STUDENT'S SOLUTIONS MANUAL

David Dubriske

University of Arkansas, Fort Smith

FINITE MATHEMATICS
AN APPLIED APPROACH

THIRD EDITION

Paula Grafton Young

Salem College

Todd Lee

Elon University

Paul E. Long

University of Arkansas

Jay Graening

University of Arkansas

PEARSON

Addison
Wesley

Boston   San Francisco   New York
London   Toronto   Sydney   Tokyo   Singapore   Madrid
Mexico City   Munich   Paris   Cape Town   Hong Kong   Montreal

Copyright © 2004 Pearson Education, Inc.
Publishing as Pearson Addison-Wesley, 75 Arlington Street, Boston, MA 02116

ISBN   0-321-17338-4

2 3 4 5 6 VHG 07 06 05

Table of Contents

Chapter 1
Applications of Linear Functions

Exercises Section 1.1

1. The given points are plotted on the graph below:

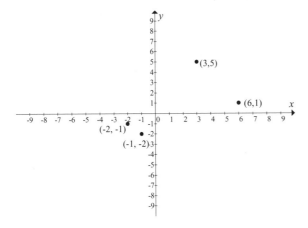

3. The given points are plotted on the graph below:

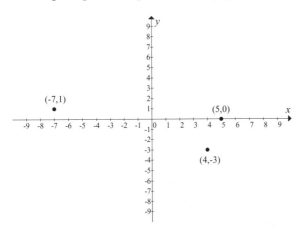

5. The given points are plotted on the graph below:

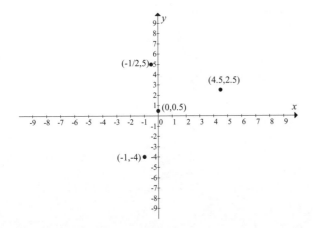

7. The given points are plotted on the graph below:

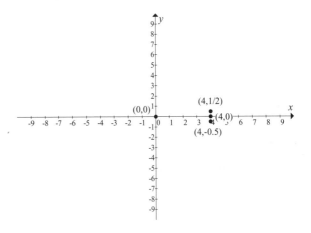

9. The coordinates are shown next to the appropriate point in the graph below:

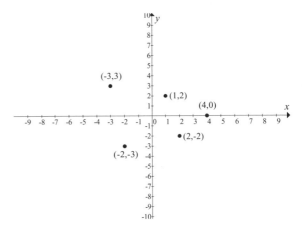

11. The data is shown below:

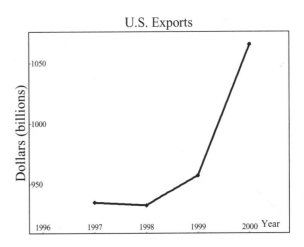

13. The data is shown below:

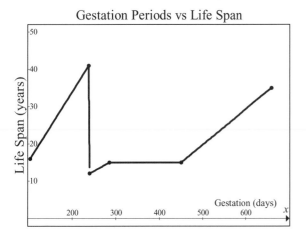

15. The data is shown below:

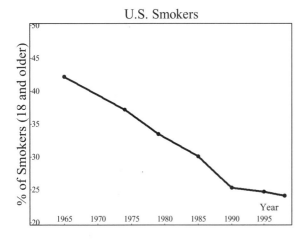

17. The Data is shown below:

19. Recognizing that the equation is a linear equation, we simply need to find two points on the line. So we will create a representative table using any two values of x we wish to use.

x	y	(x, y)
0	$0 - 4 = -4$	$(0, -4)$
4	$4 - 4 = 0$	$(4, 0)$

Plotting the points and connecting them with a smooth curve we see the following graph:

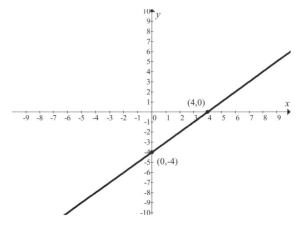

21. Recognizing that the equation is a linear equation, we simply need to find two points on the line. So we will create a representative table using any two values of x we wish to use.

x	y	(x, y)
-1	$5(-1) = -5$	$(-1, -5)$
1	$5(1) = 5$	$(1, 5)$

Plotting the points and connecting them with a smooth curve we see the following graph:

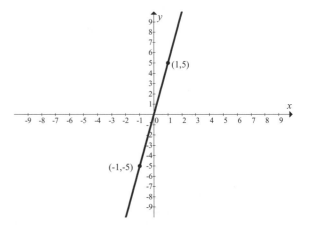

23. Recognizing that the equation is a linear equation, we simply need to find two points on the line. So we will create a representative table using any two values of x we wish to use.

x	y	(x, y)
0	$-2(0) + 7 = 7$	$(0, 7)$
3	$-2(3) + 7 = 1$	$(3, 1)$

Plotting the points and connecting them with a smooth curve we see the following graph:

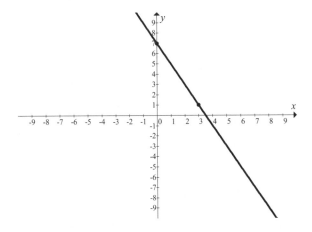

25. Recognizing that the equation is a linear equation, we simply need to find two points on the line. So we will create a representative table using any two values of x we wish to use.

x	y	(x, y)
-5	$(-5) + 5 = 0$	$(-5, 0)$
0	$(0) + 5 = 5$	$(0, 5)$

Plotting the points and connecting them with a smooth curve we see the following graph:

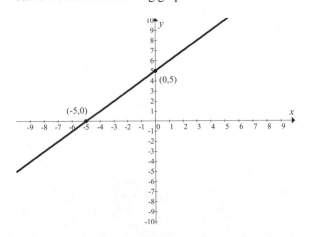

27. Because the exponent on the variable is one, this is a linear function. Therefore, we only need to plot two points. Notice the representative table below:

x	$f(x) = -2x$	$(x, f(x))$
-1	$f(-1) = -2(-1) = 2$	$(-1, 2)$
2	$f(2) = -2(2) = -4$	$(2, -4)$

Plotting the points in the plane and connecting them with a smooth curve we see the following graph:

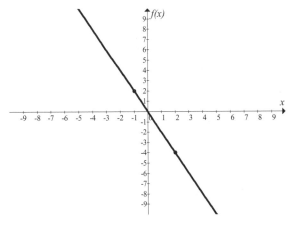

29. The exponent on the independent variable is two, the graph will be a parabola. We need to plot several points to get the basic shape. To do this we will simply choose more values of t. However, we still create the same representative table below:

t	$P(t) = t^2 - 3$	$(t, P(t))$
-2	$P(-2) = (-2)^2 - 3 = 1$	$(-2, 1)$
-1	$P(-1) = (-1)^2 - 3 = -2$	$(-1, -2)$
0	$P(0) = (0)^2 - 3 = -3$	$(0, -3)$
1	$P(1) = (1)^2 - 3 = -2$	$(1, -2)$
2	$P(2) = (2)^2 - 3 = 1$	$(2, 1)$

Plotting the points in the plane, and connecting them we a smooth curve, we graph the parabola at the top of the next page..

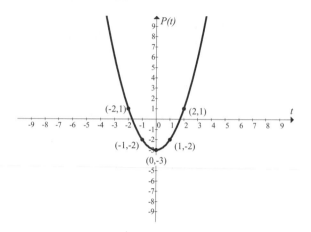

31. Once again, we have a linear function. The representative table is shown below:

p	$Z(p) = p - 1$	$(p, Z(p))$
-2	$Z(-2) = (-2) - 1 = -3$	$(-2, -3)$
5	$Z(5) = (5) - 1 = 4$	$(5, 4)$

Plotting the points and connecting them with a smooth curve, we see the graph of the line below:

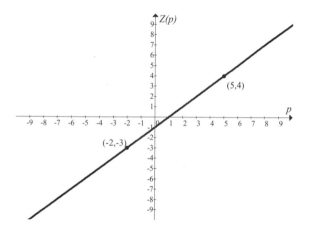

For Problems 33–37, I will be using the standard graphing window on the calculator.

```
WINDOW
 Xmin=-10
 Xmax=10
 Xscl=1
 Ymin=-10
 Ymax=10
 Yscl=1
 Xres=1█
```

33. Entering $y = x^2 + 1$ into the graphing calculator, shows the following graph:

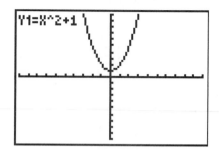

35. Entering $y = x - 6$ into the graphing calculator, shows the following graph:

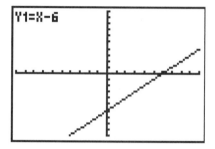

37. Entering $y = (x - 1)^3$ into the graphing calculator, shows the following graph:

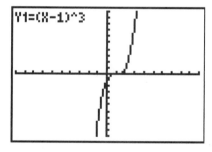

39. The independent variable in this function is thousands of miles x. This application implies that $x \geq 0$. It also would not make sense to have a negative tread depth, so $f(x) \geq 0$. This implies that $\left(1 - \frac{x}{40}\right) \geq 0 \Rightarrow x \leq 40$. So an appropriate domain for this function would be $0 \leq x \leq 40$. Creating a representative table using appropriate values from the domain we see:

x	$f(x)$	$(x, f(x))$
0	$f(0) = 2\left(1 - \frac{0}{40}\right) = 2$	$(0, 2)$
20	$f(20) = 2\left(1 - \frac{20}{40}\right) = 1$	$(20, 1)$
40	$f(40) = 2\left(1 - \frac{40}{40}\right) = 0$	$(40, 0)$

Using the data from the previous page, we see the graph of the function below:

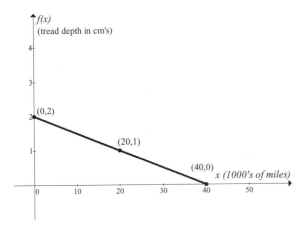

41. The independent variable is price p in hundreds of dollars. Since price must be positive, $p \geq 0$. It also makes sense to have a positive number of stereos sold. This means that $\left(16 - \frac{p^2}{4}\right) \geq 0 \Rightarrow -8 \leq p \leq 8$. However since price is positive, and appropriate domain would be $0 \leq p \leq 8$. Create a representative table:

p	$S(p) = 30\left(16 - \frac{p^2}{4}\right)$	$(p, S(p))$
0	$S(0) = 30\left(16 - \frac{0^2}{4}\right) = 480$	$(0, 480)$
2	$S(2) = 30\left(16 - \frac{2^2}{4}\right) = 450$	$(2, 450)$
6	$S(6) = 30\left(16 - \frac{6^2}{4}\right) = 210$	$(6, 210)$
8	$S(8) = 30\left(16 - \frac{8^2}{4}\right) = 0$	$(8, 0)$

Plotting the points and connecting them with a smooth curve we get the graph of the function:

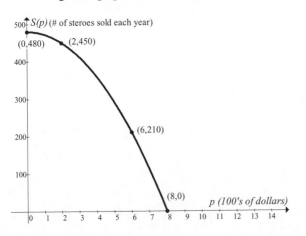

43. The independent variable in this application is the number of shifts needed each week x. Since there are eight students available to work one shift a day, there is a maximum number of 40 shifts per week. The appropriate domain for this function is $0 \leq x \leq 40$.

Create a representative table using appropriate points from the domain.

x	$C(x) = 80 + 40x$	$(x, C(x))$
0	$C(0) = 80 + 40(0) = 80$	$(0, 80)$
20	$C(20) = 80 + 40(20) = 880$	$(20, 880)$
40	$C(40) = 80 + 40(40) = 1680$	$(40, 1680)$

Plotting the points and connecting them with a smooth curve, we get the graph of the function:

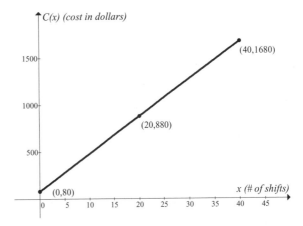

Exercises Section 1.2

1.

a. The slope is $m = \dfrac{-1-5}{6-4} = \dfrac{-6}{2} = -3$

b. The graph is:

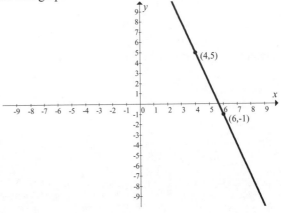

3.

a. The slope is $m = \dfrac{5-5}{-1-\frac{1}{3}} = \dfrac{0}{\frac{-4}{3}} = 0$

b. The graph is:

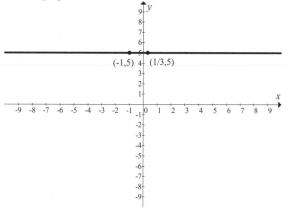

5.

a. The slope is $m = \dfrac{5.8-3.7}{1-\frac{7}{8}} = \dfrac{2.1}{\frac{1}{8}} = 16.8$

b. The graph is:

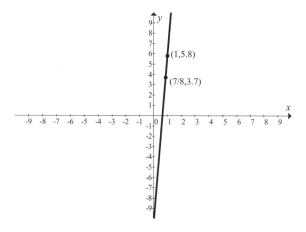

7.

a. The slope is $m = \dfrac{5-(-2)}{3-3} = \dfrac{7}{0} = No\ Slope$

b. The graph is displayed at the top of the next column.

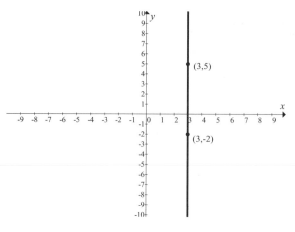

9.

a. The slope is $m = \dfrac{3-1}{3-1} = \dfrac{2}{2} = 1$

b. The graph is:

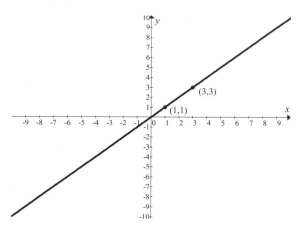

11.

a. To begin with we will use point-slope to find the slope-intercept form of the line. Substituting the appropriate slope and coordinates we see the equation: $y-5 = -2(x-2)$. Now solve for y to get the slope-intercept form.

$$y-5 = -2x+4$$
$$y-5+5 = -2x+4+5$$

Slope-intercept Form: $y = -2x+9$

Once the slope-intercept form of the line is known, we isolate the constant term to get the general form of the line. Do this by subtracting the x term from both sides of the equation.

$$y = -2x+9$$
$$2x+y = -2x+9+2x$$

General Form: $2x+y = 9$

b. The graph is shown below:

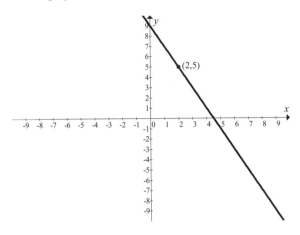

13.

a. To begin with we will use point-slope to find the slope-intercept form of the line. Substituting the appropriate slope and coordinates we see the equation:

$$y - 5.7 = \tfrac{2}{3}(x - 0).$$

Now solve for y to get the slope-intercept form.

$$y - 5.7 = \tfrac{2}{3}x$$
$$y - 5.7 + 5.7 = \tfrac{2}{3}x + 5.7$$

Slope-intercept Form: $y = \tfrac{2}{3}x + 5.7$

Once the slope-intercept form of the line is known, we isolate the constant term to get the general form of the line. Do this by subtracting the x term from both sides of the equation.

$$y = \tfrac{2}{3}x + 5.7$$
$$-\tfrac{2}{3}x + y = \tfrac{2}{3}x + 5.7 - \tfrac{2}{3}x$$

General Form: $\tfrac{-2}{3}x + y = 5.7 \Rightarrow -20x + 30y = 171$

b. The graph is shown at the top of the next column.

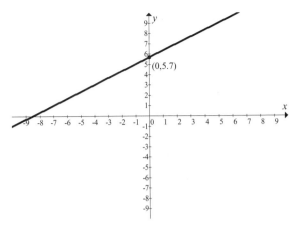

15.

a. To begin, find the slope of the line passing through the two points. $m = \dfrac{-1-4}{6-3} = \dfrac{-5}{3}$. Now use point-slope to find the slope-intercept form of the line. Substitute the appropriate slope and one of the known coordinates to get the equation:

$$y - 4 = \tfrac{-5}{3}(x - 3).$$

We solve for y to put the equation in slope–intercept form.

$$y - 4 = \tfrac{-5}{3}x + 5$$
$$y - 4 + 4 = \tfrac{-5}{3}x + 5 + 4$$

Slope-intercept Form: $y = \tfrac{-5}{3}x + 9$

Once the slope-intercept form of the line is known, we isolate the constant term to get the general form of the line. Do this by subtracting the x term from both sides of the equation as shown:

$$y = \tfrac{-5}{3}x + 9$$
$$\tfrac{5}{3}x + y = \tfrac{-5}{3}x + 9 + \tfrac{5}{3}x$$

General Form: $\tfrac{5}{3}x + y = 9 \Rightarrow 5x + 3y = 27$

The solution is continued on the next page.

b. The graph is shown below:

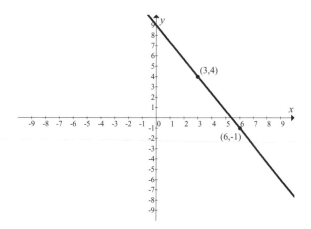

17.

a. To begin, find the slope of the line passing through the two points. $m = \dfrac{3-1}{-2-0.3} = \dfrac{-20}{23}$. Now use point-slope to find the slope-intercept form of the line. Substitute the appropriate slope and one of the known coordinates to get the equation shown at the top of the next column:

$$y - 1 = \tfrac{-20}{23}(x - 0.3).$$

Now solve for y to get the slope-intercept form.

$$y - 1 = \tfrac{-20}{23}x + \tfrac{6}{23}$$
$$y - 1 + 1 = \tfrac{-20}{23}x + \tfrac{6}{23} + 1$$

Slope-intercept Form: $y = \tfrac{-20}{23}x + \tfrac{29}{23}$

Once the slope-intercept form of the line is known, we isolate the constant term to get the general form of the line. Do this by subtracting the x term from both sides of the equation.

$$y = \tfrac{-20}{23}x + \tfrac{29}{23}$$
$$\tfrac{20}{23}x + y = \tfrac{-20}{23}x + \tfrac{29}{23} + \tfrac{20}{23}x$$

General Form: $\tfrac{20}{23}x + y = \tfrac{29}{23} \Rightarrow 20x + 23y = 29$

b. The graph is shown at the top of the next column.

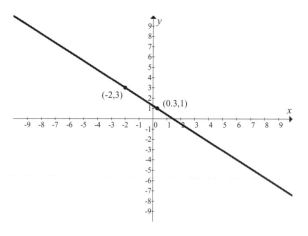

19.

a. To begin, find the slope of the line passing through the two points. $m = \dfrac{150 - 300}{23 - 17} = -25$. Now use point-slope to find the slope-intercept form of the line. Substitute the appropriate slope and one of the known coordinates to get the equation:

$$y - 300 = -25(x - 17).$$

Now solve for y to get the slope-intercept form.

$$y - 300 = -25x + 425$$
$$y - 300 + 300 = -25x + 425 + 300$$

Slope-intercept Form: $y = -25x + 725$

Once the slope-intercept form of the line is known, we isolate the constant term to get the general form of the line. Do this by subtracting the x term from both sides of the equation.

$$y = -25x + 725$$
$$25x + y = -25x + 725 + 25x$$

General Form: $25x + y = 725$

The solution is continued on the next page.

b. The graph is shown below:

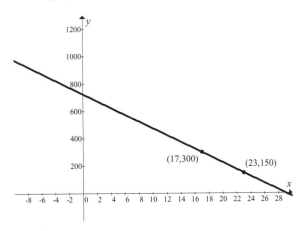

21.

a. Since the x-intercept is the point $(1.6, 0)$ and the y-intercept is the point $(0, 4.3)$, the slope of the line passing through the two points is $m = \dfrac{0 - (4.3)}{1.6 - 0} = \dfrac{-43}{16}$. Now use the slope-intercept form of the line to find the equation. Substitute the appropriate slope and the y-intercept into the equation:

$$y = \left(\tfrac{-43}{16}\right)x + (4.3).$$

Now solve for y to get the slope-intercept form.

Slope-intercept Form: $y = \tfrac{-43}{16}x + \tfrac{43}{10}$

Once the slope-intercept form of the line is known, we isolate the constant term to get the general form of the line. Do this by subtracting the x term from both sides of the equation.

$$y = \tfrac{-43}{16}x + \tfrac{43}{10}$$
$$\tfrac{43}{16}x + y = \tfrac{-43}{16}x + \tfrac{43}{10} + \tfrac{43}{16}x$$

General Form: $\tfrac{43}{16}x + y = \tfrac{43}{10} \Rightarrow 215x + 80y = 344$

b. The graph is shown at the top of the next column.

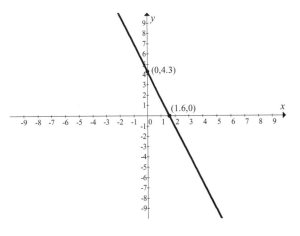

23. The equation of the horizontal line is $y = 4$
The equation of the vertical line is $x = \tfrac{-1}{2}$

25. The equation of the horizontal line is $y = -8.6$
The equation of the vertical line is $x = 1.2$

Note: There are many solutions to 27–31. The following are arbitrary choices.

27. Parallel lines have the same slope, so simply change the constant term in the equation to get a parallel equation.

Original line: $4x - 7y = 6$

$$4x - 7y = -56$$
Parallel Lines: $4x - 7y = -28$
$$4x - 7y = 28$$

The graph of each equation is shown below:

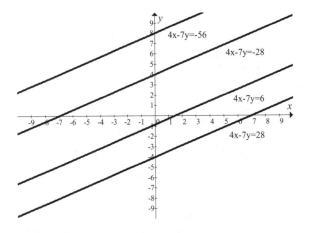

29. Parallel lines have the same slope, so simply change the constant term in the equation to get a parallel equation.

Original line: $x = 5$

$$x = -8$$

Parallel Lines: $x = -4$

$$x = 2$$

The graph of each equation is shown at the top of the next page.

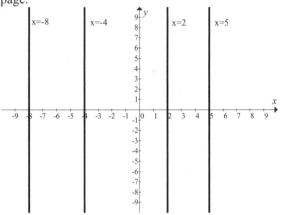

31. Parallel lines have the same slope, so simply change the constant term in the equation to get a parallel equation.

Original line: $-2x + 4y = -3$

$$-2x + 4y = -25$$

Parallel Lines: $-2x + 4y = 16$

$$-2x + 4y = 32$$

The graph of each equation is shown below:

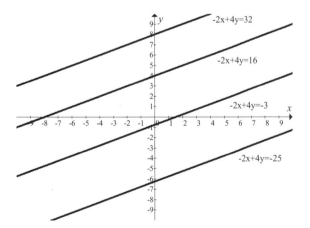

33. The given line is in slope-intercept form $y = -3x - 7$. Therefore the slope of the line is $m_1 = -3$. Find the desired equation using point-slope form:

$$y - 2 = -3(x - 1.3)$$

Thus the equation parallel to the given line is:

$$y - 2 = -3x + 3.9 \Rightarrow y = -3x + 5.9$$

35. Solve $2x + 3y = 6$ for y to find the slope of the line.

$$3y = -2x + 6 \Rightarrow y = \tfrac{-2}{3}x + 2.$$

Therefore the slope of the line is $m_1 = \tfrac{-2}{3}$. Since the point that we have is the y-intercept, find the desired equation using slope-intercept form:

$$y = \tfrac{-2}{3}x - 3.5$$

Thus the equation parallel to the given line is:

$$y = \tfrac{-2}{3}x - 3.5$$

37.
a. Since the equation is in slope-intercept form, change the coefficient on the x-term to create a line with the same y-intercept.
$$y = -2x - 4$$
New equations: $y = x - 4$
$$y = 2x - 4$$

b. Change the y-coefficient of an equation in slope-intercept form to find the equation of line that has the same x-intercept.
$$-y = -x - 4 \Rightarrow y = x + 4$$
New equations: $2y = -x - 4 \Rightarrow y = \tfrac{-1}{2}x - 2$
$$\tfrac{1}{2}y = -x - 4 \Rightarrow y = -2x - 8$$

39. First solve the equation for y to put the equation in slope-intercept form:

$$x - 2y = 5 \Rightarrow y = \tfrac{1}{2}x - \tfrac{5}{2}$$

a. Since the equation is now in slope-intercept form, change the coefficient on the x-term to create a line with the same y-intercept. The new equations are shown at the top of the next page.

$$y = -2x - \tfrac{5}{2}$$

New equations: $y = -x - \tfrac{5}{2}$

$$y = 5x - \tfrac{5}{2}$$

b. Change the y-coefficient of an equation in slope-intercept form to find the equation of line that has the same x-intercept.

$$-y = \tfrac{1}{2}x - \tfrac{5}{2} \Rightarrow \ y = \tfrac{-1}{2}x + \tfrac{5}{2}$$

New equations: $\tfrac{-1}{2}y = \tfrac{1}{2}x - \tfrac{5}{2} \Rightarrow \ y = -x + 5$

$$\tfrac{1}{2}y = \tfrac{1}{2}x - \tfrac{5}{2} \Rightarrow \ y = x - 5$$

41. First solve the equation for y to put the equation in slope-intercept form:

$$7x + 6y = 0 \Rightarrow y = \tfrac{7}{6}x$$

a, b. Notice that this line goes through the origin. That means that the x-intercept and the y-intercept are the same point. Therefore, simply change the x-coefficient to get the new lines with the same x-intercept and y-intercept.

$$y = -2x$$

New Equations: $y = x$

$$y = 2x$$

43. Since $\dfrac{\Delta y}{\Delta x} = m \Rightarrow \Delta y = m \cdot \Delta x$ (provided $\Delta x \neq 0$).

The slope of the line is $m = 3$. So if:

a. $\Delta x = 1$ then $\Delta y = 3 \cdot 1 = 3$

b. $\Delta x = 2$ then $\Delta y = 3 \cdot 2 = 6$

c. $\Delta x = 5$ then $\Delta y = 3 \cdot 5 = 15$

45. Since $\dfrac{\Delta y}{\Delta x} = m \Rightarrow \Delta y = m \cdot \Delta x$ (provided $\Delta x \neq 0$).

The slope of the line is $m = -2$. So if:

a. $\Delta x = 1$ then $\Delta y = -2 \cdot 1 = -2$

b. $\Delta x = 2$ then $\Delta y = -2 \cdot 2 = -4$

c. $\Delta x = 5$ then $\Delta y = -2 \cdot 5 = -10$

47. Since $\dfrac{\Delta y}{\Delta x} = m \Rightarrow \Delta y = m \cdot \Delta x$ (provided $\Delta x \neq 0$).

The slope of the line is $m = \tfrac{-2}{5}$. So if:

a. $\Delta x = 1$ then $\Delta y = \tfrac{-2}{5} \cdot 1 = \tfrac{-2}{5}$

b. $\Delta x = 2$ then $\Delta y = \tfrac{-2}{5} \cdot 2 = \tfrac{-4}{5}$

c. $\Delta x = 5$ then $\Delta y = \tfrac{-2}{5} \cdot 5 = -2$

49.

a.

$$S(5) = 800(5) + 6000 = 10,000$$

$$S(20) = 800(20) + 6000 = 22,000$$

The paper had 10,000 subscribers after five weeks, and 22,000 subscribers after 20 weeks.

b. Yes, the subscriptions are increasing by 800 subscriptions per week (the slope of the linear function).

c. The slope is the number of new subscribers per week after the start of the advertising campaign. The y-intercept is the number of subscribers that were already subscribing to the paper when the advertising campaign started.

d.

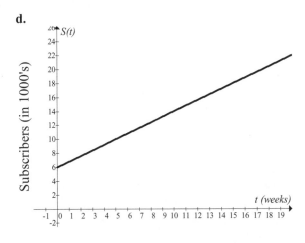

e. $S(t) = 18,000 \Rightarrow 800t + 6000 = 18,000$ Solve the linear equation by isolating the variable.

$$800t + 6000 = 18,000$$

$$800t = 12,000 \Rightarrow t = 15$$

It will take 15 weeks for subscriptions to reach 18,000.

51.

a. Since August 31st corresponds to $t = 0$, September 15th corresponds to $t = 15$ and September 20th corresponds to $t = 20$. Substitute the appropriate values of t into the function.

$$P(15) = -2(15) + 100 = 70$$
$$P(20) = -2(20) + 100 = 60$$

The agent predicts that there will be 70,000 grasshoppers per acre on September 15th, and 60,000 grasshoppers per acre on September 20th.

b. According to the model, the grasshopper population is decreasing by 2 thousand grasshoppers per acre each day.

c. The slope of the graph is the rate in which the population of grasshoppers per acre is decreasing each day. The *y*-intercept is the estimated population of grasshoppers per acre that were present on August 31st.

d. The domain of this function is $0 \le t \le 50$ because for values of t larger than fifty, the estimated population of grasshoppers is negative.

The graph of the function on this domain is shown at the top of the next page.

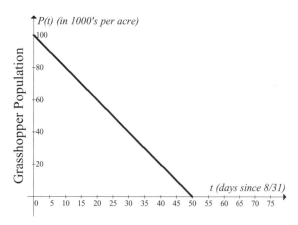

e. Since $P(t) = 20 \Rightarrow -2t + 100 = 20$, solve the linear equation for t. $-2t + 100 = 20 \Rightarrow 2t = 80 \Rightarrow t = 40$
It will take 40 days for the grasshopper population to reach 20,000 per acre.

53.

a. The corresponding rise in temperature will be $\frac{5}{9}°$ Celsius for each increase of $1°$ Fahrenheit.

b. $C = \frac{5}{9}(80) - \left(\frac{160}{9}\right) = 26\frac{2}{3}$. The corresponding Celsius temperature will be $26\frac{2}{3}° \approx 26.67°$ Celsius.

c. The original equation is $C = \left(\frac{5}{9}\right)F - \left(\frac{160}{9}\right)$.

$$9C = 5F - 160$$
$$9C + 160 = 5F$$
$$\tfrac{9}{5}C + 32 = F \Rightarrow F = \tfrac{9}{5}C + 32$$

d. The corresponding rise in temperature will be $\frac{9}{5}°$ Fahrenheit for each increase of $1°$ Celsius

55.

a. The graph $y = M(x)$ is shown below:

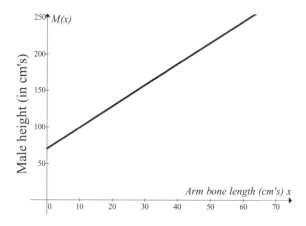

b. Substitute $x = 46$ in to the function.

$$M(46) = 2.9(46) + 70.6 = 204.$$

If a male measures 46 centimeters in length from elbow to shoulder, then he is estimated to be 204 centimeters tall.

c. Now $W(x) = 180$. Set the equation equal to 180 and solve for x.

$$2.8x + 71.5 = 180 \Rightarrow 2.8x = 108.5 \Rightarrow x = 38.75$$

A female 180 centimeters tall, should measure 38.75 centimeters from elbow to shoulder.

57.

a. Substitute $x = 11$ which corresponds to the year 2006 into the function. $S(11) = 0.91(11) + 28.96 = 38.97$
In 2006, the company sales should reach 38,970 cases of soft drinks.

b. Solve the equation $0.91x + 28.96 = 50$ for x.

$0.91x = 21.04 \Rightarrow x \approx 23.12$.

Add this to 1995 to get the answer.

The company should reach 50,000 cases in sales in the year 2018 if this trend continues.

c. The graph is shown at the top of the next page.

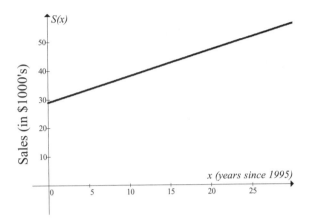

Exercises Section 1.3

1. For a function to be linear, it must have a constant rate of change. The increase in yogurt outlets in the first year is 10, followed by an increase of 20 outlets in the second year. Thus the rate of change for this relation is increasing each year. This relation *is not* modeled by a linear function.

3. For a function to be linear, it must have a constant rate of change. The increase in yogurt outlets in the first year is 20, followed by an increase of 20 outlets in the second year, and each year after that. Thus the rate of change for this relation is constant each year. This relation *is* modeled by a linear function.

5. For a function to be linear, it must have a constant rate of change. The initial number of outlets is 60. The number of outlets does not change after that, giving the relation a constant rate of change of zero each year. This relation *is* modeled by a linear function.

7. $C(x) = 3x + 20$

a. The marginal cost is the additional cost associated with the production of an additional item. For linear functions, marginal cost is the slope of the line. Thus the marginal cost is $3 per item.

b. The fixed cost is a cost that is incurred regardless of number of units produced. For linear functions, fixed cost is the constant term (the coefficient without a variable). Thus, the fixed cost is $20.

c. To find the total cost of producing the first 20 items, evaluate the cost function at $x = 20$.

$$C(20) = 3(20) + 20 = 80.$$

Thus the total cost for producing the first 20 items is $80.

d. Average total cost is the total cost of production divided by the number of items produced. Mathematically it is
$\dfrac{C(x)}{x} = \dfrac{3x + 20}{x}$. Remember that you have to perform the operations in the numerator before dividing. To find the average cost of producing the first 20 items, evaluate the average cost function at $x = 20$.

$$AC = \frac{3(20) + 20}{20} = \frac{80}{20} = 4.$$

So the average cost of producing the first 20 items is $4.

Repeat this process for 100 and 200 items.

$$AC = \frac{3(100) + 20}{100} = \frac{320}{100} = 3.20$$
$$AC = \frac{3(200) + 20}{200} = \frac{620}{200} = 3.10$$

The average cost of producing the first 100 items is $3.20 and the average cost of producing the first 200 items is $3.10.

9. $C(x) = 3.2x + 1680$

a. The marginal cost is the additional cost associated with the production of an additional item. For linear functions, marginal cost is the slope of the line. Thus the marginal cost is $3.20 per item.

b. The fixed cost is a cost that is incurred regardless of number of units produced. For linear functions, fixed cost is the constant term (the coefficient without a variable). Thus, the fixed cost is $1680.

c. To find the total cost of producing the first 20 items, evaluate the cost function at $x = 20$.

$$C(20) = 3.2(20) + 1680 = 1744.$$

Thus the total cost for producing the first 20 items is $1744.

d. Average total cost is the total cost of production divided by the number of items produced. Mathematically it is $\dfrac{C(x)}{x} = \dfrac{3.2x+1680}{x}$. Remember that you have to perform the operations in the numerator before dividing. To find the average cost of producing the first 20 items, evaluate the average cost function at $x = 20$.

$$AC = \frac{3.2(20)+1680}{20} = \frac{1744}{20} = 87.20.$$

The average cost of producing the first 20 items is $87.20.

Repeat this process for 100 and 200 items.

$$AC = \frac{3.2(100)+1680}{100} = \frac{2000}{100} = 20$$

$$AC = \frac{3.2(200)+1680}{200} = \frac{2320}{200} = 11.60$$

The average cost of producing the first 100 items is $20 and the average cost of producing the first 200 items is $11.60.

11. $C(x) = 1.6x + 5000$

a. The marginal cost is the additional cost associated with the production of an additional item. For linear functions, marginal cost is the slope of the line. Thus the marginal cost is $1.60 per item.

b. The fixed cost is a cost that is incurred regardless of number of units produced. For linear functions, fixed cost is the constant term (the coefficient without a variable). Thus, the fixed cost is $5000.

c. To find the total cost of producing the first 20 items, evaluate the cost function at $x = 20$.

$$C(20) = 1.6(20) + 5000 = 5032.$$

Thus the total cost for producing the first 20 items is $4000.

d. Average total cost is the total cost of production divided by the number of items produced. Mathematically it is

$$\frac{C(x)}{x} = \frac{50x+3000}{x}.$$

Remember that you have to perform the operations in the numerator before dividing. To find the average cost of producing the first 20 items, evaluate the average cost function at $x = 20$.

$$AC = \frac{1.6(20)+5000}{20} = \frac{5032}{20} = 251.60.$$

So the average cost of producing the first 20 items is $251.60.

Repeat this process for 100 and 200 items.

$$AC = \frac{1.6(100)+5000}{100} = \frac{5160}{100} = 51.60$$

$$AC = \frac{1.6(200)+5000}{200} = \frac{5320}{200} = 26.60$$

The average cost of producing the first 100 items is $51.60 and the average cost of producing the first 200 items is $26.60.

13. Graph each of the functions by plotting x along the horizontal axis and $C(x) = 8x + 75$ along the vertical axis. We organize the data in the table below:

x	$y_1 = \frac{C(x)}{x}$	$y_2 = MC$	(x, y_1)	(x, y_2)
1	$\frac{8(1)+75}{1} = 83$	8	$(1, 83)$	$(1, 8)$
5	$\frac{8(5)+75}{5} = 23$	8	$(5, 23)$	$(5, 8)$
10	$\frac{8(10)+75}{10} = 15.5$	8	$(10, 15.5)$	$(10, 8)$

The graphs are shown below:

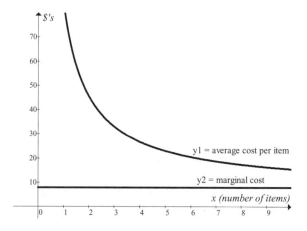

The solution is continued on the next page.

Notice that the marginal cost per item stays a constant $8 per item regardless of production. This implies that the additional cost of each item is $8. However, the average cost per item falls as output is increased. This is due to the fixed cost of production. As output continues to increase, the average fixed cost will get closer and closer to $8 per item.

15. Remember that fixed costs are the constant term of a linear cost function that takes the form $C(x) = mx + b$. So $b = 4000$. Also, it is given that when $x = 20$, $C(20) = 10,000$. Putting these two pieces of information together we know that
$m(20) + 4000 = 10,000$.

Solve this equation for m by subtracting 4000 from each side of the equation then dividing each side by 20.

$20m + 4000 = 10,000 \Rightarrow 20m = 6000 \Rightarrow m = 300$

Substitute the values of $m = 20$ and $b = 4000$ into the general form of the cost function to get the solution.

$C(x) = 300x + 4000$.

17. Remember that marginal costs are the slope of a linear cost function that takes the form $C(x) = mx + b$. Thus, $m = 25$. Also, it is given that when $x = 40$, $C(40) = 4000$.
Putting these two pieces of information together we know that $25(40) + b = 4000$. Solve this equation for b by subtracting 1000 from each side of the equation.

$25(40) + b = 4000 \Rightarrow 1000 + b = 4000 \Rightarrow b = 3000$

Substitute the values of $m = 25$ and $b = 3000$ into the general form of the cost function to get the solution.

$C(x) = 25x + 3000$.

19. We have two ordered pairs that represent this cost function; the first is $(10, 2000)$ and the second is $(15, 2700)$. Find the marginal cost by finding the slope between these two points $m = \dfrac{2700 - 2000}{15 - 10} = 140$.
Now find the equation of the cost function using point-slope form of the line as shown at the top of the next column.

$y - 2000 = 140(x - 10) \Rightarrow y - 2000 = 140x - 1400$

Add 2000 to both sides of the equation to get

$y = 140x + 600$.

Thus, the cost function that will give total cost for producing x refrigerators is $C(x) = 140x + 600$.

21. We have two ordered pairs that represent this cost function; the first is $(50, 1000)$ and the second is $(60, 1200)$. Find the marginal cost by finding the slope between these two points $m = \dfrac{1200 - 1000}{60 - 50} = 20$.

Now find the equation of the cost function using point-slope form of the line.

$y - 1000 = 20(x - 50) \Rightarrow y - 1000 = 20x - 1000$

Add 1000 to both sides of the equation to get

$y = 20x + 0$.

Thus the cost function that will give total cost for producing x ladies' coats is $C(x) = 20x$.

23. We have two ordered pairs that represent this cost function; the first is $(15, 900)$ and the second is $(30, 1560)$. Find the marginal cost by finding the slope between these two points $m = \dfrac{1560 - 900}{30 - 15} = 44$.

Now find the equation of the cost function using point-slope form of the line.

$y - 900 = 44(x - 15) \Rightarrow y - 900 = 44x - 660$

Add 900 to both sides of the equation to get

$y = 44x + 240$.

Thus, the cost function that will give total cost for producing x sweepers is $C(x) = 44x + 240$.

25. Since 25% of people skip breakfast each day, the slope of the linear function that models the number of people y who skip breakfast from a group of x people is $m = .25$. Since 25% of zero people is still zero, the y-intercept is zero. Therefore, the linear function that models the number of people y who skip breakfast from a group of x people is $y = .25x$.

27.

a. To model a linear function it is necessary to have two points that represent the function. Let x = the number of years since 2000. Since the enrollment is increasing at 600 students per year and the initial population of students is 12,000, when $x = 1$, $y = 12,000 + 600 = 12,600$ and when $x = 2$, $y = 12,600 + 600 = 13,200$.

So the two ordered pairs that are needed are $(1, 12,600)$ and $(2, 13,200)$.

Use these two points to find the slope of the linear function

$$m = \frac{13,200 - 12,600}{2 - 1} = 600.$$

Now use the point-slope formula to find the equation.

$$y - 12,600 = 600(x - 1) \Rightarrow$$
$$y - 12,600 = 600x - 600$$

Add 12,600 to both sides of the equation to get the linear function that models the number of students y enrolled in the college x years after 2000.

$$y = 600x + 12,000.$$

b. To graph the function plot the y-intercept and then use the slope to increase the value of y 600 units for every one unit increase in the value of x. The graph is shown below:

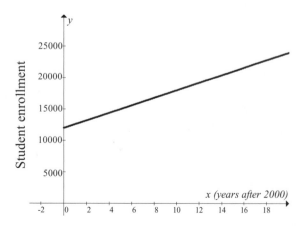

c. The slope of the linear function represents the increase in the number of students each year since 2000. The y-intercept of the function is the value of the function when $x = 0$. Thus the student enrollment in the year 2000 of 12,000 students is the y-intercept.

d. To find out the number of years it will take for enrollment to reach $y = 20,000$. Substitute 20,000 for y into the function and solve for x.

$20,000 = 600x + 12,000$ Subtract 12,000 from both sides of the equation and then divide each side by 600.

$$20,000 = 600x + 12,000 \Rightarrow$$
$$8000 = 600x \Rightarrow$$
$$x = 13.33.$$

Since the enrollment figures are taken on a yearly basis, this means that it will take 14 years for enrollment to reach 20,000 students or the year 2014.

29.

a. Using the methods done in previous problems we notice that the function decreases in value $3000 for each year, hence $m = -3000$ and the y-intercept of the function is $25,000. Using slope-intercept the linear function is $y = -3000x + 25,000$. Where the variable y is the value of the car, and the variable x is the number of years after the initial purchase.

b. The slope is the rate of depreciation of the value of the car after it's initial purchase. The y-intercept is the initial value of the car (the purchase price).

c. Substitute the value $y = 5000$ into the function and solve for x to get the number of years it will take for the car to depreciate to $5000.

$$5000 = -3000x + 25,000 \Rightarrow$$
$$-20000 = -3000x \Rightarrow$$
$$x = 6.67$$

The value of the car will be $5000 approximately 6.67 years after it is purchased.

Likewise to find when the car will be worth $1500, Substitute $y = 1500$ into the function and sole for x.

$$1500 = -3000x + 25,000 \Rightarrow$$
$$-23,500 = -3000x \Rightarrow$$
$$x = 7.83$$

The value of the car depreciates to $1500 approximately 7.83 years after it was purchased.

d. Graph the function by plotting the y-intercept and using slope depreciate the value 3000 dollars for each year after the car was purchased. The graph is shown below:

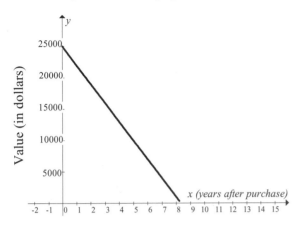

31.
a. Marginal cost is $20 dollars per radio and fixed cost is $2000, using slope-intercept formula the cost function is $y = 20x + 2000$ where y is the total cost of producing x clock radios. The graph is shown below:

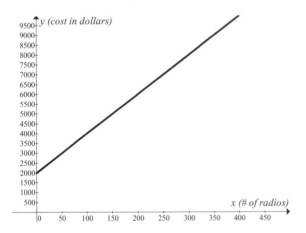

b. Substitute the value $x = 150$ into the function to get $y = 20(150) + 2000 = 5000$. The total cost of producing 150 radios is $5000.

c. The marginal cost of a linear cost function is the slope of the cost function. Since slope is constant over the entire line, then the marginal cost of producing the 201st radio is $20.

d. Find the total cost of producing 200 radios and divide that by 200. So average cost of producing the first 200 radios is $\dfrac{20(200) + 2000}{200} = 30$.

The average cost of producing 200 radios is 30 dollars per radio.

e. Proceeding in the same method of part d, substitute 500, 1000, and 2000 respectively into the equation. The results are shown below:

$$\frac{20(500) + 2000}{500} = 24$$

$$\frac{20(1000) + 2000}{1000} = 22$$

$$\frac{20(2000) + 2000}{2000} = 21$$

The average cost of producing 500 radios is $24 dollars per radio. The average cost of producing 1000 radios is $22 dollars per radio. The average cost of producing 2000 radios is $21 dollars per radio.

33.
a. The marginal cost of each inspection is $1,500 per year and the fixed cost of monitoring is $25,000. Using slope-intercept formula, the linear function is $y = 1500x + 25000$ where y is the total cost of treatment, and x is the number of inspections.

b. Substitute $x = 50$ into the function to get:
$y = 1500(50) + 25000 = 100000$.

The total cost of 50 inspections is $100,000.

c. The additional cost of the 51st inspection is the marginal cost of $1,500.

d. Find the total cost of 50 inspections and divide that by 50. So average cost of 50 inspections is:

$$\frac{1500(50) + 25000}{50} = 2000.$$

The average cost of 50 inspections is $2,000.

Likewise for 100 and 200 on site inspections

$$\frac{1500(100) + 25000}{100} = 1750$$

$$\frac{1500(200) + 25000}{200} = 1625$$

The average cost of 100 on-site inspections is $1,750 and the average cost of 200 on-site inspections is $1,625.

e. The graph of the function is shown below

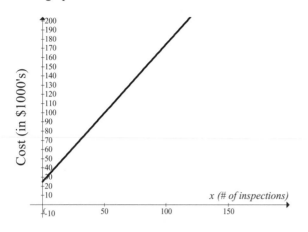

35.
a. Fixed monthly income will be $1500 per month. This is the amount if you make zero dollars in sales and is equivalent to the y-intercept. The slope of this function is 4% of total sales in the month. Using the slope-intercept formula, the function is $y = .04x + 1500$ where y is total income per month and x is dollars in sales.

b. The graph of the function is shown below:

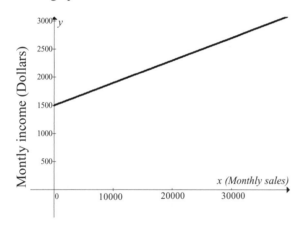

c. Substitute $x = 20,000$ into the function to get
$y = .04(20,000) + 1500 = 2300$. Your monthly income would be $2300.

d. Because you make 4% of all sales. The additional income that you would earn would be $0.04.

e. Substitute $y = 3000$ into the function and solve for x to find the dollar amount of sales.
$3000 = .04x + 1500 \Rightarrow$
$1500 = .04x \Rightarrow$
$x = 37,500$

If you make $37,500 in sales you will earn a monthly income of $3000.

37.
a. Let y be the number of apartments rented, and let x be the rent per month. The slope of the line that will model this relationship is $m = \dfrac{60 - 80}{800 - 700} = \dfrac{-1}{5}$. Use the point-slope formula to find the linear equation

$y - 80 = \frac{-1}{5}(x - 700) \Rightarrow$
$y - 80 = \frac{-1}{5}x + 140 \Rightarrow$
$y = \frac{-1}{5}x + 220$

The linear function that models occupancy as a function of monthly rent is $y = \frac{-1}{5}x + 220$. The domain of this function is $700 \le x \le 1100$.

The graph of the function is shown below:

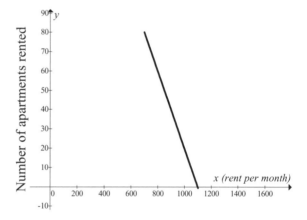

b. To check the answer, substitute $700 and $800 into the function to see if you get the correct number of apartments rented.

$y = \frac{-1}{5}(700) + 220 = 80$
$y = \frac{-1}{5}(800) + 220 = 60$

The function is accurate.

c. Substitute $760 into the function to get
$y = \frac{-1}{5}(760) + 220 = 68$.

At a monthly rent of $760 the predicted occupancy is 68 apartments.

39.
a. Let x be the number of years since 2000 and y be sales in millions. The slope for this function is

$$m = \frac{3.6 - 2.7}{2000 - 1998} = .45.$$

Since $x = 0$ corresponds to the year 2000 the y-intercept is 3.6. Using slope-intercept form, the equation of the function is $y = 0.45x + 3.6$.

b. Substitute $x = 5$ the number of years since 2000 into the function to get $y = 0.45(5) + 3.6 = 5.85$. The sales will be $5.85 million in 2005.

c. Substitute $y = 10$ into the function and solve for x.

$$10 = .45x + 3.6 \Rightarrow$$
$$.45x = 6.4 \Rightarrow x = 14.2$$

This tells us that in the year 2015 sales will top $10 million.

d. The linear model probably does not give a good approximation to sales figures for years that far away from the current year because of the complexity of financial markets.

41.
a. Let x be the number of years since 1960 and let y be the amount of paper and paperboard waste in millions of tons. Since the year 1997 corresponds to the value $x = 37$, use the points given find the slope $m = \frac{38.5 - 5.4}{37 - 0} = .895$.
Now the slope-intercept formula yields $y = .895x + 5.4$. This equation approximates the amount of waste for a given year.

b. The slope 0.895 is the increase in paper waste generated each year since 1960.

c. Substitute $x = 36$ into the equation to check the accuracy $y = .895(36) + 5.4 = 37.6$. The model approximates that 37.6 million tons of waste was generated in 1996, where the actual data tells us that 38.0 million tons were generated. The model underestimates the amount of waste generated in 1996 by 0.4 million tons.

d. Substitute $x = 38$ into the equation to check the accuracy $y = .895(38) + 5.4 = 39.4$. The model approximates that 39.4 million tons of waste was generated in 1996, where the actual data tells us that 38.2 million tons were generated. The model overestimates the amount of waste generated in 1998 by 1.2 million tons.

43.
a. To find the average growth per year, find the slope using the two endpoints of the data

$$m = \frac{220.2 - 151.5}{1998 - 1980} = 3.817.$$

The average growth between of waste between 1980 and 1998 is 3.817 million tons per year.

b. Let x be the number of years since 1980, and y be the waste in millions of tons. Therefore the point $(0, 151.5)$ is the y-intercept. Use slope-intercept form to find the linear function that approximates y, the waste generated in any year beyond 1980:

The function is $y = 3.817x + 151.5;\ 0 \le x$.

Substitute $x = 15$ into the function to get

$$y = 3.817(15) + 151.5 = 208.8$$

The model estimates that 208.8 million tons will be generated in 1995 and the data shows 211.4 million tons of waste was actually generated, so the model underestimates the value by 2.6 million tons.

c. Substitute $x = 30$ into the function to get $y = 3.817(30) + 151.5 = 266.0$.

The model predicts that 266 million tons will be generated in the year 2010.

d. Substitute $y = 300$ into the function to get

$$300 = 3.817x + 151.5.$$

Solve the equation for x.

$$300 = 3.817x + 151.5 \Rightarrow$$
$$148.5 = 3.817x \Rightarrow x = 38.9$$

This model predicts that 2019 will be the first year in which the amount of waste generated will be greater than 300 million tons. $(1980 + 39 = 2019)$

45.
a. The amount to be depreciated each year is $5000.
$$\frac{100,000}{20} = 5000$$

Making this value negative gives us the slope of the depreciation function.

b. Use slope-intercept with the initial value as the y-intercept to get $y = -5000x + 100,000$

Where y is the book value of the building, and x is the number of years since the building was purchased.

c. The graph is shown at the top of the next page. Notice the domain is $0 \le x \le 20$ the time it takes the building to depreciate to zero.

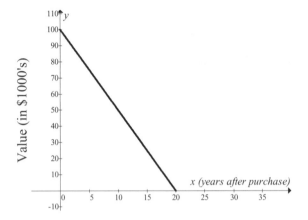

47.
a. The amount to be depreciated each year is $3500.
$$\frac{80,000 - 10,000}{20} = 3500$$

Making this value negative gives us the slope of the depreciation function.

b. Use slope-intercept with the initial value as the y-intercept to get
$$y = -3500x + 80,000, \ 0 \le x \le 20$$

Where y is the book value of the machine, and x is the number of years since purchased.

c. The graph is shown below: Notice the domain is $0 \le x \le 20$ the time it takes the machine to depreciate to $10,000.

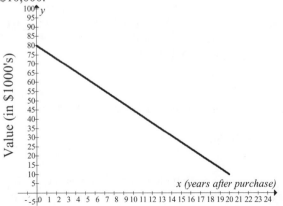

49.
a. The amount to be depreciated each year is $20,000.
$$\frac{550,000 - 50000}{25} = 20,000$$

Making this value negative gives us the slope of the depreciation function.

b. Use slope-intercept with the initial value as the y-intercept to get $y = -20,000x + 550,000, 0 \le x \le 25$

Where y is the book value of the apartment building, and x is the number of years since purchased.

c. The graph is shown below. Notice the domain is $0 \le x \le 25$.

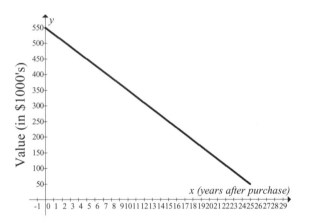

51.
a. When $p = 20$ we can find quantity demanded and quantity supplied by substituting into demand and supply functions.

$$S(20) = 6(20) = 120$$
$$D(20) = 160 - 6(20) = 40$$

Supply exceeds demand at this price which means there is a surplus of 80 items at the price of $20.

b. Substitute $D(p) = 100$ into the demand function and solve for p.

$$100 = 160 - 6(p) \Rightarrow$$
$$6p = 60 \Rightarrow p = 10$$

A price of $10 will support a demand of 100 items.

The solution is continued on the next page.

At a price of $10, the supply and demand function show

$$S(10) = 6(10) = 60$$
$$D(10) = 160 - 6(10) = 100$$

Demand exceeds supply at this price which means there is a shortage of 40 items at the price of $10.

c. The graphs are shown below:

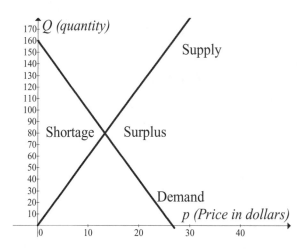

53.
a. Find the slope using the two relations given in the problem. We see $m = \dfrac{9-12}{6-4} = \dfrac{-3}{2}$. Now point-slope formula yields:

$$y - 9 = \tfrac{-3}{2}(x - 6) \Rightarrow$$
$$y = -1.5x + 18$$

Letting price be x the demand function is

$$D(p) = -1.5p + 18$$

b. Substitute $p = 5.25$ into the demand function

$$D(5.25) = -1.5(5.25) + 18 = 10.1$$ the demand for soybeans will be 10.1 million bushels at this price.

c. Substitute $D(p) = 7$ and solve for price.

$$7 = -1.5p + 18 \Rightarrow 1.5p = 11 \Rightarrow p = 7.33.$$ A price of $7.33 will support a demand of 7 million bushels.

d. At a price of zero dollars, the demand for soybeans will be 18 million bushels. This is not realistic. Demand would be much higher if soybeans were free in order to help alleviate food shortages.

Exercises Section 1.4

1. Since the slopes of the two lines are different, the pair of equations will have one point in common.

2. Since the slopes of the two lines are the same and the y-intercepts are the same, these two lines will have all points in common.

3. Since the slopes of the lines are the same and the y-intercepts are different, these two lines will have no points in common.

5. Since the slopes of the lines are different, these two lines will have one point in common.

7.
a. Since the slopes of the two lines are different, this system has only one solution.

b.
$$2x + y = 3$$
$$x - y = 2$$

Solve the second equation for x
$$x - y = 2 \Rightarrow x = 2 + y$$

Now substitute x into the first equation
$$2(2 + y) + y = 3$$

$$4 + 2y + y = 3$$
$$4 + 3y = 3$$
$$3y = -1$$
$$y = \frac{-1}{3}$$

Now substitute the value of y back into the original equation to find x.
$$x - \left(\tfrac{-1}{3}\right) = 2$$
$$x + \tfrac{1}{3} = 2$$
$$x = \frac{5}{3}$$

The point of intersection will be $\left(\tfrac{5}{3}, \tfrac{-1}{3}\right)$.

9.
a. Since the slopes of the two lines are the same, and the y-intercepts are the same, this system has infinitely many solutions.

b.
$2x + y = 3$
$x + 0.5y = 1.5$
Let x be the parameter and solve either of the equations for y to find the solution. One method would be to subtract

$2x$ from both sides of equation 1.
$y = -2x + 3$

The solution is:

$x =$ any real number .
$y = -2x + 3$

11.
a. Since the slopes of the two lines are different, this system has only one solution.

b.
$x + 2y = 0$
$x - y = 0$

Solve equation 2 for x by adding y to both sides.
$x = y$

Now substitute the value of x back into the original equation to find y.
$y + 2y = 0$
$y = 0$

Substitute the value of y back into one of the original equations $x - (0) = 0$

The point of intersection will be $(0,0)$.

13.
a. Since the slopes of the two lines are the same and the y-intercepts are the same, this system has infinitely many solutions.

b.
$x + 2y = 0$
$2x + 4y = 0$

Let y be the parameter, solve either of the equations for x to find the solution. One method would be to subtract $2y$ from both sides of equation 1.
$x = -2y$

The solution is:

Let y be any real number.
$x = -2y$

15.
a. The slopes of the two lines are equal, but the y-intercepts are different. This means that there are no solutions.

b. Since there is no solution, there are no points of intersection.

17.
a. Since the slopes of the two lines are different, this system has only one solution.

b.
$0.5x \quad = 6$
$x + 3y = 4$

To solve this system first solve equation 1 for x by multiplying both sides by 2
$0.5x \quad = 6$
$x = 12$

Now substitute the value of x back into equation 2 in order to find y.

$(12) + 3y = 4$ subtract 12 from both sides
$3y = -8$ then divide by 3
$y = \frac{-8}{3}$.

The point of intersection is $\left(12, \frac{-8}{3}\right)$.

19.
a. Since the slopes of the two lines are different, this system has only one solution.

b.
$0.5x + 2y = 3$
$x - y = 1$

To solve this system first solve equation 2 for x by adding y to both sides
$x = y + 1$
The solution is continued on the next page.

Now substitute the value of x back into equation 1 and solve for y .

$0.5(y+1)+2y=3$
$.5y+0.5+2y=3$
$0.5+2.5y=3$
$2.5y=2.5$ subtract 0.5 from both sides
$y=1$.

Now substitute the value of y back into the original equation to get x .
$x-(1)=1$ add one to both sides.
$x=2$

The point of intersection is $(2,1)$.

21.
a. Since the slopes of the two lines are different, this system has only one solution.

b.
$0.2x-y=0$
$x+0.6y=0$

To solve this system first solve equation 2 for x by adding $0.6y$ to both sides
$x=0.6y$

Now substitute the value of x back into equation 1 in order to find y .
$0.2(0.6y)-y=0$
$y=0$.

Now substitute the value of y back into the original equation to get x .
$x-0.6(0)=0$ add one to both sides.
$x=0$

The point of intersection is $(0,0)$.

23.
a. Since the slopes of the two lines are different, this system has only one solution.

b.
$x-2y=3$
$2x-y=6$

To solve this system first solve equation 1 for x by adding $2y$ to both sides
$x=2y+3$

Now substitute the value of x back into equation 2 in order to find y .
$2(2y+3)-y=6$
$4y+6-y=6$ Subtract 6 from both sides
$3y=0$ Divide by 3
$y=0$.

Now substitute the value of y back into the original equation to get x .
$x-2(0)=3$
$x=3$

The point of intersection is $(3,0)$.

25. To solve the system using a graphing calculator you first must solve each equation for y

Equation 1   Equation 2
$2x+3y=150$ $1.8x+3.5y=167$
$3y=150-2x$ $3.5y=167-1.8x$
$y=\dfrac{(150-2x)}{3}$ $y=\dfrac{(167-1.8x)}{3.5}$

Notice that the equations are not simplified. The calculator does not need to see pleasant looking equations so there is no need to simplify them here.

Type equation 1 into Y1 of your graphing calculator and type equation 2 into Y2 of your graphing calculator and hit the graph button.

Depending on your window you may not have the picture of the intersection, so you must change your window.

Change the window to look like

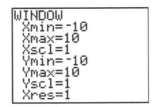

Now hit the graph button

Now hit 2nd Trace to get the calculate screen and select 5: intersection.

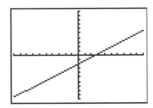

The solution to the problem is the following:

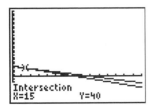

This time the slopes are equal and so is the y-intercept. There are infinitely many solutions. To double check hit 2nd graph to pull up the table of values.

X	Y₁	Y₂
1	-1.25	-1.25
2	-.5	-.5
3	.25	.25
4	1	1
5	1.75	1.75
6	2.5	2.5
7	3.25	3.25
X=1		

These points are examples of solutions to the problem.

27. To solve the system using a graphing calculator you first must solve each equation for y

Equation 1   Equation 2

$$0.3x - 0.4y = 0.8 \qquad 6x - 8y = 16$$

$$-0.4y = 0.8 - 0.3x \qquad -8y = 16 - 6x$$

$$y = \frac{(0.8 - 0.3x)}{(-0.4)} \qquad y = \frac{(16 - 6x)}{(-8)}$$

Notice that the equations are not simplified. The calculator does not need to see pleasant looking equations so there is no need to simplify them here.

Type equation 1 into Y1 of your graphing calculator and type equation 2 into Y2 of your graphing calculator.

29.
a. The break-even quantity is the quantity where revenue equals costs.

$$R(x) = C(x)$$
$$50x = 20x + 900 \text{ subtract } 20x \text{ from both sides}$$
$$30x = 900 \qquad \text{divide by 30}$$
$$x = 30$$

The break-even quantity is 30 items.

b. To find the break-even point substitute $x = 30$ into either function. $R(30) = 50(30) = 1500$

So the break-even point is $(30, 1500)$

c. The graph is shown at the top of the next page.

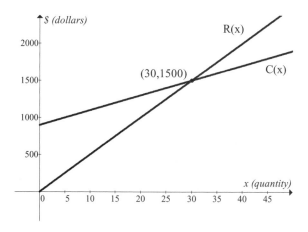

31.

a. The break-even quantity is the quantity where revenue equals costs.

$$R(x) = C(x)$$
$20x = 10x + 2500$ subtract $10x$ from both sides
$10x = 2500$ divide by 10
$x = 250$

The break-even quantity is 250 items.

b. To find the break-even point substitute $x = 250$ into either function. $R(250) = 20(250) = 5000$

So the break-even point is $(250, 5000)$

c. The graph is shown below:

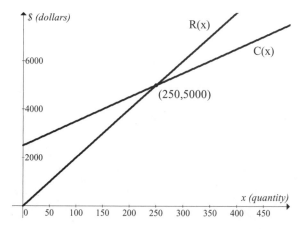

33.

a. The break-even quantity is the quantity where revenue equals costs.

$$R(x) = C(x)$$
$15x = 8x + 490$ subtract $8x$ from both sides
$7x = 490$ divide by 7
$x = 70$

The break-even quantity is 70 items.

b. To find the break-even point substitute $x = 70$ into either function. $R(70) = 15(70) = 1050$

So the break-even point is $(70, 1050)$

c. The graph is shown below:

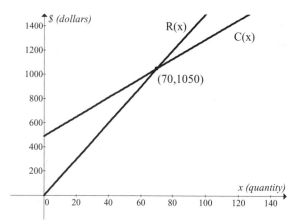

35.

a. The break-even quantity is the quantity where revenue equals costs.

$$R(x) = C(x)$$
$3x = 2x + 12$ subtract $2x$ from both sides
$x = 12$

The break-even quantity is 12 items.

b. To find the break-even point substitute $x = 12$ into either function. $R(12) = 3(12) = 36$

So the break-even point is $(12, 36)$

c. The profit function is revenue minus cost so

$$P(x) = R(x) - C(x)$$
$$P(x) = 3x - (2x + 12)$$
$$P(x) = x - 12$$

d. The graphs are shown below:

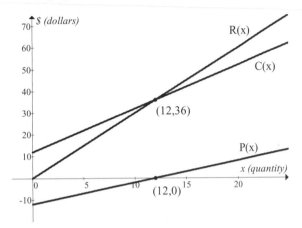

Notice the break-even quantity is the quantity where the profit function is zero.

37.
a. The break-even quantity is the quantity where revenue equals costs.

$$R(x) = C(x)$$
$$20x = 12x + 490 \quad \text{subtract } 12x \text{ from both sides}$$
$$8x = 490 \quad \text{Divide by 8}$$
$$x = 61.25$$

The break-even quantity is 61.25 items.

b. To find the break-even point substitute $x = 61.25$ into either function. $R(61.25) = 20(61.25) = 1225$

So the break-even point is $(61.25, 1225)$

c. The profit function is revenue minus cost so:

$$P(x) = R(x) - C(x)$$
$$P(x) = 20x - (12x + 490)$$
$$P(x) = 8x - 490$$

d. The graphs are shown at the top of the next column.

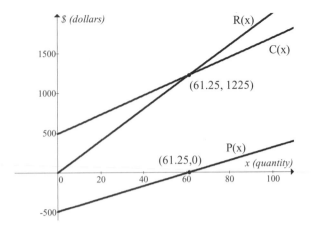

Notice the break-even quantity is the quantity where the profit function is zero.

39.
a. Revenue is price times quantity so $R(x) = 10.8x$
Cost is variable cost plus fixed cost so

$$C(x) = 8.36x + 8000$$

Profit is revenue minus cost so
$$P(x) = 10.8x - (8.36x + 8000)$$
$$P(x) = 2.44x - 8000$$

b.
To break-even $P(x) = 0$

$$2.44x - 8000 = 0 \quad \text{Add 8000 to both sides}$$
$$2.44x = 8000 \quad \text{Divide by 2.44}$$
$$x = 3278.6$$

To break-even approximately 3279 C.D.'s must be sold each month.

c. The graphs are shown below:

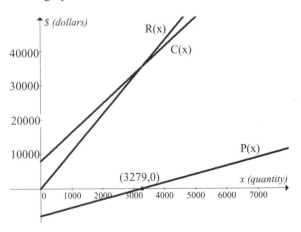

d. Since both quantities are below the break-even quantity, a loss will be incurred by the company.

41.
a.

$$D(p) = 15 - 0.5p; S(p) = p + 12$$

To find equilibrium price, set demand equal to supply and solve for p.

$15 - 0.5p = p + 12$ add $0.5p$ to both sides
$15 = 1.5p + 12$ subtract 12 from both sides
$1.5p = 3$ divide both sides by 1.5
$p = 2$

The equilibrium price is $p_0 = \$2$.

b.
To find equilibrium quantity substitute $p_0 = \$2$ into the demand or supply function.

$$D(2) = 15 - 0.5(2) = 14$$
$$S(2) = 2 + 12 = 14$$

c. The graphs are shown below:

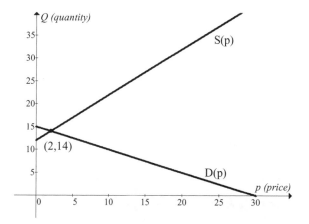

43.
a.

$$D(p) = 400 - 10p; S(p) = p + 180$$

To find equilibrium price, set demand equal to supply and solve for p.

$400 - 10p = p + 180$ add $10p$ to both sides
$400 = 11p + 180$ subtract 180 from both sides
$11p = 220$ divide both sides by 11
$p = 20$

The equilibrium price is $p_0 = \$20$.

b.
To find equilibrium quantity substitute $p_0 = \$20$ into the demand or supply function.

$$D(20) = 400 - 10(20) = 200$$
$$S(20) = (20) + 180 = 200$$

The equilibrium quantity is $q_0 = 200$.

c. The graphs are shown at the top of the next column.

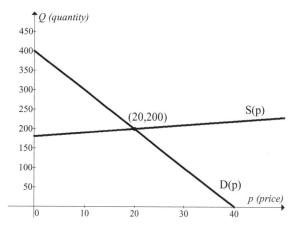

45.
a.

$$D(p) = 50 - 0.5p; S(p) = p - 12$$

To find equilibrium price, set demand equal to supply and solve for p.

$50 - 0.5p = p - 12$ add $0.5p$ to both sides
$50 = 1.5p - 12$ add 12 to both sides
$1.5p = 62$ divide both sides by 1.5
$p = 41.33$

The equilibrium price is $p_0 = \$41.33$.

b.
To find equilibrium quantity substitute $p_0 = \$41.33$ into the demand or supply function.

$$D(41.33) = 50 - 0.5(41.33) = 29.3$$
$$S(41.33) = (41.33) - 12 = 29.3$$

The equilibrium quantity is $q_0 = 29.3$. This means that the equilibrium quantity is 2,930 units of software.

47.

a. The fixed cost for the car is $20,000 and the maintenance cost is $0.40 per mile. Thus the cost function is

$$C(x) = .4x + 20,000.$$

b. Since the car was rented for 160 days, the company gets a fixed revenue of $(160)(50) = 8000$ for the year. The company also gets a variable revenue of $0.50 per mile. So the revenue function is $R(x) = .5x + 8000$.

c. To break-even revenue must equal cost.

$.5x + 8000 = .4x + 20,000$ subtract $.4x$ from both sides
$0.1x + 8000 = 20,000$ subtract 8000 from both sides
$0.1x = 12,000$ Divide by 0.1
$x = 120,000$.

The car must be driven 120,000 miles to break-even during the first year.

d. The graph is shown below:

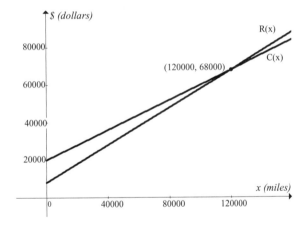

49.

a. The fixed cost associated with producing the soft-drink container is $50,000. The direct cost of the container is $1.50 per container. Thus the cost function is

$C(x) = 1.50x + 50,000$ where x is number containers produced. The revenue function is price times quantity. Since the containers are being sold at $2 a bottle, the revenue function is $R(x) = 2x$

b. Profit is revenue minus cost so the profit function is

$P(x) = 2x - (1.50x + 50,000)$ which yields

$P(x) = 0.5x - 50,000$.

The graphs of the three functions are shown at the top of the next column.

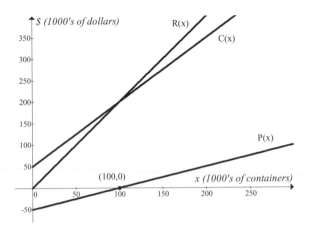

c. If the sales department is correct in their estimates, then the company will operate at a loss of $20,000

$P(60,000) = 0.5(60,000) - 50,000 = -20,000$ the first year if they sell 60,000 containers.

d. To find the break-even point set revenue equal to cost.

$R(x) = C(x)$ and solve for x
$2x = 1.5x + 50,000$ subtract $1.5x$
$.5x = 50,000$ divide by 0.5
$x = 100,000$

The company would have to sell 100,000 containers to break-even.

e. For $x = 20,000$ Total cost is

$$C(20,000) = 1.50(20,000) + 50,000 = 80,000$$

The total cost is $80,000 dollars.

To find average cost of producing 20,000 units divide total cost by the number of units $\dfrac{80,000}{20,000} = 4$

Average cost is $4 per unit for the first 20,000 units.
For $x = 40,000$ Total cost is

$$C(40,000) = 1.50(40,000) + 50,000 = 110,000$$

The total cost is $110,000 dollars.

To find average cost of producing 40,000 units divide total cost by the number of units $\dfrac{110,000}{40,000} = 2.75$

Average cost is $275 per unit for the first 40,000 units.

f. Marginal cost is the additional cost of producing the next additional unit, so

$$C(20,001) - C(20,000) = 80,001.50 - 80,000 = 1.50.$$

The marginal cost is $1.50 per unit. This is also equal to the slope of the cost curve. We will use this information in subsequent problems.

51.
a. For the computer method of printing, the fixed cost is $300. Since it cost $25 to print 1000 copies, the linear cost function for the computer printing method is

$C(x) = 25x + 300$ where x is 1000's of copies printed.

For the offset process, the fixed cost are $500 dollars and Since it cost $20 to print 1000 copies, the linear cost function is $C(x) = 20x + 500$

b. The graphs are shown below:

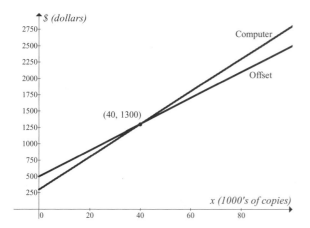

c. From the graph we see that the quantity is 40,000 copies where cost for both printing process equal $1300.

d. For any number of copies below: 40,000 (hence 30,000) it would be more advantageous to use the computer method. However, for any number of copies above 40,000 (hence 70,000) it would be more advantageous to use the offset method.

53.
a. The cost of brand A is simply what you paid for it since there is not service contract the total cost of ownership is $C(x) = 500$.

For brand B, the fixed cost is $350. Adding the service contract, there is a variable cost of 35 dollars per year so the total cost of ownership is $C(x) = 35x + 350$

b. The graphs are shown at the top of the next column.

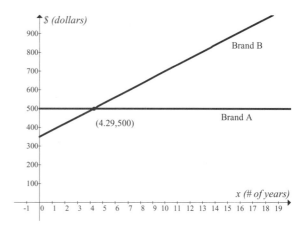

c. Set the two cost functions equal to each other and solve for x.

$$500 = 35x + 350 \quad \text{subtract 350 from both sides}$$
$$150 = 35x \quad \text{divide by 35}$$
$$x = 4.29.$$

It will take approximately 4.29 years for the two brands to have equal costs.

d. Brand A cost $500 at the five-year mark, while brand B cost 525 at the five-year mark. Brand A costs less if you own the stereo for five years.

55. Let x be the number of years after 2000. Finding a linear sales function for A-Mart and B-Mart is the first step.

A-Mart had a growth of $1.1 million per year since 2000 and the total sales of $36.3 million in 2000 is the y-intercept so the sales function for A-Mart is $S(x) = 1.1x + 36.3$ where sales are in millions of dollars.

B-Mart had a growth of $2.3 million per year since 2000 and the total sales of $16.7 million is the y-intercept so the sales function for B-Mart is $S(x) = 2.3x + 16.7$

Setting the sales function equal to each other we see
$1.1x + 36.3 = 2.3x + 16.7$ Solve for x.
$$36.3 = 1.2x + 16.7$$
$$19.6 = 1.2x$$
$$x = 16.3$$

B-Mart will overtake A-Mart in 17 years or the year 2017.

57. Let x be the number of years since 2000.
The city of Rumford can model the amount of recycled waste paper by using the average increase of two tons per year. Starting with the 80 tons they recycled in 2000 the linear population model of Rumford is $R(x) = 2x + 80$.

The solution is continued on the next page.

The city of Galeton can model the amount of recycled waste paper by using the average increase of three tons per year. Starting with the 38 tons they recycled in 2000 the linear population model of Rumford is $G(x) = 3x + 38$.

Set the two equations equal to each other and solve for x.

$3x + 38 = 2x + 80$
$x + 38 = 80$
$x = 42$

The two cities will be recycling the same amount in 42 years, or the year 2042.

59. The horizontal boundary line that passes through $(0,6)$ has the equation $y = 6$. Likewise the horizontal boundary line that passes through $(0,2)$ has the equation $y = 2$. The vertical boundary line that passes through the point $(8,0)$ has the equation $x = 8$.

The fourth boundary line passes through the points $(0,8);(6,0)$. The slope of this line is $m = \dfrac{0-8}{6-0} = \dfrac{-4}{3}$.

Using slope-intercept formula the equation of the boundary line is $y = \dfrac{-4}{3}x + 8$

The boundary equations are:

$y = 6$, $y = 2$, $x = 8$, $y = \dfrac{-4}{3}x + 8$

We can find two of the corner points by inspection. The corner points that are formed by the intersection of the horizontal and vertical boundary lines are $(8,2)$ and $(8,6)$. To find the last two corner points substitute $y = 6$ into $y = \dfrac{-4}{3}x + 8$ to get $6 = \dfrac{-4}{3}x + 8 \Rightarrow \dfrac{4}{3}x = 2 \Rightarrow x = \dfrac{3}{2}$ so the corner point is $\left(\dfrac{3}{2}, 6\right)$.

Likewise substitute $y = 2$ into $y = \dfrac{-4}{3}x + 8$ to get $2 = \dfrac{-4}{3}x + 8 \Rightarrow \dfrac{4}{3}x = 6 \Rightarrow x = \dfrac{9}{2}$ so the corner point is $\left(\dfrac{9}{2}, 2\right)$.

The four corner points are:

$\left(\dfrac{3}{2}, 6\right), \left(\dfrac{9}{2}, 2\right), (8,2)$ and $(8,6)$.

61. The y-axis is one of the boundary lines, and it as the equation $x = 0$. The increasing boundary line that passes through $(-2,0)$ and $(4,4)$ has slope $m = \dfrac{4-0}{4-(-2)} = \dfrac{2}{3}$.

Using the point-slope formula the equation is $y - (0) = \dfrac{2}{3}\left(x - (-2)\right)$. Distribution gives the boundary line equation of $y = \dfrac{2}{3}x + \dfrac{4}{3}$ The decreasing boundary line that passes through $(0,6)$ and $(4,4)$ has slope $m = \dfrac{6-4}{0-4} = \dfrac{-1}{2}$. Using the slope-intercept formula the equation is $y = \dfrac{-1}{2}x + 6$.

The equations of the boundary lines are

$x = 0$, $y = \dfrac{2}{3}x + \dfrac{4}{3}$, $y = \dfrac{-1}{2}x + 6$

One of the corner points is already labeled. It is $(4,4)$ To find the other corner points substitute $x = 0$ into $y = \dfrac{2}{3}x + \dfrac{4}{3}$ to get $y = \dfrac{2}{3}(0) + \dfrac{4}{3} = \dfrac{4}{3}$. The corner point is $\left(0, \dfrac{4}{3}\right)$. To find the second corner point substitute $x = 0$ into $y = \dfrac{-1}{2}x + 6$ to get $y = \dfrac{-1}{2}(0) + 6 = 6$. The corner point is $(0,6)$.

The three corner points are, $(0,6)$ $\left(0, \dfrac{4}{3}\right)$ and $(4,4)$.

63. The y-axis is one boundary line and it has the equation $x = 0$. The increasing boundary line that passes through $(-2,0)$ and $(0,1)$ has slope $m = \dfrac{1-0}{0-(-2)} = \dfrac{1}{2}$. Using slope intercept form the equation of the boundary line is $y = \dfrac{1}{2}x + 1$

The decreasing boundary line that passes through $(10,0)$ and $(0,6)$ has slope $m = \dfrac{6-0}{0-10} = \dfrac{-3}{5}$. Using slope intercept form the equation of the boundary line is $y = \dfrac{-3}{5}x + 6$. The decreasing boundary line that passes through $(5,0)$ and $(0,8)$ has slope $m = \dfrac{8-0}{0-5} = \dfrac{-8}{5}$. Using slope intercept form the equation of the boundary line is $y = \dfrac{-8}{5}x + 8$.

The solution is continued on the next page.

The four boundary line equations are

$$y = \tfrac{-8}{5}x + 8, \ y = \tfrac{-3}{5}x + 6, \ y = \tfrac{1}{2}x + 1, \ x = 0.$$

We can choose find two corner points by inspection. The two corner points on the y-axis are $(0,1)$ and $(0,6)$

To find the next corner points subtract

$y = \tfrac{-8}{5}x + 8$ from $y = \tfrac{-3}{5}x + 6$ to get

$0 = x - 2 \Rightarrow x = 2$. Substitute the value of x into one of the boundary lines yields $y = \tfrac{-3}{5}(2) + 6 = \tfrac{24}{5}$. The

boundary point is $\left(2, \tfrac{24}{5}\right)$. To find the final corner point

subtract $y = \tfrac{-8}{5}x + 8$ from $y = \tfrac{1}{2}x + 1$ to get

$0 = \tfrac{21}{10}x - 7 \Rightarrow x = \tfrac{10}{3}$. Substitute the value of x into one

of the boundary lines yields $y = \tfrac{1}{2}\left(\tfrac{10}{3}\right) + 1 = \tfrac{8}{3}$. The

boundary point is $\left(\tfrac{10}{3}, \tfrac{8}{3}\right)$.

The four boundary points are:

$$\left(\tfrac{10}{3}, \tfrac{8}{3}\right), \left(2, \tfrac{24}{5}\right), (0,1) \text{ and } (0,6).$$

Exercises Section 1.5

1. Create the table as shown. We have two points.

n	x	y	x*x	x*y
1	1	2	1	2
2	3	7	9	21
Sum	4	9	10	23

Let m and b represent the slope and y-intercept of the least squares regression line. The system of equations that give us these values is:

$2b + 4m = 9$
$4b + 10m = 23$

To solve this system multiply the first equation by -2 and add it to the second equation to get

$2m = 5$ which implies that $m = \tfrac{5}{2}$.

Substitute this value back into one of the original equations to get $2b + 4\left(\tfrac{5}{2}\right) = 9$

Solve this equation for b.

$2b + 10 = 9$ subtract 10
$2b = -1$ divide by 2
$b = \tfrac{-1}{2}$

Therefore the least squares regression line for these data points is $y = \tfrac{5}{2}x - \tfrac{1}{2}$.

b. The graph of the function and the data points are shown below:

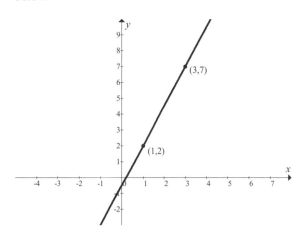

Notice that with two points the least squares regression line gives the exact fit of the line to the data points.

3. Create the table as shown. We have two points.

n	x	y	x*x	x*y
1	-1	3	1	-3
2	2	-1	4	-2
Sum	1	2	5	-5

Let m and b represent the slope and y-intercept of the least squares regression line. The system of equations that give us these values is

$2b + m = 2$
$b + 5m = -5$

To solve this system multiply the first equation by $\tfrac{-1}{2}$ and add it to the second equation to get

$\tfrac{9}{2}m = -6$ which implies that $m = \tfrac{-4}{3}$.

Substitute this value back into one of the original equations to get $2b + \left(\tfrac{-4}{3}\right) = 2$

Solve this equation for b.

$2b + \left(\tfrac{-4}{3}\right) = 2$ add $\tfrac{4}{3}$
$2b = \tfrac{10}{3}$ divide by 2
$b = \tfrac{5}{3}$

Therefore the least squares regression line for these data points is $y = \tfrac{-4}{3}x + \tfrac{5}{3}$.

b. The graph of the function and the data points are shown below:

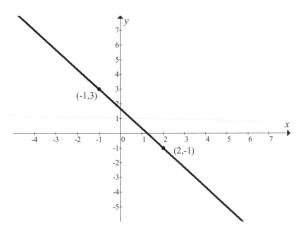

Notice that with two points the least squares regression line gives the exact fit of the line to the data points.

5. Create the table as shown. We have three points.

n	x	y	x*x	x*y
1	1	1	1	1
2	2	4	4	8
3	3	8	9	24
Sum	6	13	14	33

Let m and b represent the slope and y-intercept of the least squares regression line. The system of equations that give us these values is

$$3b + 6m = 13$$
$$6b + 14m = 33$$

To solve this system multiply the first equation by -2 and add it to the second equation to get

$2m = 7$ which implies that $m = \frac{7}{2}$.

Substitute this value back into one of the original equations to get $3b + 6\left(\frac{7}{2}\right) = 13$

Solve this equation for b.

$3b + 21 = 13$ subtract 21
$3b = -8$ divide by 3
$b = \frac{-8}{3}$

Therefore the least squares regression line for these data points is $y = \frac{7}{2}x - \frac{8}{3}$.

b. The graph of the function and the data points are shown at the top of the next column.

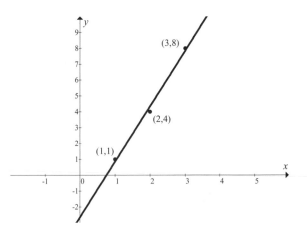

Notice that with three points the least squares regression line does not give the exact fit of the line to the data points.

7. Create the table as shown. We have four points.

n	x	y	x*x	x*y
1	1	1	1	1
2	2	3	4	6
3	3	6	9	18
4	4	5	16	20
Sum	10	15	30	45

Let m and b represent the slope and y-intercept of the least squares regression line. The system of equations that give us these values is

$$4b + 10m = 15$$
$$10b + 30m = 45$$

To solve this system multiply the first equation by $\frac{-10}{4}$ and add it to the second equation to get

$5m = \frac{15}{2}$ which implies that $m = \frac{3}{2}$.

Substitute this value back into one of the original equations to get $4b + 10\left(\frac{3}{2}\right) = 15$

Solve this equation for b.

$4b + 15 = 15$ subtract 15
$4b = 0$
$b = 0$

Therefore the least squares regression line for these data points is $y = \frac{3}{2}x$.

b. The graph of the function and the data points are shown at the top of the next page.

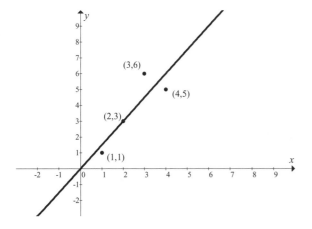

9.

a.

The scatter diagram and the regression line are plotted below:

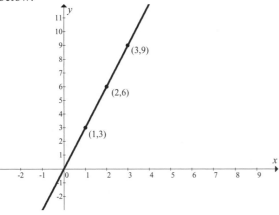

b. Create the table as shown. We have three points.

n	x	y	x*x	x*y	y*y
1	2	6	4	12	36
2	1	3	1	3	9
3	3	9	9	27	81
Sum	6	18	14	42	126

Let m and b represent the slope and y-intercept of the least squares regression line. The system of equations that give us these values is

$3b + 6m = 18$

$6b + 14m = 42$

To solve this system multiply the first equation by -2 and add it to the second equation to get

$2m = 6$ which implies that $m = 3$.

Substitute this value back into one of the original equations to get $3b + 6(3) = 18$

The solution is continued at the top of the next column.

Solve this equation for b.

$3b + 18 = 18$ subtract 18

$4b = 0$ divide by 4

$b = 0$

Therefore the least squares regression line for these data points is $y = 3x$.

c. To find the correlation coefficient use the formula:

$$r = \frac{n\left(\sum xy\right) - \left(\sum x\right)\left(\sum y\right)}{\sqrt{n\left(\sum x^2\right) - \left(\sum x\right)^2}\sqrt{n\left(\sum y^2\right) - \left(\sum y\right)^2}}$$

and substitute the values from the table.

$$r = \frac{3(42) - (6)(18)}{\sqrt{3(14) - (6)^2}\sqrt{3(126) - (18)^2}}$$

$r = 1$.

The correlation coefficient is one.

11.

a.

The scatter diagram and the regression line are plotted on the graph below:

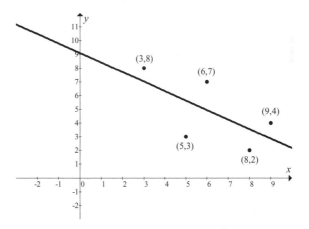

b. Create the table as shown. We have five points.

n	x	y	x*x	x*y	y*y
1	3	8	9	24	64
2	5	3	25	15	9
3	6	7	36	42	49
4	8	2	64	16	4
5	9	4	81	36	16
Sum	31	24	215	133	142

Let m and b represent the slope and y-intercept of the least squares regression line. Using the information in the table on the previous page, the system of equations that gives us these values is:

$5b + 31m = 24$
$31b + 215m = 133$

To solve this system multiply the first equation by $\frac{-31}{5}$ and add it to the second equation to get

$22.8m = -15.8$ which implies that $m = -0.693$.

Substitute this value back into one of the original equations to get $5b + 31(-0.693) = 24$

Solve this equation for b.
$5b - 21.482 = 24$ add 21.482
$5b = 45.483$ divide by 5
$b = 9.097$

Therefore the least squares regression line for these data points is $y = -0.693x + 9.097$.

c. To find the correlation coefficient use the formula:

$$r = \frac{n\left(\sum xy\right) - \left(\sum x\right)\left(\sum y\right)}{\sqrt{n\left(\sum x^2\right) - \left(\sum x\right)^2}\sqrt{n\left(\sum y^2\right) - \left(\sum y\right)^2}}$$

and substitute the values from the table:

$$r = \frac{5(133) - (31)(24)}{\sqrt{5(215) - (31)^2}\sqrt{5(142) - (24)^2}}$$

$r = -0.64$

The correlation coefficient is $r = -0.64$.

13.
a.
The scatter diagram and the regression line are plotted at the top of the next column.

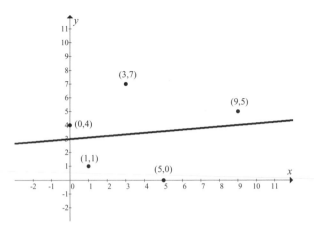

b. Create the table as shown. We have five points.

n	x	y	x*x	x*y	y*y
1	0	4	0	0	16
2	5	0	25	0	0
3	9	5	81	45	25
4	3	7	9	21	49
5	1	1	1	1	1
Sum	**18**	**17**	**116**	**67**	**91**

Let m and b represent the slope and y-intercept of the least squares regression line. The system of equations that give us these values is

$5b + 18m = 17$
$18b + 116m = 67$

To solve this system multiply the first equation by $\frac{-18}{5}$ and add it to the second equation to get equation at the top of the next column.

$51.2m = 5.8$ which implies that $m = 0.113$.

Substitute this value back into one of the original equations to get $5b + 18(0.113) = 17$

Solve this equation for b.
$5b + 2.039 = 17$ subtract 2.039
$5b = 14.961$ divide by 5
$b = 2.992$

Therefore the least squares regression line for these data points is $y = 0.113x + 2.992$.

c. To find the correlation coefficient use the formula:

$$r = \frac{n\left(\sum xy\right) - \left(\sum x\right)\left(\sum y\right)}{\sqrt{n\left(\sum x^2\right) - \left(\sum x\right)^2}\sqrt{n\left(\sum y^2\right) - \left(\sum y\right)^2}}$$

and substitute the values from the table:

$$r = \frac{5(67) - (18)(17)}{\sqrt{5(116) - (18)^2}\sqrt{5(91) - (17)^2}}$$

$$r = 0.141$$

The correlation coefficient is $r = 0.141$.

15. There should be a very high positive correlation between the number of accidents and the number of automobiles on the highway. The denser the traffic, the better chance an accident will occur. The correlation is probably above 0.9.

17. There should be almost no correlation between the heights of persons and having bank accounts and the amount of money in their bank account. However, since height is positive, and we like to think that most bank accounts are positive as well, the correlation coefficient will probably end up being a between 0.1 and 0.2.

19. In general there should be a very high negative correlation. Most old cars have depreciated in value over the years. There are a few classics (like a mint condition '67 Hemi-Barracuda) that might weaken this negative correlation, but it still should be in the -0.9 to -0.8 range.

21.
a. Using techniques done in the first part of this section, or using a spreadsheet we can find the least squares regression line. The regression line and the scatter plot are shown below:

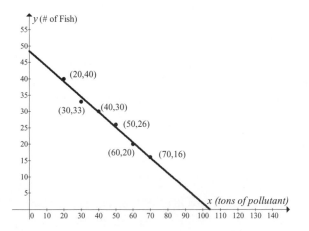

Create a chart like the following:

n	x	y	x*x	x*y	y*y
1	20	40	400	800	1600
2	30	33	900	990	1089
3	40	30	1600	1200	900
4	50	26	2500	1300	676
5	60	20	3600	1200	400
6	70	16	4900	1120	256
Sum	**270**	**165**	**13900**	**6610**	**4921**

The least squares regression line is calculated to be $y = -0.466x + 48.457$.

b. Using the formula from earlier or a spreadsheet, the correlation coefficient is $r = -0.995$.

c. The slope is the key to this solution. According to the linear model, we would expect -0.466 thousand black bass or 466 fish for each ton of pollutant that is introduced into the water.

d. Substitute $x = 65$ into the least squares regression line to get $y = -0.466(65) + 48.457 = 18.2$ The population of black bass should be 18,200 if there is 65 tons of pollutants present.

e. Substitute 35 into the least squares regression line to get $35 = -0.466x + 48.457$ Solve this equation for x. First subtract 48.457 from both sides to get $-13.457 = -0.466x$ now divide by -0.466 $x = 28.9$. We can expect 28.9 tons of pollution will result in a population of 35,000 fish..

23.
a. Using techniques done in the first part of this section, or using a spreadsheet we can find the least squares regression line. The regression line and the scatter plot are shown below:

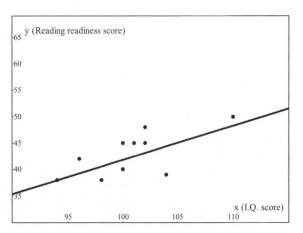

b. Create a chart like the one below:

n	x	y	x*x	x*y	y*y
1	102	45	10404	4590	2025
2	110	50	12100	5500	2500
3	102	48	10404	4896	2304
4	98	38	9604	3724	1444
5	100	40	10000	4000	1600
6	101	45	10201	4545	2025
7	96	42	9216	4032	1764
8	100	45	10000	4500	2025
9	104	39	10816	4056	1521
10	94	38	8836	3572	1444
Sum	**1007**	**430**	**101581**	**43415**	**18652**

The least squares regression line is calculated to be $y = 0.647x - 22.891$. Plot this line on the graph on the previous page.

c. Using the formula from earlier or a spreadsheet, the correlation coefficient is $r = 0.675$.

25.
a. Create a table like the one shown below:

n	x	y	x*x	x*y	y*y
1	17	19	289	323	361
2	19	25	361	475	625
3	21	33	441	693	1089
4	23	57	529	1311	3249
5	25	71	625	1775	5041
6	28	113	784	3164	12769
7	32	123	1024	3936	15129
8	38	252	1444	9576	63504
9	39	260	1521	10140	67600
10	41	293	1681	12013	85849
Sum	**283**	**1246**	**8699**	**43406**	**255216**

Let m and b represent the slope and y-intercept of the least squares regression line. The system of equations that give us these values is
$$10b + 283m = 1246$$
$$283b + 8699m = 43406$$

To solve this system multiply the first equation by -28.3 and add it to the second equation to get
$690.1m = 8144.2$ which implies that $m = 11.801$.

Substitute this value back into one of the original equations to get $10b + 283(11.801) = 1246$

Solve this equation for b.
$$10b + 3339.818 = 1246$$
$$10b = -2093.818 \qquad \text{divide by 5}$$
$$b = -209.382$$

Therefore the least squares regression line for these data points is $y = 11.801x - 209.382$. where x is the diameter of the tree in inches and y is volume in hundreds of board feet that can be retrieved from each tree.

To find the correlation coefficient use the formula:

$$r = \frac{n\left(\sum xy\right) - \left(\sum x\right)\left(\sum y\right)}{\sqrt{n\left(\sum x^2\right) - \left(\sum x\right)^2}\sqrt{n\left(\sum y^2\right) - \left(\sum y\right)^2}}$$

and substitute the values from the table as shown:

$$r = \frac{10(43406) - (283)(1246)}{\sqrt{10(8699) - (283)^2}\sqrt{10(255216) - (1246)^2}}$$

$$r = 0.981$$

The correlation coefficient is $r = 0.981$.

b. The rate of change between board feet and the inches in diameter of the tree is the slope of the regression line. For each additional inch in diameter, the regression model predicts an increase in 11.801 hundred feet of boards, or 1,180.1 additional board feet.

c. Substitute $x = 45$ into the regression equation to get
$$y = 11.801(45) - 209.382 = 321.663.$$
The regression model predicts that a tree with a diameter of 45 inches will produce 321.663 hundred feet of boards or 32,166 feet of boards.

d. To find out how large a tree was in diameter that produced 250 hundred (25000) board feet, substitute $y = 250$ into the regression equation and solve for x.

$$250 = 11.801(x) - 209.382$$

$$459.382 = 11.801(x)$$

$$x = 38.93.$$

The regression model predicts that the tree that produces 25000 feet of boards is 38.93 inches in diameter.

27.
a. Create a table like the ones shown below for each set of data.

Current Dollars

n	x	y	x*x	x*y	y*y
1	0	4.88	0	0	23.8144
2	5	5.94	25	29.7	35.2836
3	10	6.75	100	67.5	45.5625
4	15	7.69	225	115.35	59.1361
5	19	9.08	361	172.52	82.4464
Sum	49	34.34	711	385.07	246.243

Using the methods shown earlier in the exercises or calculator or software regression packages we find the regression equation for current dollars to be $y = 0.210x + 4.81$, where y is average hourly earnings in current dollars and x is the year since 1980.

To find the correlation coefficient use the formula and substitute the values from the table:

$$r = \frac{5(385.07) - (49)(34.34)}{\sqrt{5(711) - (49)^2}\sqrt{5(246.243) - (34.34)^2}}$$

$$r = 0.991$$

The correlation coefficient between hourly average earnings in current dollars and the year since 1980 is $r = 0.991$.

Constant Dollars

n	x	y	x*x	x*y	y*y
1	0	5.7	0	0	32.49
2	5	5.39	25	26.95	29.0521
3	10	5.07	100	50.7	25.7049
4	15	4.97	225	74.55	24.7009
5	19	5.39	361	102.41	29.0521
Sum	49	26.52	711	254.61	141

Using the methods shown earlier in the exercises or calculator or software regression packages we find the regression equation for constant dollars to be $y = -0.0229x + 5.53$, where y is average hourly earnings in constant dollars and x is the year since 1980.

To find the correlation coefficient use the formula and substitute the values from the table as shown at the top of the next column.

$$r = \frac{5(254.61) - (49)(26.52)}{\sqrt{5(711) - (49)^2}\sqrt{5(141) - (26.52)^2}}$$

$$r = -0.599.$$

The correlation coefficient between hourly average earnings in constant dollars and the year since 1980 is $r = 0.991$.

b. Plot the year since 1980 on the x-axis and then draw the regressions lines on the same coordinate plane. The graph is shown below:.

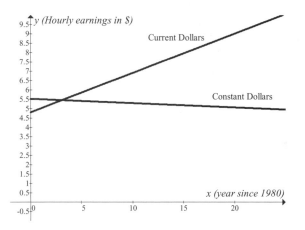

c. Substitute $x = 25$ into the appropriate regression equation to get

Current Dollars: $y = 0.210(25) + 4.81 = 10.06$

Constant Dollars: $y = -0.0229(25) + 5.53 = 4.96$

Subtract the two numbers to find the difference between current dollar and constant dollar average hourly earnings in 2005. We see that the difference is $10.84 - 4.96 = 5.88$ or there is a difference of $5.88 between current dollar and constant dollar average hourly earnings.

29.
a. Create a chart like the following.

n	x	y	x*x	x*y	y*y
1	0	79	0	0	6241
2	5	135	25	675	18225
3	9	135	81	1215	18225
4	10	163	100	1630	26569
5	12	279	144	3348	77841
Sum	36	791	350	6868	147101

Let m and b represent the slope and y-intercept of the least squares regression line. Using the data in the chart on the previous page, the system of equations that give us these values is

$5b + 36m = 791$
$36b + 350m = 6868$

To solve this system multiply the first equation by -7.2 and add it to the second equation to get
$90.8m = 1172.8$ which implies that $m = 12.916$.
Substitute this value back into one of the original equations
to get $5b + 36(12.916) = 791$

Solve the equation in the previous column for b.
$5b + 464.976 = 791$
$5b = 326.024$
$b = 65.2$

The least squares regression line is calculated to be
$y = 12.916x + 65.2$, where y is number of millions of dollars spent on magazine advertising for drugs and remedies, and x is number of years since 1980.

Using the correlation coefficient formula

$$r = \frac{5(6868) - (36)(791)}{\sqrt{5(350) - (36)^2}\,\sqrt{5(147101) - (791)^2}}$$

The correlation coefficient is $r = 0.83$.

b. Substitute $x = 30$ into the regression equation to get
$y = 12.916(30) + 65.2 = 452.68$. According to the regression equation there will be $452.68 million spent on magazine advertising for drugs and remedies.

31.
a. Create a table like the ones shown below for each set of data.

Gross waste generated (in millions of tons)

n	x	y	x*x	x*y	y*y
1	0	87.8	0	0	7708.84
2	10	121.9	100	1219	14859.61
3	20	151.5	400	3030	22952.25
4	25	164.4	625	4110	27027.36
5	27	178.1	729	4808.7	31719.61
6	30	205.2	900	6156	42107.04
7	35	211.4	1225	7399	44689.96
8	38	220.2	1444	8367.6	48488.04
Sum	185	1340.5	5423	35090.3	239552.7

Let m and b represent the slope and y-intercept of the least squares regression line. The system of equations that give us these values is
$8b + 185m = 1340.5$
$185b + 5423m = 35090.3$

Using the methods shown earlier in the exercises or calculator or software regression packages we find the regression equation gross waste generated to be
$y = 3.57x + 84.925$, where the variable y is the amount of waste generated in millions of tons and x is the year since 1960.

To find the correlation coefficient use the formula, and substitute the values from the table.

$$r = \frac{8(35181.3) - (185)(1343.1)}{\sqrt{8(5423) - (185)^2}\,\sqrt{8(240658.8) - (1343.1)^2}}$$

$r = 0.989$.

The correlation coefficient between tons of waste generated and years since 1960 is $r = 0.989$.

Materials recovered (in millions of tons)

n	x	y	x*x	x*y	y*y
1	0	5.9	0	0	34.81
2	10	8.6	100	86	73.96
3	20	14.5	400	290	210.25
4	25	16.4	625	410	268.96
5	27	20.4	729	550.8	416.16
6	30	33.6	900	1008	1128.96
7	35	54.9	1225	1921.5	3014.01
8	38	62.2	1444	2363.6	3868.84
Sum	185	216.5	5423	6629.9	9015.95

Let m and b represent the slope and y-intercept of the least squares regression line. The system of equations that give us these values is
$8b + 185m = 216.5$
$185b + 5423m = 6629.9$

Using the methods shown earlier in the exercises or calculator or software regression packages we find the regression equation waste generated to be
$y = 1.418x - 5.727$, where the variable y is amount of waste recovered in millions of tons, and x is the year since 1960.

The solution is continued on the next page.

To find the correlation coefficient use the formula, and substitute the values from the table as shown below:

$$r = \frac{8(6629.9) - (185)(216.5)}{\sqrt{8(5423) - (185)^2}\sqrt{8(9015.95) - (216.5)^2}}$$

$$r = 0.854$$

The correlation coefficient between amount of waste recovered and the number of years since 1960 is $r = 0.854$.

b. The solution to this problem is found by setting the equation you found in part (a) for waste recovery equal to .5 times the equation you found in part (b) for waste generation. In other words solve

$$1.418x - 5.727 = .5(3.574x + 84.626)$$

$$1.418x - 5.727 = 1.787x + 42.313$$

$$.369x = -48.04$$

$$x = -130.2$$

If you could use this model to predict into the past as well as the future, it predicts that the last time we recovered half as much waste as we generated was the year 1830. An intuitive solution to the previous problem would be to recognize that the slope for the equation that models the generation of waste is larger than the slope that models the recovery of waste. This means that instead of recovering a greater percentage of generated waste each year, we are actually recovering a smaller percentage of waste each year.

c. Substitute $x = 50$ into each regression equation.

The waste generation regression equation predicts that the amount waste generated in 2010 will be

$$y = 3.574(50) + 84.626 = 263.326 \text{ million tons.}$$

The waste recovery regression equation predicts that the amount of waste recovered in 2010 will be

$$y = 1.418(50) - 5.727 = 65.173 \text{ million tons.}$$

Chapter 1 Summary Exercises

1. Substitute $y = 0$ into the equation to find the x-intercept $3x - 5(0) = 30$. Solving this equation for x by dividing by 3 yields $x = 10$. The x-intercept is the point $(10, 0)$.

Substitute $x = 0$ into the equation to find the y-intercept $3(0) - 5y = 30$. Solving this equation for y by dividing by -5 yields $y = -6$. The y-intercept is the point $(0, -6)$.

3. The horizontal line that passes through the point $(-3, 8)$ will have the equation $y = 8$.

The vertical line that passes through the point $(-3, 8)$ will have the equation $x = -3$

5. Substitute the expression $(x + 2)$ into the equation for x to find the y-coordinate.

$$y = 3x + 4 \qquad \text{substitute } x = (x + 2)$$

$$y = 3(x + 2) + 4 \text{ simplify the right hand side}$$

$$y = 3x + 6 + 4$$

$$y = 3x + 10.$$

7. $2y = 5x + 10 \Rightarrow y = \frac{5}{2}x + 5$ The slope of the line that is parallel to $2y = 5x + 10$ is $m = \dfrac{5}{2}$. Using point-slope formula, the equation of the line that passes through $(0, 0)$ with slope $m = \dfrac{5}{2}$ is

$$y - 0 = \tfrac{5}{2}(x - 0)$$

$$y = \tfrac{5}{2}x.$$

9. Add the two equations together to eliminate the y variable.

$$3x + 6y = 13$$
$$\underline{+5x - 6y = 11}$$
$$8x + 0y = 24$$

Solve this equation for x.

$$x = \frac{24}{8} = 3.$$

Now substitute the value of x back into one of the original equations to get the value for y as shown at the top of the next page.

$$3(3) + 6y = 13$$
$$6y = 4$$
$$y = \frac{2}{3}.$$

The point of intersection is $\left(3, \frac{2}{3}\right)$.

11. The slopes of the two equations are equal. The y-intercept of $5x + 3y = 9$ is $(0,3)$ while the y-intercept of $10x + 6y = 18$ is $(0,3)$ as well. Therefore the lines are equivalent, thus the common points are all the points that satisfy $5x + 3y = 9$.

13.

a. Substitute $x = 60$ into the cost function $C(60) = 7.9(60) + 2520 = 2994$. The total cost of producing 60 items is $2994.

b. Take the total cost of producing 60 items in found in part (a) and divide by 60. $\dfrac{2994}{60} = 49.9$ The average cost of producing the first 60 items is $49.90.

c. The marginal cost is the additional cost of producing the 61st item. The marginal cost is $7.90. (the slope of the linear cost function)

15.

a. Profit is revenue minus cost.
$$P(x) = R(x) - C(x)$$
$$P(x) = 65x - (25x + 4000)$$
$$P(x) = 40x - 4000$$

b. The break-even point is when $P(x) = 0$.
$$40x - 4000 = 0$$
$$40x = 4000$$
$$x = 100$$

The company will break-even when it sells 100 items.

c. The functions are graphed at the top of the next column.

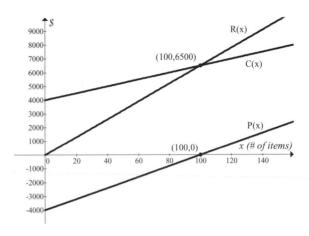

17. The first thing we notice is the y-intercept is $(0,4)$ so that eliminates choices (c), (d), and (e). Now using $(0,4)$ and $(2,8)$ we see the slope of the relation is $m = \dfrac{8-4}{2-0} = 2$. This eliminates answer (a), and reconfirms the correct choice of answer (b) $y = 2x + 4$.

Sample Test Answers

1. The points are plotted below:

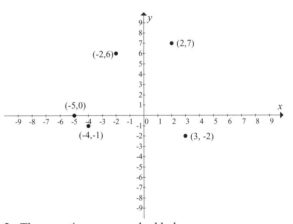

2. The equations are graphed below:

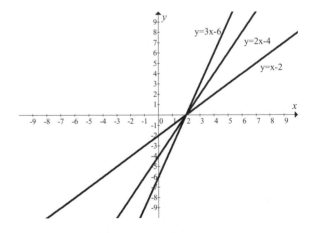

3. Use the slope-intercept formula to find the equation. The slope is the per unit change or $m = 3.7$. The equation of the line will be $y = 3.7x - 2.8$.

4. Let x be the number of items and let y be the total cost of production. The two points associated with this relation are $(50, 650)$ and $(70, 714)$. The slope of the line is $m = \dfrac{714 - 650}{70 - 50} = 3.2$. Now use the point slope formula to find the equation of the line.

$$y - 650 = 3.2(x - 50)$$
$$y - 650 = 3.2x - 160$$
$$y = 3.2x + 490$$

5. Let x be the number of years from the current year and let y be the population of the town.. If the town is growing at a rate of 250 per year, the slope of the line is $m = 250$.

The current population (population when $x = 0$) is 2300, which is the y-intercept. Using the slope-intercept formula, the linear function that will give the approximate population x years from now is

$$y = 250x + 2300.$$

6. The equation of the vertical line that passes through the point $(-2, 7)$ will be $x = -2$.

7. Solve the equation for y.

$$x - 2y = 9$$
$$-2y = 9 - x$$
$$y = \frac{9 - x}{-2}$$
$$y = \tfrac{1}{2}x - \tfrac{9}{2}.$$

The equation is now in slope-intercept form. The slope is $m = \tfrac{1}{2}$ and the y-intercept is $\left(0, \tfrac{-9}{2}\right)$.

To find the x-intercept set $y = 0$ in the original equation.

$$x - 2(0) = 9$$
$$x = 9.$$

The x-intercept is $(0, 9)$

8. The two points on the line are $(-2, 0)$ and $(0, 6.4)$.

The slope of the line is $m = \dfrac{0 - 6.4}{-2 - 0} = 3.2$. Use the slope and the y-intercept in the slope-intercept formula to get the equation $y = 3.2x + 6.4$.

9. Let x be the price of the special, and y be the number of specials sold. Use the points $(4.75, 78)$ and $(5.50, 63)$.

The slope of the line is $m = \dfrac{63 - 78}{5.50 - 4.75} = -20$.

Use the point-slope formula to find the equation of the line. The calculations are shown below:

$$y - 78 = -20(x - 4.75)$$
$$y - 78 = -20x + 95$$
$$y = -20x + 173.$$

The equation that will model the number of lunch specials y sold for a given price x is $y = -20x + 173$.

10. Let x be the number of years after the purchase of the tractor, and let y be the value of the tractor. Two points associated with this line are $(0, 5600)$ and $(10, 2100)$.

The slope of the line is $m = \dfrac{5600 - 2100}{0 - 10} = -350$.

Use the slope and the y-intercept to find the equation of the line that models the value of the tractor as a function of the number of years from the date of purchase. The function is $y = -350x + 5600$.

11. The slopes of the two equations are different; therefore there is only one intersection point.

12.
Substitution method

Solve $x + y = 5$ for x.
$$x = 5 - y$$
Substitute x into $2x - y = 12$
$$2(5 - y) - y = 12$$
$$10 - 2y - y = 12$$
$$-3y = 2$$
$$y = \tfrac{-2}{3}$$

Now substitute the value of y back into one of the original equations to get x as shown at the top of the next page.

$x + \left(\frac{-2}{3}\right) = 5$

$x = \frac{17}{3}$

The solution is $\left(\frac{17}{3}, \frac{-2}{3}\right)$

Elimination method

Add the two equations together to eliminate y

$x + y = 5$

$\underline{+2x - y = 12}$

$3x + 0y = 17$

Solve for x

$3x = 17$

$x = \frac{17}{3}$

Substitute x back into one of the original equations to find y.

$\frac{17}{3} + y = 5$

$y = \frac{-2}{3}$.

The solution is $\left(\frac{17}{3}, \frac{-2}{3}\right)$

13. Solve both equations for y.

Equation 1

$0.2x + 0.8y = 3$

$0.8y = -0.2x + 3$

$y = \frac{-1}{4}x + \frac{15}{4}$

Equation 2

$x + 4y = 15$

$4y = -x + 15$

$y = \frac{-1}{4}x + \frac{15}{4}$

The slope of both equations is $m = \frac{-1}{4}$. The y-intercept of both equations is $\left(0, \frac{15}{4}\right)$. Therefore there are infinitely many solutions. All points given by the equation $y = \frac{-1}{4}x + \frac{15}{4}$ will solve the system.

14. Revenue is price times quantity. The revenue function is $R(x) = 98x$.

The cost function is direct cost plus fixed cost. The Cost function is $C(x) = 73x + 40,000$.

Profit is revenue minus cost. The profit function will be

$P(x) = R(x) - C(x)$

$P(x) = 98x - (73x + 40,000)$

$P(x) = 25x - 40,000$

To find the break-even point set $P(x) = 0$ and solve for x.

$25x - 40,000 = 0$

$25x = 40,000$

$x = 1600$

The company will break-even if they sell 1600 calculators.

15. To find the equilibrium price and quantity set supply equal to demand and solve for p as shown:

$S(p) = D(p)$

$0.03p - 9 = 17.5 - 0.015p$ Solve for p.

$0.045p - 9 = 17.5$

$0.045p = 26.5$

$p = 588.89$

Substitute the price back into supply or demand to get the equilibrium quantity.

$S(588.89) = 0.03(588.89) - 9 = 8.67$.

Equilibrium is reached when the price is \$588.89. At this price the annual supply is 8.67 quilts, and the annual demand for quilts is 8.67.

16. Let x start with year 1, be the year the mutual fund was invested, and let y be the return (in percentage terms) achieved by the mutual fund $(y = 4 \Rightarrow 4\%\ return)$.

a. Using the above variables four data points will be $(1,8); (2,9); (3,12); (4,14)$ over the first four years of the fund. These points are plotted on the axis below:

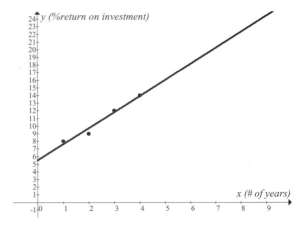

b. Create a chart like the following using the data points.

n	x	y	x*x	x*y	y*y
1	1	8	1	8	64
2	2	9	4	18	81
3	3	12	9	36	144
4	4	14	16	56	196
Sum	10	43	30	118	485

Let m and b represent the slope and y-intercept of the least squares regression line. The system of equations that give us these values is

$$4b + 10m = 43$$
$$10b + 30m = 118$$

To solve the system on the previous page multiply the first equation by $\frac{-10}{4}$ and add it to the second equation to get

$5m = 10.5$ which implies that $m = 2.1$.

Substitute this value back into one of the original equations to get $4b + 10(2.1) = 43$

Solve this equation for b.

$4b + 21 = 43$ subtract 21

$4b = 22$ divide by 4

$b = 5.5$

Therefore the least squares regression line for these data points is $y = 2.1x + 5.5$.

c. To find the correlation coefficient use the formula from the text and substitute the appropriate values from the table.

$$r = \frac{4(118) - (10)(43)}{\sqrt{4(30) - (10)^2}\sqrt{4(485) - (43)^2}}$$

$$r = 0.984$$

The correlation coefficient is $r = 0.984$.

d. Substitute $x = 5$ into the regression equation to get

$y = 2.1(5) + 5.5 = 16$.

The regression model predicts a 16% return on investments for the fifth year.

e. Substitute $y = 20$ into the regression equation and solve for x.

$2.1x + 5.5 = (20)$

$2.1x = 14.5$

$x = 6.9$

The regression model predicts that in the 7th year the fund will return 20% on the investment.

Chapter 2
Systems of Linear Equations

Exercises Section 2.1

1. The coefficient matrix of the system is:

$$\begin{bmatrix} 1 & 3 \\ 2 & -1 \end{bmatrix}$$

The augmented matrix for the system is:

$$\begin{bmatrix} 1 & 3 & | & 5 \\ 2 & -1 & | & -4 \end{bmatrix}$$

3. The coefficient matrix of the system is:

$$\begin{bmatrix} 1 & 3 & 0 \\ 2 & 1 & -1 \\ 1 & -1 & 1 \end{bmatrix}$$

The augmented matrix for the system is:

$$\begin{bmatrix} 1 & 3 & 0 & | & 0 \\ 2 & 1 & -1 & | & 0 \\ 1 & -1 & 1 & | & 0 \end{bmatrix}$$

5. The coefficient matrix of the system is:

$$\begin{bmatrix} 1 & 2 & 3 & 1 \\ 1 & 1 & -1 & 0 \\ 3 & 5 & 0 & 2 \end{bmatrix}$$

The augmented matrix for the system is:

$$\begin{bmatrix} 1 & 2 & 3 & 1 & | & 5 \\ 1 & 1 & -1 & 0 & | & 3 \\ 3 & 5 & 0 & 2 & | & -2 \end{bmatrix}$$

7. The coefficient matrix of the system is:

$$\begin{bmatrix} 5 & 3 & 2 \\ 1 & -2 & 0 \\ 3 & 1 & 1 \\ 1 & 0 & 4 \end{bmatrix}$$

The augmented matrix for the system is:

$$\begin{bmatrix} 5 & 3 & 2 & | & 0 \\ 1 & -2 & 0 & | & 0 \\ 3 & 1 & 1 & | & 1 \\ 1 & 0 & 4 & | & 2 \end{bmatrix}$$

9. To create the system, use the left side of the augmented matrix as the coefficients and the right side as your constants. The system is:

$$2x \quad\quad = 2$$
$$x + 3y = 1$$

11. To create the system, use the left side of the augmented matrix as the coefficients and the right side as your constants. The system is:

$$x \quad\quad = -8$$
$$\quad +y = \quad 7$$

13. To create the system, use the left side of the augmented matrix as the coefficients and the right side as your constants. The system is:

$$x \quad\quad = 18$$
$$\quad y \quad = 25$$
$$\quad\quad z = 30$$

15. Let x be the number of shares of Datafix, and y be the number of shares of Rocktite. The information can be placed into the following table:

	Shares of Datafix	Shares of Rocktite	Total
Price per share	$30	$20	$4000
Dividends per share	$2	$1	$220

The solution is continued at the top of the next page.

The system of equations that will model the information on the previous page is:

$$30x + 20y = 4000$$
$$2x + y = 220$$

17. Let x be the number of tons of coal used and y be the number of thousands of cubic feet of natural gas used. The information can be placed into the following table:

	Coal	Natural gas	Total
Megawatt-hours of electricity	3	4.50	183
Pounds of sulfur dioxide per hour	60	1	990

The system of equations that will model the information is:

$$3x + 4.50y = 183$$
$$60x + y = 990$$

19. Let x be the number of wooden pens, y be the number of silver pens, and z be the number of gold pens.

The information can be found placed in the following table:

	Wooden	Silver	Gold	Total
Grinder time	1	0.5	3	12000
Bonder time	2	2	2.5	9600

The system of equations that will model this information is:

$$x + 0.5y + 3z = 12000$$
$$2x + 2y + 2.5z = 9600$$

21. Let x be the number of skilled workers, y be the number of semiskilled workers, and z be the number of supervisors.

To satisfy the production schedule:

$$x + y = 140$$

The remaining information can be placed in the table at the top of the next column.

	Skilled	Semi-skilled	Supervisors	Total
Payroll	12	9	15	1560
Training	3	5	1	588

The equations that will model this information are:

$$12x + 9y + 15z = 1560$$
$$3x + 5y + z = 588$$

Put together the three equations to get the system of equations that will solve this problem. The system is:

$$x + y = 140$$
$$12x + 9y + 15z = 1560$$
$$3x + 5y + z = 588$$

23. Let A, B, C, and D be the number of pounds of the respective ingredients.

The amount of ingredients must total 200 pounds so the first equation is:

$$A + B + C + D = 200$$

The other equations can be found using the information in the following table:

	A	B	C	D	Total
Units of Vitamin A	2	1	3	1	300
Units of Protein	4	6	2	2	600

The equations that will model this information are:

$$2A + B + 3C + D = 300$$
$$4A + 6B + 2C + 2D = 600$$

Combine the three equations to get the system of equations that will solve this problem. The system is:

$$A + B + C + D = 200$$
$$2A + B + 3C + D = 300$$
$$4A + 6B + 2C + 2D = 600$$

25. Let s be the acres of soybeans, w be the acres of wheat, c be the acres of corn, and b be the acres of barley.

The information given in the problem can be organized in the table at the top of the next page.

	Soybean	Wheat	Corn	Barley	Total
profit	25	50	40	30	10 mil.
labor	5	6	8	3	50,000

The system of equations that will solve this problem is:

$$25s + 50w + 40c + 30b = 10,000,000$$
$$5s + 6w + 8c + 3b = 50,000$$

27. Let x be the number of workers on the first shift, y be the number of workers on the second shift, and z be the number of workers on the third shift.

Since the first and third shift each has to have the same number of workers, the first equation is:

$x = z$ or
$x - z = 0$

Since 10 workers must overlap the first and second shift, the second equation is:

$x + y = 10$

And finally since eight workers must overlap the second and third shift, the third equation is:

$y + z = 8$

Combine the three equations to get the system that solves this problem. The system is shown below:

$$x \quad - z = 0$$
$$x + y \quad = 10$$
$$y + z = 8$$

29. Let A be the number of highway miles in Alma, B be the number of highway miles in Bedford, and C be the number of highway miles in Cooksville.

Since the highway department needs to build twice as many miles in the Alma district then in the Bedford district, the first equation is:

$A = 2B$, or:
$A - 2B = 0$

The total budget (in millions of dollars) the highway department has to spend will give us the second equation:

$2A + 0.5B + C = 120$

Combine these two equations to get the system of equations that will model this problem. The system is:

$$2A + 0.5B + C = 120$$
$$A - 2B = 0$$

31. Let x be the number of four-ply tires and y be the number of radial tires.

Since 600 tires were bought total, the first equation is:

$x + y = 600$

The cost associated with each type of tire gives must equal the total amount spent, so the second equation is:

$40x + 50y = 28,500$

Finally since the manager wants to buy three times as many steel-belted radial tires as four-ply nylon tires, the third equation is:

$3x = y$, or:
$3x - y = 0$

Combining the three equations will produce the system of equations that will model this problem. The system is:

$$x + y = 600$$
$$40x + 50y = 28,500$$
$$3x - y = 0$$

33. Let x be the amount of 8% gasohol, and y be the amount of 13% gasohol.

This is a percent mixture problem and the information can be organized in the following table:

	8% mixture	13% mixture	10% mixture
Gallons	x	y	2000
% mixture	0.08	0.13	.10
Totals	$0.08x$	$0.13y$	200

To get the totals, multiply down each column. To get the system, add across the gallons row, and across the totals row.

The system of equations that will model this problem is:

$$x + y = 2000$$
$$0.08x + 0.13y = 200$$

35. Let x be the amount of the 21% solution, and y be the amount of the 14% solution.

Sometimes it is helpful to organize mixture information in the following chart:

	%mixture	Amount	totals
21% mixture	0.21	x	$0.21x$
14%mixture	0.14	y	$0.14y$
18%mixture	0.18	2	0.36

The chart is found by multiplying across the rows to get the totals. To find the equations needed add down the amount column and the totals column.

Since two liters are needed, the first equation is:

$x + y = 2$

The solution mixture requires that:

$0.21x + 0.14y = 0.36$

Combine these two equations to get the system of equations that models this problem. The system is:

$$x + \quad y = \quad 2$$
$$0.21x + 0.14y = 0.36$$

Exercises Section 2.2

1. The original matrix is

$$\begin{bmatrix} 1 & 1 & 1 & | & 2 \\ 0 & 2 & 0 & | & 4 \\ 0 & 1 & 3 & | & 2 \end{bmatrix}$$

$$-\frac{1}{2}R_2 \rightarrow R_2$$

$$\begin{bmatrix} 1 & 1 & 1 & | & 2 \\ 0 & -1 & 0 & | & -2 \\ 0 & 1 & 3 & | & 2 \end{bmatrix}$$

3. The original matrix is displayed at the top of the next column.

$$\begin{bmatrix} 1 & 4 & 0 & | & 5 \\ 0 & 2 & 2 & | & 4 \\ 0 & 1 & -1 & | & 0 \end{bmatrix}$$

Make the two in row 2, column 2 a one.

$$\frac{1}{2}R_2 \rightarrow R_2$$

$$\begin{bmatrix} 1 & 4 & 0 & | & 5 \\ 0 & 1 & 1 & | & 2 \\ 0 & 1 & -1 & | & 0 \end{bmatrix}$$

Now make the other elements in column 2 zero

$$-4R_2 + R_1 \rightarrow R_1$$

$$\begin{bmatrix} 1 & 0 & -4 & | & -3 \\ 0 & 1 & 1 & | & 2 \\ 0 & 1 & -1 & | & 0 \end{bmatrix}$$

$$-R_2 + R_3 \rightarrow R_3$$

$$\begin{bmatrix} 1 & 0 & -4 & | & -3 \\ 0 & 1 & 1 & | & 2 \\ 0 & 0 & -2 & | & -2 \end{bmatrix}$$

5. Since the element in row 1, column 1 is already a one, first make the element in row 2, column 2 a one.

The original matrix is:

$$\begin{bmatrix} 1 & 2 & | & 4 \\ 0 & 2 & | & 2 \end{bmatrix}$$

$$\frac{1}{2}R_2 \rightarrow R_2$$

$$\begin{bmatrix} 1 & 2 & | & 4 \\ 0 & 1 & | & 1 \end{bmatrix}$$

The solution is continued on the next page.

Now make the remaining element in column 2 a zero.

$$-2R_2 + R_1 \rightarrow R_1$$

$$\begin{bmatrix} 1 & 0 & | & 2 \\ 0 & 1 & | & 1 \end{bmatrix}$$

The matrix is in reduced row echelon form.

7. Since the element in row 1, column 1 is already a one with zeros in the other rows, and the element in row 2, column 2 is a one, make the other elements in column 2 zeros.

The original matrix is:

$$\begin{bmatrix} 1 & 2 & 1 & | & 7 \\ 0 & 1 & 3 & | & 4 \\ 0 & -1 & 0 & | & 0 \end{bmatrix}$$

$$-2R_2 + R_1 \rightarrow R_1$$

$$\begin{bmatrix} 1 & 0 & -5 & | & -1 \\ 0 & 1 & 3 & | & 4 \\ 0 & -1 & 0 & | & 0 \end{bmatrix}$$

$$R_2 + R_3 \rightarrow R_3$$

$$\begin{bmatrix} 1 & 0 & -5 & | & -1 \\ 0 & 1 & 3 & | & 4 \\ 0 & 0 & 3 & | & 4 \end{bmatrix}$$

Now make the element in row 3, column 3 a one.

$$\frac{1}{3}R_2 \rightarrow R_2$$

$$\begin{bmatrix} 1 & 0 & -5 & | & -1 \\ 0 & 1 & 3 & | & 4 \\ 0 & 0 & 1 & | & \frac{4}{3} \end{bmatrix}$$

Now make the remaining elements in column 3 zero.

$$5R_3 + R_1 \rightarrow R_1$$

$$\begin{bmatrix} 1 & 0 & 0 & | & \frac{17}{3} \\ 0 & 1 & 3 & | & 4 \\ 0 & 0 & 1 & | & \frac{4}{3} \end{bmatrix}$$

$$-3R_3 + R_2 \rightarrow R_2$$

$$\begin{bmatrix} 1 & 0 & 0 & | & \frac{17}{3} \\ 0 & 1 & 0 & | & 0 \\ 0 & 0 & 1 & | & \frac{4}{3} \end{bmatrix}$$

The matrix is in reduced row echelon form.

9. The augmented matrix is:

$$\begin{bmatrix} 1 & 1 & 1 & | & 4 \\ 0 & 1 & -1 & | & -1 \\ 1 & 3 & 1 & | & 6 \end{bmatrix}$$

Begin by pivoting on row 1, column 1. Since element 1, 1 is already a one; make the remaining elements in column 1 a zero.

$$-R_1 + R_3 \rightarrow R_3$$

$$\begin{bmatrix} 1 & 1 & 1 & | & 4 \\ 0 & 1 & -1 & | & -1 \\ 0 & 2 & 0 & | & 2 \end{bmatrix}$$

Next pivot on row 2, column 2. Since element 2, 2 is already a one, make the remaining elements in the column a zero.

$$-R_2 + R_1 \rightarrow R_1$$

$$\begin{bmatrix} 1 & 0 & 2 & | & 5 \\ 0 & 1 & -1 & | & -1 \\ 0 & 2 & 0 & | & 2 \end{bmatrix}$$

$$-2R_2 + R_3 \rightarrow R_3$$

$$\begin{bmatrix} 1 & 0 & 2 & | & 5 \\ 0 & 1 & -1 & | & -1 \\ 0 & 0 & 2 & | & 4 \end{bmatrix}$$

The solution is continued at the top of the next page.

Now pivot on row 3, column 3. We need to make element 3, 3 a one first, and then make the remaining elements in column 3 a zero.

$$\frac{1}{2}R_3 \rightarrow R_3$$

$$\begin{bmatrix} 1 & 0 & 2 & | & 5 \\ 0 & 1 & -1 & | & -1 \\ 0 & 0 & 1 & | & 2 \end{bmatrix}$$

$$-2R_3 + R_1 \rightarrow R_1$$

$$\begin{bmatrix} 1 & 0 & 0 & | & 1 \\ 0 & 1 & -1 & | & -1 \\ 0 & 0 & 1 & | & 2 \end{bmatrix}$$

$$R_3 + R_2 \rightarrow R_2$$

The reduced matrix is:

.

$$\begin{bmatrix} 1 & 0 & 0 & | & 1 \\ 0 & 1 & 0 & | & 1 \\ 0 & 0 & 1 & | & 2 \end{bmatrix}$$

Now that the matrix is in reduced row echelon form, the solution is:

$$x = 1;\ y = 1;\ z = 2\ .$$

11. The augmented matrix is

$$\begin{bmatrix} 0 & 1 & -3 & | & -11 \\ 1 & -1 & 1 & | & 5 \\ 3 & 0 & 1 & | & 10 \end{bmatrix}$$

Begin by pivoting on row 1, column 1. Element 1, 1 must be equal to one. Switch row 1 and row 2 to do this.

$$R_1 \leftrightarrow R_2$$

$$\begin{bmatrix} 1 & -1 & 1 & | & 5 \\ 0 & 1 & -3 & | & -11 \\ 3 & 0 & 1 & | & 10 \end{bmatrix}$$

Now make the remaining elements in column 1 zero.

$$-3R_1 + R_3 \rightarrow R_3$$

$$\begin{bmatrix} 1 & -1 & 1 & | & 5 \\ 0 & 1 & -3 & | & -11 \\ 0 & 3 & -2 & | & -5 \end{bmatrix}$$

Next pivot on row 2, column 2, since element 2, 2 is already equal to one, make the remaining elements in column 2 zero.

$$R_2 + R_1 \rightarrow R_1$$

$$\begin{bmatrix} 1 & 0 & -2 & | & -6 \\ 0 & 1 & -3 & | & -11 \\ 0 & 3 & -2 & | & -5 \end{bmatrix}$$

$$-3R_2 + R_3 \rightarrow R_3$$

$$\begin{bmatrix} 1 & 0 & -2 & | & -6 \\ 0 & 1 & -3 & | & -11 \\ 0 & 0 & 7 & | & 28 \end{bmatrix}$$

Next pivot on row 3, column 3, begin the pivoting process by making element 3, 3 equal to one.

$$\frac{1}{7}R_3 \rightarrow R_3$$

$$\begin{bmatrix} 1 & 0 & -2 & | & -6 \\ 0 & 1 & -3 & | & -11 \\ 0 & 0 & 1 & | & 4 \end{bmatrix}$$

Now make the remaining elements in column 3 zero.

$$2R_3 + R_1 \rightarrow R_1$$

$$\begin{bmatrix} 1 & 0 & 0 & | & 2 \\ 0 & 1 & -3 & | & -11 \\ 0 & 0 & 1 & | & 4 \end{bmatrix}$$

The solution is continued on the next page.

$3R_3 + R_2 \rightarrow R_2$

$$\begin{bmatrix} 1 & 0 & 0 & | & 2 \\ 0 & 1 & 0 & | & 1 \\ 0 & 0 & 1 & | & 4 \end{bmatrix}$$

Now that the matrix is in reduced row echelon form, the solution is:

$x = 2, y = 1, z = 4$.

13. The augmented matrix is:

$$\begin{bmatrix} 2 & 4 & 2 & | & 5 \\ 1 & 1 & 1 & | & \frac{3}{2} \\ 2 & 1 & 4 & | & 4 \end{bmatrix}$$

Begin by pivoting on row 1, column 1. Element 1, 1 must be equal to one. Switch row 1 and row 2 to do this.

$R_1 \leftrightarrow R_2$

$$\begin{bmatrix} 1 & 1 & 1 & | & \frac{3}{2} \\ 2 & 4 & 2 & | & 5 \\ 2 & 1 & 4 & | & 4 \end{bmatrix}$$

Now make the remaining elements in column 1 zero.

$-2R_1 + R_2 \rightarrow R_2$

$$\begin{bmatrix} 1 & 1 & 1 & | & \frac{3}{2} \\ 0 & 2 & 0 & | & 2 \\ 2 & 1 & 4 & | & 4 \end{bmatrix}$$

$-2R_1 + R_3 \rightarrow R_3$

$$\begin{bmatrix} 1 & 1 & 1 & | & \frac{3}{2} \\ 0 & 2 & 0 & | & 2 \\ 0 & -1 & 2 & | & 1 \end{bmatrix}$$

Next pivot on row 2, column 2, begin the pivoting process by making element 2, 2 equal to 1.

$\dfrac{1}{2}R_2 \rightarrow R_2$

The equivalent matrix is:

$$\begin{bmatrix} 1 & 1 & 1 & | & \frac{3}{2} \\ 0 & 1 & 0 & | & 1 \\ 0 & -1 & 2 & | & 1 \end{bmatrix}$$

Now make the remaining elements in column 2 zero.

$-R_2 + R_1 \rightarrow R_1$

$$\begin{bmatrix} 1 & 0 & 1 & | & \frac{1}{2} \\ 0 & 1 & 0 & | & 1 \\ 0 & -1 & 2 & | & 1 \end{bmatrix}$$

$R_2 + R_3 \rightarrow R_3$

$$\begin{bmatrix} 1 & 0 & 1 & | & \frac{1}{2} \\ 0 & 1 & 0 & | & 1 \\ 0 & 0 & 2 & | & 2 \end{bmatrix}$$

Next pivot on row 3, column 3, begin the pivoting process by making element 3, 3 equal to 1.

$\dfrac{1}{2}R_3 \rightarrow R_3$

$$\begin{bmatrix} 1 & 0 & 1 & | & \frac{1}{2} \\ 0 & 1 & 0 & | & 1 \\ 0 & 0 & 1 & | & 1 \end{bmatrix}$$

Now make the remaining element in column 3 zero.

$-R_3 + R_1 \rightarrow R_1$

$$\begin{bmatrix} 1 & 0 & 0 & | & \frac{-1}{2} \\ 0 & 1 & 0 & | & 1 \\ 0 & 0 & 1 & | & 1 \end{bmatrix}$$

Now that the matrix is in reduced row echelon form, the solution is:

$x = \frac{-1}{2}; y = 1; z = 1$

15. The augmented matrix is

$$\begin{bmatrix} 1 & 2 & 1 & 3 \\ 3 & 3 & 3 & 7 \\ 2 & 1 & 2 & 1 \end{bmatrix}$$

Begin by pivoting on row 1, column 1. Element 1, 1 is equal to one, therefore make the remaining elements in column 1 zero.

$$-3R_1 + R_2 \rightarrow R_2$$

$$\begin{bmatrix} 1 & 2 & 1 & 3 \\ 0 & -3 & 0 & -2 \\ 2 & 1 & 2 & 1 \end{bmatrix}$$

$$-2R_1 + R_3 \rightarrow R_3$$

$$\begin{bmatrix} 1 & 2 & 1 & 3 \\ 0 & -3 & 0 & -2 \\ 0 & -3 & 0 & -5 \end{bmatrix}$$

Next pivot on row 2, column 2, start the pivoting process by making element 2, 2 equal to one.

$$-\frac{1}{3}R_2 \rightarrow R_2$$

$$\begin{bmatrix} 1 & 2 & 1 & 3 \\ 0 & 1 & 0 & \frac{2}{3} \\ 0 & -3 & 0 & -5 \end{bmatrix}$$

Now make the remaining elements in column 2 zero.

$$-2R_2 + R_1 \rightarrow R_1$$

$$\begin{bmatrix} 1 & 0 & 1 & \frac{5}{3} \\ 0 & 1 & 0 & \frac{2}{3} \\ 0 & -3 & 0 & -5 \end{bmatrix}$$

$$3R_2 + R_3 \rightarrow R_3$$

The equivalent matrix is shown at the top of the next column.

$$\begin{bmatrix} 1 & 0 & 1 & \frac{5}{3} \\ 0 & 1 & 0 & \frac{2}{3} \\ 0 & 0 & 0 & -3 \end{bmatrix}$$

We see that the last row is inconsistent. We can stop the Gauss-Jordan process, the system has no solution.

17. The augmented matrix is:

$$\begin{bmatrix} 1 & 3 & 8 \\ 2 & -1 & 4 \end{bmatrix}$$

Begin by pivoting on row 1, column 1. Since element 1, 1 is equal to one, make the remaining element column 1 zero.

$$-2R_1 + R_2 \rightarrow R_2$$

$$\begin{bmatrix} 1 & 3 & 8 \\ 0 & -7 & -12 \end{bmatrix}$$

Next pivot on row 2 column2, to begin the pivoting process make element 2, 2 equal to 1.

$$-\frac{1}{7}R_2 \rightarrow R_2$$

$$\begin{bmatrix} 1 & 3 & 8 \\ 0 & 1 & \frac{12}{7} \end{bmatrix}$$

Now make the remaining element in column 2 zero.

$$-3R_2 + R_1 \rightarrow R_1$$

The reduced matrix is:

$$\begin{bmatrix} 1 & 0 & \frac{20}{7} \\ 0 & 1 & \frac{12}{7} \end{bmatrix}$$

Now that the matrix is in reduced row echelon form, the solution is:

$$x = \frac{20}{7}; y = \frac{12}{7}.$$

19. The augmented matrix is

$$\begin{bmatrix} 3 & 1 & | & -1 \\ 1 & -2 & | & 3 \\ 4 & -1 & | & 2 \\ 8 & 5 & | & -6 \end{bmatrix}$$

Begin by pivoting on row 1, column 1. Element 1, 1 must be equal to one. Switch row 1 and row 2 to do this.

$$R_1 \leftrightarrow R_2$$

$$\begin{bmatrix} 1 & -2 & | & 3 \\ 3 & 1 & | & -1 \\ 4 & -1 & | & 2 \\ 8 & 5 & | & -6 \end{bmatrix}$$

Next make the remaining elements in column 1 zero. In order to save space, I will do the following operations in one matrix. If you chose to combine steps, be very careful in your bookkeeping.

Perform the following three operations:

$$-3R_1 + R_2 \rightarrow R_2$$
$$-4R_1 + R_3 \rightarrow R_3$$
$$-8R_1 + R_4 \rightarrow R_4$$

$$\begin{bmatrix} 1 & -2 & | & 3 \\ 0 & 7 & | & -10 \\ 0 & 7 & | & -10 \\ 0 & 21 & | & -30 \end{bmatrix}$$

Next pivot on row 2, column 2, start the pivoting process by making element 2, 2 equal to 1.

$$\frac{1}{7}R_2 \rightarrow R_2$$

$$\begin{bmatrix} 1 & -2 & | & 3 \\ 0 & 1 & | & \frac{-10}{7} \\ 0 & 7 & | & -10 \\ 0 & 21 & | & -30 \end{bmatrix}$$

Next make the remaining elements in column 2 zero. Again, I will perform all three row operations in one step.

$$2R_2 + R_1 \rightarrow R_1$$
$$-7R_2 + R_3 \rightarrow R_3$$
$$-21R_2 + R_4 \rightarrow R_4$$

$$\begin{bmatrix} 1 & 0 & | & \frac{1}{7} \\ 0 & 1 & | & \frac{-10}{7} \\ 0 & 0 & | & 0 \\ 0 & 0 & | & 0 \end{bmatrix}$$

Now that the matrix is in reduced row echelon form, the solution is:

$$x = \tfrac{1}{7}; \; y = \tfrac{-10}{7}.$$

21. The augmented matrix is

$$\begin{bmatrix} 1 & 0 & 1 & | & -2 \\ 1 & 1 & 1 & | & -2 \\ 3 & 2 & 2 & | & -3 \\ 2 & 1 & 0 & | & 2 \end{bmatrix}$$

Begin by pivoting on row 1, column 1. Since element 1, 1 is equal to one, make the remaining elements column 1 zero by performing the following 3 row operations:

$$-R_1 + R_2 \rightarrow R_2$$
$$-3R_1 + R_3 \rightarrow R_3$$
$$-2R_1 + R_4 \rightarrow R_4$$

$$\begin{bmatrix} 1 & 0 & 1 & | & -2 \\ 0 & 1 & 0 & | & 0 \\ 0 & 2 & -1 & | & 3 \\ 0 & 1 & -2 & | & 6 \end{bmatrix}$$

Next pivot on row 2, column 2, since element 2, 2 is already equal to 1, make the remaining elements in column 2 zero by performing the following row operations:

$$-2R_2 + R_3 \rightarrow R_3$$
$$-R_2 + R_4 \rightarrow R_4$$

The matrix is displayed at the top of the next page.

$$\begin{bmatrix} 1 & 0 & 1 & -2 \\ 0 & 1 & 0 & 0 \\ 0 & 0 & -1 & 3 \\ 0 & 0 & -2 & 6 \end{bmatrix}$$

Next pivot on row 3, column 3, by making element 3, 3 equal to one.

$$-R_3 \rightarrow R_3$$

$$\begin{bmatrix} 1 & 0 & 1 & -2 \\ 0 & 1 & 0 & 0 \\ 0 & 0 & 1 & -3 \\ 0 & 0 & -2 & 6 \end{bmatrix}$$

Now make the remaining elements in column 3 zero by performing the following row operations:

$$-R_3 + R_1 \rightarrow R_1$$
$$2R_3 + R_4 \rightarrow R_4$$

$$\begin{bmatrix} 1 & 0 & 0 & 1 \\ 0 & 1 & 0 & 0 \\ 0 & 0 & 1 & -3 \\ 0 & 0 & 0 & 0 \end{bmatrix}$$

Now that the matrix is in reduced row echelon form, the solution is:

$$x = 1; \; y = 0; \; z = -3.$$

23. The augmented matrix is:

$$\begin{bmatrix} 1 & 1 & -3 & 2 \\ 2 & 0 & -1 & 1 \\ 3 & -2 & 0 & 0 \\ 6 & -1 & -3 & 5 \end{bmatrix}$$

Pivot on row 1, column 1, since element 1, 1 is already a one, make the other elements in column 1 zero by performing the following operations:

$$-2R_1 + R_2 \rightarrow R_2$$
$$-3R_1 + R_3 \rightarrow R_3$$
$$-6R_1 + R_4 \rightarrow R_4$$

$$\begin{bmatrix} 1 & 1 & -3 & 2 \\ 0 & -2 & 5 & -3 \\ 0 & -5 & 9 & -6 \\ 0 & -7 & 15 & -7 \end{bmatrix}$$

Next pivot on row 2, column 2, begin the pivoting process by making element 2, 2 equal to one.

$$-\frac{1}{2}R_2 \rightarrow R_2$$

$$\begin{bmatrix} 1 & 1 & -3 & 2 \\ 0 & 1 & \frac{-5}{2} & \frac{3}{2} \\ 0 & -5 & 9 & -6 \\ 0 & -7 & 15 & -7 \end{bmatrix}$$

Now make the remaining elements in column 2 zero by performing the following row operations:

$$-R_2 + R_1 \rightarrow R_1$$
$$5R_2 + R_3 \rightarrow R_3$$
$$7R_2 + R_4 \rightarrow R_4$$

$$\begin{bmatrix} 1 & 0 & \frac{-1}{2} & \frac{1}{2} \\ 0 & 1 & \frac{-5}{2} & \frac{3}{2} \\ 0 & 0 & \frac{-7}{2} & \frac{3}{2} \\ 0 & 0 & \frac{-5}{2} & \frac{7}{2} \end{bmatrix}$$

Next pivot on row 3, column 3 by making element 3, 3 equal to 1

$$-\frac{2}{7}R_3 \rightarrow R_3$$

The matrix is:

$$\begin{bmatrix} 1 & 0 & \frac{-1}{2} & \frac{1}{2} \\ 0 & 1 & \frac{-5}{2} & \frac{3}{2} \\ 0 & 0 & 1 & \frac{-3}{7} \\ 0 & 0 & \frac{-5}{2} & \frac{7}{2} \end{bmatrix}$$

The solution is continued on the next page.

We notice that row 3 and 4 of the previous matrix are not multiples of each other. Performing the following row operations:

$$\frac{5}{2}R_3 + R_4 \to R_4$$

$$\begin{bmatrix} 1 & 0 & \frac{-1}{2} & \Big| & \frac{1}{2} \\ 0 & 1 & \frac{-5}{2} & \Big| & \frac{3}{2} \\ 0 & 0 & 1 & \Big| & \frac{-3}{7} \\ 0 & 0 & 0 & \Big| & \frac{17}{7} \end{bmatrix}$$

Row four is inconsistent. Therefore the system has no solution.

25. The augmented matrix is

$$\begin{bmatrix} 1 & 1 & 3 & \Big| & 0 \\ \frac{1}{2} & -1 & \frac{-7}{2} & \Big| & 0 \\ 1 & 2 & 0 & \Big| & 0 \end{bmatrix}$$

Pivot on row 1, column 1, since element 1, 1 is already a one, make the other elements in column 1 zero by performing the following operations:

$$-\frac{1}{2}R_1 + R_2 \to R_2$$

$$-R_1 + R_3 \to R_3$$

$$\begin{bmatrix} 1 & 1 & 3 & \Big| & 0 \\ 0 & \frac{-3}{2} & -5 & \Big| & 0 \\ 0 & 1 & -3 & \Big| & 0 \end{bmatrix}$$

Next pivot on row 2, column 2 by making element 2, 2 equal to 1.

$$-\frac{2}{3}R_2 \to R_2$$

$$\begin{bmatrix} 1 & 1 & 3 & \Big| & 0 \\ 0 & 1 & \frac{10}{3} & \Big| & 0 \\ 0 & 1 & -3 & \Big| & 0 \end{bmatrix}$$

Next make the remaining elements in column 2 zero by performing the following row operations at the top of the next column.

$$-R_2 + R_1 \to R_1$$

$$-R_2 + R_3 \to R_3$$

$$\begin{bmatrix} 1 & 0 & \frac{-1}{3} & \Big| & 0 \\ 0 & 1 & \frac{10}{3} & \Big| & 0 \\ 0 & 0 & \frac{-19}{3} & \Big| & 0 \end{bmatrix}$$

From this point it should be clear that $z = 0$, which in turn means that $y = 0$, and $x = 0$. The only solution to this system is:

$$x = 0;\ y = 0;\ z = 0$$

27. Enter the augmented matrix into you graphing calculator. The augment matrix is:

$$\begin{bmatrix} 1 & 1 & 1 & -1 & \Big| & 1 \\ 2 & \frac{1}{2} & 0 & \frac{2}{3} & \Big| & \frac{5}{12} \\ -1 & -1 & \frac{1}{5} & 0 & \Big| & \frac{7}{20} \\ 1 & \frac{3}{4} & 1 & 4 & \Big| & \frac{17}{8} \end{bmatrix}$$

Pivot on row 1, column 1 making the remaining elements zero by performing the proper row operations. The calculator should show the following:

$$\begin{bmatrix} 1 & 1 & 1 & -1 & \Big| & 1 \\ 0 & \frac{-3}{2} & -2 & \frac{8}{3} & \Big| & \frac{-19}{12} \\ 0 & 0 & \frac{6}{5} & -1 & \Big| & \frac{27}{20} \\ 0 & \frac{-1}{4} & 0 & 5 & \Big| & \frac{9}{8} \end{bmatrix}$$

Next pivot on row 2, column 2 by performing the operation $\frac{-2}{3}R_2 \to R_2$, and then making the remaining elements zero by performing the proper row operations:

The calculator should show the following:

$$\begin{bmatrix} 1 & 1 & 1 & -1 & \Big| & 1 \\ 0 & 1 & \frac{4}{3} & \frac{16}{9} & \Big| & \frac{19}{18} \\ 0 & 0 & \frac{6}{5} & -1 & \Big| & \frac{27}{20} \\ 0 & 0 & \frac{1}{3} & \frac{41}{9} & \Big| & \frac{25}{18} \end{bmatrix}$$

The solution is continued on the next page.

Next pivot on row 3, column 3 by performing the operation $\frac{5}{6} R_3 \to R_3$, and then making the remaining elements zero by performing the proper row operations. The calculator should show the following:

$$\begin{bmatrix} 1 & 1 & 1 & -1 & 1 \\ 0 & 1 & \frac{4}{3} & \frac{16}{9} & \frac{19}{18} \\ 0 & 0 & 1 & \frac{-5}{6} & \frac{9}{8} \\ 0 & 0 & 0 & \frac{29}{6} & \frac{73}{72} \end{bmatrix}$$

Next pivot on row 3, column 3 by performing the operation $\frac{6}{29} R_4 \to R_4$, and then making the remaining elements above the main diagonal zero by performing the proper row operations. The calculator should show the following:

$$\begin{bmatrix} 1 & 0 & 0 & 0 & \frac{56}{261} \\ 0 & 1 & 0 & 0 & \frac{-53}{174} \\ 0 & 0 & 1 & 0 & \frac{1357}{1044} \\ 0 & 0 & 0 & 1 & \frac{73}{348} \end{bmatrix}$$

The matrix is in reduced row echelon form, and the solution to the system is

$$w = \tfrac{56}{261}; x = \tfrac{-53}{174}; y = \tfrac{1357}{1044}; x = \tfrac{73}{348}$$

29. Enter the augmented matrix into you graphing calculator. The augment matrix is:

$$\begin{bmatrix} 0.2 & -0.82 & 0.03 & -1.96 \\ 0.02 & 0.14 & -0.6 & -2.82 \\ 0.47 & 0.22 & 0.62 & 3.92 \end{bmatrix}$$

Pivot on row 1, column 1 first, then pivot on row 2, column 2, and then pivot on row 3, column 3. The final matrix should look like the following:

$$\begin{bmatrix} 1 & 0 & 0 & .1095 \\ 0 & 1 & 0 & 2.611 \\ 0 & 0 & 1 & 5.313 \end{bmatrix}$$

The matrix is in reduced row echelon form, and the solution is

$$x = .1095; y = 2.611; z = 5.313$$

31. Enter the augmented matrix into your calculator. The augmented matrix is:

$$\begin{bmatrix} 1 & -1 & 2 & 1 & 9 \\ 0.3 & 0 & 0.52 & -1 & 1.62 \\ 0.6 & 2 & -1 & 0 & 3 \\ 0.2 & 0.43 & 0.6 & 1 & 2.15 \end{bmatrix}$$

Pivot on row 1, column 1 to get the following matrix:

$$\begin{bmatrix} 1 & -1 & 2 & 1 & 9 \\ 0 & .3 & -.08 & -1.3 & -1.08 \\ 0 & 2.6 & -2.2 & -0.6 & -2.4 \\ 0 & 0.63 & 0.2 & 0.8 & .35 \end{bmatrix}$$

Multiply row 2 by the reciprocal of element 2, 2 $\left(\frac{1}{0.3}\right)$ to get row 2, column 2 ready to pivot. After performing the pivot operation the following matrix is obtained.

$$\begin{bmatrix} 1 & 0 & \frac{26}{15} & \frac{-10}{3} & \frac{27}{5} \\ 0 & 1 & \frac{-4}{15} & \frac{-13}{3} & \frac{-18}{5} \\ 0 & 0 & \frac{-113}{75} & \frac{32}{3} & \frac{174}{25} \\ 0 & 0 & \frac{46}{125} & \frac{353}{100} & \frac{1309}{500} \end{bmatrix}$$

Multiply row 3 by the reciprocal of element 3, 3 $\left(\frac{-75}{113}\right)$ to get row 3, column 3 ready to pivot. After performing the pivot operation the following matrix is obtained.

$$\begin{bmatrix} 1 & 0 & 0 & \frac{1010}{113} & \frac{1515}{113} \\ 0 & 1 & 0 & \frac{-703}{113} & \frac{-546}{113} \\ 0 & 0 & 1 & \frac{-800}{113} & \frac{-522}{113} \\ 0 & 0 & 0 & 6.135 & 4.317 \end{bmatrix}$$

Multiply row 4 by the reciprocal of element 4, 4 to get row 4 column 4 ready to pivot. After performing the pivot operation the following matrix is obtained.

$$\begin{bmatrix} 1 & 0 & 0 & 0 & 7.116 \\ 0 & 1 & 0 & 0 & -0.4534 \\ 0 & 0 & 1 & 0 & 0.3631 \\ 0 & 0 & 0 & 1 & 0.704 \end{bmatrix}$$

The matrix on the previous page is in reduced row echelon form and the solution is:

$$w \approx 7.12; \ x \approx -0.45; \ y \approx 0.36; \ z \approx 0.70 \ .$$

33. The set up to this problem can be found in exercise 17 out of chapter 2.1.

The system is:

$$3x + 4.50y = 183$$
$$60x + \quad y = 990$$

Where x is tons of coal used, and y is thousands of cubic feet of gas.

The augmented matrix from the system is:

$$\begin{bmatrix} 3 & 4.50 & | & 183 \\ 60 & 1 & | & 990 \end{bmatrix}$$

Pivot on row 1, column 1, to make the element 1, 1 equal to one by performing the operation:

$$\frac{1}{3} R_1 \rightarrow R_1$$

$$\begin{bmatrix} 1 & 1.5 & | & 61 \\ 60 & 1 & | & 990 \end{bmatrix}$$

Now make the remaining elements in column 1 zero by performing the row operation:

$$-60 R_1 + R_2 \rightarrow R_2$$

$$\begin{bmatrix} 1 & 1.5 & | & 61 \\ 0 & -89 & | & -2670 \end{bmatrix}$$

Pivot on row 2, column 2 by performing the operation:

$$\frac{-1}{89} R_2 \rightarrow R_2$$

$$\begin{bmatrix} 1 & 1.5 & | & 61 \\ 0 & 1 & | & 30 \end{bmatrix}$$

Now make the remaining element in column 2 zero by performing the row operation:

$$-1.5 R_2 + R_1 \rightarrow R_1$$

$$\begin{bmatrix} 1 & 0 & | & 16 \\ 0 & 1 & | & 30 \end{bmatrix}$$

The electrical plant should use 16 tons of coal per hour and 30 thousand cubic feet of gas per hour to produce the desired amount of electricity while exactly meeting the maximum pollution allowance.

35. Let x be the number of Q6 models, y be the number of Q10 models, and z be the number of Q12 models.

The printer budget yields the equation:

$$800x + 1500y + 3000z = 11,400$$

The maintenance budget constraint yields the equation:

$$20x + 40y + 120z = 380$$

The Q6 prints 6 pages per minute or 360 pages per hour. The Q10 prints 10 pages per minute or 600 pages per hour. The Q12 prints 12 pages per minute or 720 pages per hour. Using these conversions, the output constraint yields the equation:

$$360x + 600y + 720z = 3720$$

The memory constraint yields the equation:

$$2x + 4y + 12z = 38$$

Putting the four equations together we get the system:

$$\begin{bmatrix} 800 & 1500 & 3000 & | & 11400 \\ 20 & 40 & 120 & | & 380 \\ 360 & 600 & 720 & | & 3720 \\ 2 & 4 & 12 & | & 38 \end{bmatrix}$$

Pivoting on row 1, column 1, we will need to perform the following operations as shown at the top of the next page.

First make the element 1, 1 equal to 1.

$$\frac{1}{800} R_1 \rightarrow R_1$$

$$\begin{bmatrix} 1 & \frac{15}{8} & \frac{15}{4} & \frac{57}{4} \\ 20 & 40 & 120 & 380 \\ 360 & 600 & 720 & 3720 \\ 2 & 4 & 12 & 38 \end{bmatrix}$$

Now make the remaining elements in column 1 zero by performing the following operations:

$$-20R_1 + R_2 \rightarrow R_2$$
$$-360R_1 + R_3 \rightarrow R_3$$
$$-2R_1 + R_4 \rightarrow R_4$$

$$\begin{bmatrix} 1 & \frac{15}{8} & \frac{15}{4} & \frac{57}{4} \\ 0 & \frac{5}{2} & 45 & 95 \\ 0 & -75 & -630 & -1410 \\ 0 & \frac{1}{4} & \frac{9}{2} & \frac{19}{2} \end{bmatrix}$$

Now pivot on row 2, column 2.

The four pivot operations should be performed in order:

$$\tfrac{2}{5} R_2 \rightarrow R_2$$
$$\tfrac{-15}{8} R_2 + R_1 \rightarrow R_1$$
$$75R_2 + R_3 \rightarrow R_3$$
$$\tfrac{-1}{4} R_2 + R_4 \rightarrow R_4$$

$$\begin{bmatrix} 1 & 0 & -30 & -57 \\ 0 & 1 & 18 & 38 \\ 0 & 0 & 720 & 1440 \\ 0 & 0 & 0 & 0 \end{bmatrix}$$

Finally pivot on row 3, column 3. The following operations should be performed in the exact order:

$$\tfrac{1}{720} R_3 \rightarrow R_3$$
$$30R_3 + R_1 \rightarrow R_1$$
$$-18R_3 + R_2 \rightarrow R_2$$

$$\begin{bmatrix} 1 & 0 & 0 & 3 \\ 0 & 1 & 0 & 2 \\ 0 & 0 & 1 & 2 \\ 0 & 0 & 0 & 0 \end{bmatrix}$$

The company should buy 3 of the Q6 printers, 2 of the Q10 printers and 2 of the Q12 printer to meet its needs.

37. Let x be the number of newspaper ads, y be the number of radio ads, and z be the number of TV ads.

The fact that the charity wants to run a total of 24 ads produces the equation:

$$x + y + z = 24$$

The condition that they run 3 times as many radio ads as TV ads yields the equation:

$$y = 3z \text{ or } y - 3z = 0$$

To meet the effective rating points of 188, the equation is:

$$7x + 8y + 9z = 188$$

Put these three equations together to get the system:

$$\begin{aligned} x + \;\; y + \;z &= \;\; 24 \\ y - 3z &= \;\;\;\; 0 \\ 7x + 8y + 9z &= 188 \end{aligned}$$

The augmented matrix for this system is:

$$\begin{bmatrix} 1 & 1 & 1 & 24 \\ 0 & 1 & -3 & 0 \\ 7 & 8 & 9 & 188 \end{bmatrix}$$

Pivot on the first row first column. The only operation that we need to perform is:

$$-7R_1 + R_3 \rightarrow R_3$$

$$\begin{bmatrix} 1 & 1 & 1 & 24 \\ 0 & 1 & -3 & 0 \\ 0 & 1 & 2 & 20 \end{bmatrix}$$

Next pivot on row 2, column 2, the operations that are required to complete this pivot are displayed at the top of the next page.

$-R_2 + R_1 \rightarrow R_1$
$-R_2 + R_3 \rightarrow R_3$

$$\begin{bmatrix} 1 & 0 & 4 & | & 24 \\ 0 & 1 & -3 & | & 0 \\ 0 & 0 & 5 & | & 20 \end{bmatrix}$$

Now pivot on row 3, column 3, the operations that we need to perform are:

$\dfrac{1}{5} R_3 \rightarrow R_3$

$-4R_3 + R_1 \rightarrow R_1$

$3R_3 + R_2 \rightarrow R_2$

$$\begin{bmatrix} 1 & 0 & 0 & | & 8 \\ 0 & 1 & 0 & | & 12 \\ 0 & 0 & 1 & | & 4 \end{bmatrix}$$

The charity should take out eight newspaper ads, 12 radio ads and four TV ads to meet their advertising needs.

39. The set up for this problem can be found in exercise 27 out of section 2.1. The system of equations is:

$x \quad - z = 0$
$x + y \quad = 10$
$\quad y + z = 8$

The augmented matrix associated with this system is:

$$\begin{bmatrix} 1 & 0 & -1 & | & 0 \\ 1 & 1 & 0 & | & 10 \\ 0 & 1 & 1 & | & 8 \end{bmatrix}$$

Pivot on the row 1, column 1 by performing the operation:

$-R_1 + R_2 \rightarrow R_2$

$$\begin{bmatrix} 1 & 0 & -1 & | & 0 \\ 0 & 1 & 1 & | & 10 \\ 0 & 1 & 1 & | & 8 \end{bmatrix}$$

Next pivot on row 2, column 2 by performing the operation at the top of the next column.

$-R_2 + R_3 \rightarrow R_3$

$$\begin{bmatrix} 1 & 0 & -1 & | & 0 \\ 0 & 1 & 1 & | & 10 \\ 0 & 0 & 0 & | & -2 \end{bmatrix}$$

The third row is inconsistent. Therefore there is no solution to this system. This means that Brenda cannot staff the three shifts to meet the given conditions.

41. The set up for this problem can be seen in exercise 31 out of section 2.1.

The system of equations that models this problem is:

$x + \quad y = \quad 600$
$40x + 50y = 28,500$
$3x \quad - y = \quad 0$

Where, x is the number of four-ply tires and y is the number of radial tires.

The augmented matrix to this system is:

$$\begin{bmatrix} 1 & 1 & | & 600 \\ 40 & 50 & | & 28500 \\ 3 & -1 & | & 0 \end{bmatrix}$$

Pivot on row 1, column 1 by performing the row operations:

$-40R_1 + R_2 \rightarrow R_2$
$-3R_1 + R_3 \rightarrow R_3$

$$\begin{bmatrix} 1 & 1 & | & 600 \\ 0 & 10 & | & 4500 \\ 0 & -4 & | & -1800 \end{bmatrix}$$

Now pivot on row 2, column 2 by performing the following row operations:

$\dfrac{1}{10} R_2 \rightarrow R_2$

$-R_2 + R_1 \rightarrow R_1$

$4R_2 + R_3 \rightarrow R_3$

The matrix is shown at the top of the next page.

$$\begin{bmatrix} 1 & 0 & | & 150 \\ 0 & 1 & | & 450 \\ 0 & 0 & | & 0 \end{bmatrix}$$

The tire store manager should purchase 150 four ply nylon tires and 450 steel-belted radial tires to meet all the use the entire budget on 600 tires.

43. Let x be the length of shorter piece of string, and y be the length of the longer piece of string. The first condition is the total length of the two pieces must equal 160 inches. This is modeled by the following equation:

$$x + y = 160$$

The longer piece must be four times longer than the shorter one which means:

$$4x = y \text{ or } 4x - y = 0$$

Combine the two equations to get the system:

$$x + y = 160$$
$$4x - y = \quad 0$$

The augmented matrix associated with this system is:

$$\begin{bmatrix} 1 & 1 & | & 160 \\ 4 & -1 & | & 0 \end{bmatrix}$$

Pivot on row 1, column 1 by performing the following operation:

$$-4R_1 + R_2 \rightarrow R_2$$

$$\begin{bmatrix} 1 & 1 & | & 160 \\ 0 & -5 & | & -640 \end{bmatrix}$$

Next pivot on row 2, column 2 by performing the following operations:

$$-\frac{1}{5}R_2 \rightarrow R_2$$
$$-R_2 + R_1 \rightarrow R_1$$

$$\begin{bmatrix} 1 & 0 & | & 32 \\ 0 & 1 & | & 128 \end{bmatrix}$$

The length of the shorter piece of string should be 32 inches long, and the length of the longer piece of string should be 128 inches long.

45. Let A be the number of Model A saws, B be the number of Model B saws and C be the number of model C saws.

Organize the information in the chart as shown:

	Model A	Model B	Model C	Totals
Assembly	1	1.5	2	200
Painting	0.5	0.5	0.5	65
Testing	0.5	0.25	0.5	26

The model that will meet the Assembly, painting and testing needs is:

$$A + \quad 1.5B + \quad 2C = 200$$
$$0.5A + \quad 0.5B + 0.5C = \quad 65$$
$$0.2A + 0.25B + 0.2C = \quad 26$$

The augmented matrix that is associated with this system is:

$$\begin{bmatrix} 1 & 1.5 & 2 & | & 200 \\ 0.5 & 0.5 & 0.5 & | & 65 \\ 0.2 & 0.25 & 0.2 & | & 26 \end{bmatrix}$$

Pivot on row 1, column 1 by performing the following operations:

$$-\frac{1}{2}R_1 + R_2 \rightarrow R_2$$
$$-\frac{1}{5}R_1 + R_3 \rightarrow R_3$$

$$\begin{bmatrix} 1 & 1.5 & 2 & | & 200 \\ 0 & -0.25 & -0.5 & | & -35 \\ 0 & -0.05 & -0.2 & | & -14 \end{bmatrix}$$

Next pivot on row 2, column 2 by performing the following operations:

$$-4R_2 \rightarrow R_2$$
$$-1.5R_2 + R_1 \rightarrow R_1$$
$$0.05R_2 + R_3 \rightarrow R_3$$

The resulting matrix is displayed at the top of the next page.

$$\begin{bmatrix} 1 & 0 & -1 & -10 \\ 0 & 1 & 2 & 140 \\ 0 & 0 & -0.1 & -7 \end{bmatrix}$$

Next pivot on row 3, column 3 by performing the following operations:

$$-10R_3 \rightarrow R_3$$
$$R_3 + R_1 \rightarrow R_1$$
$$-2R_2 + R_3 \rightarrow R_3$$

$$\begin{bmatrix} 1 & 0 & 0 & 60 \\ 0 & 1 & 0 & 0 \\ 0 & 0 & 1 & 70 \end{bmatrix}$$

The company should produce 60 Model A chain saws, no Model B chain saws and 70 Model C chain saws, to fully utilize all labor hours each week.

47. Let A, B, C be the total net income of Company A, Company B and Company C respectively. Let x be the consolidated income of company A, y be the consolidated income of company B, and z be the consolidated income of company C.

Company A receives an own net income of $A - 0.05A$, and receives $0.20B$ and $0.10C$ from company B and C. Thus the consolidated income of company A is:

$$x = 0.95A + 0.20B + 0.10C$$

Company B receives an own net income of $B - 0.20B - 0.40B$ and receives $0.30C$ from company C. The consolidated income of company B is:

$$y = 0.40B + 0.30C$$

Company C receives an own net income of $C - 0.10C - 0.30C$, and receives $0.05A$ and $0.40B$ from company A and company B respectively. Giving company C a consolidated income of:

$$z = 0.05A + 0.40B + 0.60C$$

The system of equations that models this problem is shown at the top of the next column.

$$x = 0.95A + 0.20B + 0.10C$$
$$y = \qquad\quad 0.40B + 0.30C$$
$$z = 0.05A + 0.40B + 0.60C$$

To solve this system simply substitute the net income before inter-company adjustments into the equations.

$$x = 0.95(1.23) + 0.20(4.6) + 0.10(3.8)$$
$$x = 2.4685 .$$

$$y = 0.40(4.6) + 0.30(3.8)$$
$$y = 2.98 .$$

$$z = 0.05(1.23) + 0.40(4.6) + 0.60(3.8)$$
$$z = 4.1815 .$$

The consolidated income of company A is \$2.4685 million, the consolidated income of company B is \$2.98 million and the consolidated income of company C is \$4.1815 million.

49. The augmented matrix associated with this system is:

$$\begin{bmatrix} n & \sum x & \sum y \\ \sum x & \sum x^2 & \sum xy \end{bmatrix}$$

First make element 1, 1 equal to one

$$\frac{1}{n}R_1 \rightarrow R_1$$

$$\begin{bmatrix} 1 & \frac{\sum x}{n} & \frac{\sum y}{n} \\ \sum x & \sum x^2 & \sum xy \end{bmatrix}$$

Now make the remaining elements in column 1 zero by performing the operation:

$$\left(-\sum x\right)R_1 + R_2 \rightarrow R_2$$

$$\begin{bmatrix} 1 & \frac{\sum x}{n} & \frac{\sum y}{n} \\ 0 & \frac{n\sum x^2 - \left(\sum x\right)^2}{n} & \frac{n\sum xy - \sum x \sum y}{n} \end{bmatrix}$$

The solution is continued on the next page.

Pivot on row 2, column 2 by performing the operation:

$$\frac{n}{n\sum x^2 - \left(\sum x\right)^2} R_2 \rightarrow R_2$$

$$\begin{bmatrix} 1 & \frac{\sum x}{n} & \Big| & \frac{\sum y}{n} \\ 0 & 1 & \Big| & \frac{n\sum xy - \sum x \sum y}{n\sum x^2 - \left(\sum x\right)^2} \end{bmatrix}$$

Make the remaining elements in column 2 zero by performing the operation:

$$\left(\frac{-\sum x}{n}\right) R_2 + R_1 \rightarrow R_1$$

$$\begin{bmatrix} 1 & 0 & \Big| & \frac{\left(\sum x\right)^2\left(\sum y\right) - \left(\sum x\right)\left(\sum xy\right)}{n\sum x^2 - \left(\sum x\right)^2} \\ 0 & 1 & \Big| & \frac{n\sum xy - \left(\sum x\right)\left(\sum y\right)}{n\sum x^2 - \left(\sum x\right)^2} \end{bmatrix}$$

The solution to the system is:

$$b = \frac{\left(\sum x\right)^2\left(\sum y\right) - \left(\sum x\right)\left(\sum xy\right)}{n\sum x^2 - \left(\sum x\right)^2}$$

$$m = \frac{n\sum xy - \left(\sum x\right)\left(\sum y\right)}{n\sum x^2 - \left(\sum x\right)^2}.$$

Exercises Section 2.3

1. The augmented matrix of the system is:

$$\begin{bmatrix} 2 & -1 & -3 & | & 5 \\ 1 & -2 & 1 & | & 3 \end{bmatrix}$$

Interchange row 1 and row 2:

$$R_1 \leftrightarrow R_2$$

$$\begin{bmatrix} 1 & -2 & 1 & | & 3 \\ 2 & -1 & -3 & | & 5 \end{bmatrix}$$

Pivot on row 1, column 1 by performing the operation at the top of the next column.

$$-2R_1 + R_2 \rightarrow R_2$$

$$\begin{bmatrix} 1 & -2 & 1 & | & 3 \\ 0 & 3 & -5 & | & -1 \end{bmatrix}$$

Next pivot on row 2, column 2 by performing:

$$\frac{1}{3} R_2 \rightarrow R_2$$

$$2R_2 + R_1 \rightarrow R_1$$

$$\begin{bmatrix} 1 & 0 & \frac{-7}{3} & | & \frac{7}{3} \\ 0 & 1 & \frac{-5}{3} & | & \frac{-1}{3} \end{bmatrix}$$

We can choose any real number for z. Once that choice has been made, the values of the other equations are

$$z = any\ number$$

$$x = \tfrac{7}{3} + \tfrac{7}{3}z$$

$$y = \tfrac{-1}{3} + \tfrac{5}{3}z$$

Check the solution by plugging it back into the original equations.

$$2\left(\tfrac{7}{3} + \tfrac{7}{3}z\right) - \left(\tfrac{-1}{3} + \tfrac{5}{3}z\right) - 3(z) = 5$$

$$\tfrac{14}{3} + \tfrac{14}{3}z + \tfrac{1}{3} - \tfrac{5}{3}z - 3z = 5$$

$$5 = 5$$

$$\left(\tfrac{7}{3} + \tfrac{7}{3}z\right) - 2\left(\tfrac{-1}{3} + \tfrac{5}{3}z\right) + (z) = 3$$

$$\tfrac{7}{3} + \tfrac{7}{3}z + \tfrac{2}{3} - \tfrac{10}{3}z + z = 3$$

$$3 = 3$$

To find three specific solutions plug in the values
$$z = 0$$

To get the solution:

$$x = \tfrac{7}{3} + \tfrac{7}{3}(0) = \tfrac{7}{3}$$

$$y = \tfrac{-1}{3} + \tfrac{5}{3}(0) = \tfrac{-1}{3}$$

$$z = \qquad (0) = 0$$

The other two solutions are displayed on the next page.

$z = 1$

To get the solution:

$x = \frac{7}{3} + \frac{7}{3}(1) = \frac{14}{3}$

$y = \frac{-1}{3} + \frac{5}{3}(1) = \frac{4}{3}$

$z = \qquad (1) = 1$

$z = 2$

To get the solution:

$x = \frac{7}{3} + \frac{7}{3}(2) = 7$

$y = \frac{-1}{3} + \frac{5}{3}(2) = 3$

$z = \qquad (2) = 2$

3. The augmented matrix of the system is:

$$\begin{bmatrix} 2 & 1 & -1 & | & 0 \\ 1 & -3 & 1 & | & 1 \\ 1 & 4 & -2 & | & -1 \end{bmatrix}$$

Pivot on row 1, column 1 by performing the operations

$R_1 \leftrightarrow R_2$

$-2R_1 + R_2 \rightarrow R_2$

$-R_1 + R_3 \rightarrow R_3$

$$\begin{bmatrix} 1 & -3 & 1 & | & 1 \\ 0 & 7 & -3 & | & -2 \\ 0 & 7 & -3 & | & -2 \end{bmatrix}$$

Pivot on row 2, column 2 by performing:

$\dfrac{1}{7} R_2 \rightarrow R_2$

$3R_2 + R_1 \rightarrow R_1$

$-7R_2 + R_3 \rightarrow R_3$

$$\begin{bmatrix} 1 & 0 & \frac{-2}{7} & | & \frac{1}{7} \\ 0 & 1 & \frac{-3}{7} & | & \frac{-2}{7} \\ 0 & 0 & 0 & | & 0 \end{bmatrix}$$

We can choose any real number for z. Once that choice has been made, the values of the other equations are

$z = \textit{any number}$

$x = \frac{1}{7} + \frac{2}{7} z$

$y = \frac{-2}{7} + \frac{3}{7} z$

Check the solution by plugging it back into the original equations.

$2\left(\frac{1}{7} + \frac{2}{7}z\right) + \left(\frac{-2}{7} + \frac{3}{7}z\right) - (z) = 0$

$\frac{2}{7} + \frac{4}{7}z + \frac{-2}{7} + \frac{3}{7}z - z = 0$

$0 = 0$

$\left(\frac{1}{7} + \frac{2}{7}z\right) - 3\left(\frac{-2}{7} + \frac{3}{7}z\right) + (z) = 1$

$\frac{1}{7} + \frac{2}{7}z + \frac{6}{7} - \frac{9}{7}z + z = 1$

$1 = 1$

$\left(\frac{1}{7} + \frac{2}{7}z\right) + 4\left(\frac{-2}{7} + \frac{3}{7}z\right) - 2(z) = -1$

$\frac{1}{7} + \frac{2}{7}z - \frac{8}{7} + \frac{12}{7}z - 2z = -1$

$-1 = -1$

To find three specific solutions plug in the values $z = 0$

To get the solution:

$x = \frac{1}{7} + \frac{2}{7}(0) = \frac{1}{7}$

$y = \frac{-2}{7} + \frac{3}{7}(0) = \frac{-2}{7}$

$z = \qquad (0) = 0$

$z = 1$

To get the solution:

$x = \frac{1}{7} + \frac{2}{7}(1) = \frac{3}{7}$

$y = \frac{-2}{7} + \frac{3}{7}(1) = \frac{1}{7}$

$z = \qquad (1) = 1$

$z = 2$

To get the solution:

$x = \frac{1}{7} + \frac{2}{7}(2) = \frac{5}{7}$

$y = \frac{-2}{7} + \frac{3}{7}(2) = \frac{4}{7}$

$z = \qquad (2) = 2$

5. The augmented matrix of the system is:

$$\begin{bmatrix} 1 & 2 & 1 & -3 & | & 4 \\ -2 & 1 & 1 & 1 & | & 2 \\ -1 & 3 & 2 & -2 & | & 6 \end{bmatrix}$$

Pivot on row 1, column 1 by performing the operations

$R_1 + R_2 \rightarrow R_2$

$R_1 + R_3 \rightarrow R_3$

$$\begin{bmatrix} 1 & 2 & 1 & -3 & | & 4 \\ 0 & 5 & 3 & -5 & | & 10 \\ 0 & 5 & 3 & -5 & | & 10 \end{bmatrix}$$

Pivot on row 2, column 2 by performing the following operations:

$\dfrac{1}{5}R_2 \rightarrow R_2$

$-2R_2 + R_1 \rightarrow R_1$

$-5R_2 + R_3 \rightarrow R_3$

$$\begin{bmatrix} 1 & 0 & \frac{-1}{5} & -1 & | & 0 \\ 0 & 1 & \frac{3}{5} & -1 & | & 2 \\ 0 & 0 & 0 & 0 & | & 0 \end{bmatrix}$$

We can choose any real number for y and any real number for z. Once that choice has been made, the values of the other equations are:

$z = $ *any number*

$y = $ *any number*

$w = \frac{1}{5}y + z$

$x = 2 - \frac{3}{5}y + z$

Check the solution by plugging it back into the original equations.

$\left(\frac{1}{5}y + z\right) + 2\left(2 - \frac{3}{5}y + z\right) + (y) - 3(z) = 4$

$\frac{1}{5}y + z + 4 - \frac{6}{5}y + 2z + y - 3z = 4$

$4 = 4$

$-2\left(\frac{1}{5}y + z\right) + \left(2 - \frac{3}{5}y + z\right) + (y) + (z) = 2$

$\frac{-2}{5}y - 2z + 2 - \frac{3}{5}y + z + y + z = 2$

$2 = 2$

$-\left(\frac{1}{5}y + z\right) + 3\left(2 - \frac{3}{5}y + z\right) + 2(y) - 2(z) = 6$

$\frac{-1}{5}y - z + 6 - \frac{9}{5}y + 3z + 2y - 2z = 6$

$6 = 6$

To find three specific solutions plug in the values

$z = 0$

$y = 0$

To get the solution:

$w = 0$

$x = 2$

$y = 0$

$z = 0$

For the second solution let:

$z = 1$

$y = 0$

To get the solution:

$w = 1$

$x = 3$

$y = 0$

$z = 1$

For the third solution let:

$z = 0$

$y = 1$

To get the solution:

$w = \frac{1}{5}$

$x = \frac{7}{5}$

$y = 1$

$z = 0$

7. The augmented matrix of the system is:

$$\begin{bmatrix} 1 & 4 & 2 & -3 & 1 \\ 3 & 1 & -2 & 0 & -11 \\ 1 & 2 & 1 & 0 & 5 \\ 0 & 1 & 0 & -3 & -8 \end{bmatrix}$$

Pivot on row 1, column 1 by performing the operations

$$-3R_1 + R_2 \rightarrow R_2$$
$$-R_1 + R_3 \rightarrow R_3$$

$$\begin{bmatrix} 1 & 4 & 2 & -3 & 1 \\ 0 & -11 & -8 & 9 & -14 \\ 0 & -2 & -1 & 3 & 4 \\ 0 & 1 & 0 & -3 & -8 \end{bmatrix}$$

Pivot on row 2, column 2 by performing:

$$R_2 \leftrightarrow R_4$$
$$-4R_2 + R_1 \rightarrow R_1$$
$$2R_2 + R_3 \rightarrow R_3$$
$$11R_2 + R_4 \rightarrow R_4$$

$$\begin{bmatrix} 1 & 0 & 2 & 9 & 33 \\ 0 & 1 & 0 & -3 & -8 \\ 0 & 0 & -1 & -3 & -12 \\ 0 & 0 & -8 & -24 & -102 \end{bmatrix}$$

Next pivot on row 3, column 3 by performing the following operations:

$$-R_3 \rightarrow R_3$$
$$-2R_3 + R_1 \rightarrow R_1$$
$$8R_3 + R_4 \rightarrow R_4$$

$$\begin{bmatrix} 1 & 0 & 0 & 3 & 9 \\ 0 & 1 & 0 & -3 & -8 \\ 0 & 0 & -1 & -3 & 12 \\ 0 & 0 & 0 & 0 & -6 \end{bmatrix}$$

The matrix is inconsistent. There is no solution to this system.

9. The augmented matrix is:

$$\begin{bmatrix} 1 & 1 & 1 & 3 \\ 2 & 2 & 2 & 6 \end{bmatrix}$$

Pivot on row 1, column 1 by performing the operation:

$$-2R_1 + R_2 \rightarrow R_2$$

$$\begin{bmatrix} 1 & 1 & 1 & 3 \\ 0 & 0 & 0 & 0 \end{bmatrix}$$

We can choose any real number for y and any real number for z . Once that choice has been made, the values of the other equations are:

$z = $ *any number*

$y = $ *any number*

$x = 3 - y - z$

To check this solution substitute the values back into the original equations as shown:

Equation 1:

$$(3 - y - z) + (y) + (z) = 3$$
$$3 = 3$$

Equation 2:

$$2(3 - y - z) + 2(y) + 2(z) = 6$$
$$6 - 2y - 2z + 2y + 2z = 6$$
$$6 = 6$$

To find three specific solutions plug in the values

$z = 0$

$y = 0$

To get the first solution:

$x = 3$

$y = 0$

$z = 0$

The solution is continued on the next page.

For the second solution let

$z = 1$

$y = 0$

To get the solution:

$x = 2$

$y = 0$

$z = 1$

For the third solution let

$z = 0$

$y = 1$

To get the solution:

$x = 2$

$y = 1$

$z = 0$

11. The augmented matrix is

$$\begin{bmatrix} 2 & 3 & 1 & | & 0 \\ 1 & -1 & 2 & | & 0 \\ 4 & 1 & 5 & | & 0 \end{bmatrix}$$

Pivot on row 1, column 1 by performing the operation:

$R_1 \leftrightarrow R_2$

$-2R_1 + R_2 \rightarrow R_2$

$-4R_1 + R_3 \rightarrow R_3$

$$\begin{bmatrix} 1 & -1 & 2 & | & 0 \\ 0 & 5 & -3 & | & 0 \\ 0 & 5 & -3 & | & 0 \end{bmatrix}$$

Pivot on row 2, column 2 by performing the operations:

$\dfrac{1}{5}R_2 \rightarrow R_2$

$R_2 + R_1 \rightarrow R_1$

$-5R_2 + R_3 \rightarrow R_3$

The new matrix is at the top of the next column.

$$\begin{bmatrix} 1 & 0 & \frac{7}{5} & | & 0 \\ 0 & 1 & \frac{-3}{5} & | & 0 \\ 0 & 0 & 0 & | & 0 \end{bmatrix}$$

We can choose any real number for z. Once that choice has been made, the values of the other variables are:

$z = $ *any number*

$x = \frac{-7}{5}z$

$y = \frac{3}{5}z$

To check this solution substitute the values back into the original equations.

Equation 1:

$2\left(\frac{-7}{5}z\right) + 3\left(\frac{3}{5}z\right) + (z) = 0$

$\frac{-14}{5}z + \frac{9}{5}z + z = 0$

$0 = 0$

Equation 2:

$\left(\frac{-7}{5}z\right) - \left(\frac{3}{5}z\right) + 2(z) = 0$

$\frac{-7}{5}z - \frac{3}{5}z + 2z = 0$

$0 = 0$

Equation 3:

$4\left(\frac{-7}{5}z\right) + \left(\frac{3}{5}z\right) + 5(z) = 0$

$\frac{-28}{5}z + \frac{3}{5}z + 5z = 0$

$0 = 0$

To find three specific solutions plug in the values

$z = 0$

To get the first solution:

$x = 0$

$y = 0$

$z = 0$

The solution is continued on the next page.

For the second solution let

$$z = 5$$

To get the solution:

$$x = -7$$
$$y = 3$$
$$z = 5$$

For the third solution let

$$z = 10$$

To get the solution:

$$x = -14$$
$$y = 6$$
$$z = 10$$

13. The augmented matrix is:

$$\begin{bmatrix} 1 & 0 & -1 & | & 0 \\ 0 & 1 & 1 & | & 0 \\ 1 & 1 & 0 & | & 0 \end{bmatrix}$$

Pivot on row 1, column 1 by performing the operation:

$$-R_1 + R_3 \rightarrow R_3$$

$$\begin{bmatrix} 1 & 0 & -1 & | & 0 \\ 0 & 1 & 1 & | & 0 \\ 0 & 1 & 1 & | & 0 \end{bmatrix}$$

Pivot on row 2, column 2 by performing the operations:

$$-R_2 + R_3 \rightarrow R_3$$

$$\begin{bmatrix} 1 & 0 & -1 & | & 0 \\ 0 & 1 & 1 & | & 0 \\ 0 & 0 & 0 & | & 0 \end{bmatrix}$$

We can choose any real number for z. Once that choice has been made, the values of the other variables are:

$$z = any\ number$$

$$x = z$$

$$y = -z$$

To check this solution substitute the values back into the original equations.

Equation 1:

$$(z) + 0(-z) - (z) = 0$$
$$0 = 0$$

Equation 2:

$$0(z) + (-z) + (z) = 0$$
$$0 = 0$$

Equation 3:

$$(z) + (-z) - 0(z) = 0$$
$$0 = 0$$

To find three specific solutions plug in the values

$$z = 0$$

To get the first solution:

$$x = 0$$
$$y = 0$$
$$z = 0$$

For the second solution let

$$z = 1$$

To get the solution:

$$x = 1$$
$$y = -1$$
$$z = 1$$

For the third solution let

$$z = -1$$

To get the solution:

$$x = -1$$
$$y = 1$$
$$z = -1$$

15. The augmented matrix is:

$$\begin{bmatrix} 1 & 0 & 1 & | & 0 \\ 0 & 1 & 0 & | & 0 \\ 2 & 0 & 2 & | & 0 \end{bmatrix}$$

Pivot on row 1, column 1 by performing the operation:

$$-2R_1 + R_3 \rightarrow R_3$$

$$\begin{bmatrix} 1 & 0 & 1 & | & 0 \\ 0 & 1 & 0 & | & 0 \\ 0 & 0 & 0 & | & 0 \end{bmatrix}$$

From the matrix we see $y = 0$. We can choose any real number for z. Once that choice has been made, the values of the other variables are:

$z = any\ number$

$x = -z$

$y = 0$

To check this solution substitute the values back into the original equations.

Equation 1:
$(-z) + 0(0) + (z) = 0$
$0 = 0$

Equation 2:
$0(-z) + (0) + 0(z) = 0$
$0 = 0$

Equation 3:
$2(-z) + 0(0) + 2(z) = 0$
$0 = 0$

To find three specific solutions plug in the values

$z = 0$

To get the first solution:

$x = 0$

$y = 0$

$z = 0$

The second solution is shown at the top of the next column.

$z = 1$

To get the solution:

$x = -1$

$y = 0$

$z = 1$

For the third solution let

$z = -1$

To get the solution:

$x = 1$

$y = 0$

$z = -1$

17. The augmented matrix for the system is:

$$\begin{bmatrix} 1 & 2 & -1 & | & 8 \\ 2 & -1 & 1 & | & 3 \\ 4 & 3 & -1 & | & 5 \end{bmatrix}$$

Pivot on row 1, column 1 by performing the operations:

$$-2R_1 + R_2 \rightarrow R_2$$
$$-4R_1 + R_3 \rightarrow R_3$$

$$\begin{bmatrix} 1 & 2 & -1 & | & 8 \\ 0 & -5 & 3 & | & -13 \\ 0 & -5 & 3 & | & -27 \end{bmatrix}$$

Pivot on row 2, column 2 by performing the operations

$$-\frac{1}{5}R_2 \rightarrow R_2$$
$$-2R_2 + R_1 \rightarrow R_1$$
$$-5R_2 + R_3 \rightarrow R_3$$

$$\begin{bmatrix} 1 & 0 & \frac{1}{5} & | & \frac{14}{5} \\ 0 & 1 & \frac{-3}{5} & | & \frac{13}{5} \\ 0 & 0 & 0 & | & -14 \end{bmatrix}$$

The third row represents an equation that is inconsistent. Therefore the system has no solution.

19. The augmented matrix for the system is

$$\begin{bmatrix} 1 & 2 & -3 & 1 & | & 1 \\ 4 & 8 & -12 & 4 & | & 4 \\ 2 & 4 & -6 & 2 & | & 2 \end{bmatrix}$$

Pivot on row 1, column 1 by performing the operations

$-4R_1 + R_2 \rightarrow R_2$
$-2R_1 + R_3 \rightarrow R_3$

$$\begin{bmatrix} 1 & 2 & -3 & 1 & | & 1 \\ 0 & 0 & 0 & 0 & | & 0 \\ 0 & 0 & 0 & 0 & | & 0 \end{bmatrix}$$

We can let x, y, and z be any number, after we make that choice, the general solution to the system is shown below:

$w = 1 - 2x + 3y - z$
$x = any\ number$
$y = any\ number$
$z = any\ number$

21. The augmented matrix for the system is:

$$\begin{bmatrix} 2 & -3 & 2 & | & 3 \\ 1 & 2 & -3 & | & 2 \\ 3 & -1 & -1 & | & 3 \end{bmatrix}$$

Pivot on row 1, column 1 by performing the operations:

$R_1 \leftrightarrow R_2$
$-2R_1 + R_2 \rightarrow R_2$
$-3R_1 + R_3 \rightarrow R_3$

$$\begin{bmatrix} 1 & 2 & -3 & | & 2 \\ 0 & -7 & 8 & | & -1 \\ 0 & -7 & 8 & | & -3 \end{bmatrix}$$

Pivot on row 2, column 2 by performing the operations at the top of the next column.

$-\dfrac{1}{7}R_2 \rightarrow R_2$

$-2R_2 + R_1 \rightarrow R_1$
$7R_2 + R_3 \rightarrow R_3$

$$\begin{bmatrix} 1 & 0 & \frac{-5}{7} & | & \frac{12}{7} \\ 0 & 1 & \frac{-8}{7} & | & \frac{1}{7} \\ 0 & 0 & 0 & | & -2 \end{bmatrix}$$

The third row represents an equation that is inconsistent. Therefore the system has no solution.

23. The augmented matrix for the system is:

$$\begin{bmatrix} 0.3 & 0.5 & 1 & | & 4 \\ 5 & -2 & 0.8 & | & 7 \\ 5.6 & -1 & 2.8 & | & 15 \end{bmatrix}$$

Pivot on row 1, column 1 by performing the operations:

$\dfrac{10}{3}R_1 \rightarrow R_1$

$-5R_1 + R_2 \rightarrow R_2$
$-5.6R_1 + R_3 \rightarrow R_3$

$$\begin{bmatrix} 1 & \frac{5}{3} & \frac{10}{3} & | & \frac{40}{3} \\ 0 & \frac{-31}{3} & \frac{-238}{15} & | & \frac{-179}{3} \\ 0 & \frac{-31}{3} & \frac{-238}{15} & | & \frac{-179}{3} \end{bmatrix}$$

Next pivot on Row 2, column 2 by performing the following operations:

$\frac{-3}{31}R_2 \rightarrow R_2$
$\frac{-5}{3}R_2 + R_1 \rightarrow R_1$
$\frac{-31}{3}R_2 + R_3 \rightarrow R_3$

The matrix is:

$$\begin{bmatrix} 1 & 0 & \frac{24}{31} & | & \frac{115}{31} \\ 0 & 1 & \frac{238}{155} & | & \frac{179}{31} \\ 0 & 0 & 0 & | & 0 \end{bmatrix}$$

We can let z be any number, after we make that choice, the general solution to the system is shown on the next page.

$x = \frac{115}{31} - \frac{24}{31}z$

$y = \frac{179}{31} - \frac{238}{155}z$

$z = any\ number$

25. The augmented matrix for the system is:

$$\begin{bmatrix} 0.3 & 5.24 & -8.61 & | & 5.2 \\ 1.2 & 20.96 & -34.44 & | & 8 \end{bmatrix}$$

Pivot on row 1, column 1 by performing the operations

$\dfrac{10}{3}R_1 \rightarrow R_1$

$-1.2R_1 + R_2 \rightarrow R_2$

$$\begin{bmatrix} 1 & \frac{262}{15} & \frac{-287}{10} & | & \frac{52}{3} \\ 0 & 0 & 0 & | & \frac{-64}{5} \end{bmatrix}$$

The second row represents an equation that is inconsistent. Therefore the system has no solution.

27. The augmented matrix for the system is:

$$\begin{bmatrix} 1 & 2 & -3 & 1 & | & 6.3 \\ 2 & -1 & 0.54 & 2.8 & | & 5 \\ 2.83 & 5.72 & 1 & -3 & | & 2 \\ 12.66 & 15.44 & -2.92 & 2.6 & | & 32.9 \end{bmatrix}$$

Enter the matrix on the previous page into a graphing calculator and pivot on row 1, column 1 by performing the operations:

$-2R_1 + R_2 \rightarrow R_2$

$-2.83R_1 + R_3 \rightarrow R_3$

$-12.66R_1 + R_3 \rightarrow R_3$

$$\begin{bmatrix} 1 & 2 & -3 & 1 & | & 6.3 \\ 0 & -5 & \frac{327}{50} & \frac{4}{5} & | & \frac{-38}{5} \\ 0 & \frac{3}{50} & \frac{949}{100} & \frac{-583}{100} & | & \frac{-15829}{1000} \\ 0 & \frac{-247}{25} & \frac{1753}{50} & \frac{-503}{50} & | & \frac{-23429}{500} \end{bmatrix}$$

Pivot on Row 2, column 2 by performing the following operations at the top of the next column.

$-\dfrac{1}{5}R_2 \rightarrow R_2$

$-2R_2 + R_1 \rightarrow R_1$

$\frac{-3}{30}R_2 + R_3 \rightarrow R_3$

$\frac{247}{25}R_2 + R_1 \rightarrow R_1$

$$\begin{bmatrix} 1 & 0 & \frac{-48}{125} & \frac{33}{25} & | & \frac{163}{50} \\ 0 & 1 & \frac{-327}{250} & \frac{-4}{25} & | & \frac{38}{25} \\ 0 & 0 & \frac{59803}{6250} & \frac{-14551}{2500} & | & \frac{-79601}{5000} \\ 0 & 0 & \frac{69178}{3125} & \frac{-14551}{1250} & | & \frac{-79601}{2500} \end{bmatrix}$$

Pivot on Row 2, column 2 by performing the following operations:

$\dfrac{6250}{59803}R_3 \rightarrow R_3$

$\dfrac{48}{125}R_3 + R_1 \rightarrow R_1$

$\frac{327}{250}R_3 + R_2 \rightarrow R_2$

$\frac{-69178}{3125}R_3 + R_4 \rightarrow R_4$

$$\begin{bmatrix} 1 & 0 & 0 & 1.0864 & | & 2.6211 \\ 0 & 1 & \frac{-327}{250} & -0.9556 & | & -0.5653 \\ 0 & 0 & 0 & -0.9083 & | & -1.6638 \\ 0 & 0 & 0 & 1.8248 & | & 4.9915 \end{bmatrix}$$

Finally pivot on row 4, column 4 by performing the operations:

$\dfrac{1}{1.8249}R_4 \rightarrow R_4$

$-1.0864R_4 + R_1 \rightarrow R_1$

$0.9556R_4 + R_2 \rightarrow R_2$

$0.9083R_4 + R_3 \rightarrow R_3$

$$\begin{bmatrix} 1 & 0 & 0 & 0 & | & -0.3505 \\ 0 & 1 & 0 & 0 & | & 1.9576 \\ 0 & 0 & 1 & 0 & | & 0 \\ 0 & 0 & 0 & 1 & | & 2.7352 \end{bmatrix}$$

This system is independent. There is one unique solution shown at the top of the next page.

$w = -0.3505$

$x = 1.9576$

$y = 0$

$z = 2.7352$

29. From equation 2 it is obvious that $y = 0$. From equation 1 we see that $x - 2z = 0 \Rightarrow x = 2z$. So by inspection, the solution to this system is:

$x = 2z$

$y = 0$

$z = any\ number$

31. From equation 1 we see that $y + 2z = 0 \Rightarrow y = -2z$. From equation 2 we see that $x + z = 0 \Rightarrow x = -z$. So by inspection, the solution to this system is:

$x = -z$

$y = -2z$

$z = any\ number$

33. Since u appears in only one equation and v appears in only one equation, we can easily find a specific solution by letting $x = y = 0$. The solution is:

$x = 0; y = 0; u = 10; v = 5$

The augmented matrix is shown below:

$$\begin{bmatrix} 2 & 2 & 1 & 0 & | & 10 \\ 1 & 1 & 0 & 1 & | & 5 \end{bmatrix}$$

Pivoting on row 2, column 1 by performing the operation:

$-2R_2 + R_1 \to R_1$

$$\begin{bmatrix} 0 & 0 & 1 & -2 & | & 0 \\ 1 & 1 & 0 & 1 & | & 5 \end{bmatrix}$$

This results in the system:

$u - 2v = 0$

$x + y \quad + v = 5$

Now if we let $y = v = 0$ we can find a second solution by inspection. The solution is:

$x = 5; y = 0; u = 0; v = 0$

35. Since u appears in only one equation and v appears in only one equation, we can easily find a specific solution by letting $x = y = 0$. The solution is:

$x = 0; y = 0; u = -6; v = 4$

The augmented matrix is:

$$\begin{bmatrix} 1 & -1 & 1 & 0 & | & 6 \\ 2 & 1 & 0 & 1 & | & 4 \end{bmatrix}$$

Pivoting on row 2, column 1 by performing the operations:

$\dfrac{1}{2} R_2 \to R_2$

$-R_2 + R_1 \to R_1$

$$\begin{bmatrix} 0 & \frac{-3}{2} & 1 & \frac{-1}{2} & | & 4 \\ 1 & \frac{1}{2} & 0 & \frac{1}{2} & | & 2 \end{bmatrix}$$

This results in the system:

$\frac{-3}{2} y + u - \frac{1}{2} v = 4$

$x + \frac{1}{2} y \quad + \frac{1}{2} v = 2$

Now if we let $y = v = 0$ we can find a second solution by inspection. The solution is:

$x = 2; y = 0; u = 4; v = 0$

37. Let x be the amount of money spent on streets, y be the amount of money spent on urban renewal, and z be the amount of money spent on parks (let the amounts be in millions of dollars).

Since 1 million dollars was allocated to the city, the first equation is:

$x + y + z = 1$

The solution is continued on the next page.

Since the board wants to spend the same amount on streets as they do on urban renewal and parks combined, the second equation is:

$x = y + z$, or $x - y - z = 0$

The system of equations that will model this city board problem is:

$x + y + z = 1$

$x - y - z = 0$

The augmented matrix associated with this system is:

$$\begin{bmatrix} 1 & 1 & 1 & | & 1 \\ 1 & -1 & -1 & | & 0 \end{bmatrix}$$

Pivot on row 1, column 1 by performing the operation:

$-R_1 + R_2 \rightarrow R_2$

$$\begin{bmatrix} 1 & 1 & 1 & | & 1 \\ 0 & -2 & -2 & | & -1 \end{bmatrix}$$

Pivot on row 2, column 2 by performing the operations:

$-\dfrac{1}{2} R_2 \rightarrow R_2$

$-R_2 + R_1 \rightarrow R_1$

$$\begin{bmatrix} 1 & 0 & 0 & | & 0.5 \\ 0 & 1 & 1 & | & 0.5 \end{bmatrix}$$

The general solution to this system is:

$x = 0.5$

$y = 0.5 - z$

$z = any\ number$

This means that the city board should spend $500,000 on streets. Then after they decide how many dollars z they want to spend on parks, they should spend $y = 0.5 - z$ dollars on urban renewal. The amount they spend on parks must be between $0 and $500,000.

39. Let x be the number of ads placed in newspapers, y be the number of ads played on the radio, and z be the number of ads shown on TV. Since a total of 30 ads need to be run, the first equation is shown at the top of the next column.

$x + y + z = 30$

The fact that twice as many radio ads are to be run as television ads, the second equation is:

$y = 2z$ or $y - 2z = 0$

Finally, since the sum of the newspaper adds and 2 times the number of radio ads must equal the total number of ads plus the number of television ads, we get the third equation

$x + 2y = 30 + z$ or $x + 2y - z = 30$

The system that models this problem is:

$x + y + z = 30$

$\quad\quad y - 2z = 0$

$x + 2y - z = 30$

The augmented matrix that represents this system is:

$$\begin{bmatrix} 1 & 1 & 1 & | & 30 \\ 0 & 1 & -2 & | & 0 \\ 1 & 2 & -1 & | & 30 \end{bmatrix}$$

Pivot on row 1, column 1 by performing the operation:

$-R_1 + R_3 \rightarrow R_3$

$$\begin{bmatrix} 1 & 1 & 1 & | & 30 \\ 0 & 1 & -2 & | & 0 \\ 0 & 1 & -2 & | & 0 \end{bmatrix}$$

Pivot on row 2, column 2 by performing the operations:

$-R_2 + R_1 \rightarrow R_1$

$-R_2 + R_3 \rightarrow R_3$

$$\begin{bmatrix} 1 & 0 & 3 & | & 30 \\ 0 & 1 & -2 & | & 0 \\ 0 & 0 & 0 & | & 0 \end{bmatrix}$$

The general solution to this system is shown at the top of the next page.

$x = 30 - 3z$

$y = 2z$

$z = $ *any number*

To meet the advertising needs of the company, the company may place between 0 and 10 TV ads. Once they have decided this number, then they will place 2 times as many radio ads, and $30 - 3z$ newspaper ads.

41. Let x be the number of QT20 printers, y be the number of QT40 printers and z be the number of QT60 printers manufactured.

The condition that they use 500 megabytes of memory each requires that:

$2x + 4y + 6z = 500$.

The condition that they use the available 560 worker assembly hours each week requires that:

$3x + 4y + 5z = 560$.

The condition that they use the 1060 minutes of testing time available each week requires that:

$5x + 8y + 11z = 1060$

The system of equations that models this problem is:

$$2x + 4y + 6z = 500$$
$$3x + 4y + 5z = 560$$
$$5x + 8y + 11z = 1060$$

The augmented matrix associated with this system is:

$$\begin{bmatrix} 2 & 4 & 6 & | & 500 \\ 3 & 4 & 5 & | & 560 \\ 5 & 8 & 11 & | & 1060 \end{bmatrix}$$

Pivot on row 1, column 1 by performing the operations:

$\dfrac{1}{2} R_1 \rightarrow R_1$

$-3R_1 + R_2 \rightarrow R_2$

$-5R_1 + R_3 \rightarrow R_3$

$$\begin{bmatrix} 1 & 2 & 3 & | & 250 \\ 0 & -2 & -4 & | & -190 \\ 0 & -2 & -4 & | & -190 \end{bmatrix}$$

Pivot on row 2, column 2 by performing the operations:

$-\dfrac{1}{2} R_2 \rightarrow R_2$

$-2R_2 + R_1 \rightarrow R_1$

$2R_2 + R_3 \rightarrow R_3$

$$\begin{bmatrix} 1 & 0 & -1 & | & 60 \\ 0 & 1 & 2 & | & 95 \\ 0 & 0 & 0 & | & 0 \end{bmatrix}$$

The general solution to this system is:

$x = 60 + z$

$y = 95 - 2z$

$z = $ *any number*

The manufacture can produce any number of type QT60 printers between $0 \leq z \leq 47$. Once the number of type QT60 printers has been decided, the manufacture can produce $95 - 2z$ type QT40 printers and $60 + z$ type QT20 printers.

43. Let w be the number of ovens sent from Warehouse 1 to Store 1, and let x be the number of ovens sent from Warehouse 1 to Store 2. Let y be the number of ovens sent from Warehouse 2 to Store 1, and z be the number of ovens sent from Warehouse 2 to Store 2.

To meet store 1's request of 30 ovens:

$w + y = 30$.

To meet store 2's request of 18 ovens:

$x + z = 18$.

The condition that twice as many ovens are sent from Warehouse 1 to Store 1 as from Warehouse 2 to Store 2 means that:

$w = 2z$, or $w - 2z = 0$.

The system of equations that models the problem is displayed at the top of the next page.

$$w \; + y + \qquad = 30$$
$$x \quad + \; z = 18$$
$$w \qquad - 2z = \; 0$$

The augmented matrix associated with this system is:

$$\begin{bmatrix} 1 & 0 & 1 & 0 & | & 30 \\ 0 & 1 & 0 & 1 & | & 18 \\ 1 & 0 & 0 & -2 & | & 0 \end{bmatrix}$$

Pivot on row 1, column 1 by performing the operation:

$$-R_1 + R_3 \rightarrow R_3$$

$$\begin{bmatrix} 1 & 0 & 1 & 0 & | & 30 \\ 0 & 1 & 0 & 1 & | & 18 \\ 0 & 0 & -1 & -2 & | & -30 \end{bmatrix}$$

Since column 2 has zeros in every element other than the pivot element 2, 2 and the pivot element is one, we can proceed to row 3.

Pivot on row 3, column 3 by performing the operations:

$$-R_3 \rightarrow R_3$$
$$-R_3 + R_1 \rightarrow R_1$$

$$\begin{bmatrix} 1 & 0 & 0 & -2 & | & 0 \\ 0 & 1 & 0 & 1 & | & 18 \\ 0 & 0 & 1 & 2 & | & 30 \end{bmatrix}$$

The general solution to the system is:

$$w = 2z$$
$$x = 18 - z$$
$$y = 30 - 2z$$
$$z = any\ number$$

Since we can't ship a negative number of ovens to a store, we can ship between $0 \leq z \leq 15$ ovens from Warehouse 2 to Store 2. After that number has been decided, the amount from the other warehouses to the stores will be as follows

They must ship $2z$ ovens from Warehouse 1 to Store 1.
They must ship $18 - z$ ovens from Warehouse 1 to Store 2.
They must ship $30 - 2z$ ovens from Warehouse 2 to Store 1.

Chapter 2 Summary Exercises

1. Coefficient matrix:

$$\begin{bmatrix} 3 & 0 & -1 \\ 0 & 1 & 0 \\ 6 & 0 & -2 \end{bmatrix}$$

Constant matrix:

$$\begin{bmatrix} 0 \\ 7 \\ 0 \end{bmatrix}$$

The augmented matrix is:

$$\begin{bmatrix} 3 & 0 & -1 & | & 0 \\ 0 & 1 & 0 & | & 7 \\ 6 & 0 & -2 & | & 0 \end{bmatrix}$$

3. Begin with the augmented matrix:

$$\begin{bmatrix} 1 & 2 & 1 & | & 0 \\ 3 & 0 & -1 & | & 2 \\ -1 & 2 & 1 & | & -2 \end{bmatrix}$$

Pivot on row one by performing the operations:

$$-3R_1 + R_2 \rightarrow R_2$$
$$R_1 + R_3 \rightarrow R_3$$

$$\begin{bmatrix} 1 & 2 & 1 & | & 0 \\ 0 & -6 & -4 & | & 2 \\ 0 & 4 & 2 & | & -2 \end{bmatrix}$$

Pivot on row 2, column 2 by performing the operations:

$$-\tfrac{1}{6} R_2 \rightarrow R_2$$
$$-2R_2 + R_1 \rightarrow R_1$$
$$-4R_2 + R_3 \rightarrow R_3$$

The matrix is shown at the top of the next page.

$$\begin{bmatrix} 1 & 0 & \frac{-1}{3} & \Big| & \frac{2}{3} \\ 0 & 1 & \frac{2}{3} & \Big| & \frac{-1}{3} \\ 0 & 0 & \frac{-2}{3} & \Big| & \frac{-2}{3} \end{bmatrix}$$

Pivot on row 3, column 3 by performing the operations:

$$\frac{-3}{2} R_3 \rightarrow R_3$$
$$\frac{1}{3} R_3 + R_1 \rightarrow R_1$$
$$\frac{-2}{3} R_3 + R_2 \rightarrow R_2$$

$$\begin{bmatrix} 1 & 0 & 0 & \Big| & 1 \\ 0 & 1 & 0 & \Big| & -1 \\ 0 & 0 & 1 & \Big| & 1 \end{bmatrix}$$

The system has exactly one solution.

5. Begin with the augmented matrix:

$$\begin{bmatrix} 1 & 2 & 1 & \Big| & 0 \\ 3 & 0 & 1 & \Big| & 2 \\ 2 & -2 & 0 & \Big| & 2 \end{bmatrix}$$

Pivot on row one by performing the operations:

$$-3R_1 + R_2 \rightarrow R_2$$
$$-2R_1 + R_3 \rightarrow R_3$$

$$\begin{bmatrix} 1 & 2 & 1 & \Big| & 0 \\ 0 & -6 & -2 & \Big| & 2 \\ 0 & -6 & -2 & \Big| & 2 \end{bmatrix}$$

Pivot on row 2, column 2 by performing the operations:

$$-\frac{1}{6} R_2 \rightarrow R_2$$
$$-2R_2 + R_1 \rightarrow R_1$$
$$6R_2 + R_3 \rightarrow R_3$$

$$\begin{bmatrix} 1 & 0 & \frac{1}{3} & \Big| & \frac{2}{3} \\ 0 & 1 & \frac{1}{3} & \Big| & \frac{-1}{3} \\ 0 & 0 & 0 & \Big| & 0 \end{bmatrix}$$

This system has infinitely many solutions.

7. The augmented matrix to this system is:

$$\begin{bmatrix} \frac{1}{2} & -3 & \Big| & 5 \\ 2 & -12 & \Big| & 18 \end{bmatrix}$$

Pivot on row 1, column 1 by performing the operations:

$$2R_1 \rightarrow R_1$$
$$-2R_1 + R_2 \rightarrow R_2$$

$$\begin{bmatrix} 1 & -6 & \Big| & 10 \\ 0 & 0 & \Big| & -2 \end{bmatrix}$$

The second row is inconsistent. The system has no solution.

9. The augmented matrix to this system is:

$$\begin{bmatrix} 1 & 2 & 1 & \Big| & 0 \\ 0 & 3 & 5 & \Big| & 1 \\ 5 & -4 & -10 & \Big| & 12 \end{bmatrix}$$

Pivot on row 1, column 1 by performing the operation:

$$-5R_1 + R_3 \rightarrow R_3$$

$$\begin{bmatrix} 1 & 2 & 1 & \Big| & 0 \\ 0 & 3 & 5 & \Big| & 1 \\ 0 & -14 & -15 & \Big| & 12 \end{bmatrix}$$

Pivot on row 2, column 2 by performing the operations:

$$\frac{1}{3} R_2 \rightarrow R_2$$
$$-2R_2 + R_1 \rightarrow R_1$$
$$14R_2 + R_3 \rightarrow R_3$$

$$\begin{bmatrix} 1 & 0 & \frac{-7}{3} & \Big| & \frac{-2}{3} \\ 0 & 1 & \frac{5}{3} & \Big| & \frac{1}{3} \\ 0 & 0 & \frac{25}{3} & \Big| & \frac{50}{3} \end{bmatrix}$$

Pivot on row 3, column 3 by performing the operations:

$$\frac{3}{25} R_3 \rightarrow R_3$$
$$\frac{7}{3} R_3 + R_1 \rightarrow R_1$$
$$\frac{-5}{3} R_3 + R_2 \rightarrow R_2$$

The matrix is shown at the top of the next page.

$$\begin{bmatrix} 1 & 0 & 0 & | & 4 \\ 0 & 1 & 0 & | & -3 \\ 0 & 0 & 1 & | & 2 \end{bmatrix}$$

The solution to the system is:

$x = 4; \ y = -3; \ z = 2$

11. The augmented matrix to this system is:

$$\begin{bmatrix} 1 & -2 & | & 13 \\ 3 & 1 & | & 4 \\ 2 & -1 & | & 11 \\ 4 & 2 & | & 2 \end{bmatrix}$$

Pivot on row 1, column 1 by performing the operations:

$-3R_1 + R_2 \rightarrow R_2$
$-2R_1 + R_3 \rightarrow R_3$
$-4R_1 + R_4 \rightarrow R_4$

$$\begin{bmatrix} 1 & -2 & | & 13 \\ 0 & 7 & | & -35 \\ 0 & 3 & | & -15 \\ 0 & 10 & | & -50 \end{bmatrix}$$

Pivot on row 2, column 2 by performing the operations:

$\frac{1}{7}R_2 \rightarrow R_2$
$2R_2 + R_1 \rightarrow R_1$
$-3R_2 + R_3 \rightarrow R_3$
$-10R_2 + R_4 \rightarrow R_4$

$$\begin{bmatrix} 1 & 0 & | & 3 \\ 0 & 1 & | & -5 \\ 0 & 0 & | & 0 \\ 0 & 0 & | & 0 \end{bmatrix}$$

The solution to this system is:

$x = 3; \ y = -5$

Even though there are more equations than unknowns, it turns out this system still has only one solution.

13. The augmented matrix to this system is:

$$\begin{bmatrix} 1 & 0 & 4 & | & 7 \\ 0 & 1 & 0 & | & 5 \\ 2 & 0 & 7 & | & 13 \end{bmatrix}$$

Pivot on row 1, column 1 by performing the operation:

$-2R_1 + R_3 \rightarrow R_3$

$$\begin{bmatrix} 1 & 0 & 4 & | & 7 \\ 0 & 1 & 0 & | & 5 \\ 0 & 0 & -1 & | & -1 \end{bmatrix}$$

Since element 2, 2 is equal to 1, and all other elements in column 2 are zero, we can proceed to row 3.

Pivot on row 3, column 3 by performing the operations:

$-R_3 \rightarrow R_3$
$-4R_3 + R_1 \rightarrow R_1$

$$\begin{bmatrix} 1 & 0 & 0 & | & 3 \\ 0 & 1 & 0 & | & 5 \\ 0 & 0 & 1 & | & 1 \end{bmatrix}$$

The solution to this system is:

$x = 3; \ y = 5; \ z = 1$

15. Let H be the amount of time studying history, E be the amount of time studying English, P be the amount of time studying psychology, and A be the amount of time studying Accounting. The fact that she spends a total of 36 hours a week studying means:

$H + E + P + A = 36$

Since she spends the same amount of time on history and psychology as she does on English and accounting means:

$H + P = E + A$ or $H - E + P - A = 0$

The fact that the amount of time spent studying psychology and accounting together is five times what she spends studying English, translates to:

$P + A = 5E$ or $-5E + P + A = 0$

The solution is continued on the next page.

The fact 16 more hours is spent studying history and accounting than English and psychology translates to:

$E + P = H + A - 16$ or
$-H + E + P - A = -16$

The system of equations that models this problem is:

$$\begin{aligned} H + E + P + A &= 36 \\ H - E + P - A &= 0 \\ -5E + P + A &= 0 \\ -H + E + P - A &= -16 \end{aligned}$$

The augmented matrix that is associated with the system is:

$$\begin{bmatrix} 1 & 1 & 1 & 1 & | & 36 \\ 1 & -1 & 1 & -1 & | & 0 \\ 0 & -5 & 1 & 1 & | & 0 \\ -1 & 1 & 1 & -1 & | & -16 \end{bmatrix}$$

Pivot on row 1, column 1 by performing the operations:

$-R_1 + R_2 \rightarrow R_2$
$R_1 + R_4 \rightarrow R_4$

$$\begin{bmatrix} 1 & 1 & 1 & 1 & | & 36 \\ 0 & -2 & 0 & -2 & | & -36 \\ 0 & -5 & 1 & 1 & | & 0 \\ 0 & 2 & 2 & 0 & | & 20 \end{bmatrix}$$

Pivot on row 2, column 2 by performing the operations:

$\frac{-1}{2} R_2 \rightarrow R_2$
$-1R_2 + R_1 \rightarrow R_1$
$5R_2 + R_3 \rightarrow R_3$
$-2R_2 + R_4 \rightarrow R_4$

$$\begin{bmatrix} 1 & 0 & 1 & 0 & | & 18 \\ 0 & 1 & 0 & 1 & | & 18 \\ 0 & 0 & 1 & 6 & | & 90 \\ 0 & 0 & 2 & -2 & | & -16 \end{bmatrix}$$

Pivot on row 3, column 3 by performing the operations at the top of the next column.

$-1R_3 + R_1 \rightarrow R_1$
$-2R_3 + R_4 \rightarrow R_4$

$$\begin{bmatrix} 1 & 0 & 0 & -6 & | & -72 \\ 0 & 1 & 0 & 1 & | & 18 \\ 0 & 0 & 1 & 6 & | & 90 \\ 0 & 0 & 0 & -14 & | & -196 \end{bmatrix}$$

Now pivot on row 4, column 4 by performing the following operations:

$\frac{-1}{14} R_4 \rightarrow R_4$
$6R_4 + R_1 \rightarrow R_1$
$-R_4 + R_2 \rightarrow R_2$
$-6R_4 + R_3 \rightarrow R_3$

$$\begin{bmatrix} 1 & 0 & 0 & 0 & | & 12 \\ 0 & 1 & 0 & 0 & | & 4 \\ 0 & 0 & 1 & 0 & | & 6 \\ 0 & 0 & 0 & 1 & | & 14 \end{bmatrix}$$

The solution is:

$H = 12; E = 4; P = 6; A = 14$

So she spends 12 hours studying history, 4 hours studying English, 6 hours studying psychology and 14 hours studying accounting.

17. The system of equations is

$$\begin{aligned} x + 2y - 3z &= 5 \\ y + z &= 4 \\ z &= 3 \end{aligned}$$

Substitute $z = 3$ into equation 2 to get:

$y + (3) = 4 \Rightarrow y = 1$

Substitute $z = 3$ and $y = 1$ into equation 1 to get:

$x + 2(1) - 3(3) = 5 \Rightarrow x = 12$.

The solution to the system is:

$x = 12; y = 1; z = 3$

19. The system of equations is

$$2x + y - 3z = 16$$
$$3y + z = 12$$
$$4z = 0$$

From inspection of the third equation it is obvious that

$z = 0$.

Substitute $z = 0$ into equation 2 to get:

$$3y + (0) = 12 \Rightarrow y = 4$$

Substitute $z = 0$ and $y = 4$ into equation 1 to get:

$$2x + (4) - 3(0) = 16 \Rightarrow x = 6.$$

The solution to the system is:

$x = 6;\ y = 4;\ z = 0$

21. Neither student is correct. A system of linear equations with more equations than unknowns can have exactly one solution, infinitely many solutions, or no solutions. Exercise 11 out of this section is an example of a system that has exactly one solution. Two see a system of equations that has no solution, consider the system:

$y = x$
$y = x + 2$
$y = x + 4$

Here is a system of three parallel lines. It's obvious that there will be no single value of y that satisfies all three equations for a given value of x.

An example of a system of equations that has infinitely many solutions would be

$y = 2x$
$2y = 4x$
$3y = 6x$

Here is a system of three dependent lines. Any value of x and y that satisfies the first equation will satisfy all three of them.

23. The augmented matrix is:

$$\begin{bmatrix} 1 & 0.2 & 1.4 & 2 \\ 0.3 & -1.4 & 0.5 & 1.4 \\ 1.2 & 0.8 & -0.2 & 5 \end{bmatrix}$$

Enter the matrix into a graphing calculator, or a computer program, and pivot along the main diagonal. The reduced matrix will be:

$$\begin{bmatrix} 1 & 0 & 0 & 4.3274 \\ 0 & 1 & 0 & -0.6341 \\ 0 & 0 & 1 & -1.5719 \end{bmatrix}$$

The solution to the system is:

$x = 4.3274$
$y = -0.6341$
$z = -1.5719$

Cumulative Review

25. We need to find the slope of the line to find the equation of the line. Since parallel lines have the same slope, solve $3x + 5y = 27$ for y.

$$3x + 5y = 27 \Rightarrow y = \tfrac{-3}{5}x + \tfrac{27}{5}$$

Now it is easy to tell that the lines should have slope:

$m = -\tfrac{3}{5}$.

Use the slope-intercept form of the line to find the equation.

$y = \tfrac{-3}{5}x + 13$.

27. The augmented matrix to the system is:

$$\begin{bmatrix} 2 & 3 & 7 \\ 3 & 4 & 9 \end{bmatrix}$$

Pivot on row 1, column 1 by performing the following operations:

$$\tfrac{1}{2}R_1 \to R_1$$
$$-3R_1 + R_2 \to R_2$$

The matrix is displayed at the top of the next page.

$$\begin{bmatrix} 1 & \frac{3}{2} & \bigm| & \frac{7}{2} \\ 0 & \frac{-1}{2} & \bigm| & \frac{-3}{2} \end{bmatrix}$$

Now pivot on row 2, column 2 by performing the operations:

$$-2R_2 \rightarrow R_2$$
$$\tfrac{-3}{2}R_2 + R_1 \rightarrow R_1$$

$$\begin{bmatrix} 1 & 0 & \bigm| & -1 \\ 0 & 1 & \bigm| & 3 \end{bmatrix}$$

The solution is:

$$x = -1;\ y = 3.$$

Sample Test Answers

1. The coefficient matrix for the system is:

$$\begin{bmatrix} 1 & 0 & 2 \\ -3 & 5 & -8 \\ 1 & 1 & 0 \end{bmatrix}$$

2. The augmented matrix for the system is:

$$\begin{bmatrix} 3 & -1 & \bigm| & 5 \\ -5 & 17 & \bigm| & -1.2 \end{bmatrix}$$

3. Let N be the number of nickels, D be the number of dimes, and Q be then number of quarters in the parking meter. The system of equations that models this problem is:

$$\begin{array}{rcr} N + \quad D + \quad Q = & 213 \\ 5N + 10D + 25Q = & 2115 \\ D - \ 2Q = & 0 \end{array}$$

4. Let S represent the number of student tickets sold, and A represent the number of adult tickets sold.

A total 9,073 tickets were sold, which means:

$$S + A = 9073.$$

Total receipts of \$129,673 means:

$$12.50S + 16.75A = 129,673.$$

The system that models this problem is:

$$\begin{array}{rcr} S + \quad A = & 9073 \\ 12.50S + 16.75A = & 129,673 \end{array}$$

5. Let m represent the number of recipes for meatballs, c represent the number of recipes for crab cakes, and d represent the number of recipes for dumplings.

Since he was asked to prepare 380 servings the first equation is:

$$20m + 10c + 30d = 380.$$

The budget constraint of \$174 means:

$$10m + 15c + 8d = 174$$

The fact that people tend to eat twice as many dumplings as crab cakes means:

$$2c = d \text{ or } 2c - d = 0.$$

Therefore the system that models this problem is:

$$\begin{array}{rcr} 20m + 10c + 30d = & 380 \\ 2c - \ d = & 0 \\ 10m + 15c + \ 8d = & 174 \end{array}$$

6. Matrices (c) and (e) are in reduced row-echelon form.

Matrices (a) and (b) are not in reduced row-echelon form because the leading one in row 3 is not to the right of the leading one in row 2. Matrix (d) is not in reduced row-echelon form because the row consisting entirely of zeros is not the last row of the matrix.

7. Complete the pivot by performing the operation:

$$-5R_1 + R_3 \rightarrow R_3$$

$$\begin{bmatrix} 1 & 1 & 3 & \bigm| & -6 \\ 0 & 2 & 4 & \bigm| & 10 \\ 0 & -7 & -14 & \bigm| & 30 \end{bmatrix}$$

8. The augmented matrix is:

$$\begin{bmatrix} 1 & -2 & -1 & | & 1 \\ 3 & 1 & 1 & | & 10 \\ 2 & -1 & -1 & | & 0 \end{bmatrix}$$

Pivot on row 1, column 1 by performing the operations:

$$-3R_1 + R_2 \rightarrow R_2$$
$$-2R_1 + R_3 \rightarrow R_3$$

$$\begin{bmatrix} 1 & -2 & -1 & | & 1 \\ 0 & 7 & 4 & | & 7 \\ 0 & 3 & 1 & | & -2 \end{bmatrix}$$

Pivot on row 2, column 2 by performing the operations:

$$\tfrac{1}{7}R_2 \rightarrow R_2$$
$$2R_2 + R_1 \rightarrow R_1$$
$$-3R_2 + R_3 \rightarrow R_3$$

$$\begin{bmatrix} 1 & 0 & \tfrac{1}{7} & | & 3 \\ 0 & 1 & \tfrac{4}{7} & | & 1 \\ 0 & 0 & \tfrac{-5}{7} & | & -5 \end{bmatrix}$$

Pivot on row 3 column 3 by performing the operations:

$$\tfrac{-7}{5}R_3 \rightarrow R_3$$
$$\tfrac{-1}{7}R_3 + R_1 \rightarrow R_1$$
$$\tfrac{-4}{7}R_3 + R_2 \rightarrow R_2$$

$$\begin{bmatrix} 1 & 0 & 0 & | & 2 \\ 0 & 1 & 0 & | & -3 \\ 0 & 0 & 1 & | & 7 \end{bmatrix}$$

The solution to the system is:

$$x = 2; \ y = -3; \ z = 7$$

9. The augmented matrix is:

$$\begin{bmatrix} 6 & -1 & | & -11 \\ 1 & 2 & | & 9 \\ 5 & 1 & | & 0 \\ 11 & 3 & | & 4 \end{bmatrix}$$

Pivot on row 1, column 1 by performing the operations:

$$R_1 \leftrightarrow R_2$$
$$-6R_1 + R_2 \rightarrow R_2$$
$$-5R_1 + R_3 \rightarrow R_3$$
$$-11R_1 + R_4 \rightarrow R_4$$

$$\begin{bmatrix} 1 & 2 & | & 9 \\ 0 & -13 & | & -65 \\ 0 & -9 & | & -45 \\ 0 & -19 & | & -95 \end{bmatrix}$$

Pivot on row 2, column 2 by performing the operations:

$$\tfrac{-1}{13}R_2 \rightarrow R_2$$
$$-2R_2 + R_1 \rightarrow R_1$$
$$9R_2 + R_3 \rightarrow R_3$$
$$19R_2 + R_4 \rightarrow R_4$$

$$\begin{bmatrix} 1 & 0 & | & -1 \\ 0 & 1 & | & 5 \\ 0 & 0 & | & 0 \\ 0 & 0 & | & 0 \end{bmatrix}$$

The solution to the system on the previous page is:

$$x = -1; \ y = 5$$

10. The augmented matrix is:

$$\begin{bmatrix} 1 & 1 & 1 & | & 213 \\ 5 & 10 & 25 & | & 2115 \\ 0 & 1 & -2 & | & 0 \end{bmatrix}$$

Pivot on row 1 column 1 by performing the operation at the top of the next page.

$-5R_1 + R_2 \rightarrow R_2$

$$\left[\begin{array}{ccc|c} 1 & 1 & 1 & 213 \\ 0 & 5 & 20 & 1050 \\ 0 & 1 & -2 & 0 \end{array}\right]$$

Pivot on row 2 column 2 by performing the operations:

$\frac{1}{5}R_2 \rightarrow R_2$

$-R_2 + R_1 \rightarrow R_1$

$-R_2 + R_3 \rightarrow R_3$

$$\left[\begin{array}{ccc|c} 1 & 0 & -3 & 3 \\ 0 & 1 & 4 & 210 \\ 0 & 0 & -6 & -210 \end{array}\right]$$

Pivot on row 3 column 3 by performing the operations:

$\frac{-1}{6}R_3 \rightarrow R_3$

$3R_3 + R_1 \rightarrow R_1$

$-4R_3 + R_2 \rightarrow R_2$

The matrix is:

$$\left[\begin{array}{ccc|c} 1 & 0 & 0 & 108 \\ 0 & 1 & 0 & 70 \\ 0 & 0 & 1 & 35 \end{array}\right]$$

The parking meter had 108 nickels, 70 dimes, and 35 quarters.

11. The augmented matrix for the system is:

$$\left[\begin{array}{ccc|c} 1 & -2 & 1 & 5 \\ 3 & 2 & 1 & 7 \end{array}\right]$$

Pivot on row 1, column 1 by performing the operation:

$-3R_1 + R_2 \rightarrow R_2$

$$\left[\begin{array}{ccc|c} 1 & -2 & 1 & 5 \\ 0 & 8 & -2 & -8 \end{array}\right]$$

Pivot on row 2, column 2 by performing the operations at the top of the next column.

$\frac{1}{8}R_2 \rightarrow R_2$

$2R_2 + R_1 \rightarrow R_1$

$$\left[\begin{array}{ccc|c} 1 & 0 & \frac{1}{2} & 3 \\ 0 & 1 & \frac{-1}{4} & -1 \end{array}\right]$$

The general solution to this system is:

$x = 3 - \frac{1}{2}z$

$y = -1 + \frac{1}{4}z$

$z = any\ number$

Three specific solutions to this system are:

$z = 0;\ x = 3;\ y = -1$

$z = 4;\ x = 1;\ y = 0$

$z = -4;\ x = 5;\ y = -2$

12. The augmented matrix is:

$$\left[\begin{array}{ccc|c} 1 & 1 & 0 & 5 \\ 0 & 1 & 1 & 2 \\ 1 & 0 & -1 & 3 \end{array}\right]$$

Pivot on row 1, column 1 by performing the operation:

$-R_1 + R_3 \rightarrow R_3$

$$\left[\begin{array}{ccc|c} 1 & 1 & 0 & 5 \\ 0 & 1 & 1 & 2 \\ 0 & -1 & -1 & -2 \end{array}\right]$$

Pivot on row 2, column 2 by performing the operations:

$-R_2 + R_1 \rightarrow R_1$

$R_2 + R_3 \rightarrow R_3$

$$\left[\begin{array}{ccc|c} 1 & 0 & -1 & 3 \\ 0 & 1 & 1 & 2 \\ 0 & 0 & 0 & 0 \end{array}\right]$$

The general solution to this system is displayed at the top of the next page.

$x = 3 + z$

$y = 2 - z$

$z = \text{any number}$

13. The augmented matrix is:

$$\begin{bmatrix} 1 & 3 & -2 & | & 5 \\ 1 & 2 & 1 & | & 7 \\ 0 & 1 & -3 & | & 2 \end{bmatrix}$$

Pivot on row 1, column 1 by performing the operation:

$-R_1 + R_2 \rightarrow R_2$

$$\begin{bmatrix} 1 & 3 & -2 & | & 5 \\ 0 & -1 & 3 & | & 2 \\ 0 & 1 & -3 & | & 2 \end{bmatrix}$$

Pivot on row 2, column 2 by performing the operations:

$-R_2 \rightarrow R_2$

$-3R_2 + R_1 \rightarrow R_1$

$-R_2 + R_3 \rightarrow R_3$

$$\begin{bmatrix} 1 & 0 & 7 & | & 11 \\ 0 & 1 & -3 & | & -2 \\ 0 & 0 & 0 & | & 4 \end{bmatrix}$$

The third row is inconsistent; therefore this system has no solution.

14. The augmented matrix is:

$$\begin{bmatrix} 1 & 1 & | & 9073 \\ 12.50 & 16.75 & | & 129,673 \end{bmatrix}$$

Pivot on row 1, column 1 by performing the operation:

$-12.50R_1 + R_2 \rightarrow R_2$

$$\begin{bmatrix} 1 & 1 & | & 9073 \\ 0 & 4.25 & | & 16,260.5 \end{bmatrix}$$

Pivot on row 2, column 2 by performing the operations:

$\frac{1}{4.25} R_2 \rightarrow R_2$

$-R_2 + R_1 \rightarrow R_1$

$$\begin{bmatrix} 1 & 0 & | & 5247 \\ 0 & 1 & | & 3826 \end{bmatrix}$$

A total of 5247 student tickets were sold, and 3826 adult tickets were sold.

15. The augmented matrix is:

$$\begin{bmatrix} 20 & 10 & 30 & | & 380 \\ 0 & 2 & -1 & | & 0 \\ 10 & 15 & 8 & | & 174 \end{bmatrix}$$

Pivot on row 1, column 1 by performing the operation:

$\frac{1}{20} R_1 \rightarrow R_1$

$-10R_1 + R_3 \rightarrow R_3$

$$\begin{bmatrix} 1 & \frac{1}{2} & \frac{3}{2} & | & 19 \\ 0 & 2 & -1 & | & 0 \\ 0 & 10 & -7 & | & -16 \end{bmatrix}$$

Pivot on row 2, column 2 by performing the operations:

$\frac{1}{2} R_2 \rightarrow R_2$

$\frac{-1}{2} R_2 + R_1 \rightarrow R_1$

$-10R_2 + R_3 \rightarrow R_3$

$$\begin{bmatrix} 1 & 0 & \frac{7}{4} & | & 19 \\ 0 & 1 & \frac{-1}{2} & | & 0 \\ 0 & 0 & -2 & | & -16 \end{bmatrix}$$

Pivot on row 3, column 3 by performing the following operations:

$\frac{-1}{2} R_3 \rightarrow R_3$

$\frac{-7}{4} R_3 + R_1 \rightarrow R_1$

$\frac{1}{2} R_3 + R_2 \rightarrow R_2$

The matrix is shown at the top of the next page.

$$\begin{bmatrix} 1 & 0 & 0 & 5 \\ 0 & 1 & 0 & 4 \\ 0 & 0 & 1 & 8 \end{bmatrix}$$

Steven should prepare 5 recipes of meatballs, 4 recipes of crab cakes, and 8 recipes of Chinese Dumplings.

Chapter 3
Matrix Algebra

Exercises Section 3.1

1. If the matrices are to be equal, then
$t + 1 = 4$, or $t = 3$ and $x - 3 = 6$, or
$x = 9$.

3. If the matrices are to be equal, then $x = x$; however, this is true for every value of x.

5. $-3A + B = -3\begin{bmatrix} 2 & 3 \\ 1 & -1 \end{bmatrix} + \begin{bmatrix} 6 & -1 \\ 0 & 2 \end{bmatrix} =$

$\begin{bmatrix} -6 & -9 \\ -3 & 3 \end{bmatrix} + \begin{bmatrix} 6 & -1 \\ 0 & 2 \end{bmatrix} = \begin{bmatrix} 0 & -10 \\ -3 & 5 \end{bmatrix}$

7. $\frac{1}{3}(A + B) = \frac{1}{3}\left(\begin{bmatrix} 2 & 3 \\ 1 & -1 \end{bmatrix} + \begin{bmatrix} 6 & -1 \\ 0 & 2 \end{bmatrix} \right) =$

$\begin{bmatrix} \frac{2}{3} & 1 \\ \frac{1}{3} & \frac{-1}{3} \end{bmatrix} + \begin{bmatrix} 2 & \frac{-1}{3} \\ 0 & \frac{2}{3} \end{bmatrix} = \begin{bmatrix} \frac{8}{3} & \frac{2}{3} \\ \frac{1}{3} & \frac{1}{3} \end{bmatrix}$

9. $\frac{1}{4}(2A + C)$

Matrix addition is not defined because matrix C is not the same dimension as matrix A.

11. $-5(C + 3C) =$

$-5\left(\begin{bmatrix} 2 & -3 & 5 \\ 0 & 0 & -2 \end{bmatrix} + 3\begin{bmatrix} 2 & -3 & 5 \\ 0 & 0 & -2 \end{bmatrix} \right) =$

$-5\left(\begin{bmatrix} 8 & -12 & 20 \\ 0 & 0 & -8 \end{bmatrix} \right) = \begin{bmatrix} -40 & 60 & -100 \\ 0 & 0 & 40 \end{bmatrix}$

13. First compute $A + B$:

$A + B = \begin{bmatrix} a & b \\ c & d \end{bmatrix} + \begin{bmatrix} e & f \\ g & h \end{bmatrix} =$

$\begin{bmatrix} a+e & b+f \\ c+g & d+h \end{bmatrix}$

Next compute $B + A$:

$B + A = \begin{bmatrix} e & f \\ g & h \end{bmatrix} + \begin{bmatrix} a & b \\ c & d \end{bmatrix} =$

$\begin{bmatrix} e+a & f+b \\ g+c & h+d \end{bmatrix}$

By the commutative property of real numbers we know that:

$a + e = e + a$
$b + f = f + b$
$c + g = g + c$
$d + h = h + d$

Therefore, $A + B = B + A$.

15.
a. The information displayed in a 2×3 matrix is:

$$\begin{array}{c} \\ P_1 \\ P_2 \end{array} \begin{array}{ccc} FC & TC & EC \\ \begin{bmatrix} 30 & 50 & 90 \\ 50 & 60 & 100 \end{bmatrix} \end{array}$$

b. The information displayed in a 3×2 matrix is:

$$\begin{array}{c} \\ FC \\ TC \\ EC \end{array} \begin{array}{cc} P_1 & P_2 \\ \begin{bmatrix} 30 & 50 \\ 50 & 60 \\ 90 & 100 \end{bmatrix} \end{array}$$

17.

a. The information displayed in a 4×1 matrix is:

$$\begin{array}{c} \text{F.} \\ \text{So.} \\ \text{Jr.} \\ \text{Sr.} \end{array} \begin{bmatrix} 300 \\ 287 \\ 250 \\ 240 \end{bmatrix}$$

b. The information displayed in a 1×4 matrix is:

$$\begin{array}{cccc} \text{F.} & \text{So.} & \text{Jr.} & \text{Sr.} \end{array}$$
$$\begin{bmatrix} 300 & 287 & 250 & 240 \end{bmatrix}$$

19. The following matrix indicates airline connections between cities. An entry of one indicates that a flight does exist and an entry of zero indicates that a flight does not exist.

$$\begin{array}{c} \\ A \\ B \\ C \\ D \\ E \end{array} \begin{array}{c} \begin{array}{ccccc} A & B & C & D & E \end{array} \\ \begin{bmatrix} 0 & 1 & 1 & 1 & 0 \\ 0 & 0 & 0 & 1 & 0 \\ 0 & 0 & 0 & 1 & 0 \\ 0 & 0 & 0 & 0 & 1 \\ 0 & 0 & 0 & 0 & 0 \end{bmatrix} \end{array}$$

21. Let a one indicate that the team in the row beat the team in the column. A dominance relation matrix showing the results of the first two days of the tournament is shown below:

$$\begin{array}{c} \\ B \\ W \\ T \\ C \end{array} \begin{array}{c} \begin{array}{cccc} B & W & T & C \end{array} \\ \begin{bmatrix} 0 & 1 & 1 & 0 \\ 0 & 0 & 1 & 0 \\ 0 & 0 & 0 & 1 \\ 1 & 1 & 0 & 0 \end{bmatrix} \end{array}$$

23.
$$\frac{3}{4}X - 5T = 3B \qquad \text{(add } 5T \text{)}$$
$$\frac{3}{4}X = 3B + 5T \qquad \text{(multiply by } \frac{4}{3} \text{)}$$
$$X = 4B + \frac{20}{3}T$$

25.
$$\frac{1}{2}A - 3B + C = 2D \qquad \text{(subtract } C \text{)}$$
$$\frac{1}{2}A - 3B = 2D - C \qquad \text{(subtract } \frac{1}{2}A \text{)}$$
$$-3B = 2D - C - \frac{1}{2}A \qquad \text{(divide by -3)}$$
$$B = \frac{1}{6}A + \frac{1}{3}C - \frac{2}{3}D$$

27.
$$A - T = 2C + 3T \qquad \text{(subtract } 3T \text{)}$$
$$A - 4T = 2C \qquad \text{(subtract } A \text{)}$$
$$-4T = -A + 2C \qquad \text{(divide by -4)}$$
$$T = \frac{1}{4}A - \frac{1}{2}C$$

29.
$$\begin{bmatrix} 2 & -3 & 1 \\ 4 & 5 & 2 \end{bmatrix} + 2X = -3\begin{bmatrix} -1 & 2 & -1 \\ 1 & 0 & 1 \end{bmatrix}$$

$$2X = \begin{bmatrix} 3 & -6 & 3 \\ -3 & 0 & -3 \end{bmatrix} - \begin{bmatrix} 2 & -3 & 1 \\ 4 & 5 & 2 \end{bmatrix}$$

$$2X = \begin{bmatrix} 1 & -3 & 2 \\ -7 & -5 & -5 \end{bmatrix}$$

$$X = \frac{1}{2}\begin{bmatrix} 1 & -3 & 2 \\ -7 & \frac{-5}{2} & -5 \end{bmatrix} = \begin{bmatrix} \frac{1}{2} & \frac{-3}{2} & 1 \\ \frac{-7}{2} & \frac{-5}{2} & \frac{-5}{2} \end{bmatrix}$$

$$X = \begin{bmatrix} \frac{1}{2} & \frac{-3}{2} & 1 \\ \frac{-7}{2} & \frac{-5}{2} & \frac{-5}{2} \end{bmatrix}$$

31. $2\begin{bmatrix} 1 & 1 \\ 2 & 0 \\ 3 & -2 \end{bmatrix} + 3X + \begin{bmatrix} 2 & 5 \\ -1 & 3 \\ -4 & -2 \end{bmatrix} = \begin{bmatrix} 0 & 0 \\ 0 & 0 \\ 0 & 0 \end{bmatrix}$

$$\begin{bmatrix} 2 & 2 \\ 4 & 0 \\ 6 & -4 \end{bmatrix} + 3X = \begin{bmatrix} 0 & 0 \\ 0 & 0 \\ 0 & 0 \end{bmatrix} - \begin{bmatrix} 2 & 5 \\ -1 & 3 \\ -4 & -2 \end{bmatrix}$$

$$3X = \begin{bmatrix} -2 & -5 \\ 1 & -3 \\ 4 & 2 \end{bmatrix} - \begin{bmatrix} 2 & 2 \\ 4 & 0 \\ 6 & -4 \end{bmatrix}$$

$$X = \frac{1}{3}\begin{bmatrix} -4 & -7 \\ -3 & -3 \\ -2 & -6 \end{bmatrix}$$

$$X = \begin{bmatrix} \frac{-4}{3} & \frac{-7}{3} \\ -1 & -1 \\ \frac{-2}{3} & -2 \end{bmatrix}$$

33. Simplify the left hand side of the matrix equation.

$$\begin{bmatrix} 2x & 1x & 1x \\ 0 & 1x & 1x \end{bmatrix} + \begin{bmatrix} 1x & 2x & 2x \\ 3x & -1x & -1x \end{bmatrix} = \begin{bmatrix} 9 & 9 & 9 \\ 9 & 0 & 0 \end{bmatrix}$$

$$\begin{bmatrix} 3x & 3x & 3x \\ 3x & 0 & 0 \end{bmatrix} = \begin{bmatrix} 9 & 9 & 9 \\ 9 & 0 & 0 \end{bmatrix}$$

This means that:
$3x = 9$

$x = 3$ is the single value of x that will satisfy the matrix equation.

35. To give the matrix in minutes, multiply it by 60 as shown below:

$$60 \begin{bmatrix} 0.06 & 0.02 \\ 0.03 & 0.05 \end{bmatrix} = \begin{bmatrix} 3.6 & 1.2 \\ 1.8 & 3.0 \end{bmatrix}$$

37. Let E represent the ending inventory matrix and P represent the purchase matrix. The equation that represents this problem is:

$$B + P - S = E$$

This implies that:

$$P = E - B + S$$

$$P = \begin{bmatrix} 10 & 8 \\ 12 & 10 \end{bmatrix} - \begin{bmatrix} 30 & 20 \\ 18 & 12 \end{bmatrix} + \begin{bmatrix} 21 & 18 \\ 22 & 15 \end{bmatrix}$$

$$P = \begin{bmatrix} 1 & 6 \\ 16 & 13 \end{bmatrix}$$

39.
a. The 8% earnings of each account will be represented by the expression

$$E = 0.08A.$$

b. The matrix is:

$$0.08A = 0.08 \begin{bmatrix} 30,000 & 25,000 \\ 50,000 & 40,000 \end{bmatrix}$$

$$0.08A = \begin{bmatrix} 2400 & 2000 \\ 4000 & 3200 \end{bmatrix}$$

c. The value at the end of the first year of the account will be the amount of the account plus the interest earned on the account. This translates to:

$$V = A + E$$
$$V = A + 0.08A$$
$$V = 1.08A$$

d. Computing the value results in the matrix:

$$V = 1.08 \begin{bmatrix} 30,000 & 25,000 \\ 50,000 & 40,000 \end{bmatrix}$$

$$V = \begin{bmatrix} 32,400 & 27,000 \\ 54,000 & 43,200 \end{bmatrix}$$

41. The matrix expression that represents the actual costs for the second year at each of the universities is:

$$C_2 = C + 0.05C = 1.05C$$

To compute the actual cost, replace C

$$C_2 = 1.05 \begin{bmatrix} 4500 & 6000 \\ 5000 & 9400 \end{bmatrix} =$$

$$C_2 = \begin{bmatrix} 4725 & 6300 \\ 5250 & 9870 \end{bmatrix}$$

43. The matrix expression that computes the actual cost of the fourth year at the college will be:

$$C_4 = C_3 + 0.05C_3$$
$$C_4 = 1.1025C + 0.05(1.1025C) =$$
$$C_4 = 1.157625C$$

To compute the actual cost, replace C with the cost matrix.

$$C_4 = 1.57625 \begin{bmatrix} 4500 & 6000 \\ 5000 & 9400 \end{bmatrix} =$$

$$C_4 = \begin{bmatrix} 5209.31 & 6945.75 \\ 5788.13 & 10881.68 \end{bmatrix}$$

45. To calculate the percentage by which one group surpassed another group you would divide the difference in the two times by the time of the inferior group. For example the percentage by which group two surpassed group 1 in the 1-8 digit category is:

$$\frac{16-14}{14} = \frac{2}{14} = .1428 \text{ or } 14.3\% .$$

Use this method to calculate the remaining percentages and the appropriate matrix that shows the percentage increase in each category over the group with no training is:

$$\begin{bmatrix} 14.3 & 28.6 \\ 37.5 & 87.5 \\ 66.7 & 116.7 \end{bmatrix}$$

47. Find the transpose by switching each row with the respective column.

$$B^T = \begin{bmatrix} 1 & 3 \\ 2 & -5 \end{bmatrix}$$

49. Find the transpose by switching each row with the respective column.

$$D^T = \begin{bmatrix} 1 & 2 & -3 & 4 \\ 0 & -1 & 2 & 1 \end{bmatrix}$$

Exercises Section 3.2

1. The linear system can be written as the matrix equation:

$$\begin{bmatrix} 1 & 2 \\ 3 & -1 \end{bmatrix}\begin{bmatrix} x \\ y \end{bmatrix} = \begin{bmatrix} 7 \\ 6 \end{bmatrix}$$

3. The linear system can be written as the matrix equation:

$$\begin{bmatrix} 1 & 1 \\ 3 & -1 \\ 1 & 5 \end{bmatrix}\begin{bmatrix} x \\ y \end{bmatrix} = \begin{bmatrix} 2 \\ 6 \\ -1 \end{bmatrix}$$

5. The matrix equation represents the system of equations is:

$$2x \qquad = 3$$
$$x + 5y = 5$$
$$4x + 6y = 2$$

7. The matrix equation represents the system of equations is:

$$x \quad\; + z + \quad\; + v = -2$$
$$2x \quad\; + 3z + 5w + v = \;\; 3$$
$$-1x + 2y + 2z + 2w \quad\;\; = \;\; 6$$

9. $AB =$

$$\begin{bmatrix} 2 & 3 \\ 1 & -2 \end{bmatrix}\begin{bmatrix} 1 & 0 \\ 2 & 3 \end{bmatrix} = \begin{bmatrix} 8 & 9 \\ -3 & -6 \end{bmatrix}$$

11. $AC =$

$$\begin{bmatrix} 2 & 3 \\ 1 & -2 \end{bmatrix}\begin{bmatrix} 2 & 1 \\ 1 & 0 \\ 3 & 2 \end{bmatrix} =$$

Multiplication does not exist because the number of columns in the left hand matrix does not equal the number of rows in the right hand matrix.

13. $AD =$

$$\begin{bmatrix} 2 & 3 \\ 1 & -2 \end{bmatrix}\begin{bmatrix} 1 & 3 & 5 & 9 \\ 2 & 0 & 1 & -3 \end{bmatrix} = \begin{bmatrix} 8 & 6 & 13 & 9 \\ -3 & 3 & 3 & 15 \end{bmatrix}$$

15. $2B + I =$

$$2\begin{bmatrix} 1 & 0 \\ 2 & 3 \end{bmatrix} + \begin{bmatrix} 1 & 0 \\ 0 & 1 \end{bmatrix} = \begin{bmatrix} 2 & 0 \\ 4 & 6 \end{bmatrix} + \begin{bmatrix} 1 & 0 \\ 0 & 1 \end{bmatrix} =$$

$$\begin{bmatrix} 3 & 0 \\ 4 & 7 \end{bmatrix}$$

17. $B^2 + 2I =$

$$\begin{bmatrix} 1 & 0 \\ 2 & 3 \end{bmatrix}\begin{bmatrix} 1 & 0 \\ 2 & 3 \end{bmatrix} + 2\begin{bmatrix} 1 & 0 \\ 0 & 1 \end{bmatrix} =$$

$$\begin{bmatrix} 1 & 0 \\ 8 & 9 \end{bmatrix} + \begin{bmatrix} 2 & 0 \\ 0 & 2 \end{bmatrix} =$$

$$\begin{bmatrix} 3 & 0 \\ 8 & 11 \end{bmatrix}$$

19. $A^2 =$

$$\begin{bmatrix} 2 & 3 \\ 1 & -2 \end{bmatrix}\begin{bmatrix} 2 & 3 \\ 1 & -2 \end{bmatrix} = \begin{bmatrix} 7 & 0 \\ 0 & 7 \end{bmatrix}$$

21. $A^2 + 3B =$

$$\begin{bmatrix} 2 & 3 \\ 1 & -2 \end{bmatrix}\begin{bmatrix} 2 & 3 \\ 1 & -2 \end{bmatrix} + 3\begin{bmatrix} 1 & 0 \\ 2 & 3 \end{bmatrix} =$$

$$\begin{bmatrix} 7 & 0 \\ 0 & 7 \end{bmatrix} + \begin{bmatrix} 3 & 0 \\ 6 & 9 \end{bmatrix} =$$

$$\begin{bmatrix} 10 & 0 \\ 6 & 16 \end{bmatrix}$$

23. The following example will show that this matrix product is not valid.

Let A and B be defined how they were in problems 9-22. Now,

$$(A - B)(A + B) =$$

$$\left(\begin{bmatrix} 2 & 3 \\ 1 & -2 \end{bmatrix} - \begin{bmatrix} 1 & 0 \\ 2 & 3 \end{bmatrix}\right)\left(\begin{bmatrix} 2 & 3 \\ 1 & -2 \end{bmatrix} + \begin{bmatrix} 1 & 0 \\ 2 & 3 \end{bmatrix}\right) =$$

$$\left(\begin{bmatrix} 1 & 3 \\ -1 & -5 \end{bmatrix}\right)\left(\begin{bmatrix} 3 & 3 \\ 3 & 1 \end{bmatrix}\right) =$$

$$\begin{bmatrix} 12 & 6 \\ -18 & -8 \end{bmatrix}$$

However,

$$A^2 - B^2 =$$

$$\begin{bmatrix} 2 & 3 \\ 1 & -2 \end{bmatrix}\begin{bmatrix} 2 & 3 \\ 1 & -2 \end{bmatrix} - \begin{bmatrix} 1 & 0 \\ 2 & 3 \end{bmatrix}\begin{bmatrix} 1 & 0 \\ 2 & 3 \end{bmatrix} =$$

$$\begin{bmatrix} 7 & 0 \\ 0 & 7 \end{bmatrix} - \begin{bmatrix} 1 & 0 \\ 8 & 9 \end{bmatrix} =$$

The example is continued at the top of the next column.

$$\begin{bmatrix} 6 & 0 \\ -8 & -2 \end{bmatrix}$$

Obviously,

$$\begin{bmatrix} 12 & 6 \\ -18 & -8 \end{bmatrix} \neq \begin{bmatrix} 6 & 0 \\ -8 & -2 \end{bmatrix}$$

Therefore, the difference of squares formula does not hold for matrices.

25. $(AB)C =$

$$\left(\begin{bmatrix} 3 & 1 \\ 2 & 5 \end{bmatrix}\begin{bmatrix} 1 & 0 \\ 2 & 3 \end{bmatrix}\right)\begin{bmatrix} 1 & 2 \\ 3 & 4 \end{bmatrix} =$$

$$\begin{bmatrix} 5 & 3 \\ 12 & 15 \end{bmatrix}\begin{bmatrix} 1 & 2 \\ 3 & 4 \end{bmatrix} =$$

$$\begin{bmatrix} 14 & 22 \\ 57 & 84 \end{bmatrix}$$

$$A(BC) =$$

$$\begin{bmatrix} 3 & 1 \\ 2 & 5 \end{bmatrix}\left(\begin{bmatrix} 1 & 0 \\ 2 & 3 \end{bmatrix}\begin{bmatrix} 1 & 2 \\ 3 & 4 \end{bmatrix}\right) =$$

$$\begin{bmatrix} 3 & 1 \\ 2 & 5 \end{bmatrix}\begin{bmatrix} 1 & 2 \\ 11 & 16 \end{bmatrix} =$$

$$\begin{bmatrix} 14 & 22 \\ 57 & 84 \end{bmatrix}$$

$$(AB)C = \begin{bmatrix} 14 & 22 \\ 57 & 84 \end{bmatrix} = A(BC)$$

27. $A(B + C) =$

$$\begin{bmatrix} 3 & 1 \\ 2 & 5 \end{bmatrix}\left(\begin{bmatrix} 1 & 0 \\ 2 & 3 \end{bmatrix} + \begin{bmatrix} 1 & 2 \\ 3 & 4 \end{bmatrix}\right) =$$

The solution is continued on the next page.

$$\begin{bmatrix} 3 & 1 \\ 2 & 5 \end{bmatrix}\begin{bmatrix} 2 & 2 \\ 5 & 7 \end{bmatrix} =$$

$$\begin{bmatrix} 11 & 13 \\ 29 & 39 \end{bmatrix}$$

$AB + AC =$

$$\begin{bmatrix} 3 & 1 \\ 2 & 5 \end{bmatrix}\begin{bmatrix} 1 & 0 \\ 2 & 3 \end{bmatrix} + \begin{bmatrix} 3 & 1 \\ 2 & 5 \end{bmatrix}\begin{bmatrix} 1 & 2 \\ 3 & 4 \end{bmatrix} =$$

$$\begin{bmatrix} 5 & 3 \\ 12 & 15 \end{bmatrix} + \begin{bmatrix} 6 & 10 \\ 17 & 24 \end{bmatrix} =$$

$$\begin{bmatrix} 11 & 13 \\ 29 & 39 \end{bmatrix}$$

$$A(B+C) = \begin{bmatrix} 11 & 13 \\ 29 & 39 \end{bmatrix} = AB + AC$$

29. There are many choices to this problem, one of the choices is:

$$A = \begin{bmatrix} 0 & 0 & 0 \\ 0 & 1 & 0 \\ 0 & 0 & 0 \end{bmatrix}; B = \begin{bmatrix} 1 & 0 & 0 \\ 0 & 2 & 0 \\ 0 & 0 & 3 \end{bmatrix}; C = \begin{bmatrix} 3 & 0 & 0 \\ 0 & 2 & 0 \\ 0 & 0 & 1 \end{bmatrix}$$

$$AB = \begin{bmatrix} 0 & 0 & 0 \\ 0 & 2 & 0 \\ 0 & 0 & 0 \end{bmatrix}$$

$$AC = \begin{bmatrix} 0 & 0 & 0 \\ 0 & 2 & 0 \\ 0 & 0 & 0 \end{bmatrix}$$

Yet $B \neq C$ and $A \neq 0$.

31.
a.

$$(2A)B = \left(2\begin{bmatrix} 1 & 2 \\ 3 & 1 \end{bmatrix} \right)\begin{bmatrix} 2 & 3 \\ 4 & 1 \end{bmatrix} =$$

The solution is continued at the top of the next column.

$$\begin{bmatrix} 2 & 4 \\ 6 & 2 \end{bmatrix}\begin{bmatrix} 2 & 3 \\ 4 & 1 \end{bmatrix} = \begin{bmatrix} 20 & 10 \\ 20 & 20 \end{bmatrix}$$

b.

$$A(2B) = \begin{bmatrix} 1 & 2 \\ 3 & 1 \end{bmatrix}\left(2\begin{bmatrix} 2 & 3 \\ 4 & 1 \end{bmatrix} \right) =$$

$$\begin{bmatrix} 1 & 2 \\ 3 & 1 \end{bmatrix}\begin{bmatrix} 4 & 6 \\ 8 & 2 \end{bmatrix} = \begin{bmatrix} 20 & 10 \\ 20 & 20 \end{bmatrix}$$

c.

$$2(AB) = 2\left(\begin{bmatrix} 1 & 2 \\ 3 & 1 \end{bmatrix}\begin{bmatrix} 2 & 3 \\ 4 & 1 \end{bmatrix} \right) =$$

$$2\begin{bmatrix} 10 & 5 \\ 10 & 10 \end{bmatrix} = \begin{bmatrix} 20 & 10 \\ 20 & 20 \end{bmatrix}$$

33.
a.

$$(kA)B = \left(k\begin{bmatrix} 1 & 2 \\ 3 & 1 \end{bmatrix} \right)\begin{bmatrix} 2 & 3 \\ 4 & 1 \end{bmatrix} =$$

$$\begin{bmatrix} k & 2k \\ 3k & k \end{bmatrix}\begin{bmatrix} 2 & 3 \\ 4 & 1 \end{bmatrix} = \begin{bmatrix} 10k & 5k \\ 10k & 10k \end{bmatrix}$$

b.

$$A(kB) = \begin{bmatrix} 1 & 2 \\ 3 & 1 \end{bmatrix}\left(k\begin{bmatrix} 2 & 3 \\ 4 & 1 \end{bmatrix} \right) =$$

$$\begin{bmatrix} 1 & 2 \\ 3 & 1 \end{bmatrix}\begin{bmatrix} 2k & 3k \\ 4k & k \end{bmatrix} = \begin{bmatrix} 10k & 5k \\ 10k & 10k \end{bmatrix}$$

c.

$$k(AB) = k\left(\begin{bmatrix} 1 & 2 \\ 3 & 1 \end{bmatrix}\begin{bmatrix} 2 & 3 \\ 4 & 1 \end{bmatrix} \right) =$$

$$k\begin{bmatrix} 10 & 5 \\ 10 & 10 \end{bmatrix} = \begin{bmatrix} 10k & 5k \\ 10k & 10k \end{bmatrix}$$

35. By the distributive property:

$$AX + BX = (A+B)X$$

37. By the property discussed in exercise 34, we have:

$$AX + 2AY = AX + A(2Y)$$

By the distributive property:

$$AX + A(2Y) = A(X + 2Y)$$

39. By the distributive property:

$$3AB + 4CB = (3A + 4C)B$$

41. By the property discussed in exercise 34:

$$2AX - 3AY = A(2X) + A(-3Y)$$

By the distributive property:

$$A(2X) + A(-3Y) = A(2X - 3Y)$$

43. Matrix B will be a row matrix that looks like:

$$B = \begin{bmatrix} -100 & 6 \end{bmatrix}$$

The product matrix that will show her income minus tuition each year will be:

$$BA = \begin{bmatrix} -100 & 6 \end{bmatrix} \begin{bmatrix} 15 & 17 \\ 160 & 130 \end{bmatrix}$$

This product is:

$$BA = \begin{bmatrix} -100 & 6 \end{bmatrix} \begin{bmatrix} 15 & 17 \\ 160 & 130 \end{bmatrix} = \begin{bmatrix} -540 & -920 \end{bmatrix}$$

45.
a. Matrix B will be the row matrix:

$$B = \begin{bmatrix} 1 & 1.5 & 0.73 \end{bmatrix}$$

The product matrix that will give the total intake of fat, sodium, and protein is:

$$BA = \begin{bmatrix} 1 & 1.5 & 0.73 \end{bmatrix} \begin{bmatrix} 2 & 830 & 2 \\ 5 & 30 & 8 \\ 0 & 1 & 2 \end{bmatrix}$$

$$BA = \begin{bmatrix} 9.5 & 875.73 & 15.46 \end{bmatrix}$$

b. Matrix C is the column matrix:

$$C = \begin{bmatrix} -3 \\ -1.5 \\ 2 \end{bmatrix}$$

The product matrix that gives the total points accumulated for each food is:

$$AC = \begin{bmatrix} 2 & 830 & 2 \\ 5 & 30 & 8 \\ 0 & 1 & 2 \end{bmatrix} \begin{bmatrix} -3 \\ -1.5 \\ 2 \end{bmatrix}$$

$$AC = \begin{bmatrix} -1247 \\ -44 \\ 2.5 \end{bmatrix}$$

47. The matrix B will be:

$$B = \begin{bmatrix} 3 & 5 & 2 \\ 2 & 4 & 6 \end{bmatrix}$$

The product matrix $BA =$

$$BA = \begin{bmatrix} 3 & 5 & 2 \\ 2 & 4 & 6 \end{bmatrix} \begin{bmatrix} 700 & 1000 \\ 500 & 800 \\ 1500 & 1200 \end{bmatrix}$$

$$BA = \begin{bmatrix} 7600 & 9400 \\ 12400 & 12400 \end{bmatrix}$$

The product matrix indicates that 7600 men and 9400 women were influenced by the ads in January. It also indicates that 12,400 men and 12,400 women were influenced by the ads in February.

b. The matrix C is:

$$C = \begin{bmatrix} 10 & 12 & 15 \\ 15 & 18 & 22 \end{bmatrix}$$

The solution is continued on the next page.

The product matrix $AC =$

$$AC = \begin{bmatrix} 700 & 1000 \\ 500 & 800 \\ 1500 & 1200 \end{bmatrix} \begin{bmatrix} 10 & 12 & 15 \\ 15 & 18 & 22 \end{bmatrix}$$

$$AC = \begin{bmatrix} 22000 & 26400 & 32500 \\ 17000 & 20400 & 25100 \\ 33000 & 39600 & 48900 \end{bmatrix}$$

The product matrix indicates that the newspapers ads generated $22,000 in revenue the first time the ad ran, $26,400 the second time the ad ran and $32,500 the third time the ad ran. Similarly, the radio ads generated $17,000, $20,400, and $25,100 from the first, second and third run respectively. The TV ads generated $33,000, $39,600, and $48900 from the first, second, and third run respectively.

49. The matrix B is the row matrix

$$B = \begin{bmatrix} 200 & 300 \end{bmatrix}.$$

The matrix C is the column matrix:

$$C = \begin{bmatrix} 12 \\ 20 \\ 1 \end{bmatrix}$$

The product matrix that will give total cost for filling this order is:

$$BAC = \begin{bmatrix} 200 & 300 \end{bmatrix} \begin{bmatrix} 0.03 & 0.05 & 0.01 \\ 0.02 & 0.03 & 0.01 \end{bmatrix} \begin{bmatrix} 12 \\ 20 \\ 1 \end{bmatrix} =$$

$$\begin{bmatrix} 12 & 19 & 5 \end{bmatrix} \begin{bmatrix} 12 \\ 20 \\ 1 \end{bmatrix} = \begin{bmatrix} 529 \end{bmatrix}$$

The total cost for filling this order will be $529.

51. The transition matrix for the experiment can be written as:

$$T = \begin{bmatrix} 0.1 & 0.9 \\ 0.4 & 0.6 \end{bmatrix}$$

The solution is continued at the top of the next column.

The first day of the experiment, Julie observed 48 mice went left, and 52 went right. This forms the matrix:

$$A = \begin{bmatrix} 48 & 52 \end{bmatrix}$$

The matrix product AT will determine how many mice will go left and right the next day.

$$AT = \begin{bmatrix} 48 & 52 \end{bmatrix} \begin{bmatrix} .1 & .9 \\ .4 & .6 \end{bmatrix} =$$

$$AT = \begin{bmatrix} 25.6 & 74.4 \end{bmatrix}$$

According to the earlier results, 26 mice should go left the next day while 74 mice should turn right.

Julie uses the second day's predictions multiplied with the transition matrix to predict which way the mice will turn on the third day.

$$\begin{bmatrix} 25.6 & 74.4 \end{bmatrix} \begin{bmatrix} .1 & .9 \\ .4 & .6 \end{bmatrix} =$$

$$\begin{bmatrix} 32.32 & 67.68 \end{bmatrix}$$

On day 3, Julie expects 32 mice to turn left and 68 mice to turn right.

To compute the forth day, use the third day's predictions like before.

$$\begin{bmatrix} 32.32 & 67.68 \end{bmatrix} \begin{bmatrix} .1 & .9 \\ .4 & .6 \end{bmatrix} =$$

$$\begin{bmatrix} 30.3 & 69.7 \end{bmatrix}$$

On the fourth day, Julie expects 30 mice to turn left and 70 mice to turn right.

53. Matrix C is:

$$C = \begin{bmatrix} 0.07 & 0.01 \\ 0.0485 & 0.025 \\ 0.0350 & 0.01 \\ 0.04 & 0.008 \\ 0.042 & 0.006 \end{bmatrix}$$

The solution is continued on the next page.

a.

$$Cost$$

$$B = \begin{bmatrix} 500,000 \\ 1,000,000 \end{bmatrix} \begin{matrix} 2000\,sq.ft. \\ high-rise \end{matrix}$$

b.

$$CB = \begin{bmatrix} 0.07 & 0.01 \\ 0.0485 & 0.025 \\ 0.0350 & 0.01 \\ 0.04 & 0.008 \\ 0.042 & 0.006 \end{bmatrix} \begin{bmatrix} 500,000 \\ 1,000,000 \end{bmatrix} =$$

$$CB = \begin{bmatrix} 45,000 \\ 49,250 \\ 27,250 \\ 28,000 \\ 27,000 \end{bmatrix} \begin{matrix} \text{Belgium} \\ \text{France} \\ \text{Japan} \\ \text{Mexico} \\ \text{U.S.} \end{matrix}$$

The entries in this matrix represent the total construction cost (in U.S. dollars) allotted to architects for a 2000 square foot building and structural engineers for a high-rise office building in the respective countries.

55. Remember that a one occupying an entry means that a flight from the row to the column exists.

Thus, the matrix M that shows all of the direct flights between the cities is:

$$\begin{matrix} & A & B & C \end{matrix}$$
$$M = \begin{matrix} A \\ B \\ C \end{matrix} \begin{bmatrix} 0 & 1 & 1 \\ 0 & 0 & 1 \\ 0 & 1 & 0 \end{bmatrix}$$

b.

$$\begin{matrix} & A & B & C \end{matrix}$$
$$M^2 = \begin{matrix} A \\ B \\ C \end{matrix} \begin{bmatrix} 0 & 1 & 1 \\ 0 & 1 & 0 \\ 0 & 0 & 1 \end{bmatrix}$$

The entries in this matrix give the number of two-stage flights from each city. We notice that it is impossible to return to city A in only two stages. Row 2 Column 2 and Row 3 Column 3 You can also make a round trip in two stages from city B to city C or from city C to city B.

c.

$$\begin{matrix} & A & B & C \end{matrix}$$
$$M + M^2 = \begin{matrix} A \\ B \\ C \end{matrix} \begin{bmatrix} 0 & 2 & 2 \\ 0 & 1 & 1 \\ 0 & 1 & 1 \end{bmatrix}$$

This matrix indicates the total number of paths that can be followed to get from one city to another in two stages or less.

57. $A(B+C) =$

$$\begin{bmatrix} a_{11} & a_{12} \\ a_{21} & a_{22} \end{bmatrix} \left(\begin{bmatrix} b_{11} & b_{12} \\ b_{21} & b_{22} \end{bmatrix} + \begin{bmatrix} c_{11} & c_{12} \\ c_{21} & c_{22} \end{bmatrix} \right)$$

$$\begin{bmatrix} a_{11} & a_{12} \\ a_{21} & a_{22} \end{bmatrix} \begin{bmatrix} b_{11}+c_{11} & b_{12}+c_{12} \\ b_{21}+c_{21} & b_{22}+c_{22} \end{bmatrix}$$

$$\begin{bmatrix} a_{11}b_{11}+a_{11}c_{11}+a_{12}b_{21}+a_{12}c_{21} & a_{11}b_{12}+a_{11}c_{12}+a_{12}b_{12}+a_{12}c_{12} \\ a_{21}b_{11}+a_{21}c_{11}+a_{22}b_{21}+a_{22}c_{21} & a_{21}b_{12}+a_{21}c_{12}+a_{22}b_{22}+a_{22}c_{22} \end{bmatrix}$$

$AB + AC =$

$$\begin{bmatrix} a_{11} & a_{12} \\ a_{21} & a_{22} \end{bmatrix} \begin{bmatrix} b_{11} & b_{12} \\ b_{21} & b_{22} \end{bmatrix} + \begin{bmatrix} a_{11} & a_{12} \\ a_{21} & a_{22} \end{bmatrix} \begin{bmatrix} c_{11} & c_{12} \\ c_{21} & c_{22} \end{bmatrix}$$

$$\begin{bmatrix} a_{11}b_{11}+a_{12}b_{21} & a_{11}b_{12}+a_{12}b_{22} \\ a_{21}b_{11}+a_{22}b_{21} & a_{21}b_{12}+a_{22}b_{22} \end{bmatrix} + \begin{bmatrix} a_{11}c_{11}+a_{12}c_{21} & a_{11}c_{12}+a_{12}c_{22} \\ a_{21}c_{11}+a_{22}c_{21} & a_{21}c_{12}+a_{22}c_{22} \end{bmatrix}$$

$$\begin{bmatrix} a_{11}b_{11}+a_{11}c_{11}+a_{12}b_{21}+a_{12}c_{21} & a_{11}b_{12}+a_{11}c_{12}+a_{12}b_{12}+a_{12}c_{12} \\ a_{21}b_{11}+a_{21}c_{11}+a_{22}b_{21}+a_{22}c_{21} & a_{21}b_{12}+a_{21}c_{12}+a_{22}b_{22}+a_{22}c_{22} \end{bmatrix}$$

$$A(B+C) = \begin{bmatrix} a_{11}b_{11}+a_{11}c_{11}+a_{12}b_{21}+a_{12}c_{21} & a_{11}b_{12}+a_{11}c_{12}+a_{12}b_{12}+a_{12}c_{12} \\ a_{21}b_{11}+a_{21}c_{11}+a_{22}b_{21}+a_{22}c_{21} & a_{21}b_{12}+a_{21}c_{12}+a_{22}b_{22}+a_{22}c_{22} \end{bmatrix} = AB + AC$$

Exercises Section 3.3

1. $AB=$

$$\begin{bmatrix} 3 & 2 \\ 1 & 1 \end{bmatrix} \begin{bmatrix} 1 & -2 \\ -1 & 3 \end{bmatrix} = \begin{bmatrix} 1 & 0 \\ 0 & 1 \end{bmatrix}$$

$BA=$

$$\begin{bmatrix} 1 & -2 \\ -1 & 3 \end{bmatrix} \begin{bmatrix} 3 & 2 \\ 1 & 1 \end{bmatrix} = \begin{bmatrix} 1 & 0 \\ 0 & 1 \end{bmatrix}$$

3. $AB=$

$$\begin{bmatrix} 1 & -1 & 0 \\ 2 & 1 & 0 \\ 1 & 1 & 2 \end{bmatrix} \begin{bmatrix} 1 & 0 & 0 \\ 2 & 1 & 1 \\ \frac{1}{2} & 3 & -1 \end{bmatrix} = \begin{bmatrix} -1 & -1 & -1 \\ 4 & 1 & 1 \\ 4 & 7 & -1 \end{bmatrix}$$

Since the product is the not the identity matrix, B is not the inverse of A.

5. The matrix equation is:

$$\begin{bmatrix} 3 & 1 \\ 1 & -1 \end{bmatrix} \begin{bmatrix} x \\ y \end{bmatrix} = \begin{bmatrix} -4 \\ 8 \end{bmatrix}$$

Multiply both sides of the equation on the left by the inverse of the coefficient matrix as shown on the next page.

$$\begin{bmatrix} 1 & 0 \\ 0 & 1 \end{bmatrix}\begin{bmatrix} x \\ y \end{bmatrix} = \begin{bmatrix} \frac{1}{4} & \frac{1}{4} \\ \frac{1}{4} & \frac{-3}{4} \end{bmatrix}\begin{bmatrix} -4 \\ 8 \end{bmatrix}$$

This results in the following matrix equation:

$$\begin{bmatrix} x \\ y \end{bmatrix} = \begin{bmatrix} 1 \\ -7 \end{bmatrix}$$

The solution is:

$$x = 1;\ y = -7$$

7. The matrix equation is:

$$\begin{bmatrix} 3 & 1 \\ 1 & -1 \end{bmatrix}\begin{bmatrix} m \\ n \end{bmatrix} = \begin{bmatrix} 0.2 \\ 0.3 \end{bmatrix}$$

Multiply both sides of the equation on the left by the inverse of the coefficient matrix to get:

$$\begin{bmatrix} 1 & 0 \\ 0 & 1 \end{bmatrix}\begin{bmatrix} m \\ n \end{bmatrix} = \begin{bmatrix} \frac{1}{4} & \frac{1}{4} \\ \frac{1}{4} & \frac{-3}{4} \end{bmatrix}\begin{bmatrix} 0.2 \\ 0.3 \end{bmatrix}$$

This results in the following matrix equation:

$$\begin{bmatrix} m \\ n \end{bmatrix} = \begin{bmatrix} 0.125 \\ -0.175 \end{bmatrix}$$

The solution is:

$$m = 0.125;\ n = -0.175$$

9. Form the augmented matrix:

$$\left[\begin{array}{cc|cc} 1 & 3 & 1 & 0 \\ 2 & 5 & 0 & 1 \end{array}\right]$$

Pivot on row 1, column 1 by performing the operations:

$$-2R_1 + R_2 \rightarrow R_2$$

$$\left[\begin{array}{cc|cc} 1 & 3 & 1 & 0 \\ 0 & -1 & -2 & 1 \end{array}\right]$$

Pivot on row 2, column 2 by performing the operations at the top of the next column.

$$-R_2 \rightarrow R_2$$
$$-3R_2 + R_1 \rightarrow R_1$$

$$\left[\begin{array}{cc|cc} 1 & 0 & -5 & 3 \\ 0 & 1 & 2 & -1 \end{array}\right]$$

This means that the inverse matrix is:

$$\begin{bmatrix} -5 & 3 \\ 2 & -1 \end{bmatrix}$$

11. Form the augmented matrix:

$$\left[\begin{array}{cc|cc} 2 & 3 & 1 & 0 \\ 1 & \frac{3}{2} & 0 & 1 \end{array}\right]$$

Pivot on row 1, column 1 by performing the operations:

$$\tfrac{1}{2}R_1 \rightarrow R_1$$
$$-R_1 + R_2 \rightarrow R_2$$

$$\left[\begin{array}{cc|cc} 1 & \frac{3}{2} & \frac{1}{2} & 0 \\ 0 & 0 & \frac{-1}{2} & 1 \end{array}\right]$$

Since there is a row of zeros to the left of the vertical line, the matrix has no inverse.

13. Form the augmented matrix:

$$\left[\begin{array}{cc|cc} 0 & 3 & 1 & 0 \\ 0 & 2 & 0 & 1 \end{array}\right]$$

Since the matrix on has a column of zeros to the left of the vertical line, the matrix can not be reduced to the identity matrix. Therefore this matrix has no inverse.

15. Form the augmented matrix:

$$\left[\begin{array}{cc|cc} -1 & 3 & 1 & 0 \\ 2 & -4 & 0 & 1 \end{array}\right]$$

Pivot on row 1, column 1 by performing the operations:

$$-R_1 \rightarrow R_1$$
$$-2R_1 + R_2 \rightarrow R_2$$

The solution is continued on the next page.

$$\left[\begin{array}{cc|cc} 1 & -3 & -1 & 0 \\ 0 & 2 & 2 & 1 \end{array}\right]$$

Pivot on row 2, column 2 by performing the operations:

$\frac{1}{2}R_2 \rightarrow R_2$

$3R_2 + R_1 \rightarrow R_1$

$$\left[\begin{array}{cc|cc} 1 & 0 & 2 & \frac{3}{2} \\ 0 & 1 & 1 & \frac{1}{2} \end{array}\right]$$

This means that the inverse matrix is:

$$\left[\begin{array}{cc} 2 & \frac{3}{2} \\ 1 & \frac{1}{2} \end{array}\right]$$

17. Form the augmented matrix:

$$\left[\begin{array}{ccc|ccc} 1 & 0 & 2 & 1 & 0 & 0 \\ 0 & 2 & 2 & 0 & 1 & 0 \\ 1 & 1 & 1 & 0 & 0 & 1 \end{array}\right]$$

Pivot on row 1, column 1 by performing the operations:

$-R_1 + R_3 \rightarrow R_3$

$$\left[\begin{array}{ccc|ccc} 1 & 0 & 2 & 1 & 0 & 0 \\ 0 & 2 & 2 & 0 & 1 & 0 \\ 0 & 1 & -1 & -1 & 0 & 1 \end{array}\right]$$

Pivot on row 2, column 2 by performing the operations:

$\frac{1}{2}R_2 \rightarrow R_2$

$-R_2 + R_3 \rightarrow R_3$

$$\left[\begin{array}{ccc|ccc} 1 & 0 & 2 & 1 & 0 & 0 \\ 0 & 1 & 1 & 0 & \frac{1}{2} & 0 \\ 0 & 0 & -2 & -1 & \frac{-1}{2} & 1 \end{array}\right]$$

Pivot on row 3, column 3 by performing the operations:

$\frac{-1}{2}R_3 \rightarrow R_3$

$-2R_3 + R_1 \rightarrow R_1$

$-R_3 + R_2 \rightarrow R_2$

The reduced matrix is:

$$\left[\begin{array}{ccc|ccc} 1 & 0 & 0 & 0 & \frac{-1}{2} & 1 \\ 0 & 1 & 0 & \frac{-1}{2} & \frac{1}{4} & \frac{1}{2} \\ 0 & 0 & 1 & \frac{1}{2} & \frac{1}{4} & \frac{-1}{2} \end{array}\right]$$

Therefore, the inverse matrix is:

$$\left[\begin{array}{ccc} 0 & \frac{-1}{2} & 1 \\ \frac{-1}{2} & \frac{1}{4} & \frac{1}{2} \\ \frac{1}{2} & \frac{1}{4} & \frac{-1}{2} \end{array}\right]$$

19. Form the augmented matrix:

$$\left[\begin{array}{ccc|ccc} 3 & 1 & 0 & 1 & 0 & 0 \\ 2 & 1 & 1 & 0 & 1 & 0 \\ 5 & 2 & 1 & 0 & 0 & 1 \end{array}\right]$$

Pivot on row 1, column 1 by performing the operations:

$\frac{1}{3}R_1 \rightarrow R_1$

$-2R_1 + R_2 \rightarrow R_2$

$-5R_1 + R_3 \rightarrow R_3$

$$\left[\begin{array}{ccc|ccc} 1 & \frac{1}{3} & 0 & \frac{1}{3} & 0 & 0 \\ 0 & \frac{1}{3} & 1 & \frac{-2}{3} & 1 & 0 \\ 0 & \frac{1}{3} & 1 & \frac{-5}{3} & 0 & 1 \end{array}\right]$$

Pivot on row 2, column 2 by performing the operations:

$3R_2 \rightarrow R_2$

$\frac{-1}{3}R_2 + R_1 \rightarrow R_1$

$\frac{-1}{3}R_2 + R_3 \rightarrow R_3$

$$\left[\begin{array}{ccc|ccc} 1 & 0 & -1 & 1 & -1 & 0 \\ 0 & 1 & 3 & -2 & 3 & 0 \\ 0 & 0 & 0 & -1 & -1 & 1 \end{array}\right]$$

Since Row 3 consists entirely of zeros to the left of the vertical line, the matrix has no inverse.

21. Form the augmented matrix:

$$\left[\begin{array}{ccc|ccc} 2 & 1 & 1 & 1 & 0 & 0 \\ 1 & 0 & 1 & 0 & 1 & 0 \\ 1 & 1 & 1 & 0 & 0 & 1 \end{array}\right]$$

Pivot on row 1, column 1 by performing the operations:

$\frac{1}{2}R_1 \rightarrow R_1$
$-R_1 + R_2 \rightarrow R_2$
$-R_1 + R_3 \rightarrow R_3$

$$\left[\begin{array}{ccc|ccc} 1 & \frac{1}{2} & \frac{1}{2} & \frac{1}{2} & 0 & 0 \\ 0 & \frac{-1}{2} & \frac{1}{2} & \frac{-1}{2} & 1 & 0 \\ 0 & \frac{1}{2} & \frac{1}{2} & \frac{-1}{2} & 0 & 1 \end{array}\right]$$

Pivot on row 2, column 2 by performing the operations:

$-2R_2 \rightarrow R_2$
$\frac{-1}{2}R_2 + R_1 \rightarrow R_1$
$\frac{-1}{2}R_2 + R_3 \rightarrow R_3$

$$\left[\begin{array}{ccc|ccc} 1 & 0 & 1 & 0 & 1 & 0 \\ 0 & 1 & -1 & 1 & -2 & 0 \\ 0 & 0 & 1 & -1 & 1 & 1 \end{array}\right]$$

Pivot on row 3, column 3 by performing the operations:

$-R_3 + R_1 \rightarrow R_1$
$R_3 + R_2 \rightarrow R_2$

$$\left[\begin{array}{ccc|ccc} 1 & 0 & 0 & 1 & 0 & -1 \\ 0 & 1 & 0 & 0 & -1 & 1 \\ 0 & 0 & 1 & -1 & 1 & 1 \end{array}\right]$$

Therefore the inverse matrix is:

$$\left[\begin{array}{ccc} 1 & 0 & -1 \\ 0 & -1 & 1 \\ -1 & 1 & 1 \end{array}\right]$$

23. Form the augmented matrix:

$$\left[\begin{array}{cccc|cccc} 1 & 0 & 1 & 0 & 1 & 0 & 0 & 0 \\ 2 & 2 & 2 & 0 & 0 & 1 & 0 & 0 \\ 2 & 0 & 0 & 1 & 0 & 0 & 1 & 0 \\ 1 & 0 & 2 & 1 & 0 & 0 & 0 & 1 \end{array}\right]$$

Pivot on row 1, column 1 by performing the operations:

$-2R_1 + R_2 \rightarrow R_2$
$-2R_1 + R_3 \rightarrow R_3$
$-R_1 + R_4 \rightarrow R_4$

$$\left[\begin{array}{cccc|cccc} 1 & 0 & 1 & 0 & 1 & 0 & 0 & 0 \\ 0 & 2 & 0 & 0 & -2 & 1 & 0 & 0 \\ 0 & 0 & -2 & 1 & -2 & 0 & 1 & 0 \\ 0 & 0 & 1 & 1 & -1 & 0 & 0 & 1 \end{array}\right]$$

Pivot on row 2, column 2 by performing the operations:

$\frac{1}{2}R_2 \rightarrow R_2$

$$\left[\begin{array}{cccc|cccc} 1 & 0 & 1 & 0 & 1 & 0 & 0 & 0 \\ 0 & 1 & 0 & 0 & -1 & \frac{1}{2} & 0 & 0 \\ 0 & 0 & -2 & 1 & -2 & 0 & 1 & 0 \\ 0 & 0 & 1 & 1 & -1 & 0 & 0 & 1 \end{array}\right]$$

Pivot on row 3, column 3 by performing the operations:

$R_3 \leftrightarrow R_4$
$-R_3 + R_1 \rightarrow R_1$
$2R_3 + R_4 \rightarrow R_4$

$$\left[\begin{array}{cccc|cccc} 1 & 0 & 0 & -1 & 2 & 0 & 0 & -1 \\ 0 & 1 & 0 & 0 & -1 & \frac{1}{2} & 0 & 0 \\ 0 & 0 & 1 & 1 & -1 & 0 & 0 & 1 \\ 0 & 0 & 0 & 3 & -4 & 0 & 1 & 2 \end{array}\right]$$

The solution is continued on the next page.

Pivot on row 4, column 4 by performing the operations:

$$\frac{1}{3}R_4 \rightarrow R_4$$
$$R_4 + R_1 \rightarrow R_1$$
$$-R_4 + R_3 \rightarrow R_3$$

$$\begin{bmatrix} 1 & 0 & 0 & 0 & | & \frac{2}{3} & 0 & \frac{1}{3} & \frac{-1}{3} \\ 0 & 1 & 0 & 0 & | & -1 & \frac{1}{2} & 0 & 0 \\ 0 & 0 & 1 & 0 & | & \frac{1}{3} & 0 & \frac{-1}{3} & \frac{1}{3} \\ 0 & 0 & 0 & 1 & | & \frac{-4}{3} & 0 & \frac{1}{3} & \frac{2}{3} \end{bmatrix}$$

Therefore the inverse matrix is:

$$\begin{bmatrix} \frac{2}{3} & 0 & \frac{1}{3} & \frac{-1}{3} \\ -1 & \frac{1}{2} & 0 & 0 \\ \frac{1}{3} & 0 & \frac{-1}{3} & \frac{1}{3} \\ \frac{-4}{3} & 0 & \frac{1}{3} & \frac{2}{3} \end{bmatrix}$$

25.

a. $I - A = \begin{bmatrix} 1 & 0 \\ 0 & 1 \end{bmatrix} - \begin{bmatrix} 2 & 3 \\ 1 & -4 \end{bmatrix} =$

$$\begin{bmatrix} -1 & -3 \\ -1 & 5 \end{bmatrix}$$

b. To find $(I - A)^{-1}$, use the answer from part (a) to create the augmented matrix:

$$\begin{bmatrix} -1 & -3 & | & 1 & 0 \\ -1 & 5 & | & 0 & 1 \end{bmatrix}$$

Pivot on row 1, column 1 by performing the operations:

$$-R_1 \rightarrow R_1$$
$$R_1 + R_2 \rightarrow R_2$$

$$\begin{bmatrix} 1 & 3 & | & -1 & 0 \\ 0 & 8 & | & -1 & 1 \end{bmatrix}$$

Pivot on row 2, column 2 by performing the operations:

$$\frac{1}{8}R_2 \rightarrow R_2$$
$$-3R_2 + R_1 \rightarrow R_1$$

$$\begin{bmatrix} 1 & 0 & | & \frac{-5}{8} & \frac{-3}{8} \\ 0 & 1 & | & \frac{-1}{8} & \frac{1}{8} \end{bmatrix}$$

Therefore,

$$(I - A)^{-1} = \begin{bmatrix} \frac{-5}{8} & \frac{-3}{8} \\ \frac{-1}{8} & \frac{1}{8} \end{bmatrix}$$

27. The matrix equation is:

$$\begin{bmatrix} 1 & 3 \\ 2 & 4 \end{bmatrix}\begin{bmatrix} x \\ y \end{bmatrix} = \begin{bmatrix} 7 \\ -3 \end{bmatrix}$$

The coefficient matrix is:

$$\begin{bmatrix} 1 & 3 \\ 2 & 4 \end{bmatrix}$$

The augmented matrix is:

$$\begin{bmatrix} 1 & 3 & | & 1 & 0 \\ 2 & 4 & | & 0 & 1 \end{bmatrix}$$

Pivot on row 1, column 1 by performing the operation:

$$-2R_1 + R_2 \rightarrow R_2$$

$$\begin{bmatrix} 1 & 3 & | & 1 & 0 \\ 0 & -2 & | & -2 & 1 \end{bmatrix}$$

Pivot on row 2, column 2 by performing the operations:

$$\frac{-1}{2}R_2 \rightarrow R_2$$
$$-3R_2 + R_1 \rightarrow R_1$$

$$\begin{bmatrix} 1 & 0 & | & -2 & \frac{3}{2} \\ 0 & 1 & | & 1 & \frac{-1}{2} \end{bmatrix}$$

Therefore the inverse matrix of the coefficient matrix is:

$$\begin{bmatrix} -2 & \frac{3}{2} \\ 1 & \frac{-1}{2} \end{bmatrix}$$

The solution is continued on the next page.

Multiply both sides of the matrix equation on the left by the inverse of the coefficient matrix to get the equation:

$$\begin{bmatrix} x \\ y \end{bmatrix} = \begin{bmatrix} -2 & \frac{3}{2} \\ 1 & \frac{-1}{2} \end{bmatrix} \begin{bmatrix} 7 \\ -3 \end{bmatrix} = \begin{bmatrix} \frac{-37}{2} \\ \frac{17}{2} \end{bmatrix}$$

The solution to the system is:

$$x = \frac{-37}{2}; \; y = \frac{17}{2}$$

29. The matrix equation is:

$$\begin{bmatrix} 5 & 2 \\ 4 & 2 \end{bmatrix} \begin{bmatrix} x \\ y \end{bmatrix} = \begin{bmatrix} 4 \\ -3 \end{bmatrix}$$

The coefficient matrix is:

$$\begin{bmatrix} 5 & 2 \\ 4 & 2 \end{bmatrix}$$

The augmented matrix is:

$$\left[\begin{array}{cc|cc} 5 & 2 & 1 & 0 \\ 4 & 2 & 0 & 1 \end{array}\right]$$

Pivot on row 1, column 1 by performing the operations:

$$\frac{1}{5} R_1 \rightarrow R_1$$
$$-4R_1 + R_2 \rightarrow R_2$$

$$\left[\begin{array}{cc|cc} 1 & \frac{2}{5} & \frac{1}{5} & 0 \\ 0 & \frac{2}{5} & \frac{-4}{5} & 1 \end{array}\right]$$

Pivot on row 2, column 2 by performing the operations:

$$\frac{5}{2} R_2 \rightarrow R_2$$
$$\frac{-2}{5} R_2 + R_1 \rightarrow R_1$$

$$\left[\begin{array}{cc|cc} 1 & 0 & 1 & -1 \\ 0 & 1 & -2 & \frac{5}{2} \end{array}\right]$$

Therefore the inverse matrix of the coefficient matrix is:

$$\begin{bmatrix} 1 & -1 \\ -2 & \frac{5}{2} \end{bmatrix}$$

Multiply both sides of the matrix equation on the left by the inverse of the coefficient matrix to get the equation:

$$\begin{bmatrix} x \\ y \end{bmatrix} = \begin{bmatrix} 1 & -1 \\ -2 & \frac{5}{2} \end{bmatrix} \begin{bmatrix} 4 \\ -3 \end{bmatrix} = \begin{bmatrix} 7 \\ \frac{-31}{2} \end{bmatrix}$$

The solution to the system is:

$$x = 7; \; y = \frac{-31}{2}$$

31. The matrix equation is:

$$\begin{bmatrix} 1 & 0 & 2 \\ 0 & 2 & 2 \\ 1 & 1 & 1 \end{bmatrix} \begin{bmatrix} x \\ y \\ z \end{bmatrix} = \begin{bmatrix} 1 \\ 3 \\ 5 \end{bmatrix}$$

The coefficient matrix is:

$$\begin{bmatrix} 1 & 0 & 2 \\ 0 & 2 & 2 \\ 1 & 1 & 1 \end{bmatrix}$$

The augmented matrix is:

$$\left[\begin{array}{ccc|ccc} 1 & 0 & 2 & 1 & 0 & 0 \\ 0 & 2 & 2 & 0 & 1 & 0 \\ 1 & 1 & 1 & 0 & 0 & 1 \end{array}\right]$$

Pivot on row 1, column 1 by performing the operation:

$$-R_1 + R_3 \rightarrow R_3$$

$$\left[\begin{array}{ccc|ccc} 1 & 0 & 2 & 1 & 0 & 0 \\ 0 & 2 & 2 & 0 & 1 & 0 \\ 0 & 1 & -1 & -1 & 0 & 1 \end{array}\right]$$

Pivot on row 2, column 2 by performing the operations:

$$\frac{1}{2} R_2 \rightarrow R_2$$
$$-R_2 + R_3 \rightarrow R_3$$

$$\left[\begin{array}{ccc|ccc} 1 & 0 & 2 & 1 & 0 & 0 \\ 0 & 1 & 1 & 0 & \frac{1}{2} & 0 \\ 0 & 0 & -2 & -1 & \frac{-1}{2} & 1 \end{array}\right]$$

The solution is continued on the next page.

Pivot on row 3, column 3 by performing the operations:

$$\tfrac{-1}{2}R_3 \to R_3$$
$$-2R_3 + R_1 \to R_1$$
$$-2R_3 + R_2 \to R_2$$

$$\begin{bmatrix} 1 & 0 & 0 & 0 & \tfrac{-1}{2} & 1 \\ 0 & 1 & 0 & \tfrac{-1}{2} & \tfrac{1}{4} & \tfrac{1}{2} \\ 0 & 0 & 0 & \tfrac{1}{2} & \tfrac{1}{4} & \tfrac{-1}{2} \end{bmatrix}$$

Therefore the inverse matrix of the coefficient matrix is:

$$\begin{bmatrix} 0 & \tfrac{-1}{2} & 1 \\ \tfrac{-1}{2} & \tfrac{1}{4} & \tfrac{1}{2} \\ \tfrac{1}{2} & \tfrac{1}{4} & \tfrac{-1}{2} \end{bmatrix}$$

Multiply both sides of the matrix equation on the left by the inverse of the coefficient matrix to get the equation shown:

$$\begin{bmatrix} x \\ y \\ z \end{bmatrix} = \begin{bmatrix} 0 & \tfrac{-1}{2} & 1 \\ \tfrac{-1}{2} & \tfrac{1}{4} & \tfrac{1}{2} \\ \tfrac{1}{2} & \tfrac{1}{4} & \tfrac{-1}{2} \end{bmatrix} \begin{bmatrix} 1 \\ 3 \\ 5 \end{bmatrix} = \begin{bmatrix} \tfrac{7}{2} \\ \tfrac{11}{4} \\ \tfrac{-5}{4} \end{bmatrix}$$

The solution to the system is:

$$x = \tfrac{7}{2}; \; y = \tfrac{11}{4}; \; z = \tfrac{-5}{4}$$

33. The matrix equation is:

$$\begin{bmatrix} 2 & 1 & 2 \\ 1 & -1 & 1 \\ 1 & 1 & 0 \end{bmatrix} \begin{bmatrix} x \\ y \\ z \end{bmatrix} = \begin{bmatrix} -5 \\ -5 \\ 2 \end{bmatrix}$$

The coefficient matrix is:

$$\begin{bmatrix} 2 & 1 & 2 \\ 1 & -1 & 1 \\ 1 & 1 & 0 \end{bmatrix}$$

The augmented matrix is displayed at the top of the next column.

$$\begin{bmatrix} 2 & 1 & 2 & 1 & 0 & 0 \\ 1 & -1 & 1 & 0 & 1 & 0 \\ 1 & 1 & 0 & 0 & 0 & 1 \end{bmatrix}$$

Pivot on row 1, column 1 by performing the operations:

$$R_1 \leftrightarrow R_2$$
$$-2R_1 + R_2 \to R_2$$
$$-R_1 + R_3 \to R_3$$

$$\begin{bmatrix} 1 & -1 & 1 & 0 & 1 & 0 \\ 0 & 3 & 0 & 1 & -2 & 0 \\ 0 & 2 & -1 & 0 & -1 & 1 \end{bmatrix}$$

Pivot on row 2, column 2 by performing the operations:

$$\tfrac{1}{3}R_2 \to R_2$$
$$R_2 + R_1 \to R_1$$
$$-2R_2 + R_3 \to R_3$$

$$\begin{bmatrix} 1 & 0 & 1 & \tfrac{1}{3} & \tfrac{1}{3} & 0 \\ 0 & 1 & 0 & \tfrac{1}{3} & \tfrac{-2}{3} & 0 \\ 0 & 0 & -1 & \tfrac{-2}{3} & \tfrac{2}{3} & 1 \end{bmatrix}$$

Pivot on row 3, column 3 by performing the operations:

$$-R_3 \to R_3$$
$$-R_3 + R_1 \to R_1$$

$$\begin{bmatrix} 1 & 0 & 0 & \tfrac{-1}{3} & \tfrac{2}{3} & 1 \\ 0 & 1 & 0 & \tfrac{1}{3} & \tfrac{-2}{3} & 0 \\ 0 & 0 & 1 & \tfrac{2}{3} & \tfrac{-1}{3} & -1 \end{bmatrix}$$

Therefore the inverse matrix of the coefficient matrix is:

$$\begin{bmatrix} \tfrac{-1}{3} & \tfrac{2}{3} & 1 \\ \tfrac{1}{3} & \tfrac{-2}{3} & 0 \\ \tfrac{2}{3} & \tfrac{-1}{3} & -1 \end{bmatrix}$$

Multiply both sides of the matrix equation on the left by the inverse of the coefficient matrix to get the equation shown at the top of the next page.

$$\begin{bmatrix} x \\ y \\ z \end{bmatrix} = \begin{bmatrix} \frac{-1}{3} & \frac{2}{3} & 1 \\ \frac{1}{3} & \frac{-2}{3} & 0 \\ \frac{2}{3} & \frac{-1}{3} & -1 \end{bmatrix} \begin{bmatrix} -5 \\ -5 \\ 2 \end{bmatrix} = \begin{bmatrix} \frac{1}{3} \\ \frac{5}{3} \\ \frac{-11}{3} \end{bmatrix}$$

The solution to the system is:

$$x = \tfrac{1}{3}; \ y = \tfrac{5}{3}; \ z = \tfrac{-11}{3}$$

35. The matrix equation is:

$$\begin{bmatrix} 3 & 3 & 1 \\ 1 & -1 & 1 \\ 3 & 0 & 1 \end{bmatrix} \begin{bmatrix} x \\ y \\ z \end{bmatrix} = \begin{bmatrix} \frac{8}{3} \\ 1 \\ -1 \end{bmatrix}$$

The coefficient matrix is:

$$\begin{bmatrix} 3 & 3 & 1 \\ 1 & -1 & 1 \\ 3 & 0 & 1 \end{bmatrix}$$

The augmented matrix is:

$$\left[\begin{array}{ccc|ccc} 3 & 3 & 1 & 1 & 0 & 0 \\ 1 & -1 & 1 & 0 & 1 & 0 \\ 3 & 0 & 1 & 0 & 0 & 1 \end{array}\right]$$

Pivot on row 1, column 1 by performing the operations:

$$R_1 \leftrightarrow R_2$$
$$-3R_1 + R_2 \rightarrow R_2$$
$$-3R_1 + R_3 \rightarrow R_3$$

$$\left[\begin{array}{ccc|ccc} 1 & -1 & 1 & 0 & 1 & 0 \\ 0 & 6 & -2 & 1 & -3 & 0 \\ 0 & 3 & -2 & 0 & -3 & 1 \end{array}\right]$$

Pivot on row 2, column 2 by performing the operations:

$$\tfrac{1}{6}R_2 \rightarrow R_2$$
$$R_2 + R_1 \rightarrow R_1$$
$$-3R_2 + R_3 \rightarrow R_3$$

$$\left[\begin{array}{ccc|ccc} 1 & 0 & \frac{2}{3} & \frac{1}{6} & \frac{1}{2} & 0 \\ 0 & 1 & \frac{-1}{3} & \frac{1}{6} & \frac{-1}{2} & 0 \\ 0 & 0 & -1 & \frac{-1}{2} & \frac{-3}{2} & 1 \end{array}\right]$$

Pivot on row 3, column 3 by performing the operations:

$$-R_3 \rightarrow R_3$$
$$\tfrac{-2}{3}R_3 + R_1 \rightarrow R_1$$
$$\tfrac{1}{3}R_3 + R_2 \rightarrow R_2$$

$$\left[\begin{array}{ccc|ccc} 1 & 0 & 0 & \frac{-1}{6} & \frac{-1}{2} & \frac{2}{3} \\ 0 & 1 & 0 & \frac{1}{3} & 0 & \frac{-1}{3} \\ 0 & 0 & 1 & \frac{1}{2} & \frac{3}{2} & -1 \end{array}\right]$$

Therefore the inverse matrix of the coefficient matrix is:

$$\begin{bmatrix} \frac{-1}{6} & \frac{-1}{2} & \frac{2}{3} \\ \frac{1}{3} & 0 & \frac{-1}{3} \\ \frac{1}{2} & \frac{3}{2} & -1 \end{bmatrix}$$

Multiply both sides of the matrix equation on the left by the inverse of the coefficient matrix to get the equation:

$$\begin{bmatrix} x \\ y \\ z \end{bmatrix} = \begin{bmatrix} \frac{1}{12} & \frac{1}{4} & \frac{1}{6} \\ \frac{5}{24} & \frac{-3}{8} & \frac{-1}{12} \\ \frac{1}{8} & \frac{3}{8} & \frac{-1}{4} \end{bmatrix} \begin{bmatrix} \frac{8}{3} \\ 1 \\ -1 \end{bmatrix} = \begin{bmatrix} \frac{-19}{18} \\ \frac{11}{9} \\ \frac{23}{6} \end{bmatrix}$$

The solution to the system is:

$$x = \tfrac{-29}{18}; \ y = \tfrac{11}{9}; \ z = \tfrac{23}{6}$$

37. The matrix equation is:

$$\begin{bmatrix} 3 & -1 & 1 \\ 1 & 0 & 3 \\ 2 & 3 & 1 \end{bmatrix} \begin{bmatrix} x \\ y \\ z \end{bmatrix} = \begin{bmatrix} 0 \\ 0 \\ 0 \end{bmatrix}$$

The coefficient matrix is:

$$\begin{bmatrix} 3 & -1 & 1 \\ 1 & 0 & 3 \\ 2 & 3 & 1 \end{bmatrix}$$

The solution is continued on the next page.

The augmented matrix is:

$$\left[\begin{array}{ccc|ccc} 3 & -1 & 1 & 1 & 0 & 0 \\ 1 & 0 & 3 & 0 & 1 & 0 \\ 2 & 3 & 1 & 0 & 0 & 1 \end{array}\right]$$

Pivot on row 1, column 1 by performing the operation:

$R_1 \leftrightarrow R_2$
$-3R_1 + R_2 \rightarrow R_2$
$-2R_1 + R_3 \rightarrow R_3$

$$\left[\begin{array}{ccc|ccc} 1 & 0 & 3 & 0 & 1 & 0 \\ 0 & -1 & -8 & 1 & -3 & 0 \\ 0 & 3 & -5 & 0 & -2 & 1 \end{array}\right]$$

Pivot on row 2, column 2 by performing the operations:

$-R_2 \rightarrow R_2$
$-3R_2 + R_3 \rightarrow R_3$

$$\left[\begin{array}{ccc|ccc} 1 & 0 & 3 & 0 & 1 & 0 \\ 0 & 1 & 8 & -1 & 3 & 0 \\ 0 & 0 & -29 & 3 & -11 & 1 \end{array}\right]$$

Pivot on row 3, column 3 by performing the operations:

$\frac{-1}{29}R_3 \rightarrow R_3$
$-3R_3 + R_1 \rightarrow R_1$
$-8R_3 + R_2 \rightarrow R_2$

$$\left[\begin{array}{ccc|ccc} 1 & 0 & 0 & \frac{9}{29} & \frac{-4}{29} & \frac{3}{29} \\ 0 & 1 & 0 & \frac{-5}{29} & \frac{-1}{29} & \frac{8}{29} \\ 0 & 0 & 1 & \frac{-3}{29} & \frac{11}{29} & \frac{-1}{29} \end{array}\right]$$

Therefore the inverse matrix of the coefficient matrix is:

$$\left[\begin{array}{ccc} \frac{9}{29} & \frac{-4}{29} & \frac{3}{29} \\ \frac{-5}{29} & \frac{-1}{29} & \frac{8}{29} \\ \frac{-3}{29} & \frac{11}{29} & \frac{-1}{29} \end{array}\right]$$

Multiply both sides of the matrix equation on the left by the inverse of the coefficient matrix to get the equation at the top of the next column.

$$\left[\begin{array}{c} x \\ y \\ z \end{array}\right] = \left[\begin{array}{ccc} \frac{9}{29} & \frac{-4}{29} & \frac{3}{29} \\ \frac{-5}{29} & \frac{-1}{29} & \frac{8}{29} \\ \frac{-3}{29} & \frac{11}{29} & \frac{-1}{29} \end{array}\right]\left[\begin{array}{c} 0 \\ 0 \\ 0 \end{array}\right] = \left[\begin{array}{c} 0 \\ 0 \\ 0 \end{array}\right]$$

The solution to the system is:

$x = 0; \ y = 0; \ z = 0$

39. The matrix equation is:

$$\left[\begin{array}{cccc} 2 & 0.6 & 0.3 & -1 \\ 0.8 & 0.2 & 4 & 1 \\ 0.36 & 0.4 & -3 & 0 \\ 1 & -1 & 1 & 0.8 \end{array}\right]\left[\begin{array}{c} w \\ x \\ y \\ z \end{array}\right] = \left[\begin{array}{c} 2.80 \\ 21.13 \\ -12.41 \\ 3.50 \end{array}\right]$$

The coefficient matrix is:

$$\left[\begin{array}{cccc} 2 & 0.6 & 0.3 & -1 \\ 0.8 & 0.2 & 4 & 1 \\ 0.36 & 0.4 & -3 & 0 \\ 1 & -1 & 1 & 0.8 \end{array}\right]$$

Enter this matrix into a graphing calculator, and use the inverse matrix function to find the inverse matrix.

The inverse of the coefficient matrix is:

$$\left[\begin{array}{cccc} 0.293 & 0.072 & 0.217 & 0.277 \\ 0.019 & 0.555 & 0.519 & -0.670 \\ 0.038 & 0.083 & -0.238 & -0.056 \\ -0.390 & 0.501 & 0.675 & 0.136 \end{array}\right]$$

Multiply both sides of the matrix equation on the left by the inverse of the coefficient matrix to get the equation:

$$\left[\begin{array}{c} w \\ x \\ y \\ z \end{array}\right] = \left[\begin{array}{cccc} 0.293 & 0.072 & 0.217 & 0.277 \\ 0.019 & 0.555 & 0.519 & -0.670 \\ 0.038 & 0.083 & -0.238 & -0.056 \\ -0.390 & 0.501 & 0.675 & 0.136 \end{array}\right]\left[\begin{array}{c} 2.80 \\ 21.13 \\ -12.41 \\ 3.50 \end{array}\right]$$

The solution is continued on the next page.

Multiplying the matrices on the previous page we get:

$$\begin{bmatrix} w \\ x \\ y \\ z \end{bmatrix} \approx \begin{bmatrix} 0.609 \\ 3.001 \\ 4.610 \\ 1.602 \end{bmatrix}$$

The solution to the system is:

$w \approx 0.609;\ x \approx 3.001;\ y \approx 4.610;\ z \approx 1.602$

41.
a. Referring to exercise 15 from section 2.1, the system of equations that models the problem is:

$$30x + 20y = 4000$$
$$2x + y = 220$$

Where the variable x is the number of stocks of Datafix and the variable y is the number of stocks of Rocktite.

b. The matrix equation is:

$$\begin{bmatrix} 30 & 20 \\ 2 & 1 \end{bmatrix} \begin{bmatrix} x \\ y \end{bmatrix} = \begin{bmatrix} 4000 \\ 220 \end{bmatrix}$$

c. The coefficient matrix is:

$$\begin{bmatrix} 30 & 20 \\ 2 & 1 \end{bmatrix}$$

Using your calculator (or methods shown earlier in this section) the inverse of the coefficient matrix is:

$$\begin{bmatrix} -0.1 & 2 \\ 0.2 & -3 \end{bmatrix}$$

Multiply both sides of the matrix equation on the left by the inverse of the coefficient matrix to get the equation:

$$\begin{bmatrix} x \\ y \end{bmatrix} = \begin{bmatrix} -0.1 & 2 \\ 0.2 & -3 \end{bmatrix} \begin{bmatrix} 4000 \\ 220 \end{bmatrix} = \begin{bmatrix} 40 \\ 140 \end{bmatrix}$$

Emily bought 40 shares of Datafix, and 140 shares of Rocktite.

43.
a. Referring to exercise #21 from section 2.1, the system of equations that models this problem is:

$$x + y = 140$$
$$12x + 9y + 15z = 1560$$
$$3x + 5y + z = 588$$

Where x is the number of skilled workers, y is the number of semiskilled workers, and z is the number of supervisors the department hires.

b. The matrix equation is:

$$\begin{bmatrix} 1 & 1 & 0 \\ 12 & 9 & 15 \\ 3 & 5 & 1 \end{bmatrix} \begin{bmatrix} x \\ y \\ z \end{bmatrix} = \begin{bmatrix} 140 \\ 1560 \\ 588 \end{bmatrix}$$

c. The coefficient matrix is:

$$\begin{bmatrix} 1 & 1 & 0 \\ 12 & 9 & 15 \\ 3 & 5 & 1 \end{bmatrix}$$

Using your calculator (or methods shown earlier in this section) the inverse of the coefficient matrix is:

$$\begin{bmatrix} 2 & \frac{1}{33} & \frac{-5}{11} \\ -1 & \frac{-1}{33} & \frac{5}{11} \\ -1 & \frac{2}{33} & \frac{1}{11} \end{bmatrix}$$

Multiply both sides of the matrix equation on the left by the inverse of the coefficient matrix to get the equation:

$$\begin{bmatrix} x \\ y \\ z \end{bmatrix} = \begin{bmatrix} 2 & \frac{1}{33} & \frac{-5}{11} \\ -1 & \frac{-1}{33} & \frac{5}{11} \\ -1 & \frac{2}{33} & \frac{1}{11} \end{bmatrix} \begin{bmatrix} 140 \\ 1560 \\ 588 \end{bmatrix} = \begin{bmatrix} 60 \\ 80 \\ 8 \end{bmatrix}$$

The department should hire 60 skilled workers, 80 semiskilled workers and 8 supervisors to meet the stipulations of the problem.

45. The budget cuts to the department will only change the constant matrix in the matrix equation. The new system of equations is:

$$20x + 5y = 200$$
$$20x + 30y = 440$$

The matrix equation is:

$$\begin{bmatrix} 20 & 5 \\ 20 & 30 \end{bmatrix} \begin{bmatrix} x \\ y \end{bmatrix} = \begin{bmatrix} 200 \\ 440 \end{bmatrix}$$

Since the coefficient matrix did not change, the inverse of the coefficient is still:

$$\begin{bmatrix} 0.06 & -0.01 \\ -0.04 & 0.04 \end{bmatrix}$$

Multiply both sides of the equation by the inverse of the coefficient matrix to get the solution:

$$\begin{bmatrix} x \\ y \end{bmatrix} = \begin{bmatrix} 0.06 & -0.01 \\ -0.04 & 0.04 \end{bmatrix} \begin{bmatrix} 200 \\ 440 \end{bmatrix} = \begin{bmatrix} 7.6 \\ 9.6 \end{bmatrix}$$

The police department could have 7.6 rookies and 9.6 sergeants under the new conditions. More than likely the budget would be the deciding factor on whether or not to round the number of officers up or down.

47. The augmented matrix is:

$$\left[\begin{array}{cc|cc} 3 & 0 & 1 & 0 \\ 0 & 1 & 0 & 1 \end{array}\right]$$

Perform the following operation to reduce the left hand side of the augmented matrix:

$$\tfrac{1}{3} R_1 \rightarrow R_1$$

$$\left[\begin{array}{cc|cc} 1 & 0 & \tfrac{1}{3} & 0 \\ 0 & 1 & 0 & 1 \end{array}\right]$$

The inverse of the matrix is:

$$\begin{bmatrix} \tfrac{1}{3} & 0 \\ 0 & 1 \end{bmatrix}$$

49. The augmented matrix is:

$$\left[\begin{array}{ccc|ccc} 3 & 0 & 0 & 1 & 0 & 0 \\ 0 & -1 & 0 & 0 & 1 & 0 \\ 0 & 0 & 5 & 0 & 0 & 1 \end{array}\right]$$

Perform the following operations to reduce the left hand side of the augmented matrix:

$$\tfrac{1}{3} R_1 \rightarrow R_1$$
$$-R_2 \rightarrow R_2$$
$$\tfrac{1}{5} R_3 \rightarrow R_3$$

$$\left[\begin{array}{ccc|ccc} 1 & 0 & 0 & \tfrac{1}{3} & 0 & 0 \\ 0 & 1 & 0 & 0 & -1 & 0 \\ 0 & 0 & 1 & 0 & 0 & \tfrac{1}{5} \end{array}\right]$$

The inverse of the matrix is:

$$\begin{bmatrix} \tfrac{1}{3} & 0 & 0 \\ 0 & -1 & 0 \\ 0 & 0 & \tfrac{1}{5} \end{bmatrix}$$

51. The augmented matrix is:

$$\left[\begin{array}{cccc|cccc} 3 & 0 & 0 & 0 & 1 & 0 & 0 & 0 \\ 0 & -6 & 0 & 0 & 0 & 1 & 0 & 0 \\ 0 & 0 & 2 & 0 & 0 & 0 & 1 & 0 \\ 0 & 0 & 0 & 5 & 0 & 0 & 0 & 1 \end{array}\right]$$

Perform the following operations to reduce the left hand side of the augmented matrix:

$$\tfrac{1}{3} R_1 \rightarrow R_1$$
$$\tfrac{-1}{6} R_2 \rightarrow R_2$$
$$\tfrac{1}{2} R_3 \rightarrow R_3$$
$$\tfrac{1}{5} R_4 \rightarrow R_4$$

$$\left[\begin{array}{cccc|cccc} 1 & 0 & 0 & 0 & \tfrac{1}{3} & 0 & 0 & 0 \\ 0 & 1 & 0 & 0 & 0 & \tfrac{-1}{6} & 0 & 0 \\ 0 & 0 & 1 & 0 & 0 & 0 & \tfrac{1}{2} & 0 \\ 0 & 0 & 0 & 1 & 0 & 0 & 0 & \tfrac{1}{5} \end{array}\right]$$

The solution is continued on the next page.

The inverse of the matrix is:

$$\begin{bmatrix} \frac{1}{3} & 0 & 0 & 0 \\ 0 & \frac{-1}{6} & 0 & 0 \\ 0 & 0 & \frac{1}{2} & 0 \\ 0 & 0 & 0 & \frac{1}{5} \end{bmatrix}$$

53. $AB = C$

To solve for A, right multiply by B^{-1}

$$ABB^{-1} = CB^{-1}$$
$$AI = CB^{-1}$$
$$A = CB^{-1}$$

55. $X - AX = B$

To solve for X, first use the distributive property.

$$(I - A)X = B$$

Now left multiply by $(I - A)^{-1}$

$$(I - A)^{-1}(I - A)X = (I - A)^{-1}B$$
$$IX = (I - A)^{-1}B$$
$$X = (I - A)^{-1}B$$

57. $AB + AZ = C$

To solve for A, first use the distributive property.

$$A(B + Z) = C$$

Now right multiply by $(B + Z)^{-1}$

$$A(B + Z)(B + Z)^{-1} = C(B + Z)^{-1}$$
$$AI = C(B + Z)^{-1}$$
$$A = C(B + Z)^{-1}$$

59.
a. We first compute AB:

$$AB = \begin{bmatrix} 2 & 3 \\ 1 & 2 \end{bmatrix}\begin{bmatrix} 3 & 4 \\ 1 & 2 \end{bmatrix} = \begin{bmatrix} 9 & 14 \\ 5 & 8 \end{bmatrix}$$

To compute the inverse of AB, the augmented matrix is:

$$\left[\begin{array}{cc|cc} 9 & 14 & 1 & 0 \\ 5 & 8 & 0 & 1 \end{array}\right]$$

Pivot on row 1, column 1 by performing the operations:

$$\tfrac{1}{9}R_1 \rightarrow R_1$$
$$-5R_1 + R_2 \rightarrow R_2$$

$$\left[\begin{array}{cc|cc} 1 & \frac{14}{9} & \frac{1}{9} & 0 \\ 0 & \frac{2}{9} & \frac{-5}{9} & 1 \end{array}\right]$$

Pivot on row 2, column 2 by performing the operations:

$$\tfrac{9}{2}R_2 \rightarrow R_2$$
$$\tfrac{-14}{9}R_2 + R_1 \rightarrow R_1$$

$$\left[\begin{array}{cc|cc} 1 & 0 & 4 & -7 \\ 0 & 1 & \frac{-5}{2} & \frac{9}{2} \end{array}\right]$$

Therefore,

$$(AB)^{-1} = \begin{bmatrix} 4 & -7 \\ \frac{-5}{2} & \frac{9}{2} \end{bmatrix}$$

b. To compute the inverse of A, the augmented matrix is:

$$\left[\begin{array}{cc|cc} 2 & 3 & 1 & 0 \\ 1 & 2 & 0 & 1 \end{array}\right]$$

Pivot on row 1, column 1 by performing the operations:

$$R_1 \leftrightarrow R_2$$
$$-2R_1 + R_2 \rightarrow R_2$$

$$\left[\begin{array}{cc|cc} 1 & 2 & 0 & 1 \\ 0 & -1 & 1 & -2 \end{array}\right]$$

Pivot on row 2, column 2 by performing the operations:

$$-R_2 \rightarrow R_2$$
$$-2R_2 + R_1 \rightarrow R_1$$

The solution is continued on the next page.

$$\begin{bmatrix} 1 & 0 & | & 2 & -3 \\ 0 & 1 & | & -1 & 2 \end{bmatrix}$$

Therefore,

$$A^{-1} = \begin{bmatrix} 2 & -3 \\ -1 & 2 \end{bmatrix}$$

To compute the inverse of B, the augmented matrix is shown below:

$$\begin{bmatrix} 3 & 4 & | & 1 & 0 \\ 1 & 2 & | & 0 & 1 \end{bmatrix}$$

Pivot on row 1, column 1 by performing the operations:

$R_1 \leftrightarrow R_2$

$-3R_1 + R_2 \to R_2$

$$\begin{bmatrix} 1 & 2 & | & 0 & 1 \\ 0 & -2 & | & 1 & -3 \end{bmatrix}$$

Pivot on row 2, column 2 by performing the operations:

$\frac{-1}{2}R_2 \to R_2$

$-2R_2 + R_1 \to R_1$

$$\begin{bmatrix} 1 & 0 & | & 1 & -2 \\ 0 & 1 & | & \frac{-1}{2} & \frac{3}{2} \end{bmatrix}$$

Therefore,

$$B^{-1} = \begin{bmatrix} 1 & -2 \\ \frac{-1}{2} & \frac{3}{2} \end{bmatrix}$$

Now we compute $B^{-1}A^{-1}$:

$$B^{-1}A^{-1} = \begin{bmatrix} 1 & -2 \\ \frac{-1}{2} & \frac{3}{2} \end{bmatrix}\begin{bmatrix} 2 & -3 \\ -1 & 2 \end{bmatrix} = \begin{bmatrix} 4 & -7 \\ \frac{-5}{2} & \frac{9}{2} \end{bmatrix}$$

c. The inverse of a matrix product, is the product of the inverses in the opposite order.

d. $\left((AB)C\right)^{-1} = C^{-1}(AB)^{-1} = C^{-1}B^{-1}A^{-1}$

61. The transpose of A is:

$$A^T = \begin{bmatrix} 2 & 1 \\ 3 & 2 \end{bmatrix}$$

To compute the inverse, the augmented matrix is:

$$\begin{bmatrix} 2 & 1 & | & 1 & 0 \\ 3 & 2 & | & 0 & 1 \end{bmatrix}$$

Pivot on row 1, column 1 by performing the operations:

$\frac{1}{2}R_1 \to R_1$

$-3R_1 + R_2 \to R_2$

$$\begin{bmatrix} 1 & \frac{1}{2} & | & \frac{1}{2} & 0 \\ 0 & \frac{1}{2} & | & \frac{-3}{2} & 1 \end{bmatrix}$$

Pivot on row 2, column 2 by performing the operations:

$2R_2 \to R_2$

$\frac{-1}{2}R_2 + R_1 \to R_1$

$$\begin{bmatrix} 1 & 0 & | & 2 & -1 \\ 0 & 1 & | & -3 & 2 \end{bmatrix}$$

Therefore,

$$\left(A^T\right)^{-1} = \begin{bmatrix} 2 & -1 \\ -3 & 2 \end{bmatrix}$$

b. We know from exercise 59, that the inverse of A is:

$$A^{-1} = \begin{bmatrix} 2 & -3 \\ -1 & 2 \end{bmatrix}.$$

Therefore,

$$\left(A^{-1}\right)^T = \begin{bmatrix} 2 & -1 \\ -3 & 2 \end{bmatrix}$$

c. In general, the inverse of the transpose is the transpose of the inverse.

63.

a. If $AB = 0$, we need to know that A is invertible to conclude that B must equal the zero matrix.

$$AB = 0$$
$$A^{-1}AB = A^{-1}0$$
$$B = 0$$

b. If $AB = 0$, then it is not possible for both matrices to have inverses. As shown in part (a), if A is invertible, then $B = 0$; if $B = 0$, then B is not invertible.

c. If A and B are both invertible, then it is not possible for $AB = 0$. This is the contrapositive of the statement made in part (b) above.

d. If $AB = 0$, then it is possible for neither A or B to have an inverse. For example let

$$A = \begin{bmatrix} 1 & 0 \\ 0 & 0 \end{bmatrix}, \text{ and } B = \begin{bmatrix} 0 & 0 \\ 0 & 1 \end{bmatrix}.$$

Exercises Section 3.4

1.

a. We first assign the numbers as follows:

$$\begin{array}{c|cc|cc|cc|cc|cc} I & H & A & V & E & & A & & J & O & B & . \\ 9\ 30 & 8 & 1 & 22 & 5 & 30 & 1 & 30 & 10 & 15 & 2 & 40\ 30 \end{array}$$

Thus the code matrix is:

$$C = \begin{bmatrix} 9 & 30 \\ 8 & 1 \\ 22 & 5 \\ 30 & 1 \\ 30 & 10 \\ 15 & 2 \\ 40 & 30 \end{bmatrix}$$

b. We encrypt the message by computing:

$$CA = \begin{bmatrix} 9 & 30 \\ 8 & 1 \\ 22 & 5 \\ 30 & 1 \\ 30 & 10 \\ 15 & 2 \\ 40 & 30 \end{bmatrix} \begin{bmatrix} 1 & 2 \\ -1 & 1 \end{bmatrix} = \begin{bmatrix} -21 & 48 \\ 7 & 17 \\ 17 & 49 \\ 29 & 61 \\ 20 & 70 \\ 13 & 32 \\ 10 & 110 \end{bmatrix}$$

The message received is:

-21, 48, 7, 17, 17, 49, 29, 61, 20, 70, 13, 32, 10, 110

c. To compute the decoding matrix we first form the augmented matrix:

$$\begin{bmatrix} 1 & 2 & | & 1 & 0 \\ -1 & 1 & | & 0 & 1 \end{bmatrix}$$

Reducing the left hand side of the augmented matrix by the pivoting process we get:

$$\begin{bmatrix} 1 & 0 & | & \frac{1}{3} & \frac{-2}{3} \\ 0 & 1 & | & \frac{1}{3} & \frac{1}{3} \end{bmatrix}$$

Thus,

$$A^{-1} = \begin{bmatrix} \frac{1}{3} & \frac{-2}{3} \\ \frac{1}{3} & \frac{1}{3} \end{bmatrix}$$

To decode the message, we perform the multiplication:

$$\begin{bmatrix} -21 & 48 \\ 7 & 17 \\ 17 & 49 \\ 29 & 61 \\ 20 & 70 \\ 13 & 32 \\ 10 & 110 \end{bmatrix} \begin{bmatrix} \frac{1}{3} & \frac{-2}{3} \\ \frac{1}{3} & \frac{1}{3} \end{bmatrix} = \begin{bmatrix} 9 & 30 \\ 8 & 1 \\ 22 & 5 \\ 30 & 1 \\ 30 & 10 \\ 15 & 2 \\ 40 & 30 \end{bmatrix}$$

This translates to the message:

I HAVE A JOB.

3.

a. We first assign the numbers as follows:

$$\begin{array}{cc|cc|cc|cc} N & O & W & A & Y & & . & \\ 14 & 15 & 30 & 23 & 1 & 25 & 40 & 30 \end{array}$$

Thus the code matrix is:

$$C = \begin{bmatrix} 14 & 15 \\ 30 & 23 \\ 1 & 25 \\ 40 & 30 \end{bmatrix}$$

b. We encrypt the message by computing:

$$CA = \begin{bmatrix} 14 & 15 \\ 30 & 23 \\ 1 & 25 \\ 40 & 30 \end{bmatrix} \begin{bmatrix} 1 & 1 \\ 1 & 2 \end{bmatrix} = \begin{bmatrix} 29 & 44 \\ 53 & 76 \\ 26 & 51 \\ 70 & 100 \end{bmatrix}$$

29, 44, 53, 76, 26, 51, 70, 100

c. To compute the decoding matrix we first form the augmented matrix:

$$\begin{bmatrix} 1 & 1 & 1 & 0 \\ 1 & 2 & 0 & 1 \end{bmatrix}$$

Reducing the left hand side of the augmented matrix by the pivoting process we get:

$$\begin{bmatrix} 1 & 0 & 2 & -1 \\ 0 & 1 & -1 & 1 \end{bmatrix}$$

Thus,

$$A^{-1} = \begin{bmatrix} 2 & -1 \\ -1 & 1 \end{bmatrix}$$

To decode the message, we perform the multiplication:

$$\begin{bmatrix} 29 & 44 \\ 53 & 76 \\ 26 & 51 \\ 70 & 100 \end{bmatrix} \begin{bmatrix} 2 & -1 \\ -1 & 1 \end{bmatrix} = \begin{bmatrix} 14 & 15 \\ 30 & 23 \\ 1 & 25 \\ 40 & 30 \end{bmatrix}$$

This translates to the message:

NO WAY.

5. First compute A^{-1}. The augmented matrix is:

$$\begin{bmatrix} 1 & 0 & 1 & 1 & 0 & 0 \\ 2 & 0 & 3 & 0 & 1 & 0 \\ 1 & 2 & 1 & 0 & 0 & 1 \end{bmatrix}$$

Pivot on row 1, column 1 by performing the operations:

$$-2R_1 + R_2 \rightarrow R_2$$
$$-R_1 + R_3 \rightarrow R_3$$

$$\begin{bmatrix} 1 & 0 & 1 & 1 & 0 & 0 \\ 0 & 0 & 1 & -2 & 1 & 0 \\ 0 & 2 & 0 & -1 & 0 & 1 \end{bmatrix}$$

In order to get a one in row 2, column 2 we perform the operations:

$$R_2 \leftrightarrow R_3$$
$$\tfrac{1}{2}R_2 \rightarrow R_2$$

$$\begin{bmatrix} 1 & 0 & 1 & 1 & 0 & 0 \\ 0 & 1 & 0 & \frac{-1}{2} & 0 & \frac{1}{2} \\ 0 & 0 & 1 & -2 & 1 & 0 \end{bmatrix}$$

Now pivot on row 3, column 3 by performing the operation:

$$-R_3 + R_1 \rightarrow R_1$$

$$\begin{bmatrix} 1 & 0 & 0 & 3 & -1 & 0 \\ 0 & 1 & 0 & \frac{-1}{2} & 0 & \frac{1}{2} \\ 0 & 0 & 1 & -2 & 1 & 0 \end{bmatrix}$$

Therefore the inverse matrix of the encryption matrix is:

$$A^{-1} = \begin{bmatrix} 3 & -1 & 0 \\ \frac{-1}{2} & 0 & \frac{1}{2} \\ -2 & 1 & 0 \end{bmatrix}$$

b. Arrange the message in a matrix. Since the decryption matrix has 3 rows, the message matrix needs to have three columns.

$$M = \begin{bmatrix} 48 & 10 & 66 \\ 71 & 60 & 91 \\ 26 & 40 & 27 \\ 115 & 60 & 155 \end{bmatrix}$$

To decode the matrix, multiply the message matrix by the inverse of the encryption matrix on the right.

$$\begin{bmatrix} 48 & 10 & 66 \\ 71 & 60 & 91 \\ 26 & 40 & 27 \\ 115 & 60 & 155 \end{bmatrix} \begin{bmatrix} 3 & -1 & 0 \\ \frac{-1}{2} & 0 & \frac{1}{2} \\ -2 & 1 & 0 \end{bmatrix} = \begin{bmatrix} 7 & 18 & 5 \\ 1 & 20 & 30 \\ 4 & 1 & 20 \\ 5 & 40 & 30 \end{bmatrix}$$

The message is:

7 18 5 | 1 20 30 | 4 1 20 | 5 40 30

G R E | A T | D A T | E .

7.

a. First compute A^{-1}. The augmented matrix is:

$$\begin{bmatrix} 2 & 0 & 0 & 1 & 0 & 0 \\ 0 & 1 & 0 & 0 & 1 & 0 \\ 0 & 0 & 3 & 0 & 0 & 1 \end{bmatrix}$$

To find the inverse simply perform the operations:

$$\tfrac{1}{2}R_1 \to R_1$$
$$\tfrac{1}{3}R_3 \to R_3$$

$$\begin{bmatrix} 1 & 0 & 0 & \frac{1}{2} & 0 & 0 \\ 0 & 1 & 0 & 0 & 1 & 0 \\ 0 & 0 & 1 & 0 & 0 & \frac{1}{3} \end{bmatrix}$$

Therefore the inverse matrix of the encryption matrix is:

$$A^{-1} = \begin{bmatrix} \frac{1}{2} & 0 & 0 \\ 0 & 1 & 0 \\ 0 & 0 & \frac{1}{3} \end{bmatrix}$$

b. Arrange the message in a matrix. Since the decryption matrix has 3 rows, the message matrix needs to have three columns.

$$M = \begin{bmatrix} 14 & 12 & 45 \\ 4 & 1 & 36 \\ 60 & 23 & 3 \\ 36 & 13 & 27 \\ 28 & 7 & 90 \\ 6 & 15 & 42 \\ 40 & 9 & 42 \\ 42 & 5 & 57 \\ 80 & 30 & 90 \end{bmatrix}$$

To decode the matrix, multiply the message matrix by the inverse of the encryption matrix on the right.

$$\begin{bmatrix} 14 & 12 & 45 \\ 4 & 1 & 36 \\ 60 & 23 & 3 \\ 36 & 13 & 27 \\ 28 & 7 & 90 \\ 6 & 15 & 42 \\ 40 & 9 & 42 \\ 42 & 5 & 57 \\ 80 & 30 & 90 \end{bmatrix} \begin{bmatrix} \frac{1}{2} & 0 & 0 \\ 0 & 1 & 0 \\ 0 & 0 & \frac{1}{3} \end{bmatrix} = \begin{bmatrix} 7 & 12 & 15 \\ 2 & 1 & 18 \\ 30 & 23 & 1 \\ 18 & 13 & 9 \\ 14 & 7 & 30 \\ 3 & 15 & 14 \\ 20 & 9 & 14 \\ 21 & 5 & 19 \\ 40 & 30 & 30 \end{bmatrix}$$

Decoding the message we see:

7 12 15 | 2 1 12 | 30 23 1 | 18 13 9 | 14 7 30

G L O | B A L | W A | R M I | N G

3 15 14 | 20 9 14 | 21 5 19 | 40 30 30

C O N | T I N | U E S | .

The message is:

Global warming continues.

For the remaining exercises, we will use a graphing calculator with matrix abilities to calculate all inverses.

9.
a. The entries in the matrix, T, indicate that the steel industry needs 0.02 units worth of steel and 0.15 units worth of electronics for every unit of steel that they produce, and that the electronics industry needs 0.10 units of steel and 0.01 units of electronics for every unit of electronics that they produce.

b. We have:

$$I - T = \begin{bmatrix} 0.98 & -0.1 \\ -0.15 & 0.99 \end{bmatrix}$$

and:

$$\left(I - T\right)^{-1} = \begin{bmatrix} \frac{825}{796} & \frac{125}{1194} \\ \frac{125}{796} & \frac{1225}{1194} \end{bmatrix}$$

c. We compute X :

$$\left(I - T\right)^{-1} D = \begin{bmatrix} \frac{825}{796} & \frac{125}{1194} \\ \frac{125}{796} & \frac{1225}{1194} \end{bmatrix} \begin{bmatrix} 500 \\ 800 \end{bmatrix} =$$

$$X = \begin{bmatrix} 602.0 \\ 899.3 \end{bmatrix}$$

It will take a production of 602 units of steel and 899.3 units of electronics to meet the demand for these products.

The internal consumption is given by TX :

$$TX = \begin{bmatrix} 0.02 & 0.10 \\ 0.15 & 0.01 \end{bmatrix} \begin{bmatrix} 602.0 \\ 899.3 \end{bmatrix} = \begin{bmatrix} 102.0 \\ 99.3 \end{bmatrix}$$

The producers will consume 102 units of steel and 99.3 units of electronics while producing this product.

11.
a. The entries in the matrix, T, indicate that each unit produced in the manufacturing sector requires 0.10 units of input from the manufacturing sector and 0.15 units from the electronics sector. Each unit produced from the electronics sector requires 0.20 units from the manufacturing sector and 0.05 units from the electronics sector.

b. We have:

$$I - T = \begin{bmatrix} 0.90 & -0.20 \\ -0.15 & 0.95 \end{bmatrix}$$

and:

$$\left(I - T\right)^{-1} = \begin{bmatrix} \frac{38}{33} & \frac{8}{33} \\ \frac{2}{11} & \frac{12}{11} \end{bmatrix}$$

c. We compute X :

$$\left(I - T\right)^{-1} D = \begin{bmatrix} \frac{38}{33} & \frac{8}{33} \\ \frac{2}{11} & \frac{12}{11} \end{bmatrix} \begin{bmatrix} 200 \\ 500 \end{bmatrix} =$$

$$X = \begin{bmatrix} 351.5 \\ 581.8 \end{bmatrix}$$

It will take a production of 351.5 units of agricultural output and 581.8 units of manufacturing output to meet the demand for these products

The internal consumption is given by TX :

$$TX = \begin{bmatrix} 0.10 & 0.20 \\ 0.15 & 0.05 \end{bmatrix} \begin{bmatrix} 351.5 \\ 581.8 \end{bmatrix} = \begin{bmatrix} 35.15 + 116.36 \\ 52.7 + 29.1 \end{bmatrix}$$

$$TX = \begin{bmatrix} 151.5 \\ 81.8 \end{bmatrix}$$

The manufacturing sector will need to input 151.5 units of the manufacturing product to meet production demands (35.1 units for manufacturing output and 116.4 units for electronics output).

The electronics sector will need to input 81.8 units to meet production needs (52.7 for manufacturing output and 29.1 for electronics output).

13.
a. The entries in the matrix, T, indicate that each unit produced in the agricultural sector requires 0.02 units of input from the agricultural sector, 0.20 units of input from the manufacturing sector and 0.15 units of input from the electronics sector. Each unit produced in the manufacturing sector requires 0.01 units of input from the agricultural sector, 0.10 units of input from the manufacturing sector and 0.08 units from the electronics sector. Each unit produced from the electronics sector requires 0.01 units of input from the agricultural sector, 0.10 units of input from the manufacturing sector and 0.05 units of input from the electronics sector.

b. We have:

$$I - T = \begin{bmatrix} 0.98 & -0.01 & -0.01 \\ -0.20 & 0.90 & -0.10 \\ -0.15 & -0.08 & 0.95 \end{bmatrix}$$

and:

$$\left(I - T\right)^{-1} = \begin{bmatrix} \frac{1694}{1653} & \frac{103}{8265} & \frac{20}{1653} \\ \frac{410}{1653} & \frac{1859}{1653} & \frac{200}{1653} \\ \frac{302}{1653} & \frac{799}{8265} & \frac{1760}{1653} \end{bmatrix}$$

c. We compute X :

$$\left(I - T\right)^{-1} D = \begin{bmatrix} \frac{1694}{1653} & \frac{103}{8265} & \frac{20}{1653} \\ \frac{410}{1653} & \frac{1859}{1653} & \frac{200}{1653} \\ \frac{302}{1653} & \frac{799}{8265} & \frac{1760}{1653} \end{bmatrix} \begin{bmatrix} 500 \\ 1000 \\ 2000 \end{bmatrix} =$$

$$X = \begin{bmatrix} 549 \\ 1490.6 \\ 2317.5 \end{bmatrix}$$

It will take a production of 549 units of agricultural output, 1490.6 units of manufacturing output, and 2317.5 units of electronics output to meet the demand for these products.

The internal consumption is given by TX :

$$TX = \begin{bmatrix} 0.02 & 0.01 & 0.01 \\ 0.20 & 0.10 & 0.10 \\ 0.15 & 0.08 & 0.05 \end{bmatrix} \begin{bmatrix} 549 \\ 1490.6 \\ 2317.5 \end{bmatrix} =$$

$$TX = \begin{bmatrix} 11+15+23 \\ 110+149+232 \\ 82+119+116 \end{bmatrix}$$

$$TX = \begin{bmatrix} 49 \\ 490.6 \\ 317.5 \end{bmatrix}$$

The agricultural sector will need to input 49 units of the agricultural product to meet production demands (11 units for the agricultural output, 15 units for the manufacturing output and 23 units for the electronics output).

The manufacturing sector will need to input 491 units of the manufacturing product to meet production demands (110 units for agricultural output, 149 units for manufacturing output and 232 units for electronics output).

The electronics sector will need to input 317 units of the electronics product to meet production needs (82 units for agricultural output, 119 units for manufacturing output and 116 for electronics output).

15.
a. The parts matrix is:

$$P = \begin{array}{c} \\ P_1 \\ P_2 \\ P_3 \end{array} \begin{array}{ccc} P_1 & P_2 & P_3 \\ \begin{bmatrix} 0 & 2 & 3 \\ 0 & 0 & 1 \\ 0 & 0 & 0 \end{bmatrix} \end{array}$$

b. The parts matrix tells us that 2 units of part 1 are needed for each unit of part 2 and 3 units of part 1 are needed for each unit of part 3. It also tells us that 1 unit of part 2 is needed for each unit of part 3.

c. We have:

$$I - P = \begin{bmatrix} 1 & -2 & -3 \\ 0 & 1 & -1 \\ 0 & 0 & 1 \end{bmatrix}$$

and

$$\left(I - P\right)^{-1} = \begin{bmatrix} 1 & 2 & 5 \\ 0 & 1 & 1 \\ 0 & 0 & 1 \end{bmatrix}$$

The production matrix for the given demand matrix is:

$$\left(I - P\right)^{-1} D = \begin{bmatrix} 1 & 2 & 5 \\ 0 & 1 & 1 \\ 0 & 0 & 1 \end{bmatrix} \begin{bmatrix} 200 \\ 100 \\ 100 \end{bmatrix} =$$

$$\left(I - P\right)^{-1} D = \begin{bmatrix} 900 \\ 200 \\ 100 \end{bmatrix}$$

In order to meet demand, we will require 900 units of part 1, 200 units of part 2 and 100 units of part 3.

17.
a. The parts matrix is:

$$P = \begin{array}{c} \\ P_1 \\ P_2 \\ P_3 \\ P_4 \end{array} \begin{array}{cccc} P_1 & P_2 & P_3 & P_4 \\ \begin{bmatrix} 0 & 3 & 2 & 0 \\ 0 & 0 & 0 & 1 \\ 0 & 1 & 0 & 2 \\ 0 & 0 & 0 & 0 \end{bmatrix} \end{array}$$

b. The parts matrix tells us 3 units of part 1 are needed to make part 2 and 2 units of part 1 are needed to make part 3. It also tells us that 1 unit of part 2 is needed to make part 4. Finally we will need 1 unit of part 3 to make part 2, and 2 units of part 3 to make part 4.

c. We have:

$$I - P = \begin{bmatrix} 1 & -3 & -2 & 0 \\ 0 & 1 & 0 & -1 \\ 0 & -1 & 1 & -2 \\ 0 & 0 & 0 & 1 \end{bmatrix}$$

Therefore,

$$(I-P)^{-1} = \begin{bmatrix} 1 & 5 & 2 & 9 \\ 0 & 1 & 0 & 1 \\ 0 & 1 & 1 & 3 \\ 0 & 0 & 0 & 1 \end{bmatrix}$$

The production matrix for the given demand matrix is:

$$(I-P)^{-1} D = \begin{bmatrix} 1 & 5 & 2 & 9 \\ 0 & 1 & 0 & 1 \\ 0 & 1 & 1 & 3 \\ 0 & 0 & 0 & 1 \end{bmatrix} \begin{bmatrix} 200 \\ 80 \\ 100 \\ 50 \end{bmatrix} =$$

$$(I-P)^{-1} D = \begin{bmatrix} 1250 \\ 130 \\ 330 \\ 50 \end{bmatrix}$$

In order to meet demand, we will require 1250 units of part 1, 130 units of part 2, 330 units of part 3 and 50 units of part 4.

19.
a. The parts matrix is:

$$P = \begin{array}{c} \\ P_1 \\ P_2 \\ P_3 \\ P_4 \end{array} \begin{array}{cccc} P_1 & P_2 & P_3 & P_4 \\ \begin{bmatrix} 0 & 2 & 0 & 0 \\ 0 & 0 & 2 & 0 \\ 0 & 0 & 0 & 3 \\ 0 & 0 & 0 & 0 \end{bmatrix} \end{array}$$

b. The parts matrix tells us 2 units of part 1 are needed to make part 2. It also tells us that 2 units of part 2 are needed to make part3. Finally we will need 3 units of part 3 to make part 4.

c. We have:

$$I - P = \begin{bmatrix} 1 & -2 & 0 & 0 \\ 0 & 1 & -2 & 0 \\ 0 & 0 & 1 & -3 \\ 0 & 0 & 0 & 1 \end{bmatrix}$$

and

$$(I-P)^{-1} = \begin{bmatrix} 1 & 2 & 4 & 12 \\ 0 & 1 & 2 & 6 \\ 0 & 0 & 1 & 3 \\ 0 & 0 & 0 & 1 \end{bmatrix}$$

The production matrix for the given demand matrix is:

$$(I-P)^{-1} D = \begin{bmatrix} 1 & 2 & 4 & 12 \\ 0 & 1 & 2 & 6 \\ 0 & 0 & 1 & 3 \\ 0 & 0 & 0 & 1 \end{bmatrix} \begin{bmatrix} 250 \\ 300 \\ 150 \\ 60 \end{bmatrix} =$$

$$(I-P)^{-1} D = \begin{bmatrix} 2170 \\ 960 \\ 330 \\ 60 \end{bmatrix}$$

In order to meet demand, we will require 2170 units of part 1, 960 units of part 2, 330 units of part 3 and 60 units of part 4.

The solution is continued on the next page.

$$(I-P)^{-1}D = \begin{bmatrix} 1 & 2 & 3 & 8 & 8 \\ 0 & 1 & 0 & 1 & 1 \\ 0 & 0 & 1 & 2 & 2 \\ 0 & 0 & 0 & 1 & 1 \\ 0 & 0 & 0 & 0 & 1 \end{bmatrix} \begin{bmatrix} 80 \\ 60 \\ 50 \\ 100 \\ 200 \end{bmatrix} =$$

$$(I-P)^{-1}D = \begin{bmatrix} 2760 \\ 360 \\ 650 \\ 300 \\ 200 \end{bmatrix}$$

In order to meet demand, we will require 2760 units of part 1, 360 units of part 2, 650 units of part 3 and 300 units of part 4, and 200 units of part 5.

21. The parts – arrow diagram corresponding to the parts matrix is:

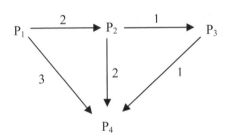

Chapter 3 Summary Exercises

1. $C + 3D =$

$$\begin{bmatrix} 9 & 1 \\ 0 & -1 \end{bmatrix} + \begin{bmatrix} 3 & -6 \\ 9 & 0 \end{bmatrix} = \begin{bmatrix} 12 & -5 \\ 9 & -1 \end{bmatrix}$$

3. The transpose of B is:

$$B^T = \begin{bmatrix} 1 & 0 & 3 \\ -2 & 4 & -1 \end{bmatrix}$$

5. The matrix product BA is:

$$BA = \begin{bmatrix} 1 & -2 \\ 0 & 4 \\ 3 & -1 \end{bmatrix} \begin{bmatrix} 2 & 0 & 1 \\ 3 & -1 & 0 \end{bmatrix} =$$

$$BA = \begin{bmatrix} -4 & 2 & 1 \\ 12 & -4 & 0 \\ 3 & 1 & 3 \end{bmatrix}$$

7. The matrix product CB is not defined because matrix C has 2 columns while matrix B has 3 rows.

9. Since matrix A does not have the same number of rows as columns, A^2 is not defined.

11. The inverse of matrix C is found by first creating the augmented matrix:

$$\left[\begin{array}{cc|cc} 9 & 1 & 1 & 0 \\ 0 & -1 & 0 & 1 \end{array} \right]$$

To save the hassle of adding fractions, since we already have a zero in row 2 column 1 we can pivot on row 2, column 2.

$$-R_2 \rightarrow R_2$$
$$-R_2 + R_1 \rightarrow R_1$$

$$\left[\begin{array}{cc|cc} 9 & 0 & 1 & 1 \\ 0 & 1 & 0 & -1 \end{array} \right]$$

Now make the element in row 1, column 1 equal to 1 by performing the operation:

$$\tfrac{1}{9} R_1 \rightarrow R_1$$

$$\left[\begin{array}{cc|cc} 1 & 0 & \frac{1}{9} & \frac{1}{9} \\ 0 & 1 & 0 & -1 \end{array} \right]$$

The inverse of matrix C is:

$$C^{-1} = \begin{bmatrix} \frac{1}{9} & \frac{1}{9} \\ 0 & -1 \end{bmatrix}$$

13. If the matrices are to be equal, we must have:

$3x = 27$, or $x = 9$

$-w = 24$, or $w = -24$

$2t = 17$, or $t = \frac{17}{2}$

$-2v + 1 = 0$, or $v = \frac{1}{2}$

15. The augmented matrix is:

$$\left[\begin{array}{ccc|ccc} 2 & 0 & 1 & 1 & 0 & 0 \\ 1 & 3 & 2 & 0 & 1 & 0 \\ 1 & 1 & 0 & 0 & 0 & 1 \end{array}\right]$$

Pivot on row 1, column 1 by performing the operations:

$R_1 \leftrightarrow R_3$

$-R_1 + R_2 \rightarrow R_2$

$-2R_1 + R_3 \rightarrow R_3$

$$\left[\begin{array}{ccc|ccc} 1 & 1 & 0 & 0 & 0 & 1 \\ 0 & 2 & 2 & 0 & 1 & -1 \\ 0 & -2 & 1 & 1 & 0 & -2 \end{array}\right]$$

Pivot on row 2, column 2 by performing the operations:

$\frac{1}{2}R_2 \rightarrow R_2$

$-R_2 + R_1 \rightarrow R_1$

$2R_2 + R_3 \rightarrow R_3$

$$\left[\begin{array}{ccc|ccc} 1 & 0 & -1 & 0 & \frac{-1}{2} & \frac{3}{2} \\ 0 & 1 & 1 & 0 & \frac{1}{2} & \frac{-1}{2} \\ 0 & 0 & 3 & 1 & 1 & -3 \end{array}\right]$$

Pivot on row 3, column 3 by performing the operations:

$\frac{1}{3}R_3 \rightarrow R_3$

$R_3 + R_1 \rightarrow R_1$

$-R_3 + R_2 \rightarrow R_2$

The resulting matrix is:

$$\left[\begin{array}{ccc|ccc} 1 & 0 & 0 & \frac{1}{3} & \frac{-1}{6} & \frac{1}{2} \\ 0 & 1 & 0 & \frac{-1}{3} & \frac{1}{6} & \frac{1}{2} \\ 0 & 0 & 1 & \frac{1}{3} & \frac{1}{3} & -1 \end{array}\right]$$

The inverse of the matrix is:

$$\left[\begin{array}{ccc} \frac{1}{3} & \frac{-1}{6} & \frac{1}{2} \\ \frac{-1}{3} & \frac{1}{6} & \frac{1}{2} \\ \frac{1}{3} & \frac{1}{3} & -1 \end{array}\right]$$

16. Let

$$A = \left[\begin{array}{ccc} 1 & 3 & 0 \\ 1 & 0 & -1 \\ 2 & 3 & -1 \end{array}\right]$$

and

$$B = \left[\begin{array}{ccc} -3 & 6 & -9 \\ 1 & -2 & 3 \\ -3 & 6 & -9 \end{array}\right]$$

There are many answers to this problem.

17. Let the entry be a one if the flight from the row to the column exists. Otherwise let the entry be a zero. The matrix that displays the flight network is displayed below:

$$\begin{array}{cccc} & w & x & y & z \\ \begin{array}{c} w \\ x \\ y \\ z \end{array} & \left[\begin{array}{cccc} 0 & 1 & 0 & 1 \\ 1 & 0 & 1 & 0 \\ 1 & 1 & 0 & 1 \\ 1 & 0 & 0 & 0 \end{array}\right] \end{array}$$

19. Multiply the message by the coding matrix to get the encrypted message.

$$CA = \left[\begin{array}{cc} 7 & 15 \\ 30 & 16 \\ 8 & 1 \\ 14 & 20 \\ 15 & 13 \\ 19 & 50 \end{array}\right] \left[\begin{array}{cc} 0 & 1 \\ 1 & 3 \end{array}\right] = \left[\begin{array}{cc} 15 & 52 \\ 16 & 78 \\ 1 & 11 \\ 20 & 74 \\ 13 & 54 \\ 50 & 169 \end{array}\right]$$

21. To decode the message from problem 18, multiply the coded message on the right by the decoding matrix

$$(CA)A^{-1} = \begin{bmatrix} 15 & 52 \\ 16 & 78 \\ 1 & 11 \\ 20 & 74 \\ 13 & 54 \\ 50 & 169 \end{bmatrix} \begin{bmatrix} -3 & 1 \\ 1 & 0 \end{bmatrix} = \begin{bmatrix} 7 & 15 \\ 30 & 16 \\ 8 & 1 \\ 14 & 20 \\ 15 & 13 \\ 19 & 50 \end{bmatrix}$$

Retrieving the message gives us:

7 15 | 30 16 | 8 1 | 14 20 | 15 13 | 19 50
G O | P | H A | N T | O M | S   !

The message sent was: Go Phantoms!

23. The matrix equation for the system is:

$$\begin{bmatrix} 2 & 1 & -1 \\ 1 & 1 & 2 \\ 3 & -1 & 2 \end{bmatrix} \begin{bmatrix} x \\ y \\ z \end{bmatrix} = \begin{bmatrix} 3 \\ 5 \\ 3 \end{bmatrix}$$

Find the inverse of the coefficient matrix by reducing the augmented matrix:

$$\begin{bmatrix} 2 & 1 & -1 & | & 1 & 0 & 0 \\ 1 & 1 & 2 & | & 0 & 1 & 0 \\ 3 & -1 & 2 & | & 0 & 0 & 1 \end{bmatrix}$$

Pivot on row 1, column 1 by performing the following operations.

$$R_1 \leftrightarrow R_2$$
$$-2R_1 + R_2 \rightarrow R_2$$
$$-3R_1 + R_3 \rightarrow R_3$$

$$\begin{bmatrix} 1 & 1 & 2 & | & 0 & 1 & 0 \\ 0 & -1 & -5 & | & 1 & -2 & 0 \\ 0 & -4 & -4 & | & 0 & -3 & 1 \end{bmatrix}$$

Pivot on row 2, column 2 by performing the following operations at the top of the next column.

$$-R_2 \rightarrow R_2$$
$$-R_2 + R_1 \rightarrow R_1$$
$$4R_2 + R_3 \rightarrow R_3$$

$$\begin{bmatrix} 1 & 0 & -3 & | & 1 & -1 & 0 \\ 0 & 1 & 5 & | & -1 & 2 & 0 \\ 0 & 0 & 16 & | & -4 & 5 & 1 \end{bmatrix}$$

Finally pivot on row 3, column 3 by performing the operations:

$$\tfrac{1}{16}R_3 \rightarrow R_3$$
$$3R_3 + R_1 \rightarrow R_1$$
$$-5R_3 + R_2 \rightarrow R_2$$

$$\begin{bmatrix} 1 & 0 & 0 & | & \tfrac{1}{4} & \tfrac{-1}{16} & \tfrac{3}{16} \\ 0 & 1 & 0 & | & \tfrac{1}{4} & \tfrac{7}{16} & \tfrac{-5}{16} \\ 0 & 0 & 1 & | & \tfrac{-1}{4} & \tfrac{5}{16} & \tfrac{1}{16} \end{bmatrix}$$

The inverse matrix is:

$$\begin{bmatrix} \tfrac{1}{4} & \tfrac{-1}{16} & \tfrac{3}{16} \\ \tfrac{1}{4} & \tfrac{7}{16} & \tfrac{-5}{16} \\ \tfrac{-1}{4} & \tfrac{5}{16} & \tfrac{1}{16} \end{bmatrix}.$$

Multiplying both sides of the matrix equation on the right by the inverse matrix we get:

$$\begin{bmatrix} 1 & 0 & 0 \\ 0 & 1 & 0 \\ 0 & 0 & 1 \end{bmatrix} \begin{bmatrix} x \\ y \\ z \end{bmatrix} = \begin{bmatrix} \tfrac{1}{4} & \tfrac{-1}{16} & \tfrac{3}{16} \\ \tfrac{1}{4} & \tfrac{7}{16} & \tfrac{-5}{16} \\ \tfrac{-1}{4} & \tfrac{5}{16} & \tfrac{1}{16} \end{bmatrix} \begin{bmatrix} 3 \\ 5 \\ 3 \end{bmatrix}$$

$$\begin{bmatrix} x \\ y \\ z \end{bmatrix} = \begin{bmatrix} 1 \\ 2 \\ 1 \end{bmatrix}$$

The solution to the system is:

$$x = 1; \ y = 2; \ z = 1$$

25. Enter the matrix into a graphing calculator, and use the matrix capabilities to find the inverse. The inverse is:

$$A^{-1} = \begin{bmatrix} \frac{5}{6} & \frac{-2}{3} \\ \frac{-1}{6} & \frac{1}{3} \end{bmatrix}$$

Cumulative Review

27. The horizontal line that contains the point $(-7,0)$ has the equation:

$$y = 0$$

29. Profit is Revenue minus Cost, so the profit function is:

$$P(x) = R(x) - C(x)$$
$$P(x) = 80x - (60x + 2100)$$
$$P(x) = 20x - 2100$$

The break-even point is when $P(x) = 0$
$$20x - 2100 = 0$$
$$20x = 2100$$
$$x = 105$$

$$R(105) = 8400 = C(105)$$

Therefore the break-even point is: $(105, 8400)$.

31. Let t be the number of additional hours of studying for the test, and let y equal the test score in percentage terms. The equation is:

$$y = 72 + 5t, \text{ where } 0 \le t \le 5.6.$$

The limits are necessary because it is not possible to score better than 100%

Sample Test Answers

1. In order for the matrices to be equal, the corresponding elements must be equal. For this to happen:

$$r = 6; t = 6; w = 0; x = 3; y = 9$$

2. $2A - B =$

$$\begin{bmatrix} 4 & 0 & 6 \\ -2 & 2 & 0 \\ 0 & 0 & 8 \end{bmatrix} - \begin{bmatrix} 1 & 2 & 3 \\ 4 & 5 & 6 \\ 7 & 8 & 9 \end{bmatrix} =$$

$$\begin{bmatrix} 3 & -2 & 3 \\ -6 & -3 & -6 \\ -7 & -8 & -1 \end{bmatrix}$$

3. Since matrix B and matrix C are not the same dimension, the operation of addition is not defined.

4.

$$A^T = \begin{bmatrix} 2 & -1 & 0 \\ 0 & 1 & 0 \\ 3 & 0 & 4 \end{bmatrix}$$

5.

$$B^2 = \begin{bmatrix} 1 & 2 & 3 \\ 4 & 5 & 6 \\ 7 & 8 & 9 \end{bmatrix}\begin{bmatrix} 1 & 2 & 3 \\ 4 & 5 & 6 \\ 7 & 8 & 9 \end{bmatrix} =$$

$$\begin{bmatrix} 30 & 36 & 42 \\ 66 & 81 & 96 \\ 102 & 126 & 150 \end{bmatrix}$$

6.

$$AB = \begin{bmatrix} 2 & 0 & 3 \\ -1 & 1 & 0 \\ 0 & 0 & 4 \end{bmatrix}\begin{bmatrix} 1 & 2 & 3 \\ 4 & 5 & 6 \\ 7 & 8 & 9 \end{bmatrix} =$$

$$\begin{bmatrix} 23 & 28 & 33 \\ 3 & 3 & 3 \\ 28 & 32 & 36 \end{bmatrix}$$

7. Since the number of rows of matrix C does not equal the number of columns of matrix C. Multiplication of C by itself is not defined.

8. The communications network matrix is:

$$
\begin{array}{c} \\ w \\ x \\ y \end{array}
\begin{array}{ccc} w & x & y \end{array}
\begin{bmatrix} 1 & 1 & 1 \\ 0 & 1 & 1 \\ 0 & 1 & 1 \end{bmatrix}
$$

9. The system of equations is:

$$2x + \qquad z = 4$$
$$3x - 2y + 5z = 17$$

10. The matrix B is the row matrix:

$$B = \begin{bmatrix} 175 & 250 \end{bmatrix}$$

The total cost of the microwave oven inventory in each store is given by:

$$\begin{bmatrix} 175 & 250 \end{bmatrix} \begin{bmatrix} 15 & 12 \\ 8 & 3 \end{bmatrix} = \begin{bmatrix} 4625 & 2850 \end{bmatrix}$$

The cost of the inventory in store 1 is $4,625 and the cost of the inventory in store 2 is $2,850.

11. Multiplying the two matrices together we see:

$$\begin{bmatrix} 3 & -1 \\ 2 & -1 \end{bmatrix} \begin{bmatrix} 1 & 1 \\ 2 & 3 \end{bmatrix} = \begin{bmatrix} 1 & 0 \\ 0 & -1 \end{bmatrix}.$$

This is not the identity matrix; therefore the two matrices are not inverses of each other.

12. The inverse is:

$$\begin{bmatrix} 3 & -7 \\ -2 & 5 \end{bmatrix}$$

13. The inverse is:

$$\begin{bmatrix} 1 & 89 & -31 \\ 0 & -20 & 7 \\ 0 & 3 & -1 \end{bmatrix}$$

14. $\quad X = (Y - B) A^{-1}$

15.

$$\begin{bmatrix} x \\ y \\ z \end{bmatrix} = \begin{bmatrix} 1 & -1 & 1 \\ 2 & -3 & 3 \\ 1 & -2 & 1 \end{bmatrix} \begin{bmatrix} 2 \\ 0 \\ 3 \end{bmatrix} = \begin{bmatrix} 5 \\ 13 \\ 5 \end{bmatrix}$$

The solution to the system is:

$$x = 5; y = 13; z = 5$$

16. The parts matrix is:

$$
\begin{array}{c} \\ P_1 \\ P_2 \\ P_3 \end{array}
\begin{array}{ccc} P_1 & P_2 & P_3 \end{array}
\begin{bmatrix} 0 & 1 & 2 \\ 0 & 0 & 3 \\ 0 & 0 & 0 \end{bmatrix}
$$

17. The entries indicate that each unit of part 2 requires 1 unit of part 1, and each unit of part 3 requires 2 units of part 1 and 3 units of part 2.

18.

$$(I - P)^{-1} = \begin{bmatrix} 1 & 1 & 5 \\ 0 & 1 & 3 \\ 0 & 0 & 1 \end{bmatrix}$$

19.

$$X = (I - P)^{-1} D$$

$$X = \begin{bmatrix} 1 & 1 & 5 \\ 0 & 1 & 3 \\ 0 & 0 & 1 \end{bmatrix} \begin{bmatrix} 100 \\ 200 \\ 300 \end{bmatrix} = \begin{bmatrix} 1800 \\ 1100 \\ 300 \end{bmatrix}$$

In order to meet the production needs, the company will need to produce 1800 units of part 1 and 1100 units of part 2, and 300 units of part 3.

20.

$$PX = \begin{bmatrix} 0 & 1 & 2 \\ 0 & 0 & 3 \\ 0 & 0 & 0 \end{bmatrix} \begin{bmatrix} 1800 \\ 1100 \\ 300 \end{bmatrix} = \begin{bmatrix} 1700 \\ 900 \\ 0 \end{bmatrix}$$

The company will consume 1700 units of part 1, and 900 units of part 2 in the production process.

Chapter 4
Linear Programming: A Graphical Approach

Exercises Section 4.1

1.

a. Let x be the number of silver pens made, and y be the number of gold pens made. The information can be organized in the following table.

	Silver	Gold	Limits
Grinder	1	3	30(60)=1800
Bonder	3	4	50(60)=3000
Number Made	x	y	
Profit	5	7	

b. The objective in this problem is to maximize profits, so the objective function should be a profit function.

$$P = 5x + 7y$$

Where the number of silver pens made is x and the number of gold pens made is y.

The profit function is subject to time constraints on the production of the pen. Specifically, it is subject to the time constraints on the grinder machine and time on the bonder machine. These constraints are formulated as:

$$x + 3y \leq 1800$$
$$3x + 4y \leq 3000$$

Finally we have the non-negativity constraints:

$$x \geq 0, y \geq 0$$

The linear programming problem is:

Maximize $\quad P = 5x + 7y$,

subject to:
$$x + 3y \leq 1800$$
$$3x + 4y \leq 3000$$
$$x \geq 0, y \geq 0$$

3.

a. Let x be the number of financial calculators made, and y be the number of scientific calculators made. The information can be organized in the following table:

	Financial	Scientific	Limits
Semiconductor	10	20	3200
Off-On switch	1	1	300
Need to make	1	0	100
Number made	x	y	
Production Steps	10	12	

b. The objective is to minimize the production steps, so the objective function is:

$$P = 10x + 12y$$

The semiconductor requirements give us the constraint:

$$10x + 20y \geq 3200$$

The off-on switch requirement gives us the constraint:

$$x + y \geq 300$$

The production requirement for the financial calculator gives us the constraint:

$$x \geq 100$$

Finally we have the non-negativity constraints:

$$x \geq 0, y \geq 0$$

The linear programming problem is:

Minimize $\quad P = 10x + 12y$,

subject to:
$$10x + 20y \geq 3200$$
$$x + \quad y \geq \quad 300$$
$$x \quad\quad \geq \quad 100$$
$$x \geq 0, y \geq 0$$

5. Let x equal the number of servings of salad, and y equal the number of servings of soup. The information can be organized in the table at the top of the next page.

	Salad	Soup	Limits
Vitamin A	4	6	10
Vitamin B	5	2	10
Servings	x	y	
mg of Fat	2	3	

The objective is to minimize the total milligrams of fat, so the objective function is:

$$F = 2x + 3y$$

The vitamin requirements give us the constraints:

$$4x + 6y \geq 10$$
$$5x + 2y \geq 10$$

Finally we have the non-negativity constraints:

$$x \geq 0, y \geq 0$$

Therefore the linear programming problem is:

Minimize $\qquad F = 2x + 3y$,

subject to:
$$4x + 6y \geq 10$$
$$5x + 2y \geq 10$$
$$x \geq 0, y \geq 0$$

7. Let w, x, y, z represent the number of times the talk show, the entertainment show, the stock market show and the national news show, are run each week.

Our objective is to maximize the number of listeners each week. The number of listeners in each week (in thousands) is given by the objective function:

$$L = 10w + 2.5x + 0.8y + 0.6z$$

The weekly cost constraint in thousands of dollars is:

$$1w + 0.8x + 0.4y + 0.3z \leq 20$$

The advertiser demands give us the constraints:

$$w \geq 5$$
$$x \geq 3$$

The station limitations give us the constraints at the top of the next column.

$$w \leq 8$$
$$x \leq 5$$
$$y \leq 3$$
$$z \leq 10$$

Finally the non-negativity constraints give us:

$$w \geq 0, x \geq 0, y \geq 0, z \geq 0$$

The linear programming problem is:

Maximize $\qquad L = 10w + 2.5x + 0.8y + 0.6z$

Subject to: $\qquad 1w + 0.8x + 0.4y + 0.3z \leq 20$
$$w \geq 5; \quad x \geq 3$$
$$w \leq 8; \quad x \leq 5$$
$$y \leq 3; \quad z \leq 10$$
$$w \geq 0, x \geq 0, y \geq 0, z \geq 0$$

9. Let x, y, z be the number of barrels of the combination of materials type A, Type B and Type C respectively.

The information can be organized in the following table:

	Type A	Type B	Type C	Limits
Quantity	x	y	z	
Carbon Black	2	3	3	12
Thinning	2	1	1	6
Expense	5	3	4	

The objective is to minimize expense, so the objective function is:

$$E = 5x + 3y + 4z$$

The requirement that 12 gallons of carbon black be used gives us the constraint:

$$2x + 3y + 3z \geq 12$$

The requirement that 6 gallons of the thinning agent be used gives us the constraint:

$$2x + y + z \geq 6$$

The non-negative constraints are:

$$x \geq 0, y \geq 0, z \geq 0$$

The solution is continued on the next page.

Using the information on the previous page, the linear programming problem is:

Minimize $\qquad E = 5x + 3y + 4z$

subject to:

$$2x + 3y + 3z \geq 12$$
$$2x + y + z \geq 6$$
$$x \geq 0, y \geq 0, z \geq 0$$

11. Let x be the number of junk bonds and y be the number of quality bonds purchased by the club. The information can be organized in the following table:

	Junk bonds	Quality Bonds	Limits
Club policy	2	-1	0
Budget	x	y	30
Rates	0.12	0.07	

The objective is to maximize the returns on the investment, so the objective function is:

$$R = 0.12x + 0.07y$$

The budget constraint is:

$$x + y \leq 30$$

The club policy constraint is:

$$2x - y \leq 0$$

The non-negative constraints are:

$$x \geq 0, y \geq 0$$

The linear programming problem is:

Maximize $\qquad R = 0.12x + 0.07y$

subject to:

$$x + y \leq 30$$
$$2x - y \leq 0$$
$$x \geq 0, y \geq 0$$

13. Let x equal the number of acres of wheat, and y equal the number of acres of soybeans. The information in the problem can be organized in the table at the top of the next column.

	Wheat	Soybeans	Limits
Quantity	x	y	
Acreage	1	1	500
Budget	200	350	25000
Income	350	525	

The objective in this problem is to maximize income, so the objective function is:

$$I = 350x + 525y$$

The acreage limitations give us the constraint:

$$x + y \leq 500$$

The Budget limitation gives us the constraint:

$$200x + 350y \leq 25,000$$

The non-negative constraints are:

$$x \geq 0, y \geq 0$$

The linear programming problem is:

Maximize $\qquad I = 350x + 525y$

subject to:

$$200x + 350y \leq 25,000$$
$$x + y \leq 500$$
$$x \geq 0, y \geq 0$$

15. Let x be the number of type A planes, and let y be the number of type B planes. The information can be organized in the following table:

	Type A	Type B	Limits
Quantity	x	y	
First-Class	50	40	500
Tourist	60	30	1000
Economy	100	80	1500
Costs	12	10	

The objective is to minimize cost, so the objective function is:

$$C = 12x + 10y$$

where cost is in thousands of dollars.

The solution is continued on the next page.

The passenger requirements for first-class, tourist-class and economy-class passengers give us the respective constraints:

$$50x + 40y \geq 500$$
$$60x + 30y \geq 1000$$
$$100x + 80y \geq 1500$$

The non-negative constraints are:

$$x \geq 0, y \geq 0.$$

The linear programming problem is:

Minimize $C = 12x + 10y$

subject to:

$$50x + 50y \geq 500$$
$$60x + 30y \geq 1000$$
$$100x + 80y \geq 1500$$
$$x \geq 0, y \geq 0$$

17. Let x be the number of Nitebrite sets, and y be the number of Solar II sets bought each month.
The information can be organized in the following table:

	Nitebrite	Solar II	Limits
Quantity	x	y	
Warehouse	1	1	300
N Demand	1		80
S Demand		1	120
Revenue	420	550	

The objective is to maximize revenue, so the objective function is:

$$R = 420x + 550y$$

The space limitation on the warehouse results in the constraint:

$$x + y \leq 300$$

The past records show that the demand for Nitebrite sets is:

$$x \geq 80$$

These records also show that the demand for Solar II sets is:

$$y \geq 120$$

The non-negative constraints are:

$$x \geq 0, y \geq 0.$$

The linear programming problem is:

Maximize: $R = 420x + 550y$

subject to:

$$x + y \leq 300$$
$$x \quad\;\; \geq 80$$
$$y \geq 120$$
$$x \geq 0, y \geq 0$$

19. Let x be the number of units of Food I, and y be the number of units of Food II.

The objective is to minimize his intake of fat, so the objective function is:

$$F = 2x + 5y$$

Gary's taste preferences result in the constraint:

$$x \geq 3y, \text{ or } x - 3y \geq 0$$

The rest of the information can be organized in the table:

	Food I	Food II	Limits
Quantity	x	y	
Carbs	5	7	105
Calories	80	110	400
Fat	2	5	

The carbohydrate limit of no more than 105 grams of carbs results in the constraint:

$$5x + 7y \leq 105$$

The need for at least 400 calories results in the constraint:

$$80x + 110y \geq 400$$

The non-negative constraints are:

$$x \geq 0, y \geq 0$$

The solution is continued on the next page.

Using the information on the previous page, the linear programming problem is:

Minimize $\qquad F = 2x + 5y$

subject to:

$$5x + 7y \leq 105$$
$$80x + 110y \geq 400$$
$$x - 3y \geq 0$$
$$x \geq 0, y \geq 0$$

Exercises Section 4.2

1. Test each point by substituting into the inequality.

a.

$$2(0) - 3(0) \leq 4$$
$$0 \leq 4$$

Yes, the point $(0,0)$ is in the feasible region.

b.

$$2(3) - 3\left(\tfrac{2}{3}\right) \leq 4$$
$$4 \leq 4$$

Yes, the point $\left(3, \tfrac{2}{3}\right)$ is in the feasible region.

c.

$$2(4) - 3(0) \leq 4$$
$$8 \leq 4$$

No, the point $(4,0)$ is not in the feasible region.

3. Test each point by substituting it into each inequality, if it fails to satisfy any inequality, it is not in the feasible region.

a. The point $(0,0)$ is not in the feasible region because
$$x > 0, y > 0$$

b. The point $(1,1)$ satisfies all four inequalities; therefore it is in the feasible region.

c. The point $(1.9, 0)$ again fails to satisfy the inequality $y > 0$, so it is not in the feasible region.

5.
a. The boundary line has the equation:

$$3x + y = 9$$

b. An appropriate test point is any point that does not lie on the boundary line. We will use the point $(0,0)$ whenever appropriate.

$$3(0) + (0) = 0 \leq 9$$

So we will choose the half plane that does contain the point $(0,0)$.

c. Since the inequality is of the form $\leq$ we include the boundary line. The graph is:

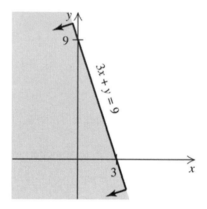

7.
a. The boundary line has the equation:

$$2x + 3y = 6$$

b. An appropriate test point is any point that does not lie on the boundary line. We will use the point $(0,0)$ whenever appropriate.

$$2(0) + 3(0) = 0 < 6$$

So we will choose the half plane that does contain the point $(0,0)$.

The solution is continued on the next page.

c. Since the inequality is of the form $<$ we do not include the boundary line. The graph is shown below:

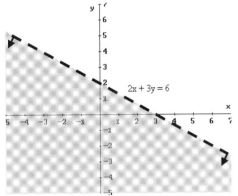

9.

a. The boundary line has the equation:

$$-2x + y = 8$$

b. An appropriate test point is any point that does not lie on the boundary line. We will use the point $(0,0)$ whenever appropriate.

$$-2(0) + (0) = 0 \le 8$$

So we will choose the half plane that does not contain the point $(0,0)$.

c. Since the inequality is of the form $\ge$ we include the boundary line. The graph is:

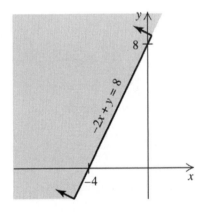

11.

a. The boundary line has the equation:

$$2x = 6, \text{ or } x = 3$$

b. An appropriate test point is any point that does not lie on the boundary line. We will use the point $(0,0)$ whenever appropriate.

$$2(0) = 0 \le 6$$

So we will choose the half plane that does not contain the point $(0,0)$.

c. Since the inequality is of the form $>$ we do not include the boundary line. The graph is:

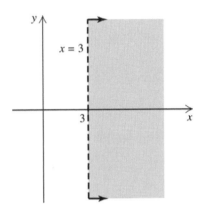

13.

a. The boundary line has the equation:

$$3x - 2y = 12$$

b. An appropriate test point is any point that does not lie on the boundary line. We will use the point $(0,0)$ whenever appropriate.

$$3(0) - 2(0) = 0 \le 12$$

So we will choose the half plane that does contain the point $(0,0)$.

c. Since the inequality is of the form $\le$ we include the boundary line. The graph is:

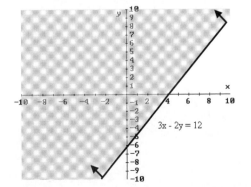

15.
a. The boundary line has the equation:

$$x - y = 0$$

b. An appropriate test point is any point that does not lie on the boundary line. We will use the point $(1, 0)$.

$$(1) - (0) = 1 \geq 0$$

So we will choose the half plane that does not contain the point $(1, 0)$.

c. Since the inequality is of the form $\leq$ we include the boundary line. The graph is:

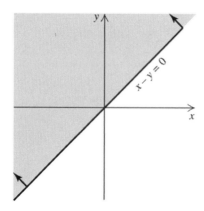

17.
a. The boundary lines are the three lines whose equations are:

$$x + 2y = 6$$
$$x = 0$$
$$y = 0$$

b. We shade in the region bounded by $x + 2y = 6$ the half plane will contain the point $(0, 0)$. We are also bounded to the half plane to the right of the line $x = 0$ and above the line $y = 0$.

The feasible region is shown at the top of the next column.

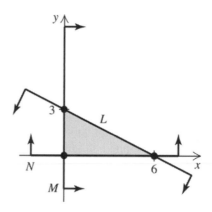

The region is a triangular region. The three corner points are labeled.

The first corner point is the intersection of the lines $x = 0$ and $y = 0$. That intersection is the point $(0, 0)$

The second corner point is the intersection of the lines $x + 2y = 6$ and $x = 0$. The lines intersect at the point $(0, 3)$

The third corner point is the intersection of the lines $x + 2y = 6$ and $y = 0$. The lines intersect at the point $(6, 0)$.

19.
a. The boundary lines are the three lines whose equations are:

$$x = y$$
$$x = 2y$$
$$x = 3$$

b. We shade in the region bounded by $x = y$ the half plane will contain the point $(2, 0)$. We shade in the region bounded by $x = 2y$ the half plane will contain the point $(0, 2)$. We are also bounded to the half plane to the left of the line $x = 3$.

The feasible region is shown at the top of the next page

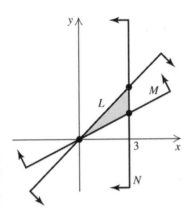

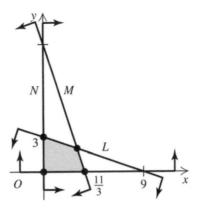

The region is a triangular region. The three corner points are labeled.

The first corner point is the intersection of the lines $x = y$ and $x = 2y$. That intersection is the point $(0,0)$

The second corner point is the intersection of the lines $x = 2y$ and $x = 3$. The lines intersect at the point $\left(3, \frac{3}{2}\right)$

The third corner point is the intersection of the lines $x = y$ and $x = 3$. The lines intersect at the point $(3,3)$.

21.
a. The boundary lines are the four lines whose equations are:

$$x + 3y = 9$$
$$3x + y = 11$$
$$x \qquad = 0$$
$$\qquad y = 0$$

b. We shade in the region bounded by $x + 3y = 9$ the half plane will contain the point $(0,0)$. We shade in the region bounded by $3x + y = 11$ the half plane will contain the point $(0,0)$. We are also bounded to the half plane to the right of the line $x = 0$ and above the line $y = 0$.

The feasible region is shown at the top of the next column.

The corner points are labeled.

The first corner point is the intersection of the lines $x = 0$ and $y = 0$. That intersection is the point $(0,0)$

The second corner point is the intersection of the lines $x = 0$ and $x + 3y = 9$. The lines intersect at the point $(0,3)$

The third corner point is the intersection of the lines $y = 0$ and $3x + y = 11$. The lines intersect at the point $\left(\frac{11}{3}, 0\right)$.

The forth corner point is the intersection of the lines $x + 3y = 9$ and $3x + y = 11$. The lines intersect that the point $(3,2)$

23.
a. The boundary lines are the five lines whose equations are:

$$x + y = 8$$
$$2x + y = 10$$
$$4x + y = 16$$
$$x \qquad = 0$$
$$\qquad y = 0$$

b. The point $(0,0)$ satisfies each of the first three inequalities, so the half-planes for the first three inequalities will contain the point $(0,0)$. We are also bounded to the half-plane to the right of the line $x = 0$ and above the line $y = 0$.

The solution is continued on the next page.

The feasible region is shown below:

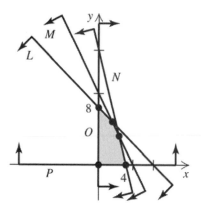

The corner points are labeled.

The first corner point is the intersection of the lines $x = 0$ and $y = 0$. That intersection is the point $(0,0)$

The second corner point is the intersection of the lines $x = 0$ and $x + y = 8$. The lines intersect at the point $(0,8)$

The third corner point is the intersection of the lines $y = 0$ and $4x + y = 16$. The lines intersect at the point $(4,0)$.

The fourth corner point is the intersection of the lines $x + y = 8$ and $2x + y = 10$. The lines intersect that the point $(2,6)$.

The fifth corner point is the intersection of the lines $4x + y = 16$ and $2x + y = 10$. The lines intersect that the point $(3,4)$.

25. The graph of the feasible region is:

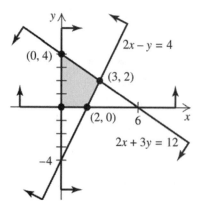

The region is the region belonging to the half-planes above $2x - y = 4$, below $2x + 3y = 12$, to the right of $x = 0$ and above $y = 0$.

The corner points of the region are the points $O(0,0)$, $A(0,4)$, $B(3,2)$, and $C(2,0)$. Both $(0,0)$ and $(1,3)$ are in the region, as they satisfy all four inequalities.

27. The graph of the feasible region is:

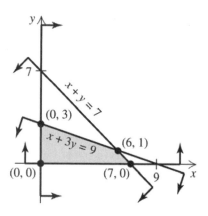

The region is the region belonging to the half-planes below $x + 3y = 9$, below $x + y = 7$, to the right of $x = 0$ and above $y = 0$.

The corner points of the region are the points $O(0,0)$, $A(0,3)$, $B(6,1)$, and $C(7,0)$. The point $(0,0)$ is in the region, as it satisfies all four inequalities. However, the point $(1,3)$ is not in the feasible region because

$$(1) + 3(3) = 10 \geq 9.$$

29. The graph of the feasible region is:

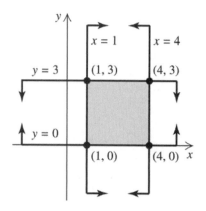

The solution is continued on the next page.

The region is the region belonging to the half-planes to the right of $x = 1$, to the left of $x = 4$, above $y = 0$ and below $y = 3$.

The corner points of the region are the points $A(1,3)$, $B(4,3)$, $C(4,0)$, and $D(1,0)$. The point $(1,3)$ is in the region, as it satisfies all four inequalities. However, the point $(0,0)$ is not in the region because $0 < 1$.

31. The graph of the feasible region is:

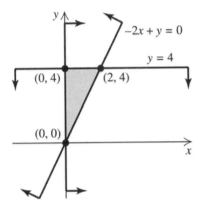

The region is the region belonging to the half-planes above $-2x + y = 0$, below $y = 4$, to the right of $x = 0$.

The corner points of the region are the points $O(0,0)$, $A(0,4)$, $B(2,4)$. Both $(0,0)$ and $(1,3)$ are in the region, as they satisfy all four inequalities.

33. The graph of the feasible region is:

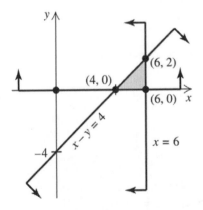

The region is the region belonging to the half-planes below $x - y = 4$, to the left of $x = 6$ and above $y = 0$.

The corner points of the region are the points, $A(4,0)$, $B(6,2)$, and $C(6,0)$. Both $(0,0)$ and $(1,3)$ are not in the region, because neither point satisfies the inequality $x - y \geq 4$.

35. Let x equal the number of days spent in Iowa, and y equal the number of days spent in Kansas. Let all dollar amounts be in thousands of dollars. The information can be organized in the table below:

	Iowa	Kansas	Limits
Expenses	0.120	0.100	18
Minimum Days	1		50
Minimum Days		1	60
Sales	3	2.5	

The linear programming problem is:

Maximize $\quad\quad S = 3x + 2.5y$
subject to:

$$0.12x + 0.10y \leq 18$$
$$x \quad\quad\quad \geq 50$$
$$y \geq 60$$

b. The graph of the feasible region is shown below:

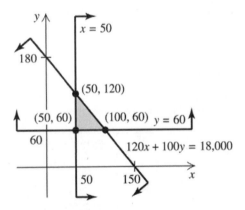

The corner points to the feasible region are:

$$A(50,120), B(100,60), \text{ and } C(50,60)$$

37. Let x be the number of kilograms of Food I, and y be the number of kilograms of Food II used in the diet. The information can be organized in the table at the top of the next page.

	Food I	Food II	Limits
Quantity	x	y	
Vitamin A	30	40	1200
Vitamin B	90	50	2970
Cost	0.80	1.10	

The linear programming problem is:

Minimize $\quad C = 0.80x + 1.10y$

subject to:

$$30x + 40y \geq 1200$$
$$90x + 50y \geq 2970$$
$$x \geq 0, y \geq 0$$

b. The feasible region is:

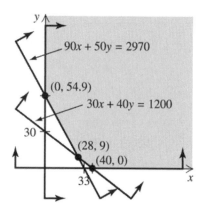

The corner points are $(0,0), (0,59.4), (28,9)$, and $(40,0)$.

39. Let x be the number of parts shipped to Sacramento, and y be the number of parts shipped to Hobbs. The information can be organized in the following table.

	Sacramento	Hobbs	Limits
Quantity	x	y	
S Demand	1		1000
H Demand		1	1200
Costs	2	3	14000
Work hours	0.2	0.25	

The linear programming problem is:

Minimize $\quad W = 0.2x + 0.25y$

subject to:

$$2x + 3y \leq 14,000$$
$$x \qquad \geq 1,000$$
$$y \geq 1,200$$

b. The feasible region is shown below:

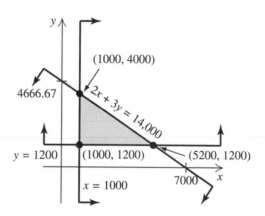

The corner points are:

$$(1000,1200), (1000,4000), \text{ and } (5200,1200)$$

41. Let x be the number of silver pens produced, and y be the number of gold pens produced. The information can be organized in the following table.

	Silver	Gold	Limits
Quantity	x	y	
Grinder	1	3	(30)(60)=1800
Bonder	3	4	(50)(60)=3000
Revenue	30	80	

The linear programming problem is:

Maximize $\quad R = 30x + 80y$

subject to:

$$x + 3y \leq 1800$$
$$3x + 4y \leq 3000$$
$$x \geq 0, y \geq 0$$

b. The feasible region is shown below:

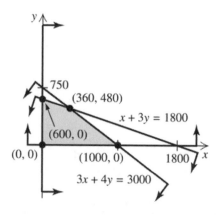

The solution is continued on the next page

The corner points of the feasible region on the previous page are:

$$(0,0), (0,600), (360,480), \text{ and } (1000,0)$$

43. Let x be equal to the number of business calculators, and y be equal to the number of scientific calculators.

The requirement that at least twice as many business calculator as scientific calculators be made translates to the inequality:

$$x \geq 2y, \text{ or } x - 2y \geq 0$$

The rest of the information can be organized in the following table:

	Business	Scientific	Limits
Quantity	x	y	
Microcircuits	10	20	320
Capacity	1		50
Revenue	50	60	

The linear programming problem is:

Maximize $R = 50x + 60y$

subject to:

$$10x + 20y \geq 320$$
$$x - 2y \geq 0$$
$$x \leq 50$$
$$x \geq 0, y \geq 0$$

b. The feasible region is:

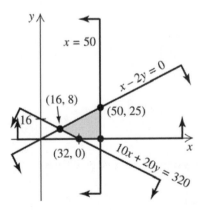

The corner points are:

$$(32,0), (16,8), (50,25), \text{ and } (50,0).$$

45. Let x be the number of college graduates who were interviewed, and y be the number of men who never attended college who were interviewed.

The information can be organized in the following table.

	College Grads.	Never Attended	Limits
Quantity	x	y	
Professor	30	35	(80)(60)=4800
Grad Student	20	45	(100)(60)=6000
Total	1	1	

The linear programming problem is:

Maximize $T = x + y$

subject to:

$$30x + 35y \leq 4800$$
$$20x + 45y \leq 6000$$
$$x \geq 0, y \geq 0$$

b. The feasible region is:

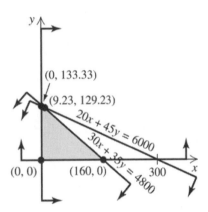

The corner points are:

$$(0,0), \left(0, \tfrac{400}{3}\right), \left(\tfrac{120}{13}, \tfrac{1680}{13}\right), \text{ and } (160,0).$$

The corner point $\left(\tfrac{120}{13}, \tfrac{1680}{13}\right)$ is not realistic since you cannot interview a fraction of a person. Again, this is analogous to the average number of interviews over several experiments.

Exercises Section 4.3

1.

a. The graph of the feasible region and the various lines $f = c$ are pictured here:

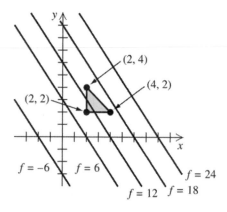

b. The minimum value of f is 10 and occurs at corner point $(2,2)$. The maximum value of f is 16 and occurs at the corner point $(4,2)$. The line $f = c$ moves in the north-east $\nearrow$ direction as the value of c increases.

c. The graph of the feasible region and the various lines $g = c$ are pictured here:

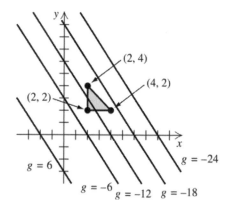

d. The minimum of g is -16 and occurs at the corner point $(4,2)$; The maximum of g is -10 and occurs at the corner point $(2,2)$. The line $g = c$ moves in the south-west $\swarrow$ direction as the value of c increases.

3.

a. The feasible region is displayed at the top of the next column.

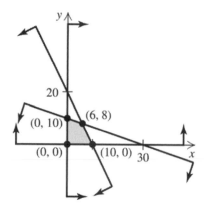

b. The corner points as shown in the above graph are:

$(0,0), (0,10), (6,8)$, and $(10,0)$

Point	$P = 4x + 4y$
$(0,0)$	$4 \cdot 0 + 4 \cdot 0 = 0$
$(0,10)$	$4 \cdot 0 + 4 \cdot 10 = 40$
$(6,8)$	$4 \cdot 6 + 4 \cdot 8 = 56$
$(10,0)$	$4 \cdot 10 + 4 \cdot 0 = 40$

c. The maximum value of P is 56 and occurs when $x = 6$ and $y = 8$.

5.

a. The feasible region is:

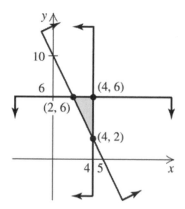

b. The corner points as shown in the above graph are:

$(2,6), (4,6)$, and $(4,2)$

The solution is continued on the next page.

Point	$P = x + 2y$
$(2,6)$	$2 + 2 \cdot 6 = 14$
$(4,6)$	$4 + 2 \cdot 6 = 16$
$(4,2)$	$4 + 2 \cdot 2 = 8$

c. The maximum value of f is 16 and occurs when $x = 4$ and $y = 6$.

7.
a. The feasible region is:

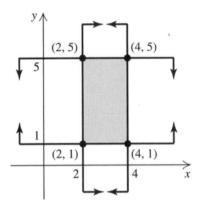

b. The corner points as shown in the above graph are:

$(2,5), (4,5), (2,1)$, and $(4,1)$

Point	$C = x - 2y$
$(2,5)$	$2 - 2 \cdot 5 = -8$
$(4,5)$	$4 - 2 \cdot 5 = -6$
$(2,1)$	$2 - 2 \cdot 1 = 0$
$(4,1)$	$4 - 2 \cdot 1 = 2$

c. The minimum value of C is -8 and occurs when $x = 2$ and $y = 5$.

9.
a. The feasible region is shown at the top of the next column.

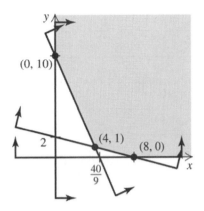

b. The corner points as shown in the above graph are:

$(0,10), (4,1)$, and $(8,0)$

Point	$C = 2x + 3y$
$(0,10)$	$2 \cdot 0 + 3 \cdot 10 = 30$
$(4,1)$	$2 \cdot 4 + 3 \cdot 1 = 11$
$(8,0)$	$2 \cdot 8 + 3 \cdot 0 = 16$

c. The minimum value of C is 11 and occurs when $x = 4$ and $y = 1$.

11.
a. The feasible region is:

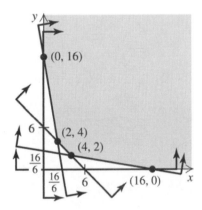

b. The corner points as shown in the above graph are:

$(0,16), (2,4), (4,2)$, and $(16,0)$

The objective function values are shown at the top of the next page.

Point	$C = 5x + 3y$
$(0,16)$	$5 \cdot 0 + 3 \cdot 16 = 48$
$(2,4)$	$5 \cdot 2 + 3 \cdot 4 = 22$
$(4,2)$	$5 \cdot 4 + 3 \cdot 2 = 26$
$(16,0)$	$5 \cdot 16 + 3 \cdot 0 = 80$

c. The minimum value of C is 22 and occurs when $x = 2$ and $y = 4$.

13.
a. The feasible region is:

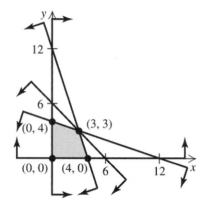

b. The corner points as shown in the above graph are:

$(0,0), (0,4), (3,3)$, and $(4,0)$

Point	$P = 8x + y$
$(0,0)$	$8 \cdot 0 + 0 = 0$
$(0,4)$	$8 \cdot 0 + 4 = 4$
$(3,3)$	$8 \cdot 3 + 3 = 27$
$(4,0)$	$8 \cdot 4 + 0 = 32$

c. The maximum value of P is 32 and occurs when $x = 4$ and $y = 0$.

15.
a. The feasible region is shown at the top of the next column. Since $y = 2$ is a constraint, the feasible region is the line segment between the points $(1,2)$ and $(4,2)$.

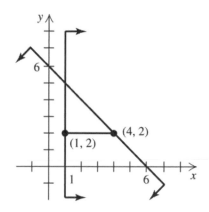

b. The corner points as shown in the above graph are:

$(1,2)$ and $(4,2)$.

Point	$f = 2x - 5y$
$(1,2)$	$2 \cdot 1 - 5 \cdot 2 = -8$
$(4,2)$	$2 \cdot 4 - 5 \cdot 2 = -2$

c. The minimum value of f is -8 and occurs when $x = 1$ and $y = 2$.

17. Consider the graph of the feasible region:

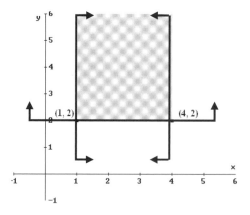

a. The function $P = x + 3y$ does not attain a maximum in the above feasible region. The region is unbounded from above. The lines that represent the different values of the function will continue to intersect the region as the function value continues to get larger. We conclude that the function does not attain a maximum in the region

b. Yes, at the point $(1,2)$, $P = 7$. This is the minimum value of the function in the feasible region.

19. Let x be the number of computers and y be the number of fax machines.

The objective is to maximize profit, so the objective function is:

$$P = 200x + 100y$$

The inventory constraint says that:

$$x + y \leq 30$$

The budget constraint implies that:

$$2000x + 800y \leq 40,800$$

The linear programming problem is:

Maximize $P = 200x + 100y$,

subject to:

$$x + \quad y \leq \quad 30$$
$$2000x + 800y \leq 40,800$$
$$x \geq 0, y \geq 0$$

The feasible region is shown below:

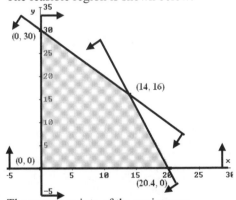

The corner points of the region are:

$$(0,0), (0,30), (14,16), \text{ and } (20.4,0)$$

Point	$P = 200x + 100y$
$(0,0)$	$200 \cdot 0 + 100 \cdot 0 = 0$
$(0,30)$	$200 \cdot 0 + 100 \cdot 30 = 3000$
$(14,16)$	$200 \cdot 14 + 100 \cdot 16 = 4400$
$(20.4,0)$	$200 \cdot 20.4 + 100 \cdot 0 = 4080$

In order to maximize profit of \$4,400 the store should have an inventory of 14 computers and 16 fax machines.

21. Let x be the number of four-ply tires and y be the number of radial tires produced.

The information can be organized in the table:

	Four-ply	Radial	Limits
Quantity	x	y	
Union	2	6	$(320)(60) = 19,200$
Total	1	1	4000
Radial		1	1600
Costs	20	30	

The linear programming problem is:

Minimize $C = 20x + 30y$
subject to:

$$2x + 6y \geq 19,200$$
$$x + \quad y \geq \quad 4000$$
$$y \geq \quad 1600$$

The feasible region is shown below:

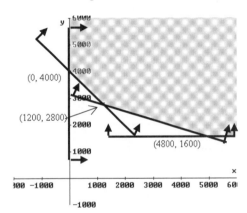

The corner points of the region are:

$$(0,4000), (1200,2800), \text{ and } (4800,1600)$$

Point	$C = 20x + 30y$
$(0,4000)$	$20 \cdot 0 + 30 \cdot 4000 = 120,000$
$(1200,2800)$	$20 \cdot 1200 + 30 \cdot 2800 = 108,000$
$(4800,1600)$	$20 \cdot 4800 + 30 \cdot 1600 = 144,000$

In order to minimize the production cost, and still meet demands, the company should produce 1200 four-ply tires and 2800 radials.

23. Let x be the number of pounds of barley, and let y be the number of pounds of corn. The information can be organized in the following table:

	Barley	Corn	Limits
Quantity	x	y	
Units of Fat	1	2	12
Barley	1		6
Corn		1	5
Units of Protein	1	1	

The objective is to maximize the units of protein in the mixture. The linear programming problem is:

Maximize $\quad P = x + y$

subject to:

$$x + 2y \leq 12$$
$$x \quad\quad \leq 6$$
$$y \leq 5$$
$$x \geq 0, y \geq 0$$

The feasible region is shown below:

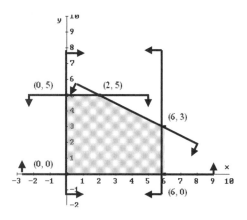

The corner points are:

$$(0,0), (0,5), (2,5), (6,3), \text{ and } (6,0)$$

Point	$P = x + y$
$(0,0)$	$0 + 0 = 0$
$(0,5)$	$0 + 5 = 5$
$(2,5)$	$2 + 5 = 7$
$(6,3)$	$6 + 3 = 9$
$(6,0)$	$6 + 0 = 6$

The solution is continued at the top of the next column.

To maximize the number of units of protein in the mixture, the processor should use 6 pounds of barley and 3 pounds of corn. The maximum number of units of protein will be 9 units.

25. Let x be equal to the number of student model desks, and y be equal to the number of secretary model desks.

The information can be organized in the following table.

	Student	Secretary	Limits
Quantity	x	y	
Woodworking	2	3	240
Finishing	3	5	390
Secretary Demand		1	40
Profit	40	50	

The linear programming problem is:

Maximize $\quad P = 40x + 50y$

subject to:

$$2x + 3y \leq 240$$
$$3x + 5y \leq 390$$
$$y \geq 40$$
$$x \geq 0, y \geq 0$$

The feasible region is shown below:

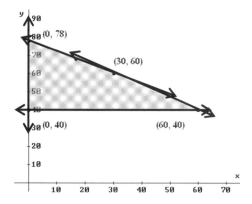

The corner points of the region are:

$$(0,40), (0,78), (30,60) \text{ and } (60,40).$$

The objective function values at the corner points are displayed in the table at the top of the next page.

Point	$P = 40x + 50y$
$(0, 40)$	$40 \cdot 0 + 50 \cdot 40 = 2000$
$(0, 78)$	$40 \cdot 0 + 50 \cdot 78 = 3900$
$(30, 60)$	$40 \cdot 30 + 50 \cdot 60 = 4200$
$(60, 40)$	$40 \cdot 60 + 50 \cdot 40 = 4400$

The company should make 60 student desks, and 40 secretary desks to maximize their profit of $4,400 per week.

27. Let x be the number of planes of type A, and y be the number of planes of type B. Also let cost be in thousands of dollars.

The information can be organized in the following table.

	Type A	Type B	Limits
Quantity	x	y	
First-Class	50	40	500
Tourist	60	30	1000
Economy	100	80	1500
Cost	12	10	

The linear programming problem is:

Minimize $\qquad C = 12x + 10y$

subject to:

$$50x + 40y \geq 500$$
$$60x + 30y \geq 1000$$
$$100x + 80y \geq 1500$$
$$x \geq 0, y \geq 0$$

The feasible region is shown below:

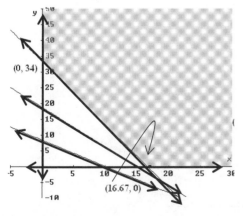

The corner points of the region are:

$(0, 34)$, and $(16.67, 0)$

Since you have to charter the whole plane, the reasonable corner point is $(17, 0)$

Point	$C = 12x + 10y$
$(0, 34)$	$12 \cdot 0 + 10 \cdot 34 = 340$
$(17, 0)$	$12 \cdot 17 + 10 \cdot 0 = 204$

The service will meet the demand, and minimize cost at $204,000 if it charters 17 type A planes and zero type B planes.

29. Let x be the number of standard models of microwave ovens that the firm makes each day, and let y be the number of deluxe models of microwave ovens that the firm makes each day.

The information can be organized in the following table:

	Standard	Deluxe	Limits
Quantity	x	y	
Cases	1	1	300
Assembly	10	15	3840
Standard orders	1		120
Deluxe orders		1	150
Profit	25	35	

The linear programming problem is:

Maximize $\qquad P = 25x + 35y$

subject to:

$$x + y \leq 300$$
$$10x + 15y \leq 3840$$
$$x \qquad \geq 120$$
$$y \geq 150$$

The feasible region is shown below:

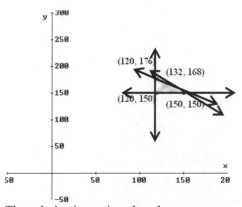

The solution is continued on the next page.

The corner points of the region on the previous page are:

$$(120,150), (120,176), (132,168) \text{ and } (150,150)$$

Point	$P = 25x + 35y$
$(120,150)$	$25 \cdot 120 + 35 \cdot 150 = 8250$
$(120,176)$	$25 \cdot 120 + 35 \cdot 176 = 9160$
$(132,168)$	$25 \cdot 132 + 35 \cdot 168 = 9180$
$(150,150)$	$25 \cdot 150 + 35 \cdot 150 = 9000$

The firm should produce 132 standard microwave ovens and 168 deluxe microwave ovens to maximize the firm's profit of $9180.

31. The set up to this problem can be found in section 4.1 exercise number 11.

The linear programming problem is:

Maximize $\quad R = 0.12x + 0.07y$

subject to:

$$x + y \le 30$$
$$2x - y \le 0$$
$$x \ge 0, y \ge 0$$

The feasible region is shown below:

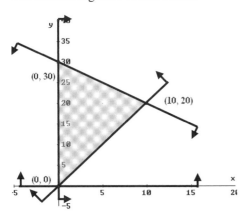

The corner points of the feasible region are:

$$(0,0), (0,30), \text{ and } (10,20)$$

Point	$P = 0.12x + 0.07y$
$(0,0)$	$0.12 \cdot 0 + 0.07 \cdot 0 = 0$
$(0,30)$	$0.12 \cdot 0 + 0.07 \cdot 30 = 2.1$
$(10,20)$	$0.12 \cdot 10 + 0.07 \cdot 20 = 2.6$

The solution is stated at the top of the next column.

If the club invests $10,000 in junk bonds, and 20,000 dollars in premium quality bonds, the club will earn a maximum return on its money of $2,600.

33. Let x be the number of mice purchased, and y be the number of rats purchased.

The objective is to maximize the number of experiments the department can run, so the objective function is:

$$N = 6x + 4y$$

The fact that it buys at least three times as many mice as rats results in the constraint:

$$x \ge 3y \text{ or } x - 3y \ge 0$$

The rest of the information can be organized in the following table.

	Mice	Rats	Limits
Quantity	x	y	
Costs	2	4	500
Space	1		<200
Limitations		1	=25
Experiments	6	4	

The linear programming problem is:

Maximize $\quad N = 6x + 4y$

subject to:

$$2x + 4y \le 500$$
$$x - 3y \ge \quad 0$$
$$x \quad \le 200$$
$$x \quad \ge 0$$
$$y = \quad 25$$

The feasible region is:

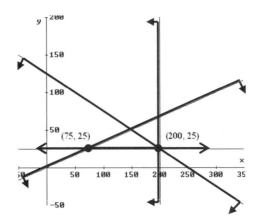

The solution is continued on the next page.

As you can see on the previous page the feasible set is the line segment between the points $(75, 25)$ and $(200, 25)$

The corner points of the region are:

$(75, 25)$, and $(200, 25)$

Point	$P = 6x + 4y$
$(75, 25)$	$6 \cdot 75 + 4 \cdot 25 = 550$
$(200, 25)$	$6 \cdot 200 + 4 \cdot 25 = 1300$

The maximum number of experiments the department can run is 1300. They will be able to do this if they purchase 200 mice and 25 rats.

Chapter 4 Summary Exercises

1. For each point substitute the values into the inequality and determine the validity of the statement.

a. The point $(0, 0)$ is not in the region because

$(0) - 2(0) = 0 \le 4$

b. The point $(6, \frac{1}{2})$ is in the region because

$(6) - 2(\frac{1}{2}) = 5 \ge 4$

c. The point $(6, \frac{1}{2})$ is in the region because

$(6) - 2(\frac{1}{2}) = 5 \ge 4$

3. The appropriate half-planes are shown below:

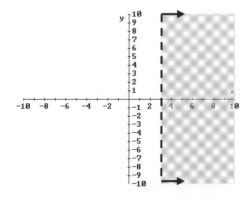

5. The appropriate half-planes are shown below:

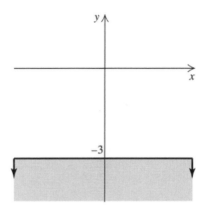

7. The corner points are:

$(10, 10), (40, 35), (30, 45)$, and $(20, 45)$

9. To find the minimum of the function over the feasible region use the following table.

Point	$f = 20x + 15y$
$(10, 10)$	$20 \cdot 10 + 15 \cdot 10 = 350$
$(40, 35)$	$20 \cdot 40 + 15 \cdot 35 = 1325$
$(30, 45)$	$20 \cdot 30 + 15 \cdot 45 = 1275$
$(20, 45)$	$20 \cdot 20 + 15 \cdot 45 = 1075$

The minimum of f occurs at the point $(10, 10)$ and has minimum value of 350.

11. Three corner points can be found by inspection of the graph. They are:

$(0, 0), (0, 2.5), (4, 0)$

To find the fourth corner point, solve the system of equations:

$5x + 2y = 20$

$x + 4y = 10$

Solve equation 2 for x to get:

$x = 10 - 4y$

Substitute the value of x into equation 1 as shown on the next page.

$$5(10-4y)+2y=20$$
$$50-20y+2y=20$$
$$-18y=-30$$
$$y=\frac{30}{18}=\frac{5}{3}$$

This means that:

$$x=10-4\left(\frac{5}{3}\right)=\frac{10}{3}$$

The fourth corner point is:

$$\left(\frac{10}{3},\frac{5}{3}\right).$$

So the corner points are:

$$\left(0,0\right),\left(0,2.5\right),\left(4,0\right), \text{ and } \left(\frac{10}{3},\frac{5}{3}\right).$$

13. To find the minimum of the function over the feasible region use the following table.

Point	$f=4x+5y$
$\left(0,0\right)$	$4\cdot0+5\cdot0=0$
$\left(0,2.5\right)$	$4\cdot0+5\cdot2.5=12.5$
$\left(4,0\right)$	$4\cdot4+5\cdot0=16$
$\left(\frac{10}{3},\frac{5}{3}\right)$	$4\cdot\frac{10}{3}+5\cdot\frac{5}{3}=\frac{65}{3}\approx21.67$

The minimum of f occurs at the point $\left(0,0\right)$ and has value zero.

15. The feasible region is shown below:

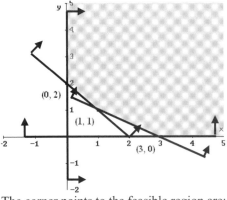

The corner points to the feasible region are:

$$\left(0,2\right),\left(1,1\right), \text{ and } \left(3,0\right)$$

To find the minimum of the function over the feasible region use the following table.

Point	$C=3x+2y$
$\left(0,2\right)$	$3\cdot0+2\cdot2=4$
$\left(1,1\right)$	$3\cdot1+2\cdot1=5$
$\left(3,0\right)$	$3\cdot3+2\cdot0=9$

The minimum value is $C=4$, and it will occur at the point $\left(0,2\right)$.

Cumulative Review

17. Using the point-slope formula.

$$y-\left(-5\right)=4\left(x-2\right)$$
$$y+5=4x-8$$
$$y=4x-13$$

19. The augmented matrix to the system is:

$$\begin{bmatrix} 2 & 3 & -5 & | & 0 \\ 4 & 2 & 1 & | & 7 \\ 1 & -3 & 0 & | & -2 \end{bmatrix}$$

Switch row 1 and row 3

$$\begin{bmatrix} 1 & -3 & 0 & | & -2 \\ 4 & 2 & 1 & | & 7 \\ 2 & 3 & -5 & | & 0 \end{bmatrix}$$

Now pivot on row 1 column 1 by performing the operations:

$$-4R_1+R_2\rightarrow R_2$$
$$-2R_1+R_3\rightarrow R_3$$

$$\begin{bmatrix} 1 & -3 & 0 & | & -2 \\ 0 & 14 & 1 & | & 15 \\ 0 & 9 & -5 & | & 4 \end{bmatrix}$$

Next pivot on row 2 column 2 by performing the operations at the top of the next page.

$\frac{1}{14} R_2 \to R_2$

$3R_2 + R_1 \to R_1$

$-9R_2 + R_3 \to R_3$

$$\begin{bmatrix} 1 & 0 & \frac{3}{14} & | & \frac{17}{14} \\ 0 & 1 & \frac{1}{14} & | & \frac{15}{14} \\ 0 & 0 & \frac{-79}{14} & | & \frac{-79}{14} \end{bmatrix}$$

Pivot on row 3 column 3 by performing the operations:

$\frac{-14}{79} R_3 \to R_3$

$\frac{-3}{14} R_3 + R_1 \to R_1$

$\frac{-1}{14} R_3 + R_2 \to R_2$

$$\begin{bmatrix} 1 & 0 & 0 & | & 1 \\ 0 & 1 & 0 & | & 1 \\ 0 & 0 & 1 & | & 1 \end{bmatrix}$$

The solution to the system is:

$x = 1; \ y = 1; \ z = 1$

21.

$$A^T = \begin{bmatrix} 1 & 2 & 3 \\ 3 & -4 & 2 \\ 0 & 1 & -2 \end{bmatrix}$$

Chapter 4 Sample Test Answers

1. The graph is shown below:

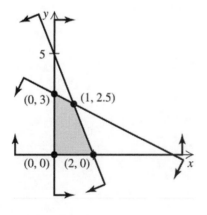

2. The graph is shown below:

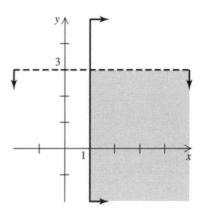

3. The corner points of the feasible region are:

$(0,0), (4,0), (3,2)$, and $\left(0, \frac{13}{5}\right)$

4. The corner points of the feasible region are:

$(25,0)$ and $(0,50)$

5. The graph of the feasible region is shown below:

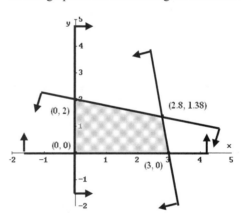

The corner points are:

$(0,0), (0,2), (3,0)$ and $\left(\frac{171}{61}, \frac{84}{61}\right)$

Point	$f = 2x + 5y$
$(0,0)$	$2 \cdot 0 + 5 \cdot 0 = 0$
$(0,2)$	$2 \cdot 0 + 5 \cdot 2 = 10$
$(3,0)$	$2 \cdot 3 + 5 \cdot 0 = 6$
$\left(\frac{171}{61}, \frac{84}{61}\right)$	$2 \cdot \frac{171}{61} + 5 \cdot \frac{84}{61} = \frac{762}{61} \approx 12.49$

The function f has a maximum value of approximately 12.49 a the point $\left(\frac{171}{61}, \frac{84}{61}\right)$.

6. The feasible region is shown below:

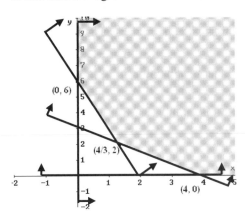

Maximize $I = 20x + 15y$

subject to:

$$30x + 15y \leq 1440$$
$$x + y \geq 30$$
$$x \geq 0, y \geq 0$$

By solving this problem we will see that Javier will earn a maximum commission of $1,440 by selling 96 unfinished bookcases a week, and zero finished bookcases.

It is clear to see that the region is not bounded. Moreover, g does not reach a minimum value over this region. The larger the value of y, the smaller the value of g.

7. The linear programming problem can be stated as:

Maximize $S = 3x + 5y$

subject to:

$$x + y \leq 36$$
$$x \geq 10$$
$$x \leq 30$$
$$y \geq 6$$
$$y \leq 20$$

The graph of the feasible region is shown below

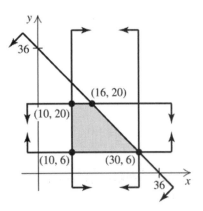

A maximum score of 148 can be achieved by answering correctly 16 true-false questions and 20 multiple choice questions.

8. Let x be the number of finished bookcases Javier sells, and y be the number of unfinished bookcases he sells. The linear programming problem is displayed at the top of the next column.

Chapter 5
Linear Programming: The Simplex Method

Exercises Section 5.1

1. The constraint is in standard form.

3. The constraint is in standard form.

5. The constraint is not in standard form.

7. The constraint is not in standard form.

In questions 9-13, we will use the following equation.

By inserting the slack variable into the inequality we get the equation:

$$x + 3y + s_1 = 9$$

9. Substitute $x = 3$ and $y = 2$ into the equation and solve for s_1.

$$(3) + 3(2) + s_1 = 9$$
$$s_1 = 0$$

11. Substitute $x = 0$ and $y = 0$ into the equation and solve for s_1.

$$(0) + 3(0) + s_1 = 9$$
$$s_1 = 9$$

13. If the slack variable is negative, and the point (x, y) satisfies the equation $x + 3y + s_1 = 9$, then the point (x, y) must lie in the half-plane satisfying the inequality $x + 3y \geq 9$ or above and to the right of the equation $x + 3y = 9$.

In questions 15-18, we will use the following equation.

By inserting the slack variable into the inequality we get the equation:

$$2x + y + s_2 = 8$$

15. Substitute $x = -2$ and $y = 7$ into the equation and solve for s_2.

$$2(-2) + (7) + s_2 = 8$$
$$s_2 = 5$$

17. The largest the slack variable can be is eight if $x = 0$ and $y = 0$. The smallest the slack variable can be is zero, so the range of the slack variable is:

$$0 \leq s_2 \leq 8$$

19. The solution to the system in which all of the non-basic variables are zero is:

$$x = 0, \ y = 2, \ s_1 = 7, \ s_2 = 0.$$

All of the variables are non-negative, so this solution is a corner point of the feasible region for the problem.

21. The solution to the system in which all of the non-basic variables are zero is:

$$x = 0, \ y = 5, \ s_1 = 0, \ s_2 = -2.$$

One of the variables is negative, so this solution is not a corner point of the feasible region for the problem.

23. The solution to the system in which all of the non-basic variables are zero is:

$$x = 0, \ y = 2, \ z = 1, \ s_1 = 0, \ s_2 = 0, \ s_3 = 5.$$

All of the variables are non-negative, so this solution is a corner point of the feasible region for the problem.

25. The system of slack equations is:

$$\begin{aligned} x + y + s_1 \quad &= 9 \\ -x + 2y \quad + s_2 &= 12 \end{aligned}$$

27. The initial augmented matrix is:

$$\begin{array}{cccc} x & y & s_1 & s_2 \end{array}$$
$$\left[\begin{array}{cccc|c} 1 & 1 & 1 & 0 & 9 \\ -1 & 2 & 0 & 1 & 12 \end{array} \right]$$

There is only one positive coefficient in the x column, so the pivot element must be the one in row 1, column 1.

The solution is continued on the next page.

If we were to pivot on the other element in column 1, we would see the resulting matrix:

$$\begin{array}{cccc} x & y & s_1 & s_2 \end{array}$$
$$\begin{bmatrix} 0 & 3 & 1 & 1 & | & 21 \\ 1 & -2 & 0 & -1 & | & -12 \end{bmatrix}$$

The negative constant on the right of row 2 indicates that the solution is not a basic feasible solution.

29. Pivoting on the proper pivot element (element 1, 1) in the x column of the initial matrix gives us the matrix:

$$\begin{array}{cccc} x & y & s_1 & s_2 \end{array}$$
$$\begin{bmatrix} 1 & 1 & 1 & 0 & | & 9 \\ 0 & 3 & 1 & 1 & | & 21 \end{bmatrix}$$

This results in the basic feasible solution:

$$x = 9,\ y = 0,\ s_1 = 0,\ s_2 = 21$$

Dividing each constant by the appropriate element in the y column we get:

$$R_1: \quad \tfrac{9}{1} = 9$$
$$R_2: \quad \tfrac{21}{3} = 7$$

Therefore the 3 in row 2, column 2 is the pivot element. Pivoting on this element we get the matrix:

$$\begin{array}{cccc} x & y & s_1 & s_2 \end{array}$$
$$\begin{bmatrix} 1 & 0 & \tfrac{2}{3} & \tfrac{-1}{3} & | & 2 \\ 0 & 1 & \tfrac{1}{3} & \tfrac{1}{3} & | & 7 \end{bmatrix}$$

This results in the basic feasible solution

$$x = 2,\ y = 7,\ s_1 = 0,\ s_2 = 0.$$

31. The system of slack equations is:

$$\begin{aligned} x + y + s_1 &= 10 \\ 2x - 2y + s_2 &= 8 \end{aligned}$$

33. The initial augmented matrix is displayed at the top of the next column.

$$\begin{array}{cccc} x & y & s_1 & s_2 \end{array}$$
$$\begin{bmatrix} 1 & 1 & 1 & 0 & | & 10 \\ 2 & -2 & 0 & 1 & | & 8 \end{bmatrix}$$

Dividing the respective constant term by each of the elements in column 1 we see

$$R_1: \quad \tfrac{10}{1} = 10$$
$$R_2: \quad \tfrac{8}{2} = 4$$

The pivot element in the x column is the two in row 2, column 1.

If we were to pivot on the other element in column 1, we would see the resulting matrix:

$$\begin{array}{cccc} x & y & s_1 & s_2 \end{array}$$
$$\begin{bmatrix} 1 & 1 & 1 & 0 & | & 10 \\ 0 & -4 & -2 & 1 & | & -12 \end{bmatrix}$$

The negative constant on the right of row 2 indicates that the solution is not a basic feasible solution.

35. The initial augmented matrix is:

$$\begin{array}{cccc} x & y & s_1 & s_2 \end{array}$$
$$\begin{bmatrix} 1 & 1 & 1 & 0 & | & 10 \\ 2 & -2 & 0 & 1 & | & 8 \end{bmatrix}$$

Pivoting on the proper pivot element (element 2, 1) in the x column of the initial matrix gives us the matrix:

$$\begin{array}{cccc} x & y & s_1 & s_2 \end{array}$$
$$\begin{bmatrix} 0 & 2 & 1 & \tfrac{-1}{2} & | & 6 \\ 1 & -1 & 0 & \tfrac{1}{2} & | & 4 \end{bmatrix}$$

This results in the basic feasible solution:

$$x = 4,\ y = 0,\ s_1 = 6,\ s_2 = 0.$$

Since there is only one positive element in column 2, the pivot element must be the element 2 in row 1, column 2. Pivoting on this element we get the matrix displayed at the top of the next page.

$$x \quad y \quad s_1 \quad s_2$$

$$\begin{bmatrix} 0 & 1 & \frac{1}{2} & \frac{-1}{4} & 3 \\ 1 & 0 & \frac{1}{2} & \frac{1}{4} & 7 \end{bmatrix}$$

This results in the basic feasible solution

$$x = 7, \; y = 3, \; s_1 = 0, \; s_2 = 0.$$

37. The system of slack equations is:

$$x + \;\; y + s_1 \quad\quad = 8$$
$$x + 2y \quad\quad + s_2 = 10$$

39. The initial augmented matrix is:

$$x \quad y \quad s_1 \quad s_2$$

$$\begin{bmatrix} 1 & 1 & 1 & 0 & 8 \\ 1 & 2 & 0 & 1 & 10 \end{bmatrix}$$

Dividing the respective constant term by each of the elements in column 1 we see

$$R_1 : \; \tfrac{8}{1} = 8$$
$$R_2 : \; \tfrac{10}{1} = 10$$

The pivot element in the x column is the one in row 1, column 1.

If we were to pivot on the other element in column 1, we would see the resulting matrix:

$$x \quad y \quad s_1 \quad s_2$$

$$\begin{bmatrix} 0 & -1 & 1 & -1 & -2 \\ 1 & 2 & 0 & 1 & 10 \end{bmatrix}$$

The negative constant on the right of row 1 indicates that the solution is not a basic feasible solution.

41. The initial augmented matrix is:

$$x \quad y \quad s_1 \quad s_2$$

$$\begin{bmatrix} 1 & 1 & 1 & 0 & 8 \\ 1 & 2 & 0 & 1 & 10 \end{bmatrix}$$

Pivoting on the proper pivot element (element 2, 1) in the x column of the initial matrix gives us the matrix displayed at the top of the next column.

$$x \quad y \quad s_1 \quad s_2$$

$$\begin{bmatrix} 1 & 1 & 1 & 0 & 8 \\ 0 & 1 & -1 & 1 & 2 \end{bmatrix}$$

This results in the basic feasible solution:

$$x = 8, \; y = 0, \; s_1 = 0, \; s_2 = 2.$$

Dividing the respective constant term by each of the elements in column 2 we see

$$R_1 : \; \tfrac{8}{1} = 8$$
$$R_2 : \; \tfrac{2}{1} = 2$$

The pivot element is the one in row 2, column 2. Pivoting on this element we get the matrix:

$$x \quad y \quad s_1 \quad s_2$$

$$\begin{bmatrix} 1 & 0 & 2 & -1 & 6 \\ 0 & 1 & -1 & 1 & 2 \end{bmatrix}$$

This results in the basic feasible solution

$$x = 6, \; y = 2, \; s_1 = 0, \; s_2 = 0.$$

43. The graph of the feasible region to the given system is:

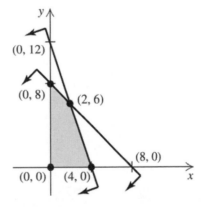

Adding slack variables to the inequalities results in the system of equations:

$$x + y + s_1 \quad\quad = 8$$
$$3x + y \quad\quad + s_2 = 12$$

The augmented matrix to the system is displayed at the top of the next page.

$$\begin{array}{cccc} x & y & s_1 & s_2 \end{array}$$

$$\left[\begin{array}{cccc|c} 1 & 1 & 1 & 0 & 8 \\ 3 & 1 & 0 & 1 & 12 \end{array}\right]$$

The basic feasible solution that this matrix represents is:

$x = 0,\ y = 0,\ s_1 = 8,\ s_2 = 12$.

For the clockwise direction the next point we wish to get to the point $x = 0, y = 8$.

To do this we need to pivot on the y-column. Using the smallest quotient method we see:

$R_1:\ \frac{8}{1} = 8$

$R_2:\ \frac{12}{1} = 12$

Since row 1 has the smallest quotient, the pivot element in column 2 is the one in row 1, column 2.

Pivoting on this element results in the matrix:

$$\begin{array}{cccc} x & y & s_1 & s_2 \end{array}$$

$$\left[\begin{array}{cccc|c} 1 & 1 & 1 & 0 & 8 \\ 2 & 0 & -1 & 1 & 4 \end{array}\right]$$

This matrix represents the basic feasible solution:

$x = 0,\ y = 8,\ s_1 = 0,\ s_2 = 4$.

The next corner point we wish to proceed to is the point $(2,6)$. To reach this point we need to pivot on the x column.

Using the smallest quotient method we see:

$R_1:\ \frac{8}{1} = 8$

$R_2:\ \frac{4}{2} = 2$

Since row 2 has the smallest quotient, the pivot element in column 1 is the two in row 2, column 1.

Pivoting on this element results in the matrix:

$$\begin{array}{cccc} x & y & s_1 & s_2 \end{array}$$

$$\left[\begin{array}{cccc|c} 0 & 1 & \frac{3}{2} & \frac{-1}{2} & 6 \\ 1 & 0 & \frac{-1}{2} & \frac{1}{2} & 2 \end{array}\right]$$

This matrix represents the basic feasible solution:

$x = 2,\ y = 6,\ s_1 = 0,\ s_2 = 0$.

The next corner point we wish to proceed to is the point $(4,0)$. To reach this point we need to pivot on the s_1 column.

Since row 1 in column 3 is the only positive entry the smallest quotient, the pivot element in column 3 is the $\frac{3}{2}$ in row 1 column 3.

Pivoting on this element results in the matrix:

$$\begin{array}{cccc} x & y & s_1 & s_2 \end{array}$$

$$\left[\begin{array}{cccc|c} 0 & \frac{2}{3} & 1 & \frac{-1}{3} & 4 \\ 1 & \frac{1}{3} & 0 & \frac{1}{3} & 4 \end{array}\right]$$

This matrix represents the basic feasible solution.

$x = 4,\ y = 0,\ s_1 = 4,\ s_2 = 0$.

The next corner point we wish to proceed to is the point $(0,0)$. To reach this point we need to pivot on the s_2 column.

Since row 2 in column 4 is the only positive entry, the pivot element in column 4 is the $\frac{1}{3}$ in row 2, column 4.

Pivoting on this element results in the matrix:

$$\begin{array}{cccc} x & y & s_1 & s_2 \end{array}$$

$$\left[\begin{array}{cccc|c} 1 & 1 & 1 & 0 & 8 \\ 3 & 1 & 0 & 1 & 12 \end{array}\right]$$

This matrix represents the basic feasible solution:

$x = 0,\ y = 0,\ s_1 = 8,\ s_2 = 12$.

To move in the counter clockwise direction, we simply reverse the procedure, first pivot on the x column to go to the corner point $(4,0)$.

Using the smallest quotient method we see:

$R_1:\ \frac{8}{1} = 8$

$R_2:\ \frac{12}{3} = 4$

The solution is continued on the next page.

From the previous page we see that row 2 has the smallest quotient, the pivot element in column 1 is the three in row 2, column 1.

Pivoting on this element results in the matrix:

$$\begin{array}{cccc} x & y & s_1 & s_2 \end{array}$$
$$\begin{bmatrix} 0 & \frac{2}{3} & 1 & \frac{-1}{3} & 4 \\ 1 & \frac{1}{3} & 0 & \frac{1}{3} & 4 \end{bmatrix}$$

This matrix represents the basic feasible solution.

$x = 4, \ y = 0, \ s_1 = 4, \ s_2 = 0$.

The next corner point we wish to proceed to is the point $(2,6)$. To reach this point we need to pivot on the y column.

Using the smallest quotient method we see:

$$R_1 : \ \frac{4}{\frac{2}{3}} = 6$$

$$R_2 : \ \frac{4}{\frac{1}{3}} = 12$$

Since row 1 has the smallest quotient, the pivot element in column 2 is the $\frac{2}{3}$ in row 1, column 2.

Pivoting on this element results in the matrix:

$$\begin{array}{cccc} x & y & s_1 & s_2 \end{array}$$
$$\begin{bmatrix} 0 & 1 & \frac{3}{2} & \frac{-1}{2} & 6 \\ 1 & 0 & \frac{-1}{2} & \frac{1}{2} & 2 \end{bmatrix}$$

This matrix represents the basic feasible solution:

$x = 2, \ y = 6, \ s_1 = 0, \ s_2 = 0$.

The next corner point we wish to proceed to is the point $(0,8)$. To reach this point we need to pivot on the s_2 column.

Since row 2 in column 4 is the only positive entry, the pivot element in column 4 is the $\frac{1}{2}$ in row 2, column 4.

Pivoting on this element results in the matrix at the top of the next column.

$$\begin{array}{cccc} x & y & s_1 & s_2 \end{array}$$
$$\begin{bmatrix} 1 & 1 & 1 & 0 & 8 \\ 2 & 0 & -1 & 1 & 4 \end{bmatrix}$$

This matrix represents the basic feasible solution:

$x = 0, \ y = 8, \ s_1 = 0, \ s_2 = 4$.

The next corner point we wish to proceed to is the point $(0,0)$. To reach this point we need to pivot on the s_1 column.

Since row 1 in column 3 is the only positive entry, the pivot element in column 3 is the one in row 1 column 3.

Pivoting on this element results in the matrix:

$$\begin{array}{cccc} x & y & s_1 & s_2 \end{array}$$
$$\begin{bmatrix} 1 & 1 & 1 & 0 & 8 \\ 3 & 1 & 0 & 1 & 12 \end{bmatrix}$$

This matrix represents the basic feasible solution:

$x = 0, \ y = 0, \ s_1 = 8, \ s_2 = 12$.

45.
a. The set of constraints for the problem is:

$$\begin{aligned} x + \ y + s_1 \quad &= \ 6 \\ 7x + 3y \quad + s_2 &= 21 \end{aligned}$$

b. The graph of the feasible region is shown below:

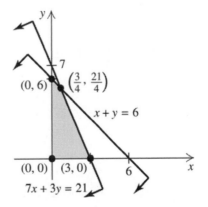

c. Using the smallest quotient method we see:

$R_1: \quad \frac{6}{1} = 6$

$R_2: \quad \frac{21}{7} = 3$

Since row 2 has the smallest quotient, the pivot element is the seven in row2 column 1.

Pivoting on this element results in the matrix

$$
\begin{array}{cccc}
x & y & s_1 & s_2 \\
\end{array}
$$
$$
\begin{bmatrix}
0 & \frac{4}{7} & 1 & \frac{-1}{7} & 3 \\
1 & \frac{3}{7} & 0 & \frac{1}{7} & 3
\end{bmatrix}
$$

This matrix represents the basic feasible solution:

$x = 3, \; y = 0, \; s_1 = 3, \; s_2 = 0$

d. Using the smallest quotient method we see:

$R_1: \quad \frac{6}{1} = 6$

$R_2: \quad \frac{21}{3} = 7$

Since row 1 has the smallest quotient, the pivot element is the one in row 1, column 2.

Pivoting on this element results in the matrix

$$
\begin{array}{cccc}
x & y & s_1 & s_2 \\
\end{array}
$$
$$
\begin{bmatrix}
1 & 1 & 1 & 0 & 6 \\
4 & 0 & -3 & 1 & 3
\end{bmatrix}
$$

This matrix represents the basic feasible solution:

$x = 0, \; y = 6, \; s_1 = 0, \; s_2 = 3$

e. First perform the pivot operation completed in step (d). Next, we will pivot on the x column.

Using the smallest quotient method we see:

$R_1: \quad \frac{6}{1} = 6$

$R_2: \quad \frac{3}{4} = \frac{3}{4}$

Since row 2 has the smallest quotient, the pivot element is the four in row 2, column 1.

Pivoting on this element results in the matrix shown at the top of the next column.

$$
\begin{array}{cccc}
x & y & s_1 & s_2 \\
\end{array}
$$
$$
\begin{bmatrix}
0 & 1 & \frac{7}{4} & \frac{-1}{4} & \frac{21}{4} \\
1 & 0 & \frac{-3}{4} & \frac{1}{4} & \frac{3}{4}
\end{bmatrix}
$$

This matrix represents the basic feasible solution:

$x = \frac{3}{4}, \; y = \frac{21}{4}, \; s_1 = 0, \; s_2 = 0$.

Exercises Section 5.2

1.
a. The tableau shows that a maximum has been obtained because all of the entries in the bottom row are non-negative.

b. The maximum value of f is 20 and is obtained when $x = 5$, and $y = 8$.

c. No surplus exists for any of the constraints.

3.
a. The tableau shows that a maximum has been obtained because all of the entries in the bottom row are non-negative.

b. The maximum value of f is $\frac{80}{5} = 16$ and is obtained when $x = \frac{8}{5}$, and $y = 0$.

c. At this point the second slack variable is equal to 16. This means that a surplus of 16 exists for the second constraint.

5.
a. The tableau shows a value of $\frac{-2}{5}$ in the last row of the y column. Thus, the objective function has not attained a maximum.

b. The value of f is 25 and is obtained when $x = 6$, $y = 0$.

c. At this point the first slack variable is equal to eight, and the second slack variable is equal to two. This means a surplus of eight exists for the first constraint, and a surplus of two exists for the second constraint.

7.

a. The feasible region is shown below:

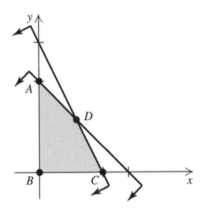

b. The value of the variables at the corner points are:

Point	x	y	f	s_1	s_2
A	0	7	42	3	0
B	0	0	0	10	7
C	5	0	40	0	2
D	3	4	48	0	0

c. The initial simplex tableau is:

$$
\begin{array}{ccccc}
x & y & s_1 & s_2 & f
\end{array}
$$
$$
\left[\begin{array}{ccccc|c}
2 & 1 & 1 & 0 & 0 & 10 \\
1 & 1 & 0 & 1 & 0 & 7 \\
\hline
-8 & -6 & 0 & 0 & 1 & 0
\end{array}\right]
$$

The basic feasible solution of the initial simplex tableau is the solution that corresponds to corner point D in part (b).

The most negative element in the last row of the tableau is negative eight. Therefore column 1 is the pivot column. Using the smallest quotient rule, we see:

$R_1: \ \frac{10}{2} = 5$

$R_2: \ \frac{7}{1} = 7$

The pivot element for the first pivot will be the 2 in row 1, column 1. Pivoting on this element results in the following tableau:

$$
\begin{array}{ccccc}
x & y & s_1 & s_2 & f
\end{array}
$$
$$
\left[\begin{array}{ccccc|c}
1 & \frac{1}{2} & \frac{1}{2} & 0 & 0 & 5 \\
0 & \frac{1}{2} & \frac{-1}{2} & 1 & 0 & 2 \\
\hline
0 & -2 & 4 & 0 & 1 & 40
\end{array}\right]
$$

The basic feasible solution of this tableau is the solution that corresponds to corner point C from part (b).

Since the negative two in column 2 is the only negative element in the last row, column 2 is the pivot column. The smallest quotient rule shows:

$R_1: \ \dfrac{5}{\frac{1}{2}} = 10$

$R_2: \ \dfrac{2}{\frac{1}{2}} = 4$

Therefore, the pivot element is the $\frac{1}{2}$ in row 2, column 2. Pivoting on this element results in the simplex tableau:

$$
\begin{array}{ccccc}
x & y & s_1 & s_2 & f
\end{array}
$$
$$
\left[\begin{array}{ccccc|c}
1 & 0 & 1 & -1 & 0 & 3 \\
0 & 1 & -1 & 2 & 0 & 4 \\
\hline
0 & 0 & 2 & 4 & 1 & 48
\end{array}\right]
$$

The basic feasible solution of this tableau is the solution that corresponds to corner point D from part (b).

All of the entries in the last row are positive. The basic feasible solution that solves the linear programming problem; i.e. the solution that results in the maximum for the objective function f is:

$$x = 3, \ y = 4, \ f = 48, \ s_1 = 0, \ s_2 = 0$$

9.

a. The feasible region is shown below:

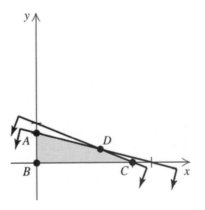

The solution is continued on the next page.

b. The value of the variables at the corner points of the feasible region on the previous page are:

Point	x	y	f	s_1	s_2
A	0	3	12	5	0
B	0	0	0	20	12
C	10	0	10	0	2
D	$\frac{20}{3}$	$\frac{4}{3}$	12	0	0

c. The initial simplex tableau is:

$$
\begin{array}{ccccc}
x & y & s_1 & s_2 & f
\end{array}
$$
$$
\left[
\begin{array}{ccccc|c}
2 & 5 & 1 & 0 & 0 & 20 \\
1 & 4 & 0 & 1 & 0 & 12 \\
\hline
-1 & -4 & 0 & 0 & 1 & 0
\end{array}
\right]
$$

The basic feasible solution of the initial simplex tableau is the solution that corresponds to corner point B in part (b).

The most negative element in the last row of the tableau is negative four. Therefore column 2 is the pivot column. The smallest quotient rule shows:

$$R_1 : \quad \frac{20}{5} = 4$$

$$R_2 : \quad \frac{12}{4} = 3$$

The pivot element for the first pivot will be the four in row 2, column 2. Pivoting on this element results in the following tableau:

$$
\begin{array}{ccccc}
x & y & s_1 & s_2 & f
\end{array}
$$
$$
\left[
\begin{array}{ccccc|c}
\frac{3}{4} & 0 & 1 & \frac{-5}{4} & 0 & 5 \\
\frac{1}{4} & 1 & 0 & \frac{1}{4} & 0 & 3 \\
\hline
0 & 0 & 0 & 1 & 1 & 12
\end{array}
\right]
$$

The basic feasible solution of this tableau is the solution that corresponds to corner point A from part (b).

All of the entries in the last row are positive. The basic feasible solution that solves the linear programming problem; i.e. the solution that results in the maximum value for the objective function f over the feasible region is:

$$x = 0, \ y = 3, \ f = 12, \ s_1 = 5, \ s_2 = 0$$

Notice from part (b) that corner point D also results in a solution that has function value $f = 12$, in fact all of the values along the boundary line that connects points A and D will result in a function value of 12.

11.
a. The feasible region is shown below:

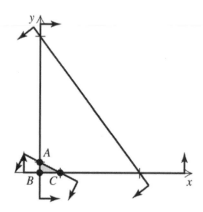

b. The value of the variables at the corner points are:

Point	x	y	f	s_1	s_2
A	0	3	3	0	111
B	0	0	0	6	120
C	6	0	30	0	96

c. The initial simplex tableau is:

$$
\begin{array}{ccccc}
x & y & s_1 & s_2 & f
\end{array}
$$
$$
\left[
\begin{array}{ccccc|c}
1 & 2 & 1 & 0 & 0 & 6 \\
4 & 3 & 0 & 1 & 0 & 120 \\
\hline
-5 & -1 & 0 & 0 & 1 & 0
\end{array}
\right]
$$

The basic feasible solution of the initial simplex tableau is the solution that corresponds to corner point B in part (b).

The most negative element in the last row of the tableau is negative five. Therefore column 1 is the pivot column. The smallest quotient rule shows:

$$R_1 : \quad \frac{6}{1} = 6$$

$$R_2 : \quad \frac{120}{4} = 30$$

The pivot element for the first pivot will be the one in row 1, column 1. Pivoting on this element results in the tableau at the top of the next page.

$$\begin{array}{ccccc} x & y & s_1 & s_2 & f \end{array}$$

$$\left[\begin{array}{ccccc|c} 1 & 2 & 1 & 0 & 0 & 6 \\ 0 & -5 & -4 & 1 & 0 & 96 \\ \hline 0 & 9 & 5 & 0 & 1 & 30 \end{array}\right]$$

The basic feasible solution of this tableau is the solution that corresponds to corner point C from part (b).

All of the entries in the last row are positive. The basic feasible solution that solves the linear programming problem; i.e. the solution that results in the maximum value for the objective function f over the feasible region is:

$$x = 6, \ y = 0, \ f = 30, \ s_1 = 0, \ s_2 = 96$$

13.
a. The feasible region is shown below:

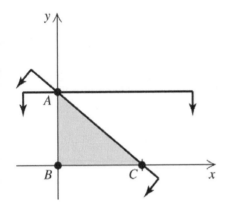

b. The value of the variables at the corner points are:

Point	x	y	f	s_1	s_2
A	0	5	5	0	0
B	0	0	0	5	5
C	5	0	10	0	5

c. The initial simplex tableau is:

$$\begin{array}{ccccc} x & y & s_1 & s_2 & f \end{array}$$

$$\left[\begin{array}{ccccc|c} 1 & 1 & 1 & 0 & 0 & 5 \\ 0 & 1 & 0 & 1 & 0 & 5 \\ \hline -2 & -1 & 0 & 0 & 1 & 0 \end{array}\right]$$

The basic feasible solution of the initial simplex tableau is the solution that corresponds to corner point B in part (b).

The most negative element in the last row of the tableau is negative two. Therefore column 1 is the pivot column.

Since the only positive element in column 1 is in row 1, the one in row 1, column 1 will be the pivot element.

Pivoting on this element results in the tableau:

$$\begin{array}{ccccc} x & y & s_1 & s_2 & f \end{array}$$

$$\left[\begin{array}{ccccc|c} 1 & 1 & 1 & 0 & 0 & 5 \\ 0 & 1 & 0 & 1 & 0 & 5 \\ \hline 0 & 1 & 2 & 0 & 1 & 10 \end{array}\right]$$

The basic feasible solution of this tableau is the solution that corresponds to corner point C from part (b).

All of the entries in the last row are positive. The basic feasible solution that solves the linear programming problem; i.e. the solution that results in the maximum value for the objective function f over the feasible region is:

$$x = 5, \ y = 0, \ f = 10, \ s_1 = 0, \ s_2 = 5$$

15.
a. The feasible region is:

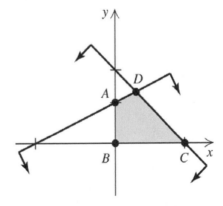

b. The value of the variables at the corner points are:

Point	x	y	f	s_1	s_2
A	0	5	15	4	0
B	0	0	0	9	10
C	9	0	27	0	19
D	$\frac{8}{3}$	$\frac{19}{3}$	27	0	0

c. The initial simplex tableau is written at the top of the next page.

$$\begin{array}{ccccc} x & y & s_1 & s_2 & f \\ \begin{bmatrix} 1 & 1 & 1 & 0 & 0 & | & 9 \\ -1 & 2 & 0 & 1 & 0 & | & 10 \\ -3 & -3 & 0 & 0 & 1 & | & 0 \end{bmatrix} \end{array}$$

The basic feasible solution of the initial simplex tableau is the solution that corresponds to corner point B in part (b).

The most negative element in the last row of the tableau is negative three. This means we can choose column 1 or column 2 as the pivot column. We choose column 1 since there is only one positive entry. Thus the pivot element is the one in row 1, column 1. Pivoting on this element results in the following tableau:

$$\begin{array}{ccccc} x & y & s_1 & s_2 & f \\ \begin{bmatrix} 1 & 1 & 1 & 0 & 0 & | & 9 \\ 0 & 3 & 1 & 1 & 0 & | & 19 \\ 0 & 0 & 3 & 0 & 1 & | & 27 \end{bmatrix} \end{array}$$

The basic feasible solution of this tableau is the solution that corresponds to corner point C from part (b).

All of the entries in the last row are positive. The basic feasible solution that solves the linear programming problem; i.e. the solution that results in the maximum value for the objective function f over the feasible region is:

$$x = 9, \ y = 0, \ f = 27, \ s_1 = 0, \ s_2 = 19$$

Notice from part (b) that corner point D also results in a solution that has function value $f = 27$, in fact all of the values along the boundary line that connects points C and D will result in a function value of 27.

17.
a. The feasible region is shown below:

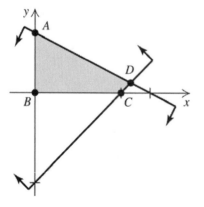

b. The value of the variables at the corner points are:

Point	x	y	f	s_1	s_2
A	0	4	-4	10	0
B	0	0	0	6	8
C	6	0	-12	0	2
D	$\frac{20}{3}$	$\frac{2}{3}$	-14	0	0

c. The initial simplex tableau is:

$$\begin{array}{ccccc} x & y & s_1 & s_2 & f \\ \begin{bmatrix} 1 & -1 & 1 & 0 & 0 & | & 6 \\ 1 & 2 & 0 & 1 & 0 & | & 8 \\ 2 & 1 & 0 & 0 & 1 & | & 0 \end{bmatrix} \end{array}$$

The basic feasible solution of the initial simplex tableau on the previous page is the solution that corresponds to corner point B in part (b).

All of the entries in the last row of the initial simplex tableau are positive. This means we are already at a maximum. The basic feasible solution that solves the linear programming problem; i.e. the solution that results in the maximum value for the objective function f over the feasible region is:

$$x = 0, \ y = 0, \ f = 0, \ s_1 = 6, \ s_2 = 8$$

19.
a. The feasible region is displayed below:

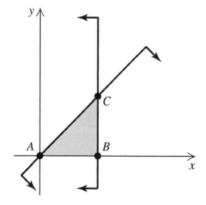

b. The value of the variables at the corner points are:

Point	x	y	f	s_1	s_2
A	0	0	0	0	2
B	2	0	6	2	0
C	2	2	8	0	0

c. In order to use the simplex method, we must convert the first constraint into the standard maximizing form. To do this simply subtract x from both sides of the inequality to get the new constraint:

$$-x + y \leq 0.$$

The initial simplex tableau is:

$$
\begin{array}{ccccc}
x & y & s_1 & s_2 & f \\
\end{array}
$$
$$
\left[
\begin{array}{ccccc|c}
-1 & 1 & 1 & 0 & 0 & 0 \\
1 & 0 & 0 & 1 & 0 & 2 \\
\hline
-3 & -1 & 0 & 0 & 1 & 0 \\
\end{array}
\right]
$$

The basic feasible solution of the initial simplex tableau is the solution that corresponds to corner point A in part (b).

The most negative element in the last row of the tableau is negative three. Therefore column 1 is the pivot column.

The only positive element in column 1 is in row 2; therefore, the pivot element for the first pivot will be the one in row 2, column 1. Pivoting on this element results in the following tableau:

$$
\begin{array}{ccccc}
x & y & s_1 & s_2 & f \\
\end{array}
$$
$$
\left[
\begin{array}{ccccc|c}
0 & 1 & 1 & 1 & 0 & 2 \\
1 & 0 & 0 & 1 & 0 & 2 \\
\hline
0 & -1 & 0 & 3 & 1 & 6 \\
\end{array}
\right]
$$

The basic feasible solution of this tableau is the solution that corresponds to corner point B from part (b).

Since the negative one in column 2 is the only negative element in the last row, column 2 is the pivot column.

Since the one in row 1 is the only positive element in column 2 this will be the pivot element.

Pivoting on this element results in the simplex tableau:

$$
\begin{array}{ccccc}
x & y & s_1 & s_2 & f \\
\end{array}
$$
$$
\left[
\begin{array}{ccccc|c}
0 & 1 & 1 & 1 & 0 & 2 \\
1 & 0 & 0 & 1 & 0 & 2 \\
\hline
0 & 0 & 1 & 4 & 1 & 8 \\
\end{array}
\right]
$$

The basic feasible solution of this tableau is the solution that corresponds to corner point C from part (b).

All of the entries in the last row of the previous tableau are positive. The basic feasible solution that solves the linear programming problem; i.e. the solution that results in the maximum for the objective function f is:

$$x = 2, \; y = 2, \; f = 8, \; s_1 = 0, \; s_2 = 0$$

21.
a. The feasible region is:

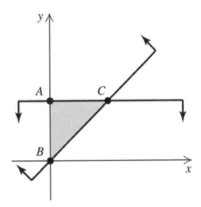

b. The value of the variables at the corner points are:

Point	x	y	f	s_1	s_2
A	0	3	6	3	0
B	0	0	0	0	3
C	3	3	21	0	0

c. In order to use the simplex method, we must convert the first constraint into the standard maximizing form. To do this simply subtract y from both sides of the inequality to get the new constraint:

$$x - y \leq 0.$$

The initial simplex tableau is displayed below:

$$
\begin{array}{ccccc}
x & y & s_1 & s_2 & f \\
\end{array}
$$
$$
\left[
\begin{array}{ccccc|c}
1 & -1 & 1 & 0 & 0 & 0 \\
0 & 1 & 0 & 1 & 0 & 3 \\
\hline
-5 & -2 & 0 & 0 & 1 & 0 \\
\end{array}
\right]
$$

The basic feasible solution of the initial simplex tableau is the solution that corresponds to corner point B in part (b).

The most negative element in the last row of the tableau is negative five. Therefore column 1 is the pivot column.

The solution is continued on the next page.

The only positive element in column 1 is in row 1; therefore, the pivot element for the first pivot will be the one in row 1, column 1. Pivoting on this element results in the following tableau:

$$x \quad y \quad s_1 \quad s_2 \quad f$$

$$\begin{bmatrix} 1 & -1 & 1 & 0 & 0 & | & 0 \\ 0 & 1 & 0 & 1 & 0 & | & 3 \\ \hline 0 & -7 & 5 & 0 & 1 & | & 0 \end{bmatrix}$$

The basic feasible solution of this tableau is the solution that corresponds to corner point B from part (b).

It appears that we have not changed points geometrically since the solution is still at the origin, but we have change basic and non-basic variables. The variable x has entered as a basic variable, and s_1 has exited. The reason for this anomaly is due to the fact that the boundary lines cross at the origin, so one of the initial slack variables already had value zero.

Since the negative seven in column 2 is the only negative element in the last row, column 2 is the new pivot column.

Since the one in row 2 is the only positive element in column 2 this will be the pivot element.

Pivoting on this element results in the simplex tableau:

$$x \quad y \quad s_1 \quad s_2 \quad f$$

$$\begin{bmatrix} 1 & 0 & 1 & 1 & 0 & | & 3 \\ 0 & 1 & 0 & 1 & 0 & | & 3 \\ \hline 0 & 0 & 5 & 7 & 1 & | & 21 \end{bmatrix}$$

The basic feasible solution of this tableau is the solution that corresponds to corner point C from part (b).

All of the entries in the last row are positive. The basic feasible solution that solves the linear programming problem; i.e. the solution that results in the maximum for the objective function f is:

$x = 3$, $y = 3$, $f = 21$, $s_1 = 0$, $s_2 = 0$.

23. Introduce the slack variables to form the system at the top of the next column.

$$\begin{aligned} x + y + s_1 \qquad\qquad &= 9 \\ 2x + y \quad + s_2 \qquad &= 12 \\ x + 4y \qquad + s_3 \quad &= 24 \\ -3x - 2y \qquad\qquad + f &= 0 \end{aligned}$$

The initial simplex tableau is:

$$x \quad y \quad s_1 \quad s_2 \quad s_3 \quad f$$

$$\begin{bmatrix} 1 & 1 & 1 & 0 & 0 & 0 & | & 9 \\ 2 & 1 & 0 & 1 & 0 & 0 & | & 12 \\ 1 & 4 & 0 & 0 & 1 & 0 & | & 24 \\ \hline -3 & -2 & 0 & 0 & 0 & 1 & | & 0 \end{bmatrix}$$

The most negative element in the last row of the tableau is negative three. Therefore column 1 is the pivot column. The smallest quotient rule shows:

$R_1: \quad \frac{9}{1} = 9$

$R_2: \quad \frac{12}{2} = 6$

$R_3: \quad \frac{24}{1} = 24$

The pivot element for the first pivot will be the two in row 2, column 1. Pivoting on this element results in the following tableau:

$$x \quad y \quad s_1 \quad s_2 \quad s_3 \quad f$$

$$\begin{bmatrix} 0 & \frac{1}{2} & 1 & \frac{-1}{2} & 0 & 0 & | & 3 \\ 1 & \frac{1}{2} & 0 & \frac{1}{2} & 0 & 0 & | & 6 \\ 0 & \frac{7}{2} & 0 & \frac{-1}{2} & 1 & 0 & | & 18 \\ \hline 0 & \frac{-1}{2} & 0 & \frac{3}{2} & 0 & 1 & | & 18 \end{bmatrix}$$

Since the element $\frac{-1}{2}$ in column 2 is the only negative element in the last row, column 2 is the pivot column. We apply the smallest quotient rule to see:

$R_1: \quad \dfrac{3}{\frac{1}{2}} = 6$

$R_2: \quad \dfrac{6}{\frac{1}{2}} = 12$

$R_3: \quad \dfrac{18}{\frac{7}{2}} = \frac{36}{7} \approx 5.14$

The solution is continued on the next page.

Therefore, the pivot element is the $\frac{7}{2}$ in row 3, column 2. Pivoting on this element results in the simplex tableau:

$$
\begin{array}{cccccc}
x & y & s_1 & s_2 & s_3 & f \\
\end{array}
$$
$$
\left[
\begin{array}{cccccc|c}
0 & 0 & 1 & \frac{-3}{7} & \frac{-1}{7} & 0 & \frac{3}{7} \\
1 & 0 & 0 & \frac{4}{7} & \frac{-1}{7} & 0 & \frac{24}{7} \\
0 & 1 & 0 & \frac{-1}{7} & \frac{2}{7} & 0 & \frac{36}{7} \\
\hline
0 & 0 & 0 & \frac{10}{7} & \frac{1}{7} & 1 & \frac{144}{7} \\
\end{array}
\right]
$$

All of the entries in the last row are positive. The basic feasible solution that solves the linear programming problem; i.e. the solution that results in the maximum value for the objective function f over the feasible region is:

$$
x = \tfrac{24}{7}, \ y = \tfrac{36}{7}, \ f = \tfrac{144}{7}, \ s_1 = \tfrac{3}{7}, \ s_2 = 0, \ s_3 = 0.
$$

25. Introduce the slack variables to form the following system:

$$
\begin{array}{rcl}
x + 2y + \ z + s_1 &=& 25 \\
3x + 2y + 2z \ + s_2 &=& 30 \\
-2x - \ y - 3z \qquad + f &=& 0
\end{array}
$$

The initial simplex tableau is:

$$
\begin{array}{cccccc}
x & y & z & s_1 & s_2 & f \\
\end{array}
$$
$$
\left[
\begin{array}{cccccc|c}
1 & 2 & 1 & 1 & 0 & 0 & 25 \\
3 & 2 & 2 & 0 & 1 & 0 & 30 \\
\hline
-2 & -1 & -3 & 0 & 0 & 1 & 0 \\
\end{array}
\right]
$$

The most negative element in the last row of the tableau is negative three. Therefore column 3 is the pivot column. The smallest quotient rule shows:

$$
R_1: \ \tfrac{25}{1} = 25
$$
$$
R_2: \ \tfrac{30}{2} = 15
$$

The pivot element for the first pivot will be the two in row 2, column 3. Pivoting on this element results in the tableau at the top of the next column.

$$
\begin{array}{cccccc}
x & y & z & s_1 & s_2 & f \\
\end{array}
$$
$$
\left[
\begin{array}{cccccc|c}
\frac{-1}{2} & 1 & 0 & 1 & \frac{-1}{2} & 0 & 10 \\
\frac{3}{2} & 1 & 1 & 0 & \frac{1}{2} & 0 & 15 \\
\hline
\frac{5}{2} & 2 & 0 & 0 & \frac{3}{2} & 1 & 45 \\
\end{array}
\right]
$$

All of the entries in the last row are positive. The basic feasible solution that solves the linear programming problem; i.e. the solution that results in the maximum value for the objective function f over the feasible region is:

$$
x = 0, \ y = 0, \ z = 15, \ f = 45, \ s_1 = 10, \ s_2 = 0.
$$

27. Introduce the slack variables to form the following system:

$$
\begin{array}{rcl}
2x + \ y + 2z + s_1 &=& 40 \\
2x + 2y + 3z \quad + s_2 &=& 50 \\
x + 3y + 2z + \qquad s_3 &=& 20 \\
-5x - 3y - 4z \qquad\qquad + f &=& 0
\end{array}
$$

The initial simplex tableau is:

$$
\begin{array}{ccccccc}
x & y & z & s_1 & s_2 & s_3 & f \\
\end{array}
$$
$$
\left[
\begin{array}{ccccccc|c}
2 & 1 & 2 & 1 & 0 & 0 & 0 & 40 \\
2 & 2 & 3 & 0 & 1 & 0 & 0 & 50 \\
1 & 3 & 2 & 0 & 0 & 1 & 0 & 20 \\
\hline
-5 & -3 & -4 & 0 & 0 & 0 & 1 & 0 \\
\end{array}
\right]
$$

The most negative element in the last row of the tableau is negative five. Therefore column 1 is the pivot column. The smallest quotient rule shows:

$$
R_1: \ \tfrac{40}{2} = 20
$$
$$
R_2: \ \tfrac{50}{2} = 25
$$
$$
R_3: \ \tfrac{20}{1} = 20
$$

We can choose to pivot on either row 1 or row 3. We choose to pivot on the one in row 3, column 1. Pivoting on this element results in the tableau displayed at the top of the next page.

$$\begin{array}{ccccccc} x & y & z & s_1 & s_2 & s_3 & f \\ \left[\begin{array}{ccccccc|c} 0 & -5 & -2 & 1 & 0 & -2 & 0 & 0 \\ 0 & -4 & -1 & 0 & 1 & -2 & 0 & 10 \\ 1 & 3 & 2 & 0 & 0 & 1 & 0 & 20 \\ \hline 0 & 12 & 6 & 0 & 0 & 5 & 1 & 100 \end{array}\right] \end{array}$$

All of the entries in the last row are positive. The basic feasible solution that solves the linear programming problem; i.e. the solution that results in the maximum value for the objective function f over the feasible region is:

$x = 20,\ y = 0,\ z = 0,\ f = 100$
$s_1 = 0,\ s_2 = 10,\ s_3 = 0$.

29. Introduce the slack variables to form the following system:

$$\begin{aligned} x - 2y + z + s_1 &= 4 \\ x - 4y + 3z + s_2 &= 12 \\ 2x + 2y + z + s_3 &= 17 \\ -2x - 2y - z + P &= 0 \end{aligned}$$

The initial simplex tableau is:

$$\begin{array}{cccccc} x & y & z & s_1 & s_2 & s_3 & P \\ \left[\begin{array}{ccccccc|c} 1 & -2 & 1 & 1 & 0 & 0 & 0 & 4 \\ 1 & -4 & 3 & 0 & 1 & 0 & 0 & 12 \\ 2 & 2 & 1 & 0 & 0 & 1 & 0 & 17 \\ \hline -2 & -2 & -1 & 0 & 0 & 0 & 1 & 0 \end{array}\right] \end{array}$$

Since there is a negative two in both column 1 and column 2 we can choose either column to be the pivot column. We choose to pivot on column 2, because there is only one positive entry in column 2.

Pivoting on the two in row 3, column 3 results in the following tableau:

$$\begin{array}{cccccc} x & y & z & s_1 & s_2 & s_3 & P \\ \left[\begin{array}{ccccccc|c} 3 & 0 & 2 & 1 & 0 & 1 & 0 & 21 \\ 5 & 0 & 5 & 0 & 1 & 2 & 0 & 46 \\ 1 & 1 & \frac{1}{2} & 0 & 0 & \frac{1}{2} & 0 & \frac{17}{2} \\ \hline 0 & 0 & 0 & 0 & 0 & 1 & 1 & 17 \end{array}\right] \end{array}$$

All of the entries in the last row are positive. The basic feasible solution that solves the linear programming problem; i.e. the solution that results in the maximum value for the objective function f over the feasible region is:

$x = 0,\ y = \frac{17}{2},\ z = 0,\ P = 17$
$s_1 = 21,\ s_2 = 46,\ s_3 = 0$.

31. Introduce the slack variables to form the following system:

$$\begin{aligned} x \quad + 2z + s_1 \quad &= 8 \\ - y + 3z + s_2 \quad &= 10 \\ -x + y - z + \quad s_3 &= 12 \\ -2x - 3y - 3z \quad + f &= 0 \end{aligned}$$

The initial simplex tableau is:

$$\begin{array}{cccccc} x & y & z & s_1 & s_2 & s_3 & f \\ \left[\begin{array}{ccccccc|c} 1 & 0 & 2 & 1 & 0 & 0 & 0 & 8 \\ 0 & -1 & 3 & 0 & 1 & 0 & 0 & 10 \\ -1 & 1 & -1 & 0 & 0 & 1 & 0 & 12 \\ \hline -2 & -3 & -3 & 0 & 0 & 0 & 1 & 0 \end{array}\right] \end{array}$$

The most negative element in the last row of the tableau is negative three. Therefore we can choose column 2 or column 3 as the pivot column. Because row 3 in column 2 is the only positive entry in the column, we will choose column 2 as the pivot column. Therefore the pivot element is the one in row 3, column 2.

Pivoting on this element results in the following tableau:

$$\begin{array}{cccccc} x & y & z & s_1 & s_2 & s_3 & f \\ \left[\begin{array}{ccccccc|c} 1 & 0 & 2 & 1 & 0 & 0 & 0 & 8 \\ -1 & 0 & 2 & 0 & 1 & 1 & 0 & 22 \\ -1 & 1 & -1 & 0 & 0 & 1 & 0 & 12 \\ \hline -5 & 0 & -6 & 0 & 0 & 3 & 1 & 36 \end{array}\right] \end{array}$$

The most negative entry in the last row now is the negative six in column 3.

The smallest quotient rule shows:

$R_1 : \ \frac{8}{2} = 4$

$R_2 : \ \frac{22}{2} = 11$

The solution is continued on the next page.

From the smallest quotient rule on the previous page, we determine that the pivot element will be the two in row 1, column 3.

Pivoting on this element results in the following tableau:

$$
\begin{array}{ccccccc}
x & y & z & s_1 & s_2 & s_3 & f \\
\end{array}
$$
$$
\left[
\begin{array}{ccccccc|c}
\frac{1}{2} & 0 & 1 & \frac{1}{2} & 0 & 0 & 0 & 4 \\
-2 & 0 & 0 & -1 & 1 & 1 & 0 & 14 \\
\frac{-1}{2} & 1 & 0 & \frac{1}{2} & 0 & 1 & 0 & 16 \\
\hline
-2 & 0 & 0 & 3 & 0 & 3 & 1 & 60 \\
\end{array}
\right]
$$

The only remaining negative entry in the last row is the negative two in column 1. This will be the pivot column. Since there is only one positive entry in column one, the pivot element will be the element in row 1, column 1.

Pivoting on this element results in the following tableau:

$$
\begin{array}{ccccccc}
x & y & z & s_1 & s_2 & s_3 & f \\
\end{array}
$$
$$
\left[
\begin{array}{ccccccc|c}
1 & 0 & 2 & 1 & 0 & 0 & 0 & 8 \\
0 & 0 & 4 & 1 & 1 & 1 & 0 & 30 \\
0 & 1 & 1 & 1 & 0 & 1 & 0 & 20 \\
\hline
0 & 0 & 4 & 5 & 0 & 3 & 1 & 76 \\
\end{array}
\right]
$$

All of the entries in the last row are positive. The basic feasible solution that solves the linear programming problem; i.e. the solution that results in the maximum value for the objective function f over the feasible region is:

$$x = 8,\ y = 20,\ z = 0,\ f = 76$$
$$s_1 = 0,\ s_2 = 30,\ s_3 = 0.$$

33. Introduce the slack variables to form the system:

$$
\begin{array}{rcl}
y + z + s_1 & = & 4 \\
y \quad\quad\quad + s_2 & = & 3 \\
z + \quad\quad s_3 & = & 2 \\
-x - 2y - 4z \quad\quad + f & = & 0 \\
\end{array}
$$

The initial simplex tableau is displayed at the top of the next column.

$$
\begin{array}{ccccccc}
x & y & z & s_1 & s_2 & s_3 & f \\
\end{array}
$$
$$
\left[
\begin{array}{ccccccc|c}
0 & 1 & 1 & 1 & 0 & 0 & 0 & 4 \\
0 & 1 & 0 & 0 & 1 & 0 & 0 & 3 \\
0 & 0 & 1 & 0 & 0 & 1 & 0 & 2 \\
\hline
-1 & -2 & -4 & 0 & 0 & 0 & 1 & 0 \\
\end{array}
\right]
$$

The most negative element in the last row of the tableau is negative four. Therefore we choose column 3 as the pivot column. However, we notice that column one has only zero entries above the negative value in the last row. This means that no matter how many pivot operations we perform, there will always be a negative in the last row of the tableau. This problem has no solution.

35. Introduce the slack variables to form the system:

$$
\begin{array}{rcl}
x + z + 2w + s_1 & = & 6 \\
2x + y + w + s_2 & = & 12 \\
x + 2z + 3w + s_3 & = & 18 \\
x + y + 2z + w + s_4 & = & 10 \\
-4x - 4y - 2z - w + f & = & 0 \\
\end{array}
$$

The initial simplex tableau is:

$$
\begin{array}{ccccccccc}
x & y & z & w & s_1 & s_2 & s_3 & s_4 & f \\
\end{array}
$$
$$
\left[
\begin{array}{ccccccccc|c}
1 & 0 & 1 & 2 & 1 & 0 & 0 & 0 & 0 & 6 \\
2 & 1 & 0 & 1 & 0 & 1 & 0 & 0 & 0 & 12 \\
1 & 0 & 2 & 3 & 0 & 0 & 1 & 0 & 0 & 18 \\
1 & 1 & 2 & 1 & 0 & 0 & 0 & 1 & 0 & 10 \\
\hline
-4 & -4 & -2 & -1 & 0 & 0 & 0 & 0 & 1 & 0 \\
\end{array}
\right]
$$

The most negative element in the last row of the tableau is negative four. We can choose column 1 or column 2 to be the pivot column. We will select column 2 to be the pivot column. Using the smallest quotient rule we see:

$$R_2 : \frac{12}{1} = 12$$
$$R_4 : \frac{10}{1} = 10$$

Therefore the pivot element is the one in row 4, column 2.

Pivoting on this element results in the tableau at the top of the next page.

$$
\begin{array}{ccccccccc}
x & y & z & w & s_1 & s_2 & s_3 & s_4 & f \\
\end{array}
$$

$$
\left[\begin{array}{ccccccccc|c}
1 & 0 & 1 & 2 & 1 & 0 & 0 & 0 & 0 & 6 \\
1 & 0 & -2 & 0 & 0 & 1 & 0 & -1 & 0 & 2 \\
1 & 0 & 2 & 3 & 0 & 0 & 1 & 0 & 0 & 18 \\
1 & 1 & 2 & 1 & 0 & 0 & 0 & 1 & 0 & 10 \\
\hline
0 & 0 & 6 & 3 & 0 & 0 & 0 & 4 & 1 & 40
\end{array}\right]
$$

All of the entries in the last row are positive. The basic feasible solution that solves the linear programming problem; i.e. the solution that results in the maximum value for the objective function f over the feasible region is:

$$x = 0,\ y = 10,\ z = 0,\ w = 0,\ f = 40$$
$$s_1 = 6,\ s_2 = 2,\ s_3 = 18,\ s_4 = 0.$$

This is not the only solution to the problem on the previous page. Had you pivoted on column 1, the simplex method would have found the solution:

$$x = 2,\ y = 8,\ z = 0,\ w = 0,\ f = 40$$
$$s_1 = 4,\ s_2 = 0,\ s_3 = 16,\ s_4 = 0.$$

37. Introduce the slack variables to form the following system:

$$
\begin{aligned}
2x_1 + \ x_2 \qquad\ \ + x_4 + 2x_5 + s_1 \qquad\qquad\quad &= 2 \\
4x_1 + \qquad\ 3x_3 + 2x_4 + \ x_5 \qquad + s_2 \qquad\quad &= 8 \\
x_1 + 2x_2 + 4x_3 + \ x_4 \qquad\qquad\qquad + s_3 \quad &= 4 \\
-2x_1 + \ x_2 - \ x_3 + 2x_4 + 3x_5 \qquad\qquad\qquad + f &= 0
\end{aligned}
$$

The initial simplex tableau is:

$$
\begin{array}{ccccccccc}
x_1 & x_2 & x_3 & x_4 & x_5 & s_1 & s_2 & s_3 & f \\
\end{array}
$$

$$
\left[\begin{array}{ccccccccc|c}
2 & 1 & 0 & 1 & 2 & 1 & 0 & 0 & 0 & 2 \\
4 & 0 & 3 & 2 & 1 & 0 & 1 & 0 & 0 & 8 \\
1 & 2 & 4 & 1 & 0 & 0 & 0 & 1 & 0 & 4 \\
\hline
-2 & 1 & -1 & 2 & 3 & 0 & 0 & 0 & 1 & 0
\end{array}\right]
$$

The most negative element in the last row of the tableau is negative two. Therefore column 1 will be the pivot column. We apply the smallest quotient rule below:

$$R_1:\ \tfrac{2}{2} = 1$$
$$R_2:\ \tfrac{8}{4} = 2$$
$$R_3:\ \tfrac{4}{1} = 4$$

Therefore the pivot element is the two in row 1, column 1.

Pivoting on this element results in the tableau:

$$
\begin{array}{ccccccccc}
x_1 & x_2 & x_3 & x_4 & x_5 & s_1 & s_2 & s_3 & f \\
\end{array}
$$

$$
\left[\begin{array}{ccccccccc|c}
1 & \frac{1}{2} & 0 & \frac{1}{2} & 1 & \frac{1}{2} & 0 & 0 & 0 & 1 \\
0 & -2 & 3 & 0 & -3 & -2 & 1 & 0 & 0 & 4 \\
0 & \frac{3}{2} & 4 & \frac{1}{2} & -1 & \frac{-1}{2} & 0 & 1 & 0 & 3 \\
\hline
0 & 2 & -1 & 3 & 5 & 1 & 0 & 0 & 1 & 2
\end{array}\right]
$$

The only negative entry in the last row is in column 3. This will be the new pivot column. Using the smallest quotient rule

Using the smallest quotient rule:

$$R_2:\ \tfrac{4}{3} \approx 1.33$$
$$R_3:\ \tfrac{3}{4} \approx 0.75$$

The pivot element will be the four in row 3, column 3.

Pivoting on this element results in the following tableau:

$$
\begin{array}{ccccccccc}
x_1 & x_2 & x_3 & x_4 & x_5 & s_1 & s_2 & s_3 & f \\
\end{array}
$$

$$
\left[\begin{array}{ccccccccc|c}
1 & \frac{1}{2} & 0 & \frac{1}{2} & 1 & \frac{1}{2} & 0 & 0 & 0 & 1 \\
0 & \frac{-25}{8} & 0 & \frac{-3}{8} & \frac{-9}{4} & \frac{-21}{8} & 1 & \frac{-3}{4} & 0 & \frac{7}{4} \\
0 & \frac{3}{8} & 1 & \frac{1}{8} & \frac{-1}{4} & \frac{-1}{8} & 0 & \frac{1}{4} & 0 & \frac{3}{4} \\
\hline
0 & \frac{19}{8} & 0 & \frac{25}{8} & \frac{19}{4} & \frac{7}{8} & 0 & \frac{1}{4} & 1 & \frac{11}{4}
\end{array}\right]
$$

All of the entries in the last row are positive. The basic feasible solution that solves the linear programming problem; i.e. the solution that results in the maximum value for the objective function f over the feasible region is:

$$x_1 = 1,\ x_2 = 0,\ x_3 = \tfrac{3}{4},\ x_4 = 0,\ f = \tfrac{11}{4}$$
$$s_1 = 0,\ s_2 = \tfrac{7}{4},\ s_3 = 0.$$

39. If we let x be the number of three speed models made each week, and y be the number of ten speed models made each week.

The conditions that assembly and painting have 80 and 100 work hours available weekly translates into the inequalities:

$$x + y \le 80$$
$$2x + y \le 100$$

The solution is continued on the next page.

The profit function is:

$$P = 80x + 60y$$

The linear programming problem becomes:

Maximize $\quad P = 80x + 60y$
subject to:

$$x + y \leq 80$$
$$2x + y \leq 100$$
$$x \geq 0, y \geq 0$$

Adding the slack variables will result in the following system:

$$x + y + s_1 \qquad = 80$$
$$2x + y \quad + s_2 \quad = 100$$
$$-80x - 60y \qquad + P = 0$$

The initial tableau is:

$$
\begin{array}{ccccc|c}
x & y & s_1 & s_2 & P & \\
1 & 1 & 1 & 0 & 0 & 80 \\
2 & 1 & 0 & 1 & 0 & 100 \\
\hline
-80 & -60 & 0 & 0 & 1 & 0
\end{array}
$$

The most negative number in the last row is in column 1. This will be the pivot column. Using the smallest quotient rule:

$R_1: \quad \frac{80}{1} = 80$

$R_2: \quad \frac{100}{2} = 50$

Therefore the pivot element will be the two in row 2, column 1.

Pivoting on this element will result in the following tableau:

$$
\begin{array}{ccccc|c}
x & y & s_1 & s_2 & P & \\
0 & \frac{1}{2} & 1 & \frac{-1}{2} & 0 & 30 \\
1 & \frac{1}{2} & 0 & \frac{1}{2} & 0 & 50 \\
\hline
0 & -20 & 0 & 40 & 1 & 4000
\end{array}
$$

The next pivot column will be column 2 since that is the only column in the last row with a negative entry.

Using the smallest quotient rule:

$R_1: \quad \dfrac{30}{\frac{1}{2}} = 60$

$R_2: \quad \dfrac{50}{\frac{1}{2}} = 100$

The next pivot element will be the $\frac{1}{2}$ in row 1, column 2. Pivoting on this element results in the following matrix:

$$
\begin{array}{ccccc|c}
x & y & s_1 & s_2 & P & \\
0 & 1 & 2 & -1 & 0 & 60 \\
1 & 0 & -1 & 1 & 0 & 20 \\
\hline
0 & 0 & 40 & 20 & 1 & 5200
\end{array}
$$

Since there are no more negative entries in the last row, we have reached a maximum value for the profit function. The maximum feasible profit per week is $5200 when 20 three-speed models and 60 ten-speed models are produced each week. With this production level, both the assembly and the painting operations will be used to their capacity.

41. If we let x be the number of true-false questions answered correctly, and y be the number of open-ended questions answered correctly on the exam.

The rules of the exam limit the number of each type of question that can be answered. These rules translate into the inequalities:

$$x \leq 20$$
$$x + y \leq 30$$

The objective is to maximize the number of points scored. The equation of the objective function is:

$$P = 5x + 8y$$

The linear programming problem becomes:

Maximize $\quad P = 5x + 8y$
subject to:

$$x \leq 20$$
$$x + y \leq 30$$
$$x \geq 0, y \geq 0$$

The solution is continued on the next page.

Adding the slack variables to the system on the previous page will result in the following system:

$$x \quad + s_1 \qquad = 20$$
$$x + y \quad + s_2 \quad = 30$$
$$-5x - 8y \qquad + P = 0$$

The initial tableau is:

$$x \quad y \quad s_1 \quad s_2 \quad P$$
$$\begin{bmatrix} 1 & 0 & 1 & 0 & 0 & 20 \\ 1 & 1 & 0 & 1 & 0 & 30 \\ -5 & -8 & 0 & 0 & 1 & 0 \end{bmatrix}$$

The most negative number in the last row is in column 2. This will be the pivot column. Since the one in row 2, column 2 is the only positive element in column 2 it will be the pivot element.

Pivoting on the appropriate pivot element results in the following tableau:

$$x \quad y \quad s_1 \quad s_2 \quad P$$
$$\begin{bmatrix} 1 & 0 & 1 & 0 & 0 & 20 \\ 1 & 1 & 0 & 1 & 0 & 30 \\ 3 & 0 & 0 & 8 & 1 & 240 \end{bmatrix}$$

Since there are no more negative entries in the last row, we have reached a maximum for the objective function. The maximum feasible number of points that can be earned on the exam is 240 and is achieved by attempting zero true-false questions and answering 30 open-ended questions correctly.

43. If we let x be the number of brand X computers, and y be the number of brand Y computers.

If we let costs be in thousands of dollars, the budget constraint can be translated into the inequality:

$$2.5x + 5y \leq 100$$

The space constraint for brand Y computers translates in the inequality:

$$y \leq 15$$

(Note: The book states that there should be no more than 15 Brand X computers. This is a misprint; the book should state that there should be no more than 15 Brand Y computers.)

The objective is to maximize the total number of computers bought. The objective function is:

$$N = x + y$$

The linear programming problem is displayed below:

Maximize $\qquad N = x + y$
subject to:
$$2.5x + 5y \leq 100$$
$$y \leq 15$$
$$x \geq 0, y \geq 0$$

Adding the slack variables will result in the following system:

$$2.5x + 5y + s_1 \qquad = 100$$
$$y \quad + s_2 \qquad = 15$$
$$-x - y \qquad + N = 0$$

The initial tableau is:

$$x \quad y \quad s_1 \quad s_2 \quad N$$
$$\begin{bmatrix} 2.5 & 5.0 & 1 & 0 & 0 & 100 \\ 0 & 1 & 0 & 1 & 0 & 15 \\ -1 & -1 & 0 & 0 & 1 & 0 \end{bmatrix}$$

Since the most negative number is the same in column 1 and column 2, we can choose either column. Since there is only one positive element in column 1, we will select column 1 as the pivot column, and pivot on the 25 in row 1, column 1.

The resulting tableau after we pivot is:

$$x \quad y \quad s_1 \quad s_2 \quad N$$
$$\begin{bmatrix} 1 & 2 & \frac{2}{5} & 0 & 0 & 40 \\ 0 & 1 & 0 & 1 & 0 & 15 \\ 0 & 1 & \frac{2}{5} & 0 & 1 & 40 \end{bmatrix}$$

Since there are no more negative entries in the last row, we have reached a maximum for the objective function. The maximum number of computers the university can purchase is 40 brand X computers and zero brand Y computers. With this purchase they will have exhausted their budget for computers. They will still have space for 15 brand Y computers.

45. Let x be the number of times the radio ads are run, y be the number of times the newspaper ads are run, and z be the number of television ads.

If the television ads can be run a maximum of 20 times and the newspaper and radio ads combined can be run a maximum of 110 times, then the inequalities that model these restrictions are:

$$z \le 20$$
$$x + y \le 110$$

The budget constraint translates into the inequality:

$$200x + 100y + 500z \le 15000$$
or,
$$2x + 1y + 5z \le 150$$

The objective is to maximize the number of people reached, so the objective function is:

$$N = 10x + 8y + 15z, \text{ where } N \text{ is in hundreds of people.}$$

The linear programming problem becomes:

Maximize $\qquad N = 10x + 8y + 15z$

subject to:
$$2x + y + 5z \le 150$$
$$z \le 20$$
$$x + y \le 110$$
$$x \ge 0, y \ge 0, z \ge 0$$

Adding the slack variables will result in the following system:

$$2x + y + 5z + s_1 = 150$$
$$z + s_2 = 20$$
$$x + y + s_3 = 110$$
$$-10x - 8y - 15z + N = 0$$

The initial tableau is:

x	y	z	s_1	s_2	s_3	N	
2	1	5	1	0	0	0	150
0	0	1	0	1	0	0	20
1	1	0	0	0	1	0	110
−10	−8	−15	0	0	0	1	0

The most negative number in the last row of the initial tableau is in column 3, so that will be the pivot column. Using the smallest quotient rule:

$$R_1 : \quad \frac{150}{5} = 30$$
$$R_2 : \quad \frac{20}{1} = 20$$

The pivot element will be the one in row 2, column 3. Pivoting on this element results in the following tableau:

x	y	z	s_1	s_2	s_3	N	
2	1	0	1	−5	0	0	50
0	0	1	0	1	0	0	20
1	1	0	0	0	1	0	110
−10	−8	0	0	15	0	1	300

Now the most negative element in the last row is in column 1. This will be the pivot column. Using the smallest quotient rule:

$$R_1 : \quad \frac{50}{2} = 25$$
$$R_3 : \quad \frac{110}{1} = 110$$

Therefore the pivot element will be the two in row 1, column 1. Pivoting on this element results in the following tableau:

x	y	z	s_1	s_2	s_3	N	
1	$\frac{1}{2}$	0	$\frac{1}{2}$	$\frac{-5}{2}$	0	0	25
0	0	1	0	1	0	0	20
1	$\frac{1}{2}$	0	$\frac{-1}{2}$	$\frac{5}{2}$	1	0	85
0	−3	0	5	−10	0	1	550

The most negative number in the last row is in column 5. This will be the pivot column. The smallest quotient rule tells us:

$$R_2 : \quad \frac{20}{1} = 20$$
$$R_3 : \quad \frac{85}{\frac{5}{2}} = 34$$

The pivot element will be the one in row 2, column 5. Pivoting on this element results in the tableau at the top of the next page.

$$\begin{bmatrix} x & y & z & s_1 & s_2 & s_3 & N & \\ 1 & \frac{1}{2} & \frac{5}{2} & \frac{1}{2} & 0 & 0 & 0 & 75 \\ 0 & 0 & 1 & 0 & 1 & 0 & 0 & 20 \\ 1 & \frac{1}{2} & \frac{-5}{2} & \frac{-1}{2} & 0 & 1 & 0 & 35 \\ \hline 0 & -3 & 10 & 5 & 0 & 0 & 1 & 750 \end{bmatrix}$$

The most negative number in the last row is in column 2. This will be the pivot column. Using the smallest quotient rule:

$$R_1: \quad \frac{75}{\frac{1}{2}} = 150$$

$$R_3: \quad \frac{35}{\frac{1}{2}} = 70$$

The pivot element will be the $\frac{1}{2}$ in row 3, column 2. Pivoting on this element results in the tableau:

$$\begin{bmatrix} x & y & z & s_1 & s_2 & s_3 & N & \\ \frac{1}{5} & 0 & 5 & 1 & 0 & -1 & 0 & 40 \\ \frac{-1}{5} & 0 & 1 & 0 & 1 & 0 & 0 & 20 \\ 1 & 1 & -5 & -1 & 0 & 2 & 0 & 70 \\ \hline 1 & 0 & -5 & 2 & 0 & 6 & 1 & 960 \end{bmatrix}$$

The only remaining negative number in the last row is in column 3. This will be the pivot column. Using the smallest quotient rule:

$$R_1: \quad \frac{40}{5} = 8$$

$$R_2: \quad \frac{20}{1} = 20$$

The pivot element will be the five in row 1, column 3. Pivoting on this element results in the tableau:

$$\begin{bmatrix} x & y & z & s_1 & s_2 & s_3 & N & \\ 0 & 0 & 1 & \frac{1}{5} & 0 & \frac{-1}{5} & 0 & 8 \\ 0 & 0 & 0 & \frac{-1}{5} & 1 & \frac{1}{5} & 0 & 12 \\ 2 & 1 & 0 & 0 & 0 & 1 & 0 & 110 \\ \hline 6 & 0 & 0 & 3 & 0 & 5 & 1 & 1000 \end{bmatrix}$$

The solution is stated at the top of the next column.

Since there are no more negative entries in the last row, we have reached a maximum for the objective function. The maximum number of people that the advertising agency can reach is 1000. They will do this if the run zero radio ads, 110 newspaper ads and 8 television ads. If they do this, they will be 12 spots under the television limit, and all of the budget will be used.

47. Let x be the number of super deluxe ovens, y be the number of deluxe ovens, and z be the number of standard ovens produce each week..

The time constraints on the assembly, painting and testing/packaging departments result in the inequalities shown below:

$$4x + 2y + z \leq 160$$
$$2x + y + z \leq 100$$
$$x + y + \tfrac{1}{2}z \leq 60$$

The profit function is:

$$P = 20x + 16y + 12z$$

The linear programming problem becomes:

Maximize $P = 20x + 16y + 12z$
subject to:

$$4x + 2y + z \leq 160$$
$$2x + y + z \leq 100$$
$$x + y + \tfrac{1}{2}z \leq 60$$
$$x \geq 0, y \geq 0, z \geq 0$$

Adding the slack variables will result in the following system:

$$4x + 2y + z + s_1 \qquad\qquad = 160$$
$$2x + y + z \quad + s_2 \qquad = 100$$
$$x + y + \tfrac{1}{2}z \qquad + s_3 \quad = 60$$
$$-20x - 16y - 12z \qquad\qquad + P = 0$$

The initial tableau is:

$$\begin{bmatrix} x & y & z & s_1 & s_2 & s_3 & P & \\ 4 & 2 & 1 & 1 & 0 & 0 & 0 & 160 \\ 2 & 1 & 1 & 0 & 1 & 0 & 0 & 100 \\ 1 & 1 & \frac{1}{2} & 0 & 0 & 1 & 0 & 60 \\ \hline -20 & -16 & -12 & 0 & 0 & 0 & 1 & 0 \end{bmatrix}$$

The most negative entry in the last row of the tableau on the previous page is the -20 in the first column. Using the smallest quotient rule:

$R_1 : \quad \frac{160}{4} = 40$

$R_2 : \quad \frac{100}{2} = 50$

$R_3 : \quad \frac{60}{1} = 60$

The pivot element will be the four in row 1, column 1. Pivoting on this element results in the tableau:

$$
\begin{array}{ccccccc|c}
x & y & z & s_1 & s_2 & s_3 & P & \\
\hline
1 & \frac{1}{2} & \frac{1}{4} & \frac{1}{4} & 0 & 0 & 0 & 40 \\
0 & 0 & \frac{1}{2} & \frac{-1}{2} & 1 & 0 & 0 & 20 \\
0 & \frac{1}{2} & \frac{1}{4} & \frac{-1}{4} & 0 & 1 & 0 & 20 \\
\hline
0 & -6 & -7 & 5 & 0 & 0 & 1 & 800
\end{array}
$$

Now the most negative element in the last row is in column 3. This will be the pivot column. Using the smallest quotient rule:

$R_1 : \quad \dfrac{40}{\frac{1}{4}} = 160$

$R_2 : \quad \dfrac{20}{\frac{1}{2}} = 40$

$R_3 : \quad \dfrac{20}{\frac{1}{4}} = 80$

We will pivot on the $\frac{1}{2}$ in row 2, column 3. Pivoting on this element results in the following tableau:

$$
\begin{array}{ccccccc|c}
x & y & z & s_1 & s_2 & s_3 & P & \\
\hline
1 & \frac{1}{2} & 0 & \frac{1}{2} & \frac{-1}{2} & 0 & 0 & 30 \\
0 & 0 & 1 & -1 & 2 & 0 & 0 & 40 \\
0 & \frac{1}{2} & 0 & 0 & \frac{-1}{2} & 1 & 0 & 10 \\
\hline
0 & -6 & 0 & -2 & 14 & 0 & 1 & 1080
\end{array}
$$

The most negative number in the last row is in column 2. This will be the pivot column. We apply the smallest quotient rule at the top of the next column.

$R_1 : \quad \dfrac{30}{\frac{1}{2}} = 60$

$R_3 : \quad \dfrac{10}{\frac{1}{2}} = 20$

The pivot element will be the $\frac{1}{2}$ in row 3, column 2. Pivoting on this element results in the tableau:

$$
\begin{array}{ccccccc|c}
x & y & z & s_1 & s_2 & s_3 & P & \\
\hline
1 & 0 & 0 & \frac{1}{2} & 0 & -1 & 0 & 20 \\
0 & 0 & 1 & -1 & 2 & 0 & 0 & 40 \\
0 & 1 & 0 & 0 & -1 & 2 & 0 & 20 \\
\hline
0 & 0 & 0 & -2 & 8 & 12 & 1 & 1200
\end{array}
$$

Now the most negative element in the last row is in column 4. This will be the pivot column. Since there is only one positive element in column 4, we will pivot on the $\frac{1}{2}$ in row 1, column 4. Pivoting on this element results in the following tableau:

$$
\begin{array}{ccccccc|c}
x & y & z & s_1 & s_2 & s_3 & P & \\
\hline
2 & 0 & 0 & 1 & 0 & -2 & 0 & 40 \\
2 & 0 & 1 & 0 & 2 & -2 & 0 & 80 \\
0 & 1 & 0 & 0 & -1 & 2 & 0 & 20 \\
\hline
4 & 0 & 0 & 0 & 8 & 8 & 1 & 1280
\end{array}
$$

Since there are no more negative entries in the last row, we have reached a maximum for the objective function. The maximum profit the company can receive is $1280 per week. To do this they need to produce zero super deluxe model toasters, 20 deluxe model toasters, and 80 standard model toasters. When producing this level of output, the company will have not used 40 hours of available assembly time, per week.

49. Let x, y, and z be the number of top stars, faded stars and high-quality local entertainers booked.

The budget constraint (in thousands of dollars) yields the inequality:

$$10x + 6y + 3z \le 50$$

The limit on the number of shows translates to the inequality:

$$x + y + z \le 6$$

The solution is continued on the next page.

The advertising constraint (in hundreds of dollars) yields the inequality:

$$5x + 4y + 2.5z \leq 40$$

The objective is to maximize the number of people in attendance. The objective function (in thousands of people) is:

$$N = 8x + 3y + 1.2z$$

The linear programming problem becomes:

Maximize $\qquad N = 8x + 3y + 1.2z$

subject to:

$$10x + 6y + 3z \leq 50$$
$$5x + 4y + 2.5z \leq 40$$
$$x + y + z \leq 6$$
$$x \geq 0, y \geq 0, z \geq 0$$

Adding the slack variables will result in the following system:

$$10x + 6y + 3z + s_1 \qquad\qquad = 50$$
$$5x + 4y + 2.5z \qquad + s_2 \qquad\quad = 40$$
$$x + y + z \qquad\qquad + s_3 \quad\; = 6$$
$$-8x - 3y - 1.2z \qquad\qquad\quad + N = 0$$

The initial tableau is:

$$
\begin{array}{c}
\begin{array}{ccccccc}
x & y & z & s_1 & s_2 & s_3 & N
\end{array} \\
\left[
\begin{array}{ccccccc|c}
10 & 6 & 3 & 1 & 0 & 0 & 0 & 50 \\
5 & 4 & 2.5 & 0 & 1 & 0 & 0 & 40 \\
1 & 1 & 1 & 0 & 0 & 1 & 0 & 6 \\
\hline
-8 & -3 & -1.2 & 0 & 0 & 0 & 1 & 0
\end{array}
\right]
\end{array}
$$

The most negative entry in the last row is in column 1, so that will be the pivot column. Using the smallest quotient rule:

$R_1: \quad \frac{50}{10} = 5$

$R_2: \quad \frac{40}{5} = 8$

$R_3: \quad \frac{6}{1} = 6$

The pivot element will be the ten in row 1, column 1. Pivoting on this element results in the tableau at the top of the next column.

$$
\begin{array}{c}
\begin{array}{ccccccc}
x & y & z & s_1 & s_2 & s_3 & N
\end{array} \\
\left[
\begin{array}{ccccccc|c}
1 & \frac{3}{5} & \frac{3}{10} & \frac{1}{10} & 0 & 0 & 0 & 5 \\
0 & 1 & 1 & \frac{-1}{2} & 1 & 0 & 0 & 15 \\
0 & \frac{2}{5} & \frac{7}{10} & \frac{-1}{10} & 0 & 1 & 0 & 1 \\
\hline
0 & \frac{9}{5} & \frac{6}{5} & \frac{4}{5} & 0 & 0 & 1 & 40
\end{array}
\right]
\end{array}
$$

Since there are no more negative entries in the last row, we have reached a maximum for the objective function. The maximum attendance that can be drawn for the fair is 40,000 people. To do this, the directors need to book 5 top stars and nobody else. They will have $1,500 left over for the advertising budget, and they will have one show time left open. However, they will have used their entire entertainment (contracting) budget.

51. Let w, x, y, and z represent the number of QP20, QP40, QP60, and QP100 printers respectively.

The memory constraints yield the inequality:

$$2w + 4x + 6y + 10z \leq 820$$

The production mix that the company wants yields the inequalities:

$$-w + 2z \leq 0$$
$$-x + y \leq 0$$
$$z \leq 20$$

The revenue function in hundreds of dollars that they expect to make is:

$$R = 5w + 6x + 10y + 20z$$

The linear programming problem becomes:

Maximize $\qquad R = 5w + 6x + 10y + 20z$

subject to:

$$2w + 4x + 6y + 10z \leq 820$$
$$-w + 2z \leq 0$$
$$-x + y \leq 0$$
$$z \leq 20$$
$$x \geq 0, y \geq 0, z \geq 0$$

The solution is continued on the next page.

Adding the slack variables to the original system on the previous page will result in the system:

$$2w + 4x + 6y + 10z + s_1 \qquad\qquad = 820$$
$$-w + \qquad\quad 2z \; + s_2 \qquad\qquad = \; 0$$
$$-x + \; y \qquad\qquad + s_3 \qquad = \; 0$$
$$z \qquad\qquad + s_4 \; = \; 20$$
$$-5w - 6x - 10y - 20z \qquad\qquad + R = \; 0$$

The initial tableau is:

$$
\begin{array}{ccccccccc|c}
w & x & y & z & s_1 & s_2 & s_3 & s_4 & R & \\
\hline
2 & 4 & 6 & 10 & 1 & 0 & 0 & 0 & 0 & 820 \\
-1 & 0 & 0 & 2 & 0 & 1 & 0 & 0 & 0 & 0 \\
0 & -1 & 1 & 0 & 0 & 0 & 1 & 0 & 0 & 0 \\
0 & 0 & 0 & 1 & 0 & 0 & 0 & 1 & 0 & 20 \\
\hline
-5 & -6 & -10 & -20 & 0 & 0 & 0 & 0 & 1 & 0
\end{array}
$$

The most negative number in the last row is in column 4, so that will be the pivot column. Using the smallest quotient rule:

$$R_1 : \quad \frac{820}{10} = 82$$
$$R_2 : \quad \frac{0}{2} = 0$$
$$R_4 : \quad \frac{20}{1} = 20$$

The pivot element will be the two in row 2, column 4. Pivoting on this element results in the following tableau:

$$
\begin{array}{ccccccccc|c}
w & x & y & z & s_1 & s_2 & s_3 & s_4 & R & \\
\hline
7 & 4 & 6 & 0 & 1 & -5 & 0 & 0 & 0 & 820 \\
\frac{-1}{2} & 0 & 0 & 1 & 0 & \frac{1}{2} & 0 & 0 & 0 & 0 \\
0 & -1 & 1 & 0 & 0 & 0 & 1 & 0 & 0 & 0 \\
\frac{1}{2} & 0 & 0 & 0 & 0 & \frac{-1}{2} & 0 & 1 & 0 & 20 \\
\hline
-15 & -6 & -10 & 0 & 0 & 10 & 0 & 0 & 1 & 0
\end{array}
$$

The most negative element in the last row is in column 1. This will be the pivot column. Using the smallest quotient rule:

$$R_1 : \quad \frac{820}{7} = 117.14$$
$$R_4 : \quad \frac{20}{\frac{1}{2}} = 40$$

Therefore the pivot element will be the $\frac{1}{2}$ in row 4, column 1. Pivoting on this element results in the following tableau:

$$
\begin{array}{ccccccccc|c}
w & x & y & z & s_1 & s_2 & s_3 & s_4 & R & \\
\hline
0 & 4 & 6 & 0 & 1 & 2 & 0 & -14 & 0 & 540 \\
0 & 0 & 0 & 1 & 0 & 0 & 0 & 1 & 0 & 20 \\
0 & -1 & 1 & 0 & 0 & 0 & 1 & 0 & 0 & 0 \\
1 & 0 & 0 & 0 & 0 & -1 & 0 & 2 & 0 & 40 \\
\hline
0 & -6 & -10 & 0 & 0 & -5 & 0 & 30 & 1 & 600
\end{array}
$$

The most negative number in the last row of the matrix is in column 3. This will be the pivot column. Using the smallest quotient rule:

$$R_1 : \quad \frac{540}{6} = 90$$
$$R_3 : \quad \frac{0}{1} = 0$$

The pivot element will be the one in row 3, column 3. Pivoting on this element results in the following tableau:

$$
\begin{array}{ccccccccc|c}
w & x & y & z & s_1 & s_2 & s_3 & s_4 & R & \\
\hline
0 & 10 & 0 & 0 & 1 & 2 & -6 & -14 & 0 & 540 \\
0 & 0 & 0 & 1 & 0 & 0 & 0 & 1 & 0 & 20 \\
0 & -1 & 1 & 0 & 0 & 0 & 1 & 0 & 0 & 0 \\
1 & 0 & 0 & 0 & 0 & -1 & 0 & 2 & 0 & 40 \\
\hline
0 & -16 & 0 & 0 & 0 & -5 & 10 & 30 & 1 & 600
\end{array}
$$

The most negative number in the last row is in column 2. This will be the pivot column. Since the ten in row 1, column 2 is the only positive element in column 2. We will pivot on this element.

Pivoting on this element results in the following tableau:

$$
\begin{array}{ccccccccc|c}
w & x & y & z & s_1 & s_2 & s_3 & s_4 & R & \\
\hline
0 & 1 & 0 & 0 & \frac{1}{10} & \frac{1}{5} & \frac{-3}{5} & \frac{-7}{5} & 0 & 54 \\
0 & 0 & 0 & 1 & 0 & 0 & 0 & 1 & 0 & 20 \\
0 & 0 & 1 & 0 & \frac{1}{10} & \frac{1}{5} & \frac{2}{5} & \frac{-7}{5} & 0 & 54 \\
1 & 0 & 0 & 0 & 0 & -1 & 0 & 2 & 0 & 40 \\
\hline
0 & 0 & 0 & 0 & \frac{8}{5} & \frac{-9}{5} & \frac{2}{5} & \frac{38}{5} & 1 & 1464
\end{array}
$$

The most negative element in the last row is in column 6. This will be the pivot column. We apply the smallest quotient rule at the top of the next page.

$$R_1: \frac{54}{\frac{1}{5}} = 270$$

$$R_3: \frac{54}{\frac{1}{5}} = 270$$

Therefore we can choose the pivot element to be in either row 1 or row 3 of the tableau on the previous page. We will choose the pivot element to be the $\frac{1}{5}$ in row 3, column 6. Pivoting on this element results in the tableau:

$$
\begin{array}{cccccccccc}
w & x & y & z & s_1 & s_2 & s_3 & s_4 & R & \\
\end{array}
$$

$$
\left[
\begin{array}{ccccccccc|c}
0 & 1 & -1 & 0 & 0 & 0 & -1 & 0 & 0 & 0 \\
0 & 0 & 0 & 1 & 0 & 0 & 0 & 1 & 0 & 20 \\
0 & 0 & 5 & 0 & \frac{1}{2} & 1 & 2 & -7 & 0 & 270 \\
1 & 0 & 5 & 0 & \frac{1}{2} & 0 & 2 & -5 & 0 & 310 \\
\hline
0 & 0 & 9 & 0 & \frac{5}{2} & 0 & 4 & -5 & 1 & 1950 \\
\end{array}
\right]
$$

The most negative element in the last row is in column 8.

The only positive element is the one in the row 2, column 8 position.

Pivoting on this element results in the following tableau:

$$
\begin{array}{cccccccccc}
w & x & y & z & s_1 & s_2 & s_3 & s_4 & R & \\
\end{array}
$$

$$
\left[
\begin{array}{ccccccccc|c}
0 & 1 & -1 & 0 & 0 & 0 & -1 & 0 & 0 & 0 \\
0 & 0 & 0 & 1 & 0 & 0 & 0 & 1 & 0 & 20 \\
0 & 0 & 5 & 7 & \frac{1}{2} & 1 & 2 & 0 & 0 & 410 \\
1 & 0 & 5 & 5 & \frac{1}{2} & 0 & 2 & 0 & 0 & 410 \\
\hline
0 & 0 & 9 & 5 & \frac{5}{2} & 0 & 4 & 0 & 1 & 2050 \\
\end{array}
\right]
$$

Since there are no more negative entries in the last row, we have reached a maximum for the objective function. The maximum revenue the company can generate next week is $205,000. This revenue will be generated if they make 410 QP20 printers, and none of the other three models.

Exercises Section 5.3

1.
a. The graph of the feasible region is shown at the top of the next column.

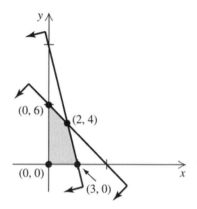

The corner points are:

$$A(3,0), B(2,4), C(0,6), \text{ and } D(0,0)$$

b. The initial tableau is shown to the linear programming problem is:

$$
\begin{array}{ccccccc}
x & y & s_1 & s_2 & f & \\
\end{array}
$$

$$
\left[
\begin{array}{ccccc|c}
1 & 1 & 1 & 0 & 0 & 6 \\
4 & 1 & 0 & 1 & 0 & 12 \\
\hline
-2 & -1 & 0 & 0 & 1 & 0 \\
\end{array}
\right]
$$

This tableau corresponds to the corner point D from part (a). Since all of the variables are non-negative we are at a corner point of the feasible region. This is the standard maximizing problem so we proceed with the simplex method as usual.

The most negative entry in the last row of the initial simplex tableau is the negative two in column 1. Applying the smallest quotient rule:

$$R_1: \frac{6}{1} = 6$$
$$R_2: \frac{12}{4} = 3$$

Therefore the four in row 2 column 1 will be the pivot element. Pivoting on this element results in the tableau:

$$
\begin{array}{ccccccc}
x & y & s_1 & s_2 & f & \\
\end{array}
$$

$$
\left[
\begin{array}{ccccc|c}
0 & \frac{3}{4} & 1 & \frac{-1}{4} & 0 & 3 \\
1 & \frac{1}{4} & 0 & \frac{1}{4} & 0 & 3 \\
\hline
0 & \frac{-1}{2} & 0 & \frac{1}{2} & 1 & 6 \\
\end{array}
\right]
$$

This tableau corresponds to the corner point A in part (a).

The solution is continued on the next page.

Now the $\frac{-1}{2}$ in column 2 is the remaining negative element in the last row. This will be the pivot column. Applying the smallest quotient rule:

$$R_1 : \frac{3}{\frac{3}{4}} = 4$$

$$R_2 : \frac{3}{\frac{1}{4}} = 12$$

Therefore the $\frac{3}{4}$ in row 1 column 2 will be the pivot element. Pivoting on this element results in the tableau:

$$
\begin{array}{ccccc}
x & y & s_1 & s_2 & f \\
\end{array}
$$
$$
\begin{bmatrix}
0 & 1 & \frac{4}{3} & \frac{-1}{3} & 0 & 4 \\
1 & 0 & \frac{-1}{3} & \frac{1}{3} & 0 & 2 \\
0 & 0 & \frac{2}{3} & \frac{1}{3} & 1 & 8 \\
\end{bmatrix}
$$

There are no more negative entries in the last row; therefore we have reached the maximum value of the objective function. The maximum value is:

$f = 8$ when $x = 2$, $y = 4$.

c. To check the solution, use the table to determine the value of the function at each corner point.

Point	$f = 2x + y$
$(0,0)$	$f = 2(0) + (0) = 0$
$(3,0)$	$f = 2(3) + (0) = 6$
$(2,4)$	$f = 2(2) + (4) = 8$
$(0,6)$	$f = 2(0) + (6) = 6$

3.
a. The graph of the feasible region is shown below:

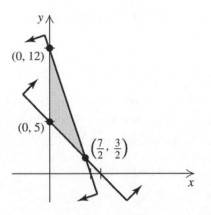

The corner points are the graph at the bottom of the previous column are:

$$A(0,12), B\left(\tfrac{7}{2}, \tfrac{1}{2}\right), \text{ and } C(0,5)$$

b. Rewrite the linear programming problem first in order to remove any greater than inequalities.

Maximize $f = 4x + 2y$
subject to:

$$-x - y \leq -5$$
$$3x + y \leq 12$$
$$x \geq 0, \; y \geq 0$$

Adding slack variables to this problem we get the following system:

$$
\begin{aligned}
-x - y + s_1 &= -5 \\
3x + y \quad + s_2 &= 12 \\
-4x - 2y \quad\quad + f &= 0
\end{aligned}
$$

The initial tableau corresponding to this linear programming problem is:

$$
\begin{array}{ccccc}
x & y & s_1 & s_2 & f \\
\end{array}
$$
$$
\begin{bmatrix}
-1 & -1 & 1 & 0 & 0 & -5 \\
3 & 1 & 0 & 1 & 0 & 12 \\
-4 & -2 & 0 & 0 & 1 & 0 \\
\end{bmatrix}
$$

Noticing that this tableau corresponds to the point $(0,0)$ which is not a corner point, we must apply Crown's rules. Locating the first negative in the column right of the vertical bar, we see that it is the negative five in the first row. Next we choose the most negative entry in the first row that is left of the vertical bar. Since both column 1 and column 2 have the same entry, we can choose either of the columns. We choose column 2 to be the pivot column. Since there is only one positive element in column 2, and the right most entry is positive as well, the one in row 2, column 2 will be the pivot element. Pivoting as usual we get:

$$
\begin{array}{ccccc}
x & y & s_1 & s_2 & f \\
\end{array}
$$
$$
\begin{bmatrix}
2 & 0 & 1 & 1 & 0 & 7 \\
3 & 1 & 0 & 1 & 0 & 12 \\
2 & 0 & 0 & 2 & 1 & 24 \\
\end{bmatrix}
$$

The solution is continued on the next page.

Looking at the matrix on the previous page, all of the entries to the right of the vertical bar are positive; we have arrived at a corner point to the feasible region. This point is $x = 0$, $y = 12$, and the function value at this corner point is $f = 24$. This corresponds to point A on the graph.

There are no more negative entries in the last row; therefore we have reached the maximum value of the objective function. The maximum value is:

$$f = 24 \text{ when } x = 0, \ y = 12.$$

c. To check the solution, use the following table to determine the value of the function at each corner point.

Point	$f = 4x + 2y$
$(0, 12)$	$f = 4(0) + 2(12) = 24$
$\left(\frac{7}{2}, \frac{3}{2}\right)$	$f = 4\left(\frac{7}{2}\right) + 2\left(\frac{3}{2}\right) = 17$
$(0, 5)$	$f = (0) + 2(5) = 10$

This confirms our solution from part (b).

5.
a. The graph of the feasible region is shown below:

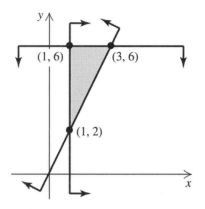

The corner points are:

$$A(1,6), B(3,6), \text{ and } C(1,2)$$

b. Rewrite the linear programming problem first in order to remove any greater than inequalities.

Maximize $f = 2x + 3y$
subject to:

$$2x - y \leq 0$$
$$-x \quad\ \leq -1$$
$$y \leq 6$$
$$x \geq 0, \ y \geq 0$$

Adding slack variables to this problem we get the following system:

$$2x - y + s_1 \qquad\qquad = 0$$
$$-x + \qquad + s_2 \qquad\ = -1$$
$$y \qquad + s_3 \quad = 6$$
$$-2x - 3y \qquad\qquad + f = 0$$

The initial tableau corresponding to this linear programming problem is:

$$
\begin{array}{cccccc}
x & y & s_1 & s_2 & s_3 & f \\
\end{array}
$$
$$
\left[
\begin{array}{cccccc|c}
2 & -1 & 1 & 0 & 0 & 0 & 0 \\
-1 & 0 & 0 & 1 & 0 & 0 & -1 \\
0 & 1 & 0 & 0 & 1 & 0 & 6 \\
\hline
-2 & -3 & 0 & 0 & 0 & 1 & 0 \\
\end{array}
\right]
$$

Noticing that this tableau corresponds to the point $(0,0)$ which is not a corner point, we must apply Crown's rules. Locating the first negative in the column right of the vertical bar, we see that it is the negative one in the second row. Next we choose the most negative entry in the third row that is left of the vertical bar. This is the negative one in column 1. Since there is only one positive element in column 1, Crown's rules indicate that we must pivot on the one in row 1. The resulting matrix is:

$$
\begin{array}{cccccc}
x & y & s_1 & s_2 & s_3 & f \\
\end{array}
$$
$$
\left[
\begin{array}{cccccc|c}
1 & \frac{-1}{2} & \frac{1}{2} & 0 & 0 & 0 & 0 \\
0 & \frac{-1}{2} & \frac{1}{2} & 1 & 0 & 0 & -1 \\
0 & 1 & 0 & 0 & 1 & 0 & 6 \\
\hline
0 & -4 & 1 & 0 & 0 & 1 & 0 \\
\end{array}
\right]
$$

The negative one in row 2 implies that we still have not reached a corner point. The point that this matrix corresponds to is $x = 0$, $y = 0$, and the function value at this corner point is $f = 0$.

Locating the first negative in the column right of the vertical bar, we see that it is the negative one in the second row. Next we choose the most negative entry in the third row that is left of the vertical bar. This is the entry $\frac{-1}{2}$ in column 2.

Since there is only one positive element in column 2, Crown's rules indicate that we must pivot on the one in row 3. The resulting matrix is displayed at the top of the next page.

$$
\begin{array}{cccccc}
x & y & s_1 & s_2 & s_3 & f \\
\end{array}
$$

$$
\left[
\begin{array}{cccccc|c}
1 & 0 & \frac{1}{2} & 0 & \frac{1}{2} & 0 & 3 \\
0 & 0 & \frac{1}{2} & 1 & \frac{1}{2} & 0 & 2 \\
0 & 1 & 0 & 0 & 1 & 0 & 6 \\
\hline
0 & 0 & 1 & 0 & 4 & 1 & 24
\end{array}
\right]
$$

Since all of the entries to the right of the vertical bar are positive, we have arrived at a corner point to the feasible region. This point is $x = 3$, $y = 6$, and the function value at this corner point is $f = 24$. This corresponds to point B on the graph.

There are no more negative entries in the last row; therefore we have reached the maximum value of the objective function. The maximum value is:

$f = 24$ when $x = 3$, $y = 6$.

c. To check the solution, use the table to determine the value of the function at each corner point.

Point	$f = 2x + 3y$
$(1,6)$	$f = 2(1) + 3(6) = 20$
$(3,6)$	$f = 2(3) + 3(6) = 24$
$(1,2)$	$f = 2(1) + 3(2) = 8$

This confirms our solution from part (b).

7.
a. The graph of the feasible region is shown below:

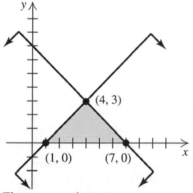

The corner points are:

$A(1,0), B(4,3),$ and $C(7,0)$

b. Rewrite the linear programming problem first in order to remove any greater than inequalities as shown at the top of the next column.

Maximize $f = -4x + 5y$
subject to:

$$
\begin{aligned}
-x + y &\le -1 \\
x + y &\le 7 \\
x \ge 0, \ y &\ge 0
\end{aligned}
$$

Adding slack variables to this problem we get the following system:

$$
\begin{aligned}
-x + y + s_1 \qquad\qquad &= -1 \\
x + y \qquad + s_2 \qquad &= 7 \\
4x - 5y \qquad\qquad + f &= 0
\end{aligned}
$$

The initial tableau corresponding to this linear programming problem is:

$$
\begin{array}{ccccc}
x & y & s_1 & s_2 & f \\
\end{array}
$$

$$
\left[
\begin{array}{ccccc|c}
-1 & 1 & 1 & 0 & 0 & -1 \\
1 & 1 & 0 & 1 & 0 & 7 \\
\hline
4 & -5 & 0 & 0 & 1 & 0
\end{array}
\right]
$$

Noticing that this tableau corresponds to the point $(0,0)$ which is not a corner point, we must apply Crown's rules. Locating the first negative in the column right of the vertical bar, we see that it is the negative one in the first row. Next we choose the most negative entry in the third row that is left of the vertical bar. This is the negative one in column 1. Since there is only one positive element in column 1, Crown's rules indicate that we must pivot on the one in row 2. The resulting matrix is:

$$
\begin{array}{ccccc}
x & y & s_1 & s_2 & f \\
\end{array}
$$

$$
\left[
\begin{array}{ccccc|c}
0 & 2 & 1 & 1 & 0 & 6 \\
1 & 1 & 0 & 1 & 0 & 7 \\
\hline
0 & -9 & 0 & -4 & 1 & -28
\end{array}
\right]
$$

Since all of the entries to the right of the vertical bar are positive, we have arrived at a corner point to the feasible region. This point is $x = 7$, $y = 0$, and the function value at this corner point is $f = -28$. This corresponds to point C on the graph.

We proceed with the simplex method. The most negative element in the last row is the negative nine in column 2. Using the smallest quotient rule, the pivot element will be the 2 in row 1 column 2. Pivoting on this element results in the matrix at the top of the next page.

$$\begin{array}{ccccc} x & y & s_1 & s_2 & f \\ \end{array}$$

$$\left[\begin{array}{ccccc|c} 0 & 1 & \frac{1}{2} & \frac{1}{2} & 0 & 3 \\ 1 & 0 & \frac{-1}{2} & \frac{1}{2} & 0 & 4 \\ \hline 0 & 0 & \frac{9}{2} & \frac{1}{2} & 1 & -1 \end{array}\right]$$

There are no more negative entries in the last row; therefore we have reached the maximum value of the objective function. The maximum value is:

$$f = -1 \text{ when } x = 4, \ y = 3.$$

c. To check the solution, use the following table to determine the value of the function at each corner point.

Point	$f = -4x + 5y$
$(1,0)$	$f = -4(1) + 5(0) = -4$
$(4,3)$	$f = -4(4) + 5(3) = -1$
$(7,0)$	$f = -4(7) + 5(0) = -28$

This confirms our solution from part (b).

9.
a. The graph of the feasible region is shown below:

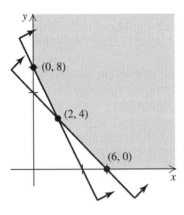

The corner points are:

$$A(0,8), B(2,4), \text{ and } C(6,0).$$

b. Rewrite the minimization problem to create the linear programming problem as shown below:

Maximize $\quad g = -f = -x - 3y$

subject to:

$$-x - y \le -6$$
$$-2x - y \le -8$$
$$x \ge 0, \ y \ge 0$$

Adding slack variables to this problem we get the following system:

$$\begin{aligned} -x - y + s_1 \qquad\qquad &= -6 \\ -2x - y \quad + s_2 \qquad &= -8 \\ x + 3y \qquad\qquad + g &= 0 \end{aligned}$$

The initial tableau corresponding to this linear programming problem is:

$$\begin{array}{ccccc} x & y & s_1 & s_2 & g \\ \end{array}$$

$$\left[\begin{array}{ccccc|c} -1 & -1 & 1 & 0 & 0 & -6 \\ -2 & -1 & 0 & 1 & 0 & -8 \\ \hline 1 & 3 & 0 & 0 & 1 & 0 \end{array}\right]$$

Noticing that this tableau corresponds to the point $(0,0)$ which is not a corner point, we must apply Crown's rules. Locating the first negative in the column right of the vertical bar, we see that it is the negative six in the first row. Next we choose the most negative entry in the first row that is left of the vertical bar. Since column 1 and column 2 each have an entry of negative 1, we can choose either column. We choose column 1, since there are no positive entries above the horizontal bar in column 1, we compute the ratios for both rows with the negative entries.

$$R_1 : \frac{-6}{-1} = 6$$
$$R_2 : \frac{-8}{-2} = 4$$

Since we are computing quotients for negative entries, we select the row with the largest quotient. This is the first row. Pivoting on the negative one in row 1 column 1 results in the following tableau:

$$\begin{array}{ccccc} x & y & s_1 & s_2 & g \\ \end{array}$$

$$\left[\begin{array}{ccccc|c} 1 & 1 & -1 & 0 & 0 & 6 \\ 0 & 1 & -2 & 1 & 0 & 4 \\ \hline 0 & 2 & 1 & 0 & 1 & -6 \end{array}\right]$$

Since all of the entries to the right of the vertical bar are positive (and above the horizontal bar), we have arrived at a corner point to the feasible region. This point is $x = 6$, $y = 0$, and the function value at this corner point is $g = -6$. This corresponds to point C on the graph.

The solution is continued on the next page.

Looking at the matrix on the previous page, we see that there are no more negative entries in the last row to the left of the vertical bar, and there are no more negative entries in the last column to the right of the vertical bar, other than the objective variable value. So we have arrived at the maximum value of the objective function g.

Since $g = -f$ the minimum value of f is

$-(-6) = 6$ and will occur at the point $x = 6,\ y = 0$

c. To check the solution, use the following table to determine the value of the function at each corner point.

Point	$f = x + 3y$
$(0,8)$	$f = (0) + 3(8) = 24$
$(2,4)$	$f = (2) + 3(4) = 14$
$(6,0)$	$f = (6) + 3(0) = 6$

This confirms our solution from part (b).

11.
a. The graph of the feasible region is:

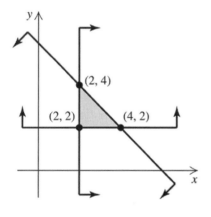

The corner points of the system are:

$A(2,2)$, $B(2,4)$, and $C(4,2)$.

b. Rewrite the minimization problem to create the linear programming problem.

Maximize $g = -f = -x - y$
subject to:

$$x + y \leq 6$$
$$-x \quad\ \leq -2$$
$$-y \leq -2$$
$$x \geq 0,\ y \geq 0$$

Adding slack variables to this problem we get the following system:

$$x + y + s_1 \qquad\qquad = 8$$
$$-x \qquad + s_2 \qquad\quad = -2$$
$$-y \qquad + s_3 \quad = -2$$
$$x + y \qquad\qquad + g = 0$$

The initial tableau corresponding to this linear programming problem is:

$$\begin{array}{cccccc} x & y & s_1 & s_2 & s_3 & g \\ \left[\begin{array}{cccccc|c} 1 & 1 & 1 & 0 & 0 & 0 & 6 \\ -1 & 0 & 0 & 1 & 0 & 0 & -2 \\ 0 & -1 & 0 & 0 & 1 & 0 & -2 \\ \hline 1 & 1 & 0 & 0 & 0 & 1 & 0 \end{array}\right] \end{array}$$

Noticing that this tableau corresponds to the point $(0,0)$ which is not a corner point, we must apply Crown's rules. Locating the first negative in the column right of the vertical bar, we see that it is the negative two in the second row. Next we choose the most negative entry in the second row that is left of the vertical bar. This is the negative one in column 1. Since there is one positive entry above the horizontal bar in column 1, we will pivot on the one in row 1 column 1. This results in the following tableau:

$$\begin{array}{cccccc} x & y & s_1 & s_2 & s_3 & g \\ \left[\begin{array}{cccccc|c} 1 & 1 & 1 & 0 & 0 & 0 & 6 \\ 0 & 1 & 1 & 1 & 0 & 0 & 4 \\ 0 & -1 & 0 & 0 & 1 & 0 & -2 \\ \hline 0 & 0 & -1 & 0 & 0 & 1 & -6 \end{array}\right] \end{array}$$

This corresponds to the point $x = 6,\ y = 0$ which is still not a corner point due to the negative in the last column above the horizontal bar. Proceeding with Crown's rule, we see that negative one is the most negative element in row 3. Using the smallest quotient rule, we will pivot on the one in row 2 column 2. This pivot results in the matrix:

$$\begin{array}{cccccc} x & y & s_1 & s_2 & s_3 & g \\ \left[\begin{array}{cccccc|c} 1 & 0 & 0 & -1 & 0 & 0 & 2 \\ 0 & 1 & 1 & 1 & 0 & 0 & 4 \\ 0 & 0 & 1 & 1 & 1 & 0 & 2 \\ \hline 0 & 0 & -1 & 0 & 0 & 1 & -6 \end{array}\right] \end{array}$$

The solution is continued on the next page.

Looking at the matrix at the bottom of the previous page, we notice that all of the entries to the right of the vertical bar are positive (and above the horizontal bar), we have arrived at a corner point to the feasible region. This point is $x = 2$, $y = 4$, and the function value at this corner point is $g = -6$. This corresponds to point B on the graph.

Since there are no more negative entries to the right of the vertical bar and above the horizontal bar, we proceed with the simplex method. The most negative entry in the last row is the negative one in column 3. Using the smallest quotient rule we will pivot on the one in row 3, column 3. The following tableau results:

$$
\begin{array}{cccccc}
x & y & s_1 & s_2 & s_3 & g \\
\end{array}
$$

$$
\left[
\begin{array}{cccccc|c}
1 & 0 & 0 & -1 & 0 & 0 & 2 \\
0 & 1 & 0 & 0 & -1 & 0 & 2 \\
0 & 0 & 1 & 1 & 1 & 0 & 2 \\
\hline
0 & 0 & 1 & 1 & 1 & 1 & -4 \\
\end{array}
\right]
$$

This point is $x = 2$, $y = 2$, and the function value at this corner point is $g = -4$. This corresponds to point A on the graph.

There are no more negative entries in the last row to the left of the vertical bar, and there are no more negative entries in the last column to the right of the vertical bar, other than the objective variable value. So we have arrived at the maximum value of the objective function g.

Since $g = -f$ the minimum value of f is $-(-4) = 4$ and will occur at the point $x = 2$, $y = 2$

c. To check the solution, use the table to determine the value of the function at each corner point.

Point	$f = x + y$
$(2,2)$	$f = (2) + (2) = 4$
$(2,4)$	$f = (2) + (4) = 6$
$(4,2)$	$f = (4) + (2) = 6$

This confirms our solution from part (b).

13.
a. The graph of the feasible region is shown at the top of the next column.

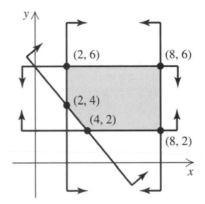

The corner points are:

$$ A(2,4), B(2,6), C(8,6), D(8,2) \text{ and } E(4,2). $$

b. Rewrite the minimization problem to create the linear programming problem.

Maximize $\qquad g = -C = -x + 2y$
subject to:

$$
\begin{aligned}
-x - y &\leq -6 \\
-x \quad\;\; &\leq -2 \\
x \quad\;\; &\leq 8 \\
-y &\leq -2 \\
y &\leq 6 \\
x \geq 0,\; y &\geq 0
\end{aligned}
$$

Adding slack variables to this problem we get the system:

$$
\begin{aligned}
-x - y + s_1 &= -6 \\
-x \qquad\quad + s_2 &= 2 \\
x \qquad\qquad + s_3 &= 8 \\
-y \qquad\qquad\quad + s_4 &= -2 \\
y \qquad\qquad\qquad + s_5 &= 6 \\
x - 2y \qquad\qquad\qquad\quad + g &= 0
\end{aligned}
$$

The initial tableau corresponding to this linear programming problem is displayed at the top of the next page.

$$\begin{array}{cccccccc}
x & y & s_1 & s_2 & s_3 & s_4 & s_5 & g \\
\end{array}$$

$$\left[\begin{array}{cccccccc|c}
-1 & -1 & 1 & 0 & 0 & 0 & 0 & 0 & -6 \\
-1 & 0 & 0 & 1 & 0 & 0 & 0 & 0 & -2 \\
1 & 0 & 0 & 0 & 1 & 0 & 0 & 0 & 8 \\
0 & -1 & 0 & 0 & 0 & 1 & 0 & 0 & -2 \\
0 & 1 & 0 & 0 & 0 & 0 & 1 & 0 & 6 \\
\hline
1 & -2 & 0 & 0 & 0 & 0 & 0 & 1 & 0
\end{array}\right]$$

Noticing that this tableau corresponds to the point $(0,0)$ which is not a corner point, we must apply Crown's rules. Locating the first negative in the column right of the vertical bar, we see that it is the negative six in the first row. Next we choose the most negative entry in the first row that is left of the vertical bar. Since there is a negative one in column 1 and column 2 we are free to choose either column. We choose to pivot on the one in row 3, column 1. Pivoting on this element results in the following tableau:

$$\begin{array}{cccccccc}
x & y & s_1 & s_2 & s_3 & s_4 & s_5 & g \\
\end{array}$$

$$\left[\begin{array}{cccccccc|c}
0 & -1 & 1 & 0 & 1 & 0 & 0 & 0 & 2 \\
0 & 0 & 0 & 1 & 1 & 0 & 0 & 0 & 6 \\
1 & 0 & 0 & 0 & 1 & 0 & 0 & 0 & 8 \\
0 & -1 & 0 & 0 & 0 & 1 & 0 & 0 & -2 \\
0 & 1 & 0 & 0 & 0 & 0 & 1 & 0 & 6 \\
\hline
0 & -2 & 0 & 0 & -1 & 0 & 0 & 1 & -8
\end{array}\right]$$

This corresponds to the point $x = 8$, $y = 0$ which is not a corner point. There is still a negative value to the right of the vertical bar in row 4. Finding the most negative entry to the left of the vertical bar in row 4, we see that column 2 will be the new pivot column. Since the one in row 5, column 2 is the only positive entry in column 2; we will pivot on that element. Pivoting results in the tableau below:

$$\begin{array}{cccccccc}
x & y & s_1 & s_2 & s_3 & s_4 & s_5 & g \\
\end{array}$$

$$\left[\begin{array}{cccccccc|c}
0 & 0 & 1 & 0 & 1 & 0 & 1 & 0 & 8 \\
0 & 0 & 0 & 1 & 1 & 0 & 0 & 0 & 6 \\
1 & 0 & 0 & 0 & 1 & 0 & 0 & 0 & 8 \\
0 & 0 & 0 & 0 & 0 & 1 & 1 & 0 & 4 \\
0 & 1 & 0 & 0 & 0 & 0 & 1 & 0 & 6 \\
\hline
0 & 0 & 0 & 0 & -1 & 0 & 2 & 1 & 4
\end{array}\right]$$

Since all of the entries to the right of the vertical bar are positive (and above the horizontal bar), we have arrived at a corner point to the feasible region. This point is $x = 8$, $y = 6$, and the function value at this corner point is $g = 4$. This corresponds to point C on the graph.

Since there are no more negative entries to the right of the vertical bar and above the horizontal bar, we proceed with the simplex method. The most negative entry in the last row is the negative one in column 5. Using the smallest quotient rule we will pivot on the one in row 2, column 5. Pivoting results in the following tableau:

$$\begin{array}{cccccccc}
x & y & s_1 & s_2 & s_3 & s_4 & s_5 & g \\
\end{array}$$

$$\left[\begin{array}{cccccccc|c}
0 & 0 & 1 & -1 & 0 & 0 & 1 & 0 & 2 \\
0 & 0 & 0 & 1 & 1 & 0 & 0 & 0 & 6 \\
1 & 0 & 0 & -1 & 0 & 0 & 0 & 0 & 2 \\
0 & 0 & 0 & 0 & 0 & 1 & 1 & 0 & 4 \\
0 & 1 & 0 & 0 & 0 & 0 & 1 & 0 & 6 \\
\hline
0 & 0 & 0 & 1 & 0 & 0 & 2 & 1 & 10
\end{array}\right]$$

This point is $x = 2$, $y = 6$, and the function value at this corner point is $g = 10$. This corresponds to point B on the graph.

There are no more negative entries in the last row to the left of the vertical bar, and there are no more negative entries in the last column to the right of the vertical bar, other than the objective variable value. So we have arrived at the maximum value of the objective function g.

Since $g = -C$ the minimum value of C is $-(10) = -10$ and will occur at the point $x = 2$, $y = 6$

c. To check the solution, use the table to determine the value of the function at each corner point.

Point	$C = x - 2y$
$(2,4)$	$f = (2) - 2(4) = -6$
$(2,6)$	$f = (2) - 2(6) = -10$
$(8,6)$	$f = (8) - 2(6) = -4$
$(8,2)$	$f = (8) - 2(2) = 4$
$(4,2)$	$f = (4) - 2(2) = 0$

This confirms our solution from part (b).

15.

a. The graph of the feasible region is shown below:

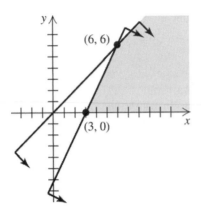

The corner points are:

$$A(3,0) \text{ , and } B(6,6)$$

b. Rewrite the minimization problem to create the linear programming problem.

Maximize $\qquad g = -f = -2x - y - 10$

subject to:

$$-x + y \le 0$$
$$-2x + y \le -6$$
$$x \ge 0, y \ge 0$$

Adding slack variables to this problem we get the system:

$$-x + y + s_1 \qquad = 0$$
$$-2x + y \quad + s_2 \qquad = -6$$
$$2x + y \qquad \quad + g = -10$$

The initial tableau corresponding to this linear programming problem is:

$$
\begin{array}{ccccc}
x & y & s_1 & s_2 & g \\
\end{array}
$$
$$
\left[
\begin{array}{ccccc|c}
-1 & 1 & 1 & 0 & 0 & 0 \\
-2 & 1 & 0 & 1 & 0 & -6 \\
\hline
2 & 1 & 0 & 0 & 1 & -10 \\
\end{array}
\right]
$$

Noticing that this tableau corresponds to the point $(0,0)$ which is not a corner point, we must apply Crown's rules.

Locating the first negative in the column right of the vertical bar, we see that it is the negative six in the second row.

Next we choose the most negative entry in the second row that is left of the vertical bar. This is the negative two in column 1. Crown's rules dictate that we pivot on the negative two in row 2, column 1. This results in the following tableau:

$$
\begin{array}{ccccc}
x & y & s_1 & s_2 & g \\
\end{array}
$$
$$
\left[
\begin{array}{ccccc|c}
0 & \frac{1}{2} & 1 & \frac{-1}{2} & 0 & 3 \\
1 & \frac{-1}{2} & 0 & \frac{-1}{2} & 0 & 3 \\
\hline
0 & 2 & 0 & 1 & 1 & -16 \\
\end{array}
\right]
$$

Since all of the entries to the right of the vertical bar are positive (and above the horizontal bar), we have arrived at a corner point to the feasible region. This point is $x = 3$, $y = 0$, and the function value at this corner point is $g = -16$. This corresponds to point B on the graph.

There are no more negative entries in the last row to the left of the vertical bar, and there are no more negative entries in the last column to the right of the vertical bar, other than the objective variable value. So we have arrived at the maximum value of the objective function g.

Since $g = -f$ the minimum value of f is

$-(-16) = 16$ and will occur at the point $x = 3$, $y = 0$

c. To check the solution, use the following table to determine the value of the function at each corner point.

Point	$f = 2x + y + 10$
$(3,0)$	$f = 2(3) + (0) + 10 = 16$
$(6,6)$	$f = 2(6) + (6) + 10 = 28$

This confirms our solution from part (b).

17.

Rewrite the minimization problem so it becomes a maximization problem.

Maximize $\qquad g = -C = -3x - 4y - z - 50$

subject to

$$x - y \qquad \le 0$$
$$-z \le -2$$
$$-x \qquad \le -4$$
$$-x - y - z \le -6$$
$$y \ge 0$$

The solution is continued on the next page.

Adding slack variables to the constraints, and then create the initial simplex tableau. The tableau is:

$$
\begin{array}{cccccccc|c}
x & y & z & s_1 & s_2 & s_3 & s_4 & g & \\
1 & -1 & 0 & 1 & 0 & 0 & 0 & 0 & 0 \\
0 & 0 & -1 & 0 & 1 & 0 & 0 & 0 & -2 \\
-1 & 0 & 0 & 0 & 0 & 1 & 0 & 0 & -4 \\
-1 & -1 & -1 & 0 & 0 & 0 & 1 & 0 & -6 \\
\hline
3 & 4 & 1 & 0 & 0 & 0 & 0 & 1 & -50
\end{array}
$$

Locating the first negative number in the right hand column in row 2, Crown's rules indicate that the pivot column will be column number 3. Since there are no positive/positive quotients, we look for negative/negative quotients. Looking for the largest ratio we see:

$$R_2 : \tfrac{-2}{-1} = 2$$

$$R_4 : \tfrac{-6}{-1} = 6$$

Therefore the pivot element will be the negative one in row 4 column 3. Pivoting on this element gives us the tableau:

$$
\begin{array}{cccccccc|c}
x & y & z & s_1 & s_2 & s_3 & s_4 & g & \\
1 & -1 & 0 & 1 & 0 & 0 & 0 & 0 & 0 \\
1 & 1 & 0 & 0 & 1 & 0 & -1 & 0 & 4 \\
-1 & 0 & 0 & 0 & 0 & 1 & 0 & 0 & -4 \\
1 & 1 & 1 & 0 & 0 & 0 & -1 & 0 & 6 \\
\hline
2 & 3 & 0 & 0 & 0 & 0 & 1 & 1 & -56
\end{array}
$$

Since there is still a negative value in the last column, we apply Crown's rules again to find the pivot element is the one in row 1 column 1. Pivoting on this element gives us the tableau:

$$
\begin{array}{cccccccc|c}
x & y & z & s_1 & s_2 & s_3 & s_4 & g & \\
1 & -1 & 0 & 1 & 0 & 0 & 0 & 0 & 0 \\
0 & 2 & 0 & -1 & 1 & 0 & -1 & 0 & 4 \\
0 & -1 & 0 & 1 & 0 & 1 & 0 & 0 & -4 \\
0 & 2 & 1 & -1 & 0 & 0 & -1 & 0 & 6 \\
\hline
0 & 5 & 0 & -2 & 0 & 0 & 1 & 1 & -56
\end{array}
$$

Since there is still a negative four in the third row of the last column, we apply Crown's rules again to find the pivot element is the two in row 2 column 2. Pivoting on this element gives us the tableau at the top of the next column.

$$
\begin{array}{cccccccc|c}
x & y & z & s_1 & s_2 & s_3 & s_4 & g & \\
1 & 0 & 0 & \tfrac{1}{2} & \tfrac{1}{2} & 0 & \tfrac{-1}{2} & 0 & 2 \\
0 & 1 & 0 & \tfrac{-1}{2} & \tfrac{1}{2} & 0 & \tfrac{-1}{2} & 0 & 2 \\
0 & 0 & 0 & \tfrac{1}{2} & \tfrac{1}{2} & 1 & \tfrac{-1}{2} & 0 & -2 \\
0 & 0 & 1 & 0 & -1 & 0 & 0 & 0 & 2 \\
\hline
0 & 0 & 0 & \tfrac{1}{2} & \tfrac{-5}{2} & 0 & \tfrac{7}{2} & 1 & -66
\end{array}
$$

Since there is still a negative two in the third row of the last column, we apply Crown's rules again to find

The most negative element in row 2 is the $\tfrac{-1}{2}$ in column 7.

The only negative/negative ratio involves the element in row 3 column 7. Pivoting on this element gives us the tableau:

$$
\begin{array}{cccccccc|c}
x & y & z & s_1 & s_2 & s_3 & s_4 & g & \\
1 & 0 & 0 & 0 & 0 & -1 & 0 & 0 & 4 \\
0 & 1 & 0 & -1 & 0 & -1 & 0 & 0 & 4 \\
0 & 0 & 0 & -1 & -1 & -2 & 1 & 0 & 4 \\
0 & 0 & 1 & 0 & -1 & 0 & 0 & 0 & 2 \\
\hline
0 & 0 & 0 & 4 & 1 & 7 & 0 & 1 & -80
\end{array}
$$

The absence of negative entries in the last column means that we have finally reached a corner point. The absence of negative entries in the last row means that we have reached the maximum of the objective function g. Since

$g = -C$ we can state that the minimum value of

$C = -(-80) = 80$, and this value occurs at the point

$x = 4, y = 4, z = 2$.

19.
Rewrite the constraints of the maximization problem.

Maximize $f = 3x + 2y + 6z$
subject to

$$x + 2y + z \le 6$$
$$y + 2z \le 4$$
$$-2x \quad - z \le -4$$
$$x \ge 0, y \ge 0, z \ge 0$$

Adding slack variables to the constraints, and then create the initial simplex tableau. The initial tableau is shown at the top of the next page.

$$
\begin{array}{ccccccc}
x & y & z & s_1 & s_2 & s_3 & f
\end{array}
$$

$$
\left[\begin{array}{ccccccc|c}
1 & 2 & 1 & 1 & 0 & 0 & 0 & 6 \\
0 & 1 & 2 & 0 & 1 & 0 & 0 & 4 \\
-2 & 0 & -1 & 0 & 0 & 1 & 0 & -4 \\
\hline
-3 & -2 & -6 & 0 & 0 & 0 & 1 & 0
\end{array}\right]
$$

The first negative number in the right hand column is the negative four in row 3. The most negative number in row four to the left of the vertical bar is the negative two in column 1. Since there is only one positive/positive quotient The pivot element will be the one in row 1 column 1. Pivoting on this element results in the following tableau:

$$
\begin{array}{ccccccc}
x & y & z & s_1 & s_2 & s_3 & f
\end{array}
$$

$$
\left[\begin{array}{ccccccc|c}
1 & 2 & 1 & 1 & 0 & 0 & 0 & 6 \\
0 & 1 & 2 & 0 & 1 & 0 & 0 & 4 \\
0 & 4 & 1 & 2 & 0 & 1 & 0 & 8 \\
\hline
0 & 4 & -3 & 3 & 0 & 0 & 1 & 18
\end{array}\right]
$$

The absence of negative values in the last column means that we have arrived at a corner point. However, the negative three in column 3 of the last row means we have not arrived at a maximum. The smallest quotient rule tells us that the pivot element will be the two in row 2 column 3. Pivoting on this element gives us the tableau:

$$
\begin{array}{ccccccc}
x & y & z & s_1 & s_2 & s_3 & f
\end{array}
$$

$$
\left[\begin{array}{ccccccc|c}
1 & \frac{3}{2} & 0 & 1 & \frac{-1}{2} & 0 & 0 & 4 \\
0 & \frac{1}{2} & 1 & 0 & \frac{1}{2} & 0 & 0 & 2 \\
0 & \frac{7}{2} & 0 & 2 & \frac{-1}{2} & 1 & 0 & 6 \\
\hline
0 & \frac{11}{2} & 0 & 3 & \frac{3}{2} & 0 & 1 & 24
\end{array}\right]
$$

The absence of negative entries in the last row means that we have reached the maximum of the objective function f. We can state that the maximum value of $f = 24$, and this value occurs at the point $x = 4, y = 0, z = 2$.

21.
Rewrite the constraints of the maximization problem as shown at the top of the next column.

Maximize $\qquad P = 3x + 4y + z$
subject to

$$3x + 2y + 4z \le 12$$
$$-x \quad\;\; - z \le -1$$
$$-y \quad\;\;\;\; \le -1$$
$$z \le 1$$
$$x \ge 0, y \ge 0, z \ge 0$$

Adding slack variables to the constraints, and then create the initial simplex tableau. The tableau is:

$$
\begin{array}{cccccccc}
x & y & z & s_1 & s_2 & s_3 & s_4 & P
\end{array}
$$

$$
\left[\begin{array}{cccccccc|c}
3 & 2 & 4 & 1 & 0 & 0 & 0 & 0 & 12 \\
-1 & 0 & -1 & 0 & 1 & 0 & 0 & 0 & -1 \\
0 & -1 & 0 & 0 & 0 & 1 & 0 & 0 & -1 \\
0 & 0 & 1 & 0 & 0 & 0 & 1 & 0 & 1 \\
\hline
-3 & -4 & -1 & 0 & 0 & 0 & 0 & 1 & 0
\end{array}\right]
$$

The first negative number in the right hand column is the negative one in row 2. The most negative number in row 2 to the left of the vertical bar is the negative one in column 1. Since there is only one positive/positive quotient The pivot element will be the three in row 1 column 1. Pivoting on this element gives us:

$$
\begin{array}{cccccccc}
x & y & z & s_1 & s_2 & s_3 & s_4 & P
\end{array}
$$

$$
\left[\begin{array}{cccccccc|c}
1 & \frac{2}{3} & \frac{4}{3} & \frac{1}{3} & 0 & 0 & 0 & 0 & 4 \\
0 & \frac{2}{3} & \frac{1}{3} & \frac{1}{3} & 1 & 0 & 0 & 0 & 3 \\
0 & -1 & 0 & 0 & 0 & 1 & 0 & 0 & -1 \\
0 & 0 & 1 & 0 & 0 & 0 & 1 & 0 & 1 \\
\hline
0 & -2 & 3 & 1 & 0 & 0 & 0 & 1 & 12
\end{array}\right]
$$

The first negative number in the right hand column is the negative one in row 3. The most negative number in row 3 to the left of the vertical bar is the negative one in column 2. The smallest positive/positive quotient tells us that the pivot element will be the $\frac{2}{3}$ in row 2, column 2. Pivoting on this element gives us the tableau at the top of the next page.

$$\begin{array}{cccccccc} x & y & z & s_1 & s_2 & s_3 & s_4 & P \end{array}$$

$$\left[\begin{array}{cccccccc|c} 1 & 0 & 1 & 0 & -1 & 0 & 0 & 0 & 1 \\ 0 & 1 & \frac{1}{2} & \frac{1}{2} & \frac{3}{2} & 0 & 0 & 0 & \frac{9}{2} \\ 0 & 0 & \frac{1}{2} & \frac{1}{2} & \frac{3}{2} & 1 & 0 & 0 & \frac{7}{2} \\ 0 & 0 & 1 & 0 & 0 & 0 & 1 & 0 & 1 \\ \hline 0 & 0 & 4 & 2 & 3 & 0 & 0 & 1 & 21 \end{array}\right]$$

The absence of negative entries in the last row means that we have reached the maximum of the objective function P. We can state that the maximum value of $P = 21$, and this value occurs at the point $x = 1, y = \frac{9}{2}, z = 0$.

23.
Rewrite the minimization problem so it becomes a maximization problem.

Maximize $\qquad g = -C = -4x - 3y - 2z$
subject to

$$-x - 2y - z \le -6$$
$$- y - 2z \le -1$$
$$2x + y + 2z \le 4$$
$$x \ge 0, y \ge 0, z \ge 0$$

Add slack variables to the constraints, and then create the initial simplex tableau. The tableau is:

$$\begin{array}{ccccccc} x & y & z & s_1 & s_2 & s_3 & g \end{array}$$

$$\left[\begin{array}{ccccccc|c} -1 & -2 & -1 & 1 & 0 & 0 & 0 & -6 \\ 0 & -1 & -2 & 0 & 1 & 0 & 0 & -1 \\ 2 & 1 & 2 & 0 & 0 & 1 & 0 & 4 \\ \hline 4 & 3 & 2 & 0 & 0 & 0 & 1 & 0 \end{array}\right]$$

The first negative number in the last column is the negative six in row 1. The most negative number in row 2 to the left of the vertical bar is the negative two in column 2. Since the only positive/positive quotient is in row 3, the pivot element will be the one in row 3, column 2. Pivoting on this element gives us the tableau:

$$\begin{array}{ccccccc} x & y & z & s_1 & s_2 & s_3 & g \end{array}$$

$$\left[\begin{array}{ccccccc|c} 3 & 0 & 3 & 1 & 0 & 2 & 0 & 2 \\ 2 & 0 & 0 & 0 & 1 & 1 & 0 & 3 \\ 2 & 1 & 2 & 0 & 0 & 1 & 0 & 4 \\ \hline -2 & 0 & -4 & 0 & 0 & -3 & 1 & -12 \end{array}\right]$$

The absence of negative entries in the last column of the matrix means that we have reached a corner point of the feasible region, but negative values in the last row to the left of the vertical bar means that we have still not reached a maximum of the objective function. Using the simplex method, we see that the most negative element in the last row is the negative four in column 3. The smallest quotient rule tells us that the pivot element will be the three in row 1, column 3. Pivoting on this element results in the tableau:

$$\begin{array}{ccccccc} x & y & z & s_1 & s_2 & s_3 & g \end{array}$$

$$\left[\begin{array}{ccccccc|c} 1 & 0 & 1 & \frac{1}{3} & 0 & \frac{2}{3} & 0 & \frac{2}{3} \\ 2 & 0 & 0 & 0 & 1 & 1 & 0 & 3 \\ 0 & 1 & 0 & \frac{-2}{3} & 0 & \frac{-1}{3} & 0 & \frac{8}{3} \\ \hline 2 & 0 & 0 & \frac{4}{3} & 0 & \frac{-1}{3} & 1 & \frac{-28}{3} \end{array}\right]$$

The only negative entry in the last row is now the $\frac{-1}{3}$ in column 6. The smallest quotient rule dictates that the pivot element will be the $\frac{2}{3}$ in row 1, column 6. Pivoting on this element results in the following tableau:

$$\begin{array}{ccccccc} x & y & z & s_1 & s_2 & s_3 & g \end{array}$$

$$\left[\begin{array}{ccccccc|c} \frac{3}{2} & 0 & \frac{3}{2} & \frac{1}{2} & 0 & 1 & 0 & 1 \\ \frac{1}{2} & 0 & \frac{-3}{2} & \frac{-1}{2} & 1 & 0 & 0 & 2 \\ \frac{1}{2} & 1 & \frac{1}{2} & \frac{-1}{2} & 0 & 0 & 0 & 3 \\ \hline \frac{5}{2} & 0 & \frac{1}{2} & \frac{3}{2} & 0 & 0 & 1 & -9 \end{array}\right]$$

The absence of negative entries in the last row means that we have reached the maximum of the objective function g. Since $g = -C$ we can state that the minimum value of $C = -(-9) = 9$, and this value occurs at the point $x = 0, y = 3, z = 0$.

25.
Rewrite the constraints of the maximization problem. Notice the trick used to write the equation as two inequalities.

Maximize $\qquad f = x + 2y + z$
subject to

$$x + y + 2z \le 12$$
$$2x + y + z \le 16$$
$$x + y \le 3$$
$$-x - y \le -3$$
$$x \ge 0, y \ge 0, z \ge 0$$

Add slack variables to the constraints on the previous page, and then create the initial simplex tableau as shown below:

$$
\begin{array}{cccccccc}
x & y & z & s_1 & s_2 & s_3 & s_4 & f \\
\end{array}
$$

$$
\left[\begin{array}{cccccccc|c}
1 & 1 & 2 & 1 & 0 & 0 & 0 & 0 & 12 \\
2 & 1 & 1 & 0 & 1 & 0 & 0 & 0 & 16 \\
1 & 1 & 0 & 0 & 0 & 1 & 0 & 0 & 3 \\
-1 & -1 & 0 & 0 & 0 & 0 & 1 & 0 & -3 \\
\hline
-1 & -2 & -1 & 0 & 0 & 0 & 0 & 1 & 0
\end{array}\right]
$$

The first negative number in the right hand column is the negative three in row 4. The most negative number in row 4 to the left of the vertical bar is the negative one in column 1 or column 2. We choose column 2. The smallest positive/positive quotient chooses the one in row 3, column 2 as the pivot element. Pivoting on this element gives us:

$$
\begin{array}{cccccccc}
x & y & z & s_1 & s_2 & s_3 & s_4 & f \\
\end{array}
$$

$$
\left[\begin{array}{cccccccc|c}
0 & 0 & 2 & 1 & 0 & -1 & 0 & 0 & 9 \\
1 & 0 & 1 & 0 & 1 & -1 & 0 & 0 & 13 \\
1 & 1 & 0 & 0 & 0 & 1 & 0 & 0 & 3 \\
0 & 0 & 0 & 0 & 0 & 1 & 1 & 0 & 0 \\
\hline
1 & 0 & -1 & 0 & 0 & 2 & 0 & 1 & 6
\end{array}\right]
$$

The absence of negative entries in the last column means that we have reached a corner point of the feasible region, but negative values in the last row to the left of the vertical bar means that we have still not reached a maximum of the objective function. Using the simplex method, we see that the most negative element in the last row is the negative one in column 3. The smallest quotient rule tells us that the pivot element will be the two in row 1 column 3. Pivoting on this element results in the tableau:

$$
\begin{array}{cccccccc}
x & y & z & s_1 & s_2 & s_3 & s_4 & f \\
\end{array}
$$

$$
\left[\begin{array}{cccccccc|c}
0 & 0 & 1 & \frac{1}{2} & 0 & \frac{-1}{2} & 0 & 0 & \frac{9}{2} \\
1 & 0 & 0 & \frac{-1}{2} & 1 & \frac{-1}{2} & 0 & 0 & \frac{17}{2} \\
1 & 1 & 0 & 0 & 0 & 1 & 0 & 0 & 3 \\
0 & 0 & 0 & 0 & 0 & 1 & 1 & 0 & 0 \\
\hline
1 & 0 & 0 & \frac{1}{2} & 0 & \frac{3}{2} & 0 & 1 & \frac{21}{2}
\end{array}\right]
$$

The absence of negative entries in the last row means that we have reached the maximum of the objective function f.

We can state that the maximum value of $f = \frac{21}{2}$, and this value occurs at the point $x = 0, y = 3, z = \frac{9}{2}$.

27.
The first constraint is an equation, so we will use the equation to eliminate one of the variables. Solving $2x - y = 3$ for y we get:

$$y = 2x - 3$$

We will substitute this value of y into the objective function and the other constraints to get:

$$C = 3x + (2x - 3) - 2z \text{ or}$$
$$C = 5x - 2z - 3$$

The second constraint becomes:

$$x + (2x - 3) - 2z \le 4 \text{ or}$$
$$3x - 2z \le 7$$

The third constraint does not involve the variable y so it will not change. The restructured problem is:

However, since $y \ge 0$ we must add the constraint:

$$2x - 3 \ge 0$$

Minimize $\qquad C = 5x - 2z - 3$
subject to

$$3x - 2z \le 7$$
$$x - z \ge 0$$
$$2x - 3 \ge 0$$
$$x \ge 0, z \ge 0$$

Now rewriting the minimization problem as a maximization problem, we get:

Maximize $\qquad g = -C = -5x + 2z + 3$
subject to

$$3x - 2z \le \;\; 7$$
$$-x + z \le \;\; 0$$
$$-2x \qquad \le -3$$
$$x \ge 0, z \ge 0$$

The initial tableau is displayed on the next page.

The initial tableau for the linear programming problem on the previous page is:

$$
\begin{array}{cccccc}
x & z & s_1 & s_2 & s_3 & g \\
\end{array}
$$

$$
\left[\begin{array}{cccccc|c}
3 & -2 & 1 & 0 & 0 & 0 & 7 \\
-1 & 1 & 0 & 1 & 0 & 0 & 0 \\
-2 & 0 & 0 & 0 & 1 & 0 & -3 \\
\hline
5 & -2 & 0 & 0 & 0 & 1 & 3
\end{array}\right]
$$

The first negative number in the last column is the negative three in row 3. The most negative number in row 2 to the left of the vertical bar is the negative two in column 1. Since the only positive/positive quotient is in row 1, the pivot element will be the three in row 1, column 1. Pivoting on this element gives us:

$$
\begin{array}{cccccc}
x & z & s_1 & s_2 & s_3 & g \\
\end{array}
$$

$$
\left[\begin{array}{cccccc|c}
1 & \frac{-2}{3} & \frac{1}{3} & 0 & 0 & 0 & \frac{7}{3} \\
0 & \frac{1}{3} & \frac{1}{3} & 1 & 0 & 0 & \frac{7}{3} \\
0 & \frac{-4}{3} & \frac{2}{3} & 0 & 1 & 0 & \frac{5}{3} \\
\hline
0 & \frac{4}{3} & \frac{-5}{3} & 0 & 0 & 1 & \frac{-26}{3}
\end{array}\right]
$$

The absence of negative entries in the last column means that we have reached a corner point of the feasible region, but negative values in the last row to the left of the vertical bar means that we have still not reached a maximum of the objective function. Using the simplex method, we see that the most negative element in the last row is $\frac{-5}{3}$ in column 3.

The smallest quotient rule tells us that the pivot element will be the $\frac{2}{3}$ in row 3, column 3. Pivoting on this element results in the tableau:

$$
\begin{array}{cccccc}
x & z & s_1 & s_2 & s_3 & g \\
\end{array}
$$

$$
\left[\begin{array}{cccccc|c}
1 & 0 & 0 & 0 & \frac{-1}{2} & 0 & \frac{3}{2} \\
0 & 1 & 0 & 1 & \frac{-1}{2} & 0 & \frac{3}{2} \\
0 & -2 & 1 & 0 & \frac{3}{2} & 0 & \frac{5}{2} \\
\hline
0 & -2 & 0 & 0 & \frac{5}{2} & 1 & \frac{-9}{2}
\end{array}\right]
$$

The only negative entry in the last row is now the negative 2 in column 2. The simplex method tells us that the one in row 2 column 2 will be the pivot element. Pivoting on this element results in the tableau at the top of the next column.

$$
\begin{array}{cccccc}
x & z & s_1 & s_2 & s_3 & g \\
\end{array}
$$

$$
\left[\begin{array}{cccccc|c}
1 & 0 & 0 & 0 & \frac{-1}{2} & 0 & \frac{3}{2} \\
0 & 1 & 0 & 1 & \frac{-1}{2} & 0 & \frac{3}{2} \\
0 & 0 & 1 & 2 & \frac{1}{2} & 0 & \frac{11}{2} \\
\hline
0 & 0 & 0 & 2 & \frac{3}{2} & 1 & \frac{-3}{2}
\end{array}\right]
$$

The absence of negative entries in the last row means that we have reached the maximum of the objective function g.

Since $g = -C$ we can state that the minimum value of

$$C = -\left(\frac{-3}{2}\right) = \frac{3}{2}.$$

From the tableau we see that $x = \frac{3}{2}, z = \frac{3}{2}$.

Don't forget that $y = 2x - 3$, so if $x = \frac{3}{2}$

$y = 2\left(\frac{3}{2}\right) - 3 = 0$. So the solution that gives the

minimum value of C is $x = \frac{3}{2}, y = 0, z = \frac{3}{2}$.

29.

Since $x_2 = 2$ we substitute that value in to the objective function and all of the constraints to get the problem:

Maximize $f = 3x_1 + 8 + x_3$

subject to

$$x_1 + 2 + x_3 \leq 8$$
$$2 + x_3 \leq 4$$
$$x_1 \geq 0, x_3 \geq 0$$

Rewrite the problem to obtain:

Maximize $f = 3x_1 + x_3 + 8$

subject to

$$x_1 + x_3 \leq 6$$
$$x_3 \leq 2$$
$$x_1 \geq 0, x_3 \geq 0$$

The initial tableau is:

$$
\begin{array}{ccccc}
x_1 & x_3 & s_1 & s_2 & f \\
\end{array}
$$

$$
\left[\begin{array}{ccccc|c}
1 & 1 & 1 & 0 & 0 & 6 \\
0 & 1 & 0 & 1 & 0 & 2 \\
\hline
-3 & -1 & 0 & 0 & 1 & 8
\end{array}\right]
$$

The solution is continued on the next page.

The problem on the previous page is a standard simplex method problem. The largest negative entry in the last row is the negative three in column 1. The only positive element in column 1 is the one in row 1. Therefore, we will pivot on the one in row 1, column 1. Pivoting results in the tableau:

$$
\begin{array}{ccccc}
x_1 & x_3 & s_1 & s_2 & f \\
\end{array}
$$
$$
\left[\begin{array}{ccccc|c}
1 & 1 & 1 & 0 & 0 & 6 \\
0 & 1 & 0 & 1 & 0 & 2 \\
\hline
0 & 2 & 3 & 0 & 1 & 26 \\
\end{array}\right]
$$

The maximum of the objective function is $f = 26$, and is obtained when $x_1 = 6$, $x_2 = 2$, $x_3 = 0$.

31.

Let x be the number of Daybrite sets stocked and sold, and let y be the number of Noglare sets. The revenue function is:

$$R = 420x + 550y$$

The warehouse restriction is:

$$x + y \leq 300.$$

The restrictions imposed by the demands for the various brands are:

$$x \geq 80$$
$$y \geq 120$$

Rewriting the appropriate constraints to form the maximization problem, we get:

Maximize $\qquad R = 420x + 550y$
subject to
$$
\begin{aligned}
x + y &\leq 300 \\
-x &\leq -80 \\
-y &\leq -120 \\
x \geq 0, y &\geq 0
\end{aligned}
$$

Adding slack variables, the initial tableau is displayed at the top of the next column.

$$
\begin{array}{cccccc}
x & y & s_1 & s_2 & s_3 & R \\
\end{array}
$$
$$
\left[\begin{array}{cccccc|c}
1 & 1 & 1 & 0 & 0 & 0 & 300 \\
-1 & 0 & 0 & 1 & 0 & 0 & -80 \\
0 & -1 & 0 & 0 & 1 & 0 & -120 \\
\hline
-420 & -550 & 0 & 0 & 0 & 1 & 0 \\
\end{array}\right]
$$

The first negative number in the last column is -80 in row 2. The only negative number to the left of the vertical bar in row 2 is the negative one in column 1. Since there is only one positive/positive ratio in column 1, the pivot element will be the one in row 1, column 1. Pivoting on this element results in the following tableau:

$$
\begin{array}{cccccc}
x & y & s_1 & s_2 & s_3 & R \\
\end{array}
$$
$$
\left[\begin{array}{cccccc|c}
1 & 1 & 1 & 0 & 0 & 0 & 300 \\
0 & 1 & 1 & 1 & 0 & 0 & 220 \\
0 & -1 & 0 & 0 & 1 & 0 & -120 \\
\hline
0 & -130 & 420 & 0 & 0 & 1 & 126000 \\
\end{array}\right]
$$

The next negative number in the last column is the -120 in row 3. The most negative number to the left of the vertical bar in row 3 is the negative one in column 2. The smallest positive/positive ratio in column 2 is row 2. Therefore, the pivot element will be the one in row 2, column 2. Pivoting on this element results in the matrix:

$$
\begin{array}{cccccc}
x & y & s_1 & s_2 & s_3 & R \\
\end{array}
$$
$$
\left[\begin{array}{cccccc|c}
1 & 0 & 0 & -1 & 0 & 0 & 80 \\
0 & 1 & 1 & 1 & 0 & 0 & 220 \\
0 & 0 & 1 & 1 & 1 & 0 & 100 \\
\hline
0 & 0 & 550 & 130 & 0 & 1 & 154,600 \\
\end{array}\right]
$$

Since there are no negatives to the right of the vertical bar, or below the horizontal bar, we have arrived at the maximum value for the objective function. The maximum revenue is \$154,600, and will occur if the company orders 80 Daybrite sets, and 220 Noglare sets.

33.

Let x be the number of half-hour speeches to civic groups, y be the number of hours spent on the phone, and z be the number of hour long trips to shopping centers. The objective is to minimize the time spent and still win. The time function is given by:

$$T = \tfrac{1}{2}x + y + z.$$

The solution is continued on the next page.

The minimum requirement of 1000 votes results in the constraint:

$$20x + 10y + 40z \geq 1000.$$

The fact that the candidate wants to make at least twice as many shopping area stops as speeches to civic groups results in the constraint:

$$2x \leq z, \text{ or } 2x - z \leq 0.$$

Since the candidate wishes to spend at least three hours on the phone, we need the constraint:

$$y \geq 3.$$

Since this is a minimization problem we will need to rewrite the objective function. If we let $g = -T = -\frac{1}{2}x - y - z$. Then finding the maximum value of g will lead us to the minimum value of T. Furthermore, we will need to rewrite all of the constraints to fit the form of the maximization problem. Once this is done, the linear programming problem is:

Maximize $\qquad g = -T = -\frac{1}{2}x - y - z$

subject to

$$-20x - 10y - 40z \leq -1000$$
$$2x - \qquad z \leq 0$$
$$-y \qquad \leq -3$$

After adding slack variables to each of the constraints, the initial tableau is stated below:

$$
\begin{array}{ccccccc}
x & y & z & s_1 & s_2 & s_3 & g
\end{array}
$$
$$
\left[
\begin{array}{ccccccc|c}
-20 & -10 & -40 & 1 & 0 & 0 & 0 & -1000 \\
2 & 0 & -1 & 0 & 1 & 0 & 0 & 0 \\
0 & -1 & 0 & 0 & 0 & 1 & 0 & -3 \\
\hline
\frac{1}{2} & 1 & 1 & 0 & 0 & 0 & 1 & 0
\end{array}
\right]
$$

Crown's rules indicate that -40 in row 1, column 3 is the pivot element. Pivoting on this element results in the tableau:

$$
\begin{array}{ccccccc}
x & y & z & s_1 & s_2 & s_3 & g
\end{array}
$$
$$
\left[
\begin{array}{ccccccc|c}
\frac{1}{2} & \frac{1}{4} & 1 & \frac{-1}{40} & 0 & 0 & 0 & 25 \\
\frac{5}{2} & \frac{1}{4} & 0 & \frac{-1}{40} & 1 & 0 & 0 & 25 \\
0 & -1 & 0 & 0 & 0 & 1 & 0 & -3 \\
\hline
0 & \frac{3}{4} & 0 & \frac{1}{40} & 0 & 0 & 1 & -25
\end{array}
\right]
$$

Since there is still a negative entry to the right of the vertical bar, Crown's rules indicate that we will pivot on the $\frac{1}{4}$ in either row 1 or row 2 of column 2. We elect to pivot on the $\frac{1}{4}$ in row 1, column 2. Pivoting on this element results in the tableau:

$$
\begin{array}{ccccccc}
x & y & z & s_1 & s_2 & s_3 & g
\end{array}
$$
$$
\left[
\begin{array}{ccccccc|c}
2 & 1 & 4 & \frac{-1}{10} & 0 & 0 & 0 & 100 \\
2 & 0 & -1 & 0 & 1 & 0 & 0 & 0 \\
2 & 0 & 4 & \frac{-1}{10} & 0 & 1 & 0 & 97 \\
\hline
\frac{-3}{2} & 0 & -3 & \frac{1}{10} & 0 & 0 & 1 & -100
\end{array}
\right]
$$

Since there are not any negative entries to the right of the vertical bar, we proceed with the simplex method. The most negative value below the horizontal bar is the negative three in column 3. The smallest quotient rule indicates that the pivot element will be the four in row 3, column 3. Pivoting on this element results in the following tableau:

$$
\begin{array}{ccccccc}
x & y & z & s_1 & s_2 & s_3 & g
\end{array}
$$
$$
\left[
\begin{array}{ccccccc|c}
0 & 1 & 0 & 0 & 0 & -1 & 0 & 3 \\
\frac{5}{2} & 0 & 0 & \frac{-1}{40} & 1 & \frac{1}{4} & 0 & \frac{97}{4} \\
\frac{1}{2} & 0 & 1 & \frac{-1}{40} & 0 & \frac{1}{4} & 0 & \frac{97}{4} \\
\hline
0 & 0 & 0 & \frac{1}{40} & 0 & \frac{3}{4} & 1 & \frac{-109}{4}
\end{array}
\right]
$$

The absence of negative entries in the bottom row indicates that a maximum for g has been found, and thus a minimum of T has been found. The tableau indicates in order to receive at least 1000 votes, the minimum amount of time spent campaigning must be $-\left(\frac{-109}{4}\right) = \frac{109}{4} = 27.25$ hours. The candidate would need to spend zero hours giving speeches, three hours on the phone, $\frac{97}{4} = 24.25$ hours visiting shopping areas.

It should be noted that if we had pivoted on the $\frac{1}{4}$ in row 2, column 2 instead of row 1, column 2 in the second pivot operation, the solution would have been the same minimum time of 27.25 hours, but would have required 4.85 half hour speeches, 19.4 hours visiting shopping areas and 3 hours on the telephone.

This parameterization of the solution to is stated at the top of the next page.

The parameterized solution to the problem on the previous page is:

$$x = 58.5 - 2t$$

$$y = 3$$

$$z = t$$

where,

$$19.4 \le t \le 24.25$$

This general solution will satisfy all of the requirements and give a minimum value of $T = 27.25$.

35.

Let x, y, and z be the number of doses of A, B, and C, respectively in a 24-hour period. The total painkilling effect is given by the function:

$$P = 2x + 2.5y + 3z.$$

The dosage restrictions are:

$$x \qquad \le 3$$

$$x + \quad z \le 5$$

$$y + z \le 7$$

The linear programming problem is:

Maximize $\qquad P = 2x + 2.5y + 3z$

subject to

$$x \qquad \le 3$$

$$x + \quad z \le 5$$

$$y + z \le 7$$

$$x \ge 0, y \ge 0, z \ge 0$$

Adding slack variables for each of the constraints, the initial tableau is:

$$
\begin{array}{ccccccc}
x & y & z & s_1 & s_2 & s_3 & P
\end{array}
$$
$$
\left[
\begin{array}{ccccccc|c}
1 & 0 & 0 & 1 & 0 & 0 & 0 & 3 \\
1 & 0 & 1 & 0 & 1 & 0 & 0 & 5 \\
0 & 1 & 1 & 0 & 0 & 1 & 0 & 7 \\
\hline
-2 & \frac{-5}{2} & -3 & 0 & 0 & 0 & 1 & 0
\end{array}
\right]
$$

The solution is continued at the top of the next column.

This is a standard simplex method problem. The most negative entry in the last row is the negative three in column 3. The smallest quotient rule indicates that the pivot element is the one in row 2, column 3. Pivoting on this element results in the tableau:

$$
\begin{array}{ccccccc}
x & y & z & s_1 & s_2 & s_3 & P
\end{array}
$$
$$
\left[
\begin{array}{ccccccc|c}
1 & 0 & 0 & 1 & 0 & 0 & 0 & 3 \\
1 & 0 & 1 & 0 & 1 & 0 & 0 & 5 \\
-1 & 1 & 0 & 0 & -1 & 1 & 0 & 2 \\
\hline
1 & \frac{-5}{2} & 0 & 0 & 3 & 0 & 1 & 15
\end{array}
\right]
$$

The only negative entry in the last row is now the $\frac{-5}{2}$ in column 2. The only positive entry above the horizontal bar in column 2 is the one in row 3. Therefore, the pivot element is the one in row 3, column 2. Pivoting on this element results in the tableau:

$$
\begin{array}{ccccccc}
x & y & z & s_1 & s_2 & s_3 & P
\end{array}
$$
$$
\left[
\begin{array}{ccccccc|c}
1 & 0 & 0 & 1 & 0 & 0 & 0 & 3 \\
1 & 0 & 1 & 0 & 1 & 0 & 0 & 5 \\
-1 & 1 & 0 & 0 & -1 & 1 & 0 & 2 \\
\hline
\frac{-3}{2} & 0 & 0 & 0 & \frac{1}{2} & \frac{5}{2} & 1 & 20
\end{array}
\right]
$$

The only negative entry in the last row is now the element in column 1. The smallest quotient rule indicates that the pivot element will be the one in row 1, column 1. Pivoting on this element results in the tableau:

$$
\begin{array}{ccccccc}
x & y & z & s_1 & s_2 & s_3 & P
\end{array}
$$
$$
\left[
\begin{array}{ccccccc|c}
1 & 0 & 0 & 1 & 0 & 0 & 0 & 3 \\
0 & 0 & 1 & -1 & 1 & 0 & 0 & 2 \\
0 & 1 & 0 & 1 & -1 & 1 & 0 & 5 \\
\hline
0 & 0 & 0 & \frac{3}{2} & \frac{1}{2} & \frac{5}{2} & 1 & 24.5
\end{array}
\right]
$$

The absence of negative values in the last row indicates that we have arrived at a maximum value for the objective function. The maximum pain-killing effect is 24.5. To achieve this level, the patient should ingest three doses of the type A painkiller, five doses of the type B painkiller and two doses of the type C painkiller.

37.
Let x, y, and z be the number of fax machine, computers, and CD players stocked and sold respectively.

The space restrictions can be translated into the following inequality:

$$x + y + z \leq 100$$

Past sales patterns indicate that:

$$x = y$$
$$z \geq 20, \text{ or } -z \leq -20$$

The objective is to maximize revenue. The revenue function (in hundreds of dollars) is:

$$R = 5x + 18y + 10z$$

Before setting up the linear programming problem, we notice that one of the constraints is an equation. Therefore we will substitute $x = y$ into the objective function and the constraints. Therefore,

$$R = 5x + 18(x) + 10z = 23x + 10z$$

and,

$$x + (x) + z \leq 100, \text{ or }$$
$$2x + z \leq 100$$

The linear programming problem will be:

Maximize $\qquad R = 23x + 10z$
subject to
$$2x + z \leq 100$$
$$-z \leq -20$$
$$x \geq 0, y \geq 0, z \geq 0$$

Adding slack variables, the initial tableau is:

$$
\begin{array}{cccccc}
x & z & s_1 & s_2 & R & \\
\left[\begin{array}{ccccc|c}
2 & 1 & 1 & 0 & 0 & 100 \\
0 & -1 & 0 & 1 & 0 & -20 \\
\hline
-23 & -10 & 0 & 0 & 1 & 0
\end{array}\right]
\end{array}
$$

The negative in the last column implies that we must use Crown's rules. Crown's rules indicate that the pivot element will be the one in row 1, column 2. Pivoting on this element results in the tableau at the top of the next column.

$$
\begin{array}{ccccccc}
x & z & s_1 & s_2 & R & \\
\left[\begin{array}{ccccc|c}
2 & 1 & 1 & 0 & 0 & 100 \\
2 & 0 & 1 & 1 & 0 & 80 \\
\hline
-3 & 0 & 10 & 0 & 1 & 1000
\end{array}\right]
\end{array}
$$

Now we can apply the simplex method. The most negative entry in the last row is the negative three in column 1. The smallest quotient rule indicates that the new pivot element will be the two in row 2, column 1. Pivoting on this element results in the tableau:

$$
\begin{array}{ccccccc}
x & z & s_1 & s_2 & R & \\
\left[\begin{array}{ccccc|c}
0 & 1 & 0 & -1 & 0 & 20 \\
1 & 0 & \frac{1}{2} & \frac{1}{2} & 0 & 40 \\
\hline
0 & 0 & \frac{23}{2} & \frac{3}{2} & 1 & 1120
\end{array}\right]
\end{array}
$$

The absence of negatives in the last row implies that we have reached a maximum for the objective function. The tableau indicates that $R = 1120$, $x = 40, z = 20$.

Remember however that $x = y$, therefore, $y = 40$. Also remember that R is in hundreds of dollars.

Therefore from the tableau, the maximum revenue that can be generated by InfoAge Communications is \$112,000. This revenue will be generated if they stock and sell 40 fax machines, 40 computers, and 20 portable CD players.

39.
Let x represent thousands of radial tires, y represent thousands of snow tires, and z represent thousands of off-road tires manufactured each day.

The union contract requires the use of at least 500 work-hours a day, or 30,000 work-minutes a day. This translates into the inequality:

$$4x + 6y + 8z \geq 30$$

The orders received by the company can be translated into the following inequalities:

$$x \geq 1$$
$$y \geq 0.4$$
$$z \geq 0.1$$

Cost function in thousands of dollars is:

$$C = 20x + 30y + 40z + 2$$

The solution is continued on the next page.

The objective is to minimize costs, to reach this objective we will need to maximize

$$g = -C = -20x - 30y - 40z - 2 \,.$$

Rewriting the constraints, the linear programming problem becomes:

Maximize $\quad g = -C = -20x - 30y - 40z - 2$

subject to

$$-4x - 6y - 8z \le -30$$
$$-x \qquad\qquad \le\ -1$$
$$-y \qquad \le -0.4$$
$$-z \le -0.1$$
$$x \ge 0,\ y \ge 0,\ z \ge 0$$

Adding the slack variables, the initial tableau becomes:

$$
\begin{array}{cccccccc}
x & y & z & s_1 & s_2 & s_3 & s_4 & g
\end{array}
$$
$$
\left[
\begin{array}{cccccccc|c}
-4 & -6 & -8 & 1 & 0 & 0 & 0 & 0 & -30 \\
-1 & 0 & 0 & 0 & 1 & 0 & 0 & 0 & -1 \\
0 & -1 & 0 & 0 & 0 & 1 & 0 & 0 & -0.4 \\
0 & 0 & -1 & 0 & 0 & 0 & 1 & 0 & -0.1 \\
\hline
20 & 30 & 40 & 0 & 0 & 0 & 0 & 1 & -2
\end{array}
\right]
$$

Crown's rules indicate that we should pivot on the negative eight in row 1, column 3. Pivoting on this element results in the tableau:

$$
\begin{array}{cccccccc}
x & y & z & s_1 & s_2 & s_3 & s_4 & g
\end{array}
$$
$$
\left[
\begin{array}{cccccccc|c}
\frac{1}{2} & \frac{3}{4} & 1 & \frac{-1}{8} & 0 & 0 & 0 & 0 & \frac{15}{4} \\
-1 & 0 & 0 & 0 & 1 & 0 & 0 & 0 & -1 \\
0 & -1 & 0 & 0 & 0 & 1 & 0 & 0 & -0.4 \\
\frac{1}{2} & \frac{3}{4} & 0 & \frac{-1}{8} & 0 & 0 & 1 & 0 & \frac{73}{20} \\
\hline
0 & 0 & 0 & 5 & 0 & 0 & 0 & 1 & -152
\end{array}
\right]
$$

Next, Crown's rules indicate that we should pivot on the $\frac{1}{2}$ in row 4, column 1. Pivoting results in the tableau stated at the top of the next column.

$$
\begin{array}{cccccccc}
x & y & z & s_1 & s_2 & s_3 & s_4 & g
\end{array}
$$
$$
\left[
\begin{array}{cccccccc|c}
0 & 0 & 1 & 0 & 0 & 0 & -1 & 0 & \frac{1}{10} \\
0 & \frac{3}{2} & 0 & \frac{-1}{4} & 1 & 0 & 2 & 0 & \frac{63}{10} \\
0 & -1 & 0 & 0 & 0 & 1 & 0 & 0 & \frac{-4}{10} \\
1 & \frac{3}{2} & 0 & \frac{-1}{4} & 0 & 0 & 2 & 0 & \frac{73}{10} \\
\hline
0 & 0 & 0 & 5 & 0 & 0 & 0 & 1 & -152
\end{array}
\right]
$$

Crown's rules now dictate that we pivot on the $\frac{3}{2}$ in row 2, column 2. Pivoting on this element results in the tableau:

$$
\begin{array}{cccccccc}
x & y & z & s_1 & s_2 & s_3 & s_4 & g
\end{array}
$$
$$
\left[
\begin{array}{cccccccc|c}
0 & 0 & 1 & 0 & 0 & 0 & -1 & 0 & 0.1 \\
0 & 1 & 0 & \frac{-1}{6} & \frac{2}{3} & 0 & \frac{4}{3} & 0 & 4.2 \\
0 & 0 & 0 & \frac{-1}{6} & \frac{2}{3} & 1 & \frac{4}{3} & 0 & 3.8 \\
1 & 0 & 0 & 0 & -1 & 0 & 0 & 0 & 1 \\
\hline
0 & 0 & 0 & 5 & 0 & 0 & 0 & 1 & -152
\end{array}
\right]
$$

The absence of negatives in the last column and last row indicate that we have found a maximum for the objective function g. The maximum value is $g = -152$ and occurs when $x = 1, y = 4.2, z = .1$ Since $g = -C$, and the x, y, and z are in thousands of tires, the minimum cost for the tire company will be $152,000. This cost will occur if they produce 1000 radial tires, 4200 snow tires, and 100 off-road tires per day. It should be noted that there are other production mixes that will satisfy all the constraints at a cost of $152,000.

41.

Let x and y represent the number of pounds of barley and corn in the mixture respectively. The objective is to maximize the total number of units of protein. The objective function is:

$$P = x + y$$

The limit on the number of units is of fat is given by:

$$x + 2y \le 12$$

The limit on the amount of each grain in the mixture is given by:

$$x \le 6$$
$$y \le 5$$

The solution is continued on the next page.

Using the constraints from the previous page, the linear programming problem is:

Maximize $\qquad P = x + y$

subject to

$$x + 2y \le 12$$
$$x \qquad \le 6$$
$$y \le 5$$
$$x \ge 0, y \ge 0$$

After adding slack variables, the initial tableau is:

$$
\begin{array}{cccccc}
x & y & s_1 & s_2 & s_3 & P \\
\end{array}
$$

$$
\left[
\begin{array}{cccccc|c}
1 & 2 & 1 & 0 & 0 & 0 & 12 \\
1 & 0 & 0 & 1 & 0 & 0 & 6 \\
0 & 1 & 0 & 0 & 1 & 0 & 5 \\
\hline
-1 & -1 & 0 & 0 & 0 & 1 & 0
\end{array}
\right]
$$

This is a standard simplex method problem. Initially we can choose column 1 or column 2 to pivot on. We choose column 1, and the pivot element will be the one in row 2, column 1. Pivoting on this element results in the following tableau:

$$
\begin{array}{cccccc}
x & y & s_1 & s_2 & s_3 & P \\
\end{array}
$$

$$
\left[
\begin{array}{cccccc|c}
0 & 2 & 1 & -1 & 0 & 0 & 6 \\
1 & 0 & 0 & 1 & 0 & 0 & 6 \\
0 & 1 & 0 & 0 & 1 & 0 & 5 \\
0 & -1 & 0 & 1 & 0 & 1 & 6
\end{array}
\right]
$$

The next pivot element is the two in row 1, column 1. Pivoting on this element results in the tableau:

$$
\begin{array}{cccccc}
x & y & s_1 & s_2 & s_3 & P \\
\end{array}
$$

$$
\left[
\begin{array}{cccccc|c}
0 & 1 & \frac{1}{2} & \frac{-1}{2} & 0 & 0 & 3 \\
1 & 0 & 0 & 1 & 0 & 0 & 6 \\
0 & 0 & \frac{-1}{2} & \frac{1}{2} & 1 & 0 & 2 \\
0 & 0 & \frac{1}{2} & \frac{1}{2} & 0 & 1 & 9
\end{array}
\right]
$$

The absence of negatives in the last row indicates that the maximum value of the objective function P has been reached. The maximum units of protein that can be put into the mixture is 9. In order to achieve this level of protein, 6 pounds of barley should be used and 3 pounds of corn should be used.

43.
Let x, y, and z represent the number of TV sets sent from El Paso to Reynosa, Tijuana and Nogales respectively. Then $60 - x$, $80 - y$, and $60 - z$ represent the number of sets sent from Mexico City to Reynosa, Tijuana, and Nogales respectively.

The total cost of shipping will be:

$$C = 2x + 3y + 4z + 2(60 - x) + 2(80 - y) + 2(60 - z)$$
$$C = 400 + y + 2z$$

The conditions on x, y, and z are:

1.   The number sent from each city must be greater than zero:

$$x \ge 0, y \ge 0, z \ge 0$$
$$60 - x \ge 0, \ 80 - y \ge 0, \ 60 - z \ge 0$$

2.   The number sent to each city must be less than 100:

$$x + y + z \le 100$$
$$(60 - x) + (80 - y) + (60 - z) \le 100$$

When we combine the two inequalities in condition two we get $x + y + z = 100$.

Therefore we can eliminate the x variable.

$$x = 100 - y - z.$$

Substituting this into the other constraints we get the linear programming problem:

Maximize $\qquad g = -C = -400 - y - 2z$

subject to

$$-y - z \le -40$$
$$y \qquad \le 80$$
$$z \le 60$$

After adding the slack variables, the initial tableau becomes:

$$
\begin{array}{cccccc}
y & z & s_1 & s_2 & s_3 & g \\
\end{array}
$$

$$
\left[
\begin{array}{cccccc|c}
-1 & -1 & 1 & 0 & 0 & 0 & -40 \\
1 & 0 & 0 & 1 & 0 & 0 & 80 \\
0 & 1 & 0 & 0 & 1 & 0 & 60 \\
\hline
1 & 2 & 0 & 0 & 0 & 1 & -400
\end{array}
\right]
$$

The solution is continued on the next page.

Crowns method indicates that either column 1 or 2 of the matrix on the previous page can be chosen as the pivot column. We choose to pivot in column 1. The pivot element will be the one in row 2, column 1. Pivoting on this element results in the tableau:

$$
\begin{array}{c c c c c c}
y & z & s_1 & s_2 & s_3 & g \\
\end{array}
$$

$$
\left[
\begin{array}{c c c c c c | c}
0 & -1 & 1 & 1 & 0 & 0 & 40 \\
1 & 0 & 0 & 1 & 0 & 0 & 80 \\
0 & 1 & 0 & 0 & 1 & 0 & 60 \\
\hline
0 & 2 & 0 & -1 & 0 & 1 & -480 \\
\end{array}
\right]
$$

Now applying the simplex method, we see that the pivot column will be column 4. The smallest quotient rule dictates that we pivot on the one in row 1, column 1. Pivoting on this element results in the following:

$$
\begin{array}{c c c c c c}
y & z & s_1 & s_2 & s_3 & g \\
\end{array}
$$

$$
\left[
\begin{array}{c c c c c c | c}
0 & -1 & 1 & 1 & 0 & 0 & 40 \\
1 & 1 & -1 & 0 & 0 & 0 & 40 \\
0 & 1 & 0 & 0 & 1 & 0 & 60 \\
\hline
0 & 1 & 1 & 0 & 0 & 1 & -440 \\
\end{array}
\right]
$$

The maximum value of the objective function is $g = -440$, when $y = 40, z = 0$. Don't forget that $x = 100 - y - z = 100 - 40 = 60$. Therefore, the minimum cost for shipping will be $440. To meet this cost you should ship 60 sets from El Paso and zero sets from Mexico City to Reynosa, 40 sets from El Paso and 40 sets from Mexico City to Tijuana, and zero from El Paso and 40 sets from Mexico City to Nogales.

45.

Let x, y, and z represent the number of boxes of books sent from Hyderabad to New Delhi, Mumbai, and Kanpur respectively. The information can be summarized in the following table:

	New Delhi	Mumbai	Kanpur
Hyderabad	x	y	z
Calcutta	$20 - x$	$15 - y$	$10 - z$

The total shipping costs will be:

$$
C = x + y + 2z + 2(20 - x) + 2(15 - y) + (10 - z)
$$
$$
C = 80 - x - y + z
$$

The constraints are:

1. $x, y, z \geq 0$

2. $x \leq 20, \ y \leq 15, \ z \leq 10$

3. $x + y + z \leq 28$

4. $(20 - x) + (15 - y) + (10 - z) \leq 40$
   or,
   $$
   -x - y - z \leq -5
   $$

To minimize the cost function we need to solve the linear programming problem stated below:

Maximize $g = -C = -80 + x + y - z$
subject to

$$
x + y + z \leq 28
$$
$$
-x - y - z \leq -5
$$
$$
x \qquad\quad \leq 20
$$
$$
y \quad \leq 15
$$
$$
z \leq 10
$$
$$
x \geq 0, y \geq 0, z \geq 0
$$

The initial tableau is:

$$
\begin{array}{c c c c c c c c c}
x & y & z & s_1 & s_2 & s_3 & s_4 & s_5 & g \\
\end{array}
$$

$$
\left[
\begin{array}{c c c c c c c c c | c}
1 & 1 & 1 & 1 & 0 & 0 & 0 & 0 & 0 & 28 \\
-1 & -1 & -1 & 0 & 1 & 0 & 0 & 0 & 0 & -5 \\
1 & 0 & 0 & 0 & 0 & 1 & 0 & 0 & 0 & 20 \\
0 & 1 & 0 & 0 & 0 & 0 & 1 & 0 & 0 & 15 \\
0 & 0 & 1 & 0 & 0 & 0 & 0 & 1 & 0 & 10 \\
\hline
-1 & -1 & 1 & 0 & 0 & 0 & 0 & 0 & 1 & -80 \\
\end{array}
\right]
$$

The first negative entry in the last column is -5 in row 2. Crown's rules allow us to choose any of the first three columns in row 2 to pivot on. We choose to pivot on column 1. The pivot element will be the one in row 3, column 1. Pivoting results in the tableau at the top of the next page.

x	y	z	s_1	s_2	s_3	s_4	s_5	g	
0	1	1	1	0	-1	0	0	0	8
0	-1	-1	0	1	1	0	0	0	15
1	0	0	0	0	1	0	0	0	20
0	1	0	0	0	0	1	0	0	15
0	0	1	0	0	0	0	1	0	10
0	-1	1	0	0	1	0	0	1	-60

The simplex method indicates that the pivot column will be column 2. The smallest quotient rule indicates that the pivot element will be the one in row 1, column 2. Pivoting on this element results in the tableau:

x	y	z	s_1	s_2	s_3	s_4	s_5	g	
0	1	1	1	0	-1	0	0	0	8
0	0	0	1	1	0	0	0	0	23
1	0	0	0	0	1	0	0	0	20
0	0	-1	-1	0	1	1	0	0	7
0	0	1	0	0	0	0	1	0	10
0	0	2	1	0	0	0	0	1	-52

The absence of negatives in the bottom row indicates that g is maximized at -52, thus C is minimized at 52, when $x = 20, y = 8, z = 0$. This means that we can minimize the total shipping cost at \$52, by having Hyderabad supply 20 boxes to New Delhi, 8 boxes to Mumbai, and zero boxes to Kanpur, while Calcutta should supply zero boxes to New Delhi, 7 boxes to Mumbai, and 10 boxes to Kanpur.

47.
a. The feasible region is shown below:

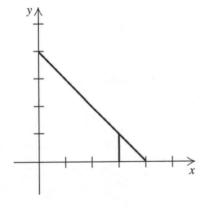

b.
First substitute the value of $x = 3$ into the constraint $x + y \le 4$ and the objective function $f = 2x + 5y$. The problem becomes:

Maximize $\quad f = 6 + 5y$
subject to: $\quad y \le 1$
$\quad\quad\quad\quad x \ge 0, y \ge 0$

The initial tableau would be:

y	s_1	f	
1	1	0	1
-5	0	1	6

The pivot element must be the one in row 1, column 1. Pivoting on this element results in the following tableau:

y	s_1	f	
1	1	0	1
0	5	1	11

The maximum value of the objective function is $f = 11$. This value occurs when $x = 3, y = 1$.

c.
Rewriting the problem replacing the equality constraint with two inequality constraints we get:

Maximize $\quad f = 2x + 5y$
subject to:
$\quad\quad x \quad\quad \le 3$
$\quad\quad -x \quad\quad \le -3$
$\quad\quad x + y \le 4$
$\quad\quad x \ge 0, y \ge 0$

The initial tableau is:

x	y	s_1	s_2	s_3	f	
1	0	1	0	0	0	3
-1	0	0	1	0	0	-3
1	1	0	0	1	0	4
-2	-5	0	0	0	1	0

The solution is continued on the next page.

Observing the matrix on the previous page, Crown's rules indicate that we pivot on one in row 1, column 1. Pivoting on this element results in the tableau:

$$
\begin{array}{cccccc|c}
x & y & s_1 & s_2 & s_3 & f & \\
\hline
1 & 0 & 1 & 0 & 0 & 0 & 3 \\
0 & 0 & 1 & 1 & 0 & 0 & 0 \\
0 & 1 & -1 & 0 & 1 & 0 & 1 \\
\hline
0 & -5 & 2 & 0 & 0 & 1 & 6
\end{array}
$$

The absence of negatives in the last column of the matrix on the previous page indicates that we have arrived at a corner point of the feasible region $x = 3, y = 0$. We have yet to reach a maximum value for the function. The simplex method indicates that we should pivot on the one in row 3, column 2. Pivoting on this element results in the tableau:

$$
\begin{array}{cccccc|c}
x & y & s_1 & s_2 & s_3 & f & \\
\hline
1 & 0 & 1 & 0 & 0 & 0 & 3 \\
0 & 0 & 1 & 1 & 0 & 0 & 0 \\
0 & 1 & -1 & 0 & 1 & 0 & 1 \\
\hline
0 & 0 & -3 & 0 & 5 & 1 & 11
\end{array}
$$

We have arrived at the corner point $x = 3, y = 1$ and the value of $f = 11$. This is the maximum value according to part b. However, the simplex tableau still indicates that we should pivot on the one in row 2, column 3. Performing this pivot we get:

$$
\begin{array}{cccccc|c}
x & y & s_1 & s_2 & s_3 & f & \\
\hline
1 & 0 & 0 & -1 & 0 & 0 & 3 \\
0 & 0 & 1 & 1 & 0 & 0 & 0 \\
0 & 1 & 0 & 1 & 1 & 0 & 1 \\
\hline
0 & 0 & 0 & 3 & 5 & 1 & 11
\end{array}
$$

The absence of negatives implies that we have arrived at a maximum for the objective function. The maximum value is $f = 11$ and occurs at $x = 3, y = 1$. This is the same solution that we arrived at in part b.

Exercises Section 5.4

1.

a. The primal matrix is:

$$
P = \begin{bmatrix}
1 & 1 & 4 \\
3 & 1 & 8 \\
\hline
2 & 1 & 0
\end{bmatrix}
$$

b. The dual matrix:

$$
D = P^T = \begin{bmatrix}
1 & 3 & 2 \\
1 & 1 & 1 \\
\hline
4 & 8 & 0
\end{bmatrix}
$$

c. The dual problem can be stated as follows:

Maximize $\qquad d = 4u + 8v$
subject to

$$u + 3v \le 2$$
$$u + v \le 1$$
$$u \ge 0, v \ge 0$$

d. The feasible set for the primal problem is shown below:

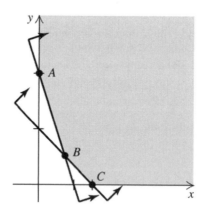

The corner points and corresponding values for the objective function f are:

$$A(0,8);\ f = 8$$
$$B(2,2);\ f = 6$$
$$C(4,0);\ f = 8$$

The solution is continued on the next page.

From the information on the previous page, the minimum value is $f = 2$ and occurs when $x = 2, y = 2$.

The feasible set for the dual problem is:

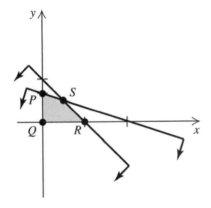

The corner points and corresponding values for the objective function d are:

$P\left(0, \frac{2}{3}\right); d = \frac{16}{3}$

$S\left(\frac{1}{2}, \frac{1}{2}\right); d = 6$

$R(1,0); d = 4$

$Q(0,0); d = 0$

The maximum value is $d = 6$ and occurs when $u = \frac{1}{2}, v = \frac{1}{2}$

e. The initial simplex tableau is:

$$
\begin{array}{ccccc|c}
u & v & s_1 & s_2 & d & \\
\hline
1 & 3 & 1 & 0 & 0 & 2 \\
1 & 1 & 0 & 1 & 0 & 1 \\
\hline
-4 & -8 & 0 & 0 & 1 & 0
\end{array}
$$

The pivot element will be the three in row 1, column 2. Pivoting on this element results in the tableau:

$$
\begin{array}{ccccc|c}
u & v & s_1 & s_2 & d & \\
\hline
\frac{1}{3} & 1 & \frac{1}{3} & 0 & 0 & \frac{2}{3} \\
\frac{2}{3} & 0 & \frac{-1}{3} & 1 & 0 & \frac{1}{3} \\
\hline
\frac{-4}{3} & 0 & \frac{8}{3} & 0 & 1 & \frac{16}{3}
\end{array}
$$

Next we pivot on the $\frac{2}{3}$ in row 2, column 1. Pivoting on this element results in the tableau at the top of the next column.

$$
\begin{array}{ccccc|c}
u & v & s_1 & s_2 & d & \\
\hline
0 & 1 & \frac{1}{2} & \frac{-1}{2} & 0 & \frac{1}{2} \\
1 & 0 & \frac{-1}{2} & \frac{3}{2} & 0 & \frac{1}{2} \\
\hline
0 & 0 & 2 & 2 & 1 & 6
\end{array}
$$

The solution to the dual problem states when $u = \frac{1}{2}, v = \frac{1}{2}$ the maximum value of the function is $d = 6$.

We can also determine the solution to the primal problem as well. The minimum value that f will take on is $f = 6$.
We can determine where this value will occur by looking at the entries in the last row of the slack variables column. Here we see that the minimum value will occur when $x = 2, y = 2$.

3.
a. The primal matrix is:

$$
P = \begin{bmatrix}
1 & 1 & 10 \\
2 & 1 & 12 \\
\hline
3 & 1 & 0
\end{bmatrix}
$$

b. The dual matrix is:

$$
D = P^T = \begin{bmatrix}
1 & 2 & 3 \\
1 & 1 & 1 \\
\hline
10 & 12 & 0
\end{bmatrix}
$$

c. The dual problem can be stated as follows:

Maximize $d = 10u + 12v$
subject to

$u + 2v \le 3$

$u + 3v \le 1$
$u \ge 0, v \ge 0$

d. The feasible set for the primal problem is shown at the top of the next page.

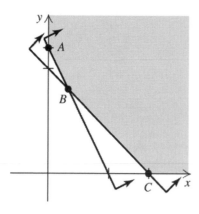

The corner points and corresponding values for the objective function C are:

$A(0,12)$; $C = 12$

$B(2,8)$; $C = 14$

$C(6,0)$; $C = 18$

The minimum value is $C = 12$ and occurs when $x = 0, y = 12$

The feasible set for the dual problem is shown below:

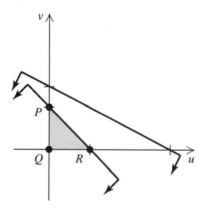

The corner points and corresponding values for the objective function d are:

$P(0,1)$; $d = 12$

$R(1,0)$; $d = 10$

$Q(0,0)$; $d = 0$

The maximum value is $d = 10$ and occurs when $u = 0, v = 1$

e. The initial simplex tableau is displayed at the top of the next column.

$$\begin{array}{ccccc} u & v & s_1 & s_2 & d \\ \end{array}$$
$$\left[\begin{array}{ccccc|c} 1 & 2 & 1 & 0 & 0 & 3 \\ 1 & 1 & 0 & 1 & 0 & 1 \\ \hline -10 & -12 & 0 & 0 & 1 & 0 \end{array}\right]$$

The pivot element will be the one in row 2, column 2. Pivoting on this element results in the tableau:

$$\begin{array}{ccccc} u & v & s_1 & s_2 & d \\ \end{array}$$
$$\left[\begin{array}{ccccc|c} -1 & 0 & 1 & -2 & 0 & 1 \\ 1 & 1 & 0 & 1 & 0 & 1 \\ \hline 2 & 0 & 0 & 12 & 1 & 12 \end{array}\right]$$

The tableau tells us that the solution to the dual problem states when $u = 0, v = 1$ then $d = 12$ is the maximum value for the function.

We can also determine the solution to the primal problem as well. The minimum value that C will take on is $C = 12$. We can determine where this value will occur by looking at the entries in the last row of the slack variables column. Here we see that the minimum value will occur when $x = 0, y = 12$.

5.
a. Since we have mixed constraints, we will have to multiply the second constraint by negative one. Now, the primal matrix is:

$$P = \left[\begin{array}{cc|c} 4 & 1 & 18 \\ -1 & -2 & -8 \\ \hline 2 & -3 & 0 \end{array}\right]$$

b. The dual matrix is:

$$D = P^T = \left[\begin{array}{cc|c} 4 & -1 & 2 \\ 1 & -2 & -3 \\ \hline 18 & -8 & 0 \end{array}\right]$$

c. The dual problem can be stated as follows:

Maximize $\quad d = 18u - 8v$

subject to

$$4u - v \le 2$$
$$u - 2v \le -3$$
$$u \ge 0, v \ge 0$$

d. The feasible set for the primal problem is shown below:

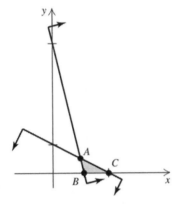

The corner points and corresponding values for the objective function f are:

$$A(4,2); \; f = 2$$
$$B(4.5,0); \; f = 9$$
$$C(8,0); \; f = 16$$

The minimum value is $f = 2$ and occurs when $x = 4, y = 2$

The feasible set for the dual problem is shown below:

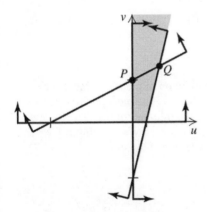

The corner points and corresponding values for the objective function d are displayed at the top of the next column.

$P(0,1.5); \; d = -12$

$Q(1,2); \; d = 2$

The maximum value is $d = 2$ and occurs when $u = 1, v = 2$

e. The initial simplex tableau is:

$$
\begin{array}{ccccc}
u & v & s_1 & s_2 & d
\end{array}
$$
$$
\left[
\begin{array}{ccccc|c}
4 & -1 & 1 & 0 & 0 & 2 \\
1 & -2 & 0 & 1 & 0 & -3 \\
\hline
-18 & 8 & 0 & 0 & 1 & 0
\end{array}
\right]
$$

Crown's rules tell us that the pivot element will be the negative two in row 2, column 2. Pivoting on this element results in the tableau:

$$
\begin{array}{ccccc}
u & v & s_1 & s_2 & d
\end{array}
$$
$$
\left[
\begin{array}{ccccc|c}
\frac{7}{2} & 0 & 1 & \frac{-1}{2} & 0 & \frac{7}{2} \\
\frac{-1}{2} & 1 & 0 & \frac{-1}{2} & 0 & \frac{3}{2} \\
\hline
-14 & 0 & 0 & 4 & 1 & -12
\end{array}
\right]
$$

At this point, the simplex algorithm can be applied. The only negative entry in the last row is in column 1. The only positive entry in column 1 above the horizontal bar is the $\frac{7}{2}$ in row 1, column 1. Pivoting on this element results in the tableau:

$$
\begin{array}{ccccc}
u & v & s_1 & s_2 & d
\end{array}
$$
$$
\left[
\begin{array}{ccccc|c}
1 & 0 & \frac{2}{7} & \frac{-1}{7} & 0 & 1 \\
0 & 1 & \frac{1}{7} & \frac{-4}{7} & 0 & 2 \\
\hline
0 & 0 & 4 & 2 & 1 & 2
\end{array}
\right]
$$

The tableau tells us that the solution to the dual problem states when $u = 1, v = 2$ then $d = 2$ is the maximum value for the function.

We can also determine the solution to the primal problem as well. The minimum value that f will take on is $f = 2$. We can determine where this value will occur by looking at the entries in the last row of the slack variables column. Here we see that the minimum value will occur when $x = 4, y = 2$.

7.

a. The primal matrix is:

$$P = \begin{bmatrix} 1 & 1 & | & 8 \\ 1 & 0 & | & 1 \\ -2 & 1 & | & 0 \end{bmatrix}$$

b. The dual matrix is:

$$D = P^T = \begin{bmatrix} 1 & 1 & | & -2 \\ 1 & 0 & | & 1 \\ 8 & 1 & | & 0 \end{bmatrix}$$

c. The dual problem can be stated as follows:

Minimize $\quad d = 8u + 1v$
subject to

$$u + v \geq -2$$
$$u \quad\;\; \geq 1$$
$$u \geq 0, v \geq 0$$

d. The feasible set for the primal problem is shown below:

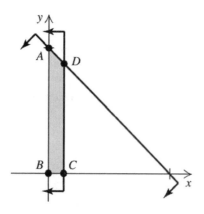

The corner points and corresponding values for the objective function f are:

$A(0,8); \; f = 8$

$B(0,0); \; f = 0$

$C(1,0); \; f = -2$

$D(1,7); \; f = 5$

The maximum value is $f = 8$ and occurs when $x = 0, y = 8$

The feasible set for the dual problem is shown at the top of the next column.

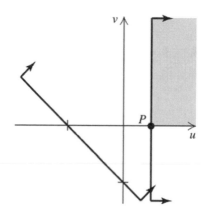

The corner points and corresponding values for the objective function d are:

$P(1,0); \; d = 8$

The minimum value is $d = 8$ and occurs when $u = 1, v = 0$

e. Since the dual problem is stated as a minimization problem, we must first rewrite the problem as a maximization problem. The problem becomes:

Maximize $\quad g = -d = -8u - v$
subject to

$$-u - v \leq \;\; 2$$
$$-u \quad\;\; \leq -1$$
$$u \geq 0, v \geq 0$$

The initial simplex tableau is:

$$\begin{array}{ccccc} u & v & s_1 & s_2 & g \end{array}$$
$$\begin{bmatrix} -1 & -1 & 1 & 0 & 0 & | & 2 \\ -1 & 0 & 0 & 1 & 0 & | & -1 \\ \hline 8 & 1 & 0 & 0 & 1 & | & 0 \end{bmatrix}$$

Crown's rules tell us to pivot on the first column. Since each entry is negative we choose the most positive ratio. The only positive ratio is the ratio out of row 2. The pivot element will be the negative one in row 2, column 1. Pivoting on this element results in the tableau:

$$\begin{array}{ccccc} u & v & s_1 & s_2 & g \end{array}$$
$$\begin{bmatrix} 0 & -1 & 1 & -1 & 0 & | & 3 \\ 1 & 0 & 0 & -1 & 0 & | & 1 \\ \hline 0 & 1 & 0 & 8 & 1 & | & -8 \end{bmatrix}$$

The solution is continued on the next page.

The absence of negatives to the right of the vertical bar, and below the horizontal bar means that we have reached the optimal solution. The maximum of g is -8, when $u = 1, v = 0$. Thus, the minimum value of d must be 8, when $u = 1, v = 0$. This is the solution to the dual problem

We can also determine the solution to the primal problem as well. The maximum value that f will take on is $f = 8$. We can determine where this value will occur by looking at the entries in the last row of the slack variables column. Here we see that the minimum value will occur when $x = 0, y = 8$.

9.

The primal matrix is:

$$P = \begin{bmatrix} 1 & 1 & 8 \\ 2 & 1 & 12 \\ \hline 1 & 2 & 0 \end{bmatrix}$$

The dual matrix is:

$$D = P^T = \begin{bmatrix} 1 & 2 & 1 \\ 1 & 1 & 2 \\ \hline 8 & 12 & 0 \end{bmatrix}$$

The dual problem is:

Maximize $d = 8u + 12v$
subject to

$$u + 2v \le 1$$
$$u + v \le 2$$
$$u \ge 0, v \ge 0$$

The initial simplex tableau for the dual problem is:

u	v	s_1	s_2	d	
1	2	1	0	0	1
1	1	0	1	0	2
-8	-12	0	0	1	0

The simplex method tells us that the pivot element is the two in row 1, column 2. Pivoting on this element results in the tableau at the top of the next column.

u	v	s_1	s_2	d	
$\frac{1}{2}$	1	$\frac{1}{2}$	0	0	$\frac{1}{2}$
$\frac{1}{2}$	0	$\frac{-1}{2}$	1	0	$\frac{3}{2}$
-2	0	6	0	1	6

Next we will pivot on the $\frac{1}{2}$ in row 1, column 1. Pivoting on this element results in the tableau:

u	v	s_1	s_2	d	
1	2	1	0	0	1
0	-1	-1	1	0	1
0	4	8	0	1	8

The tableau tells us that the solution to the dual problem states when $u = 1, v = 0$ then $d = 8$ is the maximum value for the function.

We can also determine the solution to the primal problem as well. The minimum value that f will take on is $f = 8$. We can determine where this value will occur by looking at the entries in the last row of the slack variables column. Here we see that the minimum value will occur when $x = 8, y = 0$.

11.

The primal matrix is:

$$P = \begin{bmatrix} 1 & 1 & 0 & 6 \\ 2 & 0 & 1 & 4 \\ 0 & 1 & 2 & 5 \\ \hline 3 & 1 & 2 & 0 \end{bmatrix}$$

The dual matrix is:

$$D = P^T = \begin{bmatrix} 1 & 2 & 0 & 3 \\ 1 & 0 & 1 & 1 \\ 0 & 1 & 2 & 2 \\ \hline 6 & 4 & 5 & 0 \end{bmatrix}$$

The dual problem is stated at the top of the next page.

Maximize $d = 6u + 4v + 5w$
subject to

$$u + 2v \quad\quad \leq 3$$
$$u \quad\quad + w \leq 1$$
$$v + 2w \leq 2$$
$$u \geq 0, v \geq 0$$

The initial simplex tableau for the dual problem is:

$$
\begin{array}{ccccccc}
u & v & w & s_1 & s_2 & s_3 & d \\
\end{array}
$$
$$
\left[\begin{array}{ccccccc|c}
1 & 2 & 0 & 1 & 0 & 0 & 0 & 3 \\
1 & 0 & 1 & 0 & 1 & 0 & 0 & 1 \\
0 & 1 & 2 & 0 & 0 & 1 & 0 & 2 \\
\hline
-6 & -4 & -5 & 0 & 0 & 0 & 1 & 0
\end{array}\right]
$$

The simplex method tells us that the pivot element is the one in row 1, column 2. Pivoting on this element results in the tableau:

$$
\begin{array}{ccccccc}
u & v & w & s_1 & s_2 & s_3 & d \\
\end{array}
$$
$$
\left[\begin{array}{ccccccc|c}
0 & 2 & -1 & 1 & -1 & 0 & 0 & 2 \\
1 & 0 & 1 & 0 & 1 & 0 & 0 & 1 \\
0 & 1 & 2 & 0 & 0 & 1 & 0 & 2 \\
\hline
0 & -4 & 1 & 0 & 6 & 0 & 1 & 6
\end{array}\right]
$$

Next we will pivot on the two in row 1, column 2. Pivoting on this element results in the tableau:

$$
\begin{array}{ccccccc}
u & v & w & s_1 & s_2 & s_3 & d \\
\end{array}
$$
$$
\left[\begin{array}{ccccccc|c}
0 & 1 & \frac{-1}{2} & \frac{1}{2} & \frac{-1}{2} & 0 & 0 & 1 \\
1 & 0 & 1 & 0 & 1 & 0 & 0 & 1 \\
0 & 0 & \frac{5}{2} & \frac{-1}{2} & \frac{1}{2} & 1 & 0 & 1 \\
\hline
0 & 0 & -1 & 2 & 4 & 0 & 1 & 10
\end{array}\right]
$$

Next we will pivot on the $\frac{5}{2}$ in row 3, column 3. Pivoting on this element results in the tableau:

$$
\begin{array}{ccccccc}
u & v & w & s_1 & s_2 & s_3 & d \\
\end{array}
$$
$$
\left[\begin{array}{ccccccc|c}
0 & 1 & 0 & \frac{2}{5} & \frac{-2}{5} & \frac{1}{5} & 0 & \frac{6}{5} \\
1 & 0 & 0 & \frac{1}{5} & \frac{4}{5} & \frac{-2}{5} & 0 & \frac{3}{5} \\
0 & 0 & 1 & \frac{-1}{5} & \frac{1}{5} & \frac{2}{5} & 0 & \frac{2}{5} \\
\hline
0 & 0 & 0 & \frac{9}{5} & \frac{21}{5} & \frac{2}{5} & 1 & \frac{52}{5}
\end{array}\right]
$$

The tableau tells us that the solution to the dual problem states when $u = \frac{3}{5}, v = \frac{6}{5}, w = \frac{2}{5}$ then $d = \frac{51}{5} = 10.4$ is the maximum value for the function.

We can also determine the solution to the primal problem as well. The minimum value that f will take on is $f = \frac{51}{5} = 10.4$. We can determine where this value will occur by looking at the entries in the last row of the slack variables column. Here we see that the minimum value will occur when $x = \frac{9}{5}, y = \frac{21}{5}, z = \frac{2}{5}$.

13.
Since we have mixed constraints, we will have to multiply the second constraint by negative one. The primal matrix is:

$$
P = \left[\begin{array}{ccc|c}
1 & 1 & 1 & 8 \\
0 & -1 & 0 & -4 \\
1 & 0 & 1 & 6 \\
\hline
1 & 1 & 3 & 0
\end{array}\right]
$$

The dual matrix is:

$$
D = P^T = \left[\begin{array}{ccc|c}
1 & 0 & 1 & 1 \\
1 & -1 & 0 & 1 \\
1 & 0 & 1 & 3 \\
\hline
8 & -4 & 6 & 0
\end{array}\right]
$$

The dual problem is:

Maximize $d = 8u - 4v + 6w$
subject to

$$u \quad\quad + w \leq 1$$
$$u - v \quad\quad \leq 1$$
$$u \quad\quad + w \leq 3$$
$$u \geq 0, v \geq 0$$

The initial simplex tableau for the dual problem is:

$$
\begin{array}{ccccccc}
u & v & w & s_1 & s_2 & s_3 & d \\
\end{array}
$$
$$
\left[\begin{array}{ccccccc|c}
1 & 0 & 1 & 1 & 0 & 0 & 0 & 1 \\
1 & -1 & 0 & 0 & 1 & 0 & 0 & 1 \\
1 & 0 & 1 & 0 & 0 & 1 & 0 & 3 \\
\hline
-8 & 4 & -6 & 0 & 0 & 0 & 1 & 0
\end{array}\right]
$$

The solution is continued on the next page.

The simplex method tells us that the pivot column of the matrix on the previous page is column 1. We can choose to pivot on either row 1, or row 2. We choose to pivot on the one in row 1, column 1. Pivoting on this element results in the tableau:

$$\begin{array}{ccccccc} u & v & w & s_1 & s_2 & s_3 & d \end{array}$$

$$\left[\begin{array}{ccccccc|c} 1 & 0 & 1 & 1 & 0 & 0 & 0 & 1 \\ 0 & -1 & -1 & -1 & 1 & 0 & 0 & 0 \\ 0 & 0 & 0 & -1 & 0 & 1 & 0 & 2 \\ \hline 0 & 4 & 2 & 8 & 0 & 0 & 1 & 8 \end{array}\right]$$

The tableau tells us that the solution to the dual problem states when $u = 1, v = 0, w = 0$ then $d = 8$ is the maximum value for the function.

We can also determine the solution to the primal problem from the dual problem. The minimum value that f will take on is $f = 8$. We can determine where this value will occur by looking at the entries in the last row of the slack variables column. Here we see that the minimum value will occur when $x = 8, y = 0, z = 0$.

15.

Since we have mixed constraints, we will have to multiply the first and second constraints by negative one. Now, the primal matrix is:

$$P = \left[\begin{array}{ccc|c} -1 & 0 & -1 & -6 \\ -1 & -1 & 0 & -4 \\ 1 & 1 & 1 & 10 \\ \hline 2 & 1 & -1 & 0 \end{array}\right]$$

The dual matrix is:

$$D = P^T = \left[\begin{array}{ccc|c} -1 & -1 & 1 & 2 \\ 0 & -1 & 1 & 1 \\ -1 & 0 & 1 & -1 \\ \hline -6 & -4 & 10 & 0 \end{array}\right]$$

The dual problem created from the dual matrix is stated at the top of the next column.

Minimize $d = -6u - 4v + 10w$
subject to

$$-u - v + w \geq 2$$
$$-v + w \geq 1$$
$$-u \quad + w \geq -1$$
$$u \geq 0, v \geq 0, w \geq 0$$

To use the simplex method we must rewrite the problem to look like a maximization problem. The maximization problem is:

Maximize $g = -d = 6u + 4v - 10w$
subject to

$$u + v - w \leq -2$$
$$v - w \leq -1$$
$$u \quad - w \leq 1$$
$$u \geq 0, v \geq 0, w \geq 0$$

The initial simplex tableau for the dual problem is:

$$\begin{array}{ccccccc} u & v & w & s_1 & s_2 & s_3 & g \end{array}$$

$$\left[\begin{array}{ccccccc|c} 1 & 1 & -1 & 1 & 0 & 0 & 0 & -2 \\ 0 & 1 & -1 & 0 & 1 & 0 & 0 & -1 \\ 1 & 0 & -1 & 0 & 0 & 1 & 0 & 1 \\ \hline -6 & -4 & 10 & 0 & 0 & 0 & 1 & 0 \end{array}\right]$$

Crown's rules tell us to pivot on the one in row 1, column 3. Pivoting on this element results in the tableau:

$$\begin{array}{ccccccc} u & v & w & s_1 & s_2 & s_3 & g \end{array}$$

$$\left[\begin{array}{ccccccc|c} -1 & -1 & 1 & -1 & 0 & 0 & 0 & 2 \\ -1 & 0 & 0 & -1 & 1 & 0 & 0 & 1 \\ 0 & -1 & 0 & -1 & 0 & 1 & 0 & 3 \\ \hline 4 & 6 & 0 & 10 & 0 & 0 & 1 & -20 \end{array}\right]$$

The simplex algorithm tells us that we have arrived at an optimal solution. The maximum value of g is -20, when $u = 0, v = 0, w = 2$. Thus, the minimum value of d is 20 when $u = 0, v = 0, w = 2$.

The solution to the primal problem is stated on the next page.

We can also determine the solution to the primal problem from the dual problem on the previous page. The maximum value that f will take on is $f = 20$. We can determine where this value will occur by looking at the entries in the last row of the slack variables column. Here we see that the minimum value will occur when $x = 10, y = 0, z = 0$.

17.

Let x and y represent the number of servings of Food A and Food B respectively. The objective is to minimize the number of milligrams of fat. The objective function is:

$$f = 2x + 3y.$$

The vitamin requirements imply that:

$$4x + 6y \geq 10$$
$$5x + 2y \geq 10$$

The primal matrix is:

$$P = \begin{bmatrix} 4 & 6 & 10 \\ 5 & 2 & 10 \\ \hline 2 & 3 & 0 \end{bmatrix}$$

The dual matrix is

$$D = P^T = \begin{bmatrix} 4 & 5 & 2 \\ 6 & 2 & 3 \\ \hline 10 & 10 & 0 \end{bmatrix}$$

The dual problem becomes:

Maximize $\quad d = 10u + 10v$
subject to

$$4u + 5v \leq 2$$
$$6u + 2v \leq 3$$
$$u \geq 0, v \geq 0$$

The initial simplex tableau for this problem is:

$$\begin{array}{ccccc} u & v & s_1 & s_2 & d \\ \end{array}$$
$$\begin{bmatrix} 4 & 5 & 1 & 0 & 0 & 2 \\ 6 & 2 & 0 & 1 & 0 & 3 \\ \hline -10 & -10 & 0 & 0 & 1 & 0 \end{bmatrix}$$

The simplex method indicates we can pivot on either column 1 or column 2. We choose to pivot on column 1. The smallest quotient rule indicates that we can pivot on either the four or the six in column 1. We choose to pivot on the four in row 1, column 1. Pivoting results in the tableau:

$$\begin{array}{ccccc} u & v & s_1 & s_2 & d \\ \end{array}$$
$$\begin{bmatrix} 1 & \frac{5}{4} & \frac{1}{4} & 0 & 0 & \frac{1}{2} \\ 0 & \frac{-11}{2} & \frac{-3}{2} & 1 & 0 & 0 \\ \hline 0 & \frac{5}{2} & \frac{5}{2} & 0 & 1 & 5 \end{bmatrix}$$

The solution to the primal problem is $f = 5$ when $x = \frac{5}{2}, y = 0$. Remember that we read off the values of the variables from the last row in the slack variable columns. The minimum number of milligrams of fat that can be consumed is five milligrams, while still meeting the vitamin requirements. This will occur when 2.5 units of Food A are consumed and zero units of Food B are consumed.

19.

Let x, y, and z represent the number of type A barrels, type B barrels, and type C barrels respectively. The objective is to minimize cost. The cost function is:

$$C = 5x + 3y + 4z$$

The order demands 12 gallons of carbon black and 6 gallons of thinning agent. This can be translated in to the inequalities:

$$2x + 4y + 3x \geq 12$$
$$2x + y + z \geq 6$$

The linear programming problem is:

Minimize $\quad C = 5x + 3y + 4z$
subject to

$$2x + 4y + 3x \geq 12$$
$$2x + y + z \geq 6$$
$$x \geq 0, y \geq 0, z \geq 0$$

The solution is continued on the next page.

The primal matrix of the system on the previous page is:

$$P = \begin{bmatrix} 2 & 4 & 3 & | & 12 \\ 2 & 1 & 1 & | & 6 \\ \hline 5 & 3 & 4 & | & 0 \end{bmatrix}$$

The dual matrix is:

$$D = P^T = \begin{bmatrix} 2 & 2 & | & 5 \\ 4 & 1 & | & 3 \\ 3 & 1 & | & 4 \\ \hline 12 & 6 & | & 0 \end{bmatrix}$$

The dual problem is:

Maximize $d = 12x + 6y$
subject to

$$2u + 2v \le 5$$
$$4u + v \le 3$$
$$3u + v \le 4$$
$$u \ge 0, v \ge 0$$

The initial tableau for the dual problem is:

$$\begin{array}{c} u \quad v \; s_1 \, s_2 \, s_3 \; d \\ \begin{bmatrix} 2 & 2 & 1 & 0 & 0 & 0 & | & 5 \\ 4 & 1 & 0 & 1 & 0 & 0 & | & 3 \\ 3 & 1 & 0 & 0 & 1 & 0 & | & 4 \\ \hline -12 & -6 & 0 & 0 & 0 & 1 & | & 0 \end{bmatrix} \end{array}$$

The simplex method indicates that the pivot element is the four in row 2, column 1. Pivoting on this element results in the tableau:

$$\begin{array}{c} u \quad v \; s_1 \; s_2 \; s_3 \; d \\ \begin{bmatrix} 0 & \frac{3}{2} & 1 & \frac{-1}{2} & 0 & 0 & | & \frac{7}{2} \\ 1 & \frac{1}{4} & 0 & \frac{1}{4} & 0 & 0 & | & \frac{3}{4} \\ 0 & \frac{1}{4} & 0 & \frac{-3}{4} & 1 & 0 & | & \frac{7}{4} \\ \hline 0 & -3 & 0 & 3 & 0 & 1 & | & 9 \end{bmatrix} \end{array}$$

The next pivot element will be the $\frac{3}{2}$ in row 1, column 2.

Pivoting on this element results in the tableau at the top of the next column.

$$\begin{array}{c} u \quad v \; s_1 \quad s_2 \;\, s_3 \; d \\ \begin{bmatrix} 0 & 1 & \frac{2}{3} & \frac{-1}{3} & 0 & 0 & | & \frac{7}{3} \\ 1 & 0 & \frac{-1}{6} & \frac{1}{3} & 0 & 0 & | & \frac{1}{6} \\ 0 & 0 & \frac{-1}{6} & \frac{-2}{3} & 1 & 0 & | & \frac{7}{6} \\ \hline 0 & 0 & 2 & 2 & 0 & 1 & | & 16 \end{bmatrix} \end{array}$$

The tableau indicates that we have arrived at a maximum value for the objective function d. The maximum value is $d = 16$ when $u = \frac{1}{6}, v = \frac{7}{3}$. Therefore the minimum value for the primal problem is $C = 16$, when $x = 2, y = 2, z = 0$.

Therefore, the minimum cost of the order will be $16. In order to meet this cost, the firm should use two barrels of type A, two barrels of type B, and zero barrels of type C.

21.
Let x represent the number of VHS tapes ordered and y represent the number of DVD's ordered. The objective is to minimize cost, and the cost function is:

$$C = 12x + 16y$$

The order requirements imply that

$$x + y \ge 250$$
$$y \ge 0.3(x + y)$$

The linear programming problem is;

Minimize $C = 12x + 16y$
subject to

$$x + y \ge 250$$
$$-0.3x + 0.7y \ge 0$$
$$x \ge 0, y \ge 0$$

The primal matrix is:

$$P = \begin{bmatrix} 1 & 1 & | & 250 \\ -0.3 & 0.7 & | & 0 \\ \hline 12 & 16 & | & 0 \end{bmatrix}$$

The dual matrix is stated at the top of the next page.

$$D = P^T = \begin{bmatrix} 1 & -0.3 & 12 \\ 1 & 0.7 & 16 \\ \hline 250 & 0 & 0 \end{bmatrix}$$

The dual problem is:

Maximize $\qquad d = 250u$
subject to
$$u - 0.3v \le 12$$
$$u + 0.7v \le 16$$
$$u \ge 0, v \ge 0$$

The initial simplex tableau is:

$$\begin{array}{ccccc} u & v & s_1 & s_2 & d \end{array}$$
$$\begin{bmatrix} 1 & -0.3 & 1 & 0 & 0 & 12 \\ 1 & 0.7 & 0 & 1 & 0 & 16 \\ \hline -250 & 0 & 0 & 0 & 1 & 0 \end{bmatrix}$$

The simplex method indicates that we should pivot on the one in row 1, column 1. Pivoting on this element results in the matrix:

$$\begin{array}{ccccc} u & v & s_1 & s_2 & d \end{array}$$
$$\begin{bmatrix} 1 & -0.3 & 1 & 0 & 0 & 12 \\ 0 & 1 & -1 & 1 & 0 & 4 \\ \hline 0 & -75 & 250 & 0 & 1 & 3000 \end{bmatrix}$$

We proceed by pivoting on the 1 in row 2, column 2. Pivoting on this element results in the matrix:

$$\begin{array}{ccccc} u & v & s_1 & s_2 & d \end{array}$$
$$\begin{bmatrix} 1 & 0 & \frac{7}{10} & \frac{3}{10} & 0 & \frac{66}{5} \\ 0 & 1 & -1 & 1 & 0 & 5 \\ \hline 0 & 0 & 175 & 75 & 1 & 3300 \end{bmatrix}$$

We have reached the optimal value for the dual problem. The function will be maximized when $u = \frac{66}{5}, v = 5$. The maximum value is $d = 3300$. Therefore costs will be minimized at $3300 when 175 VHS tapes are ordered, and 75 DVD's are ordered.

23.
Let x, y, and z represent the number of standard, deluxe, and super deluxe microwave ovens produced each week respectively.

The cost function is:

$$C = 100x + 110y + 120z$$

The constraints made by the contracts for each type of microwave oven imply that:

$$x + y + z \ge 300$$
$$x \qquad \ge 80$$
$$\qquad y \qquad \ge 50$$
$$\qquad z \ge 90$$

The linear programming problem is:

Minimize $\qquad C = 100x + 110y + 120z$
subject to
$$x + y + z \ge 300$$
$$x \qquad \ge 80$$
$$\qquad y \qquad \ge 50$$
$$\qquad z \ge 90$$
$$x \ge 0, y \ge 0, z \ge 0$$

The primal matrix is:

$$P = \begin{bmatrix} 1 & 1 & 1 & 300 \\ 1 & 0 & 0 & 80 \\ 0 & 1 & 0 & 50 \\ 0 & 0 & 1 & 90 \\ \hline 100 & 110 & 120 & 0 \end{bmatrix}$$

The dual matrix is:

$$D = P^T = \begin{bmatrix} 1 & 1 & 0 & 0 & 100 \\ 1 & 0 & 1 & 0 & 110 \\ 1 & 0 & 0 & 1 & 120 \\ \hline 300 & 80 & 50 & 90 & 0 \end{bmatrix}$$

The dual problem is stated at the top of the next page.

Maximize $d = 300t + 80u + 50v + 90w$
subject to

$$t + u \qquad\qquad \le 100$$
$$t \qquad + v \qquad \le 110$$
$$t \qquad\qquad + w \le 120$$
$$t \ge 0, u \ge 0, v \ge 0, w \ge 0$$

The initial simplex tableau for the dual problem is:

$$
\begin{array}{cccccccc|c}
t & u & v & w & s_1 & s_2 & s_3 & d & \\
\hline
1 & 1 & 0 & 0 & 1 & 0 & 0 & 0 & 100 \\
1 & 0 & 1 & 0 & 0 & 1 & 0 & 0 & 110 \\
1 & 0 & 0 & 1 & 0 & 0 & 1 & 0 & 120 \\
\hline
-300 & -80 & -50 & -90 & 0 & 0 & 0 & 1 & 0
\end{array}
$$

Using the simplex algorithm, the first pivot element will be the one in row 1, column 1. Pivoting on this element results in the tableau:

$$
\begin{array}{cccccccc|c}
t & u & v & w & s_1 & s_2 & s_3 & d & \\
\hline
1 & 1 & 0 & 0 & 1 & 0 & 0 & 0 & 100 \\
0 & -1 & 1 & 0 & -1 & 1 & 0 & 0 & 10 \\
0 & -1 & 0 & 1 & -1 & 0 & 1 & 0 & 20 \\
\hline
0 & 220 & -50 & -90 & 300 & 0 & 0 & 1 & 30000
\end{array}
$$

Next we are required to pivot on the one in row 3, column 4. Pivoting on this element results in the tableau:

$$
\begin{array}{cccccccc|c}
t & u & v & w & s_1 & s_2 & s_3 & d & \\
\hline
1 & 1 & 0 & 0 & 1 & 0 & 0 & 0 & 100 \\
0 & -1 & 1 & 0 & -1 & 1 & 0 & 0 & 10 \\
0 & -1 & 0 & 1 & -1 & 0 & 1 & 0 & 20 \\
\hline
0 & 130 & -50 & 0 & 210 & 0 & 90 & 1 & 31800
\end{array}
$$

Next we are required to pivot on the one in row 2, column 3. Pivoting on this element results in the tableau:

$$
\begin{array}{cccccccc|c}
t & u & v & w & s_1 & s_2 & s_3 & d & \\
\hline
1 & 1 & 0 & 0 & 1 & 0 & 0 & 0 & 100 \\
0 & -1 & 1 & 0 & -1 & 1 & 0 & 0 & 10 \\
0 & -1 & 0 & 1 & -1 & 0 & 1 & 0 & 20 \\
\hline
0 & 80 & 0 & 0 & 160 & 50 & 90 & 1 & 32300
\end{array}
$$

The solution to the dual problem is $d = 32,300$, when $t = 100, u = 0, v = 10, w = 20$. Therefore the solution to the primal problem is $C = 32,300$ when $x = 160, y = 50, z = 90$.

Costs are minimized at \$32,300 when 160 standard models are produced, 50 deluxe models are produced, and 90 super deluxe models are produced.

Chapter 5 Summary Exercises

1. The pivot column will be the second column. Using the smallest quotient rule we see:

$$R_1 : \tfrac{2}{2} = 1$$
$$R_2 : \tfrac{6}{2} = 2$$
$$R_3 : \tfrac{5}{1} = 5$$

Therefore the first pivot element will be the two in row 1, column 2.

3. The solution for this matrix is:
$$x = 0, y = 0, z = 0, P = 0, s_1 = 16, s_2 = 16, s_3 = 10$$

The absence of negative values for the basic variables indicates that this solution does represent a corner point of the feasible region.

5. Introducing slack variables the inequality constraints become a system of slack equations

$$x + y + s_1 \qquad\quad = 5$$
$$3x + 2y \qquad + s_2 = 12$$

7. The initial simplex tableau is:

$$
\begin{array}{ccccc|c}
x & y & s_1 & s_2 & f & \\
\hline
1 & 1 & 1 & 0 & 0 & 5 \\
3 & 2 & 0 & 1 & 0 & 12 \\
\hline
-4 & -3 & 0 & 0 & 1 & 0
\end{array}
$$

The most negative value in the last row is the negative four in column 1. Therefore column 1 is the pivot column.

The solution is continued on the next page.

Applying the smallest quotient rule we see:

$R_1 : \frac{5}{1} = 5$

$R_2 : \frac{12}{3} = 4$

Therefore the pivot element will be the three in row 2, column 1.

Pivoting on this element results in the tableau:

$$
\begin{array}{ccccc}
x & y & s_1 & s_2 & f
\end{array}
$$
$$
\left[\begin{array}{cccc|c}
0 & \frac{1}{3} & 1 & \frac{-1}{3} & 0 & 1 \\
1 & \frac{2}{3} & 0 & \frac{1}{3} & 0 & 4 \\
\hline
0 & \frac{-1}{3} & 0 & \frac{4}{3} & 1 & 16
\end{array}\right]
$$

The basic feasible solution that this tableau represents is:

$x = 4, y = 0, f = 16, s_1 = 1, s_2 = 0$

9. The primal matrix is:

$$
P = \left[\begin{array}{cc|c}
1 & 1 & 8 \\
2 & 1 & 12 \\
\hline
3 & 1 & 0
\end{array}\right]
$$

The dual matrix is:

$$
D = P^T = \left[\begin{array}{cc|c}
1 & 2 & 3 \\
1 & 1 & 1 \\
\hline
8 & 12 & 0
\end{array}\right]
$$

The dual problem is:

Maximize $\qquad d = 8u + 12v$
subject to
$$u + 2v \le 3$$
$$u + v \le 1$$
$$u \ge 0, v \ge 0$$

The initial simplex tableau for the dual problem is stated at the top of the next column.

$$
\begin{array}{ccccc}
u & v & s_1 & s_2 & d
\end{array}
$$
$$
\left[\begin{array}{cccc|c}
1 & 2 & 1 & 0 & 0 & 3 \\
1 & 1 & 0 & 1 & 0 & 1 \\
\hline
-8 & -12 & 0 & 0 & 1 & 0
\end{array}\right]
$$

The simplex method indicates that the first pivot element will be the one in row 2, column 2. Pivoting on this element results in the tableau:

$$
\begin{array}{ccccc}
u & v & s_1 & s_2 & d
\end{array}
$$
$$
\left[\begin{array}{cccc|c}
-1 & 0 & 1 & -2 & 0 & 1 \\
1 & 1 & 0 & 1 & 0 & 1 \\
\hline
4 & 0 & 0 & 12 & 1 & 12
\end{array}\right]
$$

The absence of negative values below the horizontal bar indicates that we have reached the maximum value for the objective function in the dual problem. The maximum value is $d = 12$ when $u = 0, v = 1$.

The Duality Theorem indicates that the solution to the primal problem is the minimum value of C is 12, and this value is reached when $x = 0, y = 12$. Remember to find the values of the variables x and y we look in the last row of the slack variable columns.

11. Let x, y, and z represent the number of batches of sugar cookies, peanut butter cookies and pecan sandies respectively. The objective is to maximize revenue, and since each batch contains 3 dozen cookies, the revenue function is:

$$R = 1.50(3x) + 1.75(3y) + 2.25(3z) \text{, or}$$
$$R = 4.50x + 5.25y + 6.75z$$

The constraints imposed by the fact that they have 10 dozen eggs or $10(12) = 120$ eggs, 300 cups of flour, and 200 cups of sugar are translated into the following inequalities:

$$2x + y \le 120$$
$$3x + \tfrac{5}{4}y + 2z \le 300$$
$$\tfrac{9}{8}x + y + \tfrac{1}{3}z \le 200$$

The linear programming problem is stated at the top of the next page.

Maximize $R = 4.50x + 5.25y + 6.75z$
subject to

$$2x + y \leq 120$$
$$3x + \tfrac{5}{4}y + 2z \leq 300$$
$$\tfrac{9}{8}x + y + \tfrac{1}{3}z \leq 200$$
$$x \geq 0, y \geq 0, z \geq 0$$

The initial simplex tableau is:

$$
\begin{array}{ccccccc|c}
x & y & z & s_1 & s_2 & s_3 & f & \\
2 & 1 & 0 & 1 & 0 & 0 & 0 & 120 \\
3 & \tfrac{5}{4} & 2 & 0 & 1 & 0 & 0 & 300 \\
\tfrac{9}{8} & 1 & \tfrac{1}{3} & 0 & 0 & 1 & 0 & 200 \\
\hline
-4.50 & -5.25 & -6.75 & 0 & 0 & 0 & 1 & 0
\end{array}
$$

The most negative value in the last row is in the third column, so this will be the pivot column. The smallest quotient rule indicates that we will pivot on the two in row 2, column 3. Pivoting on this element results in the following tableau:

$$
\begin{array}{ccccccc|c}
x & y & z & s_1 & s_2 & s_3 & f & \\
2 & 1 & 0 & 1 & 0 & 0 & 0 & 120 \\
\tfrac{3}{2} & \tfrac{5}{8} & 1 & 0 & \tfrac{1}{2} & 0 & 0 & 150 \\
\tfrac{5}{8} & \tfrac{19}{24} & 0 & 0 & \tfrac{-1}{6} & 1 & 0 & 150 \\
\hline
\tfrac{45}{8} & \tfrac{-33}{32} & 0 & 0 & \tfrac{27}{28} & 0 & 1 & \tfrac{2025}{2}
\end{array}
$$

The next pivot element will be the one in row 1, column 2. Pivoting on this element results in the tableau:

$$
\begin{array}{ccccccc|c}
x & y & z & s_1 & s_2 & s_3 & f & \\
2 & 1 & 0 & 1 & 0 & 0 & 0 & 120 \\
\tfrac{1}{4} & 0 & 1 & \tfrac{-5}{8} & \tfrac{1}{2} & 0 & 0 & 75 \\
\tfrac{-23}{24} & 0 & 0 & \tfrac{-19}{24} & \tfrac{-1}{6} & 1 & 0 & 55 \\
\hline
\tfrac{15}{16} & 0 & 0 & \tfrac{33}{32} & \tfrac{27}{8} & 0 & 1 & \tfrac{4545}{2}
\end{array}
$$

The absence of negative values to below the horizontal bar indicates we have reached an optimal value for the objective function R. The maximum revenue that the Fibonacci Math Club can generate is $1136.25. They will generate this revenue by making zero sugar cookies, 120 peanut butter cookies, and 75 pecan sandies. They will have 55 cups of sugar left over.

13. An example of such a problem is:

Minimize $f = 5x + 3y - 20$
subject to

$$2x + y \geq 4$$
$$x + y \geq 3$$
$$x \geq 0, y \geq 0$$

When we solve this problem with the simplex method, we maximize $g = -f = -5x - 3y + 20$ and the initial value of g is 20. As we apply Crown's rules and eventually the Simplex algorithm, the value of g remains positive, and the value of f remains negative throughout the problem.

In a minimization problem, if the objective function is negative at all possible corner points the entry in the lower right corner will be positive throughout the algorithm. If the objective function becomes positive, the entry will be negative if the basic solution to the tableau corresponds to a point at which the value of f is positive.

In a maximization problem, the lower right hand entry always corresponds to the value of the objective function at the point determined by the basic solution of the simplex method. That entry will remain positive if all corner points return positive values for the objective function.

15. Using Excel's solver, the minimum value of f is nine. The minimum occurs when $x = 0, y = 3, z = 0$.

Cumulative Review

17. To find the inverse of A we reduce the augmented matrix:

$$
\left[\begin{array}{cc|cc}
1 & 2 & 1 & 0 \\
0 & 2 & 0 & 1
\end{array}\right]
$$

First we need pivot on the two in row 2, column 2. This results in the matrix:

$$
\left[\begin{array}{cc|cc}
1 & 0 & 1 & -1 \\
0 & 1 & 0 & \tfrac{1}{2}
\end{array}\right]
$$

Thus the inverse of A is:

$$
A^{-1} = \begin{bmatrix} 1 & -1 \\ 0 & \tfrac{1}{2} \end{bmatrix}
$$

The solution is continued on the next page.

We place the coded message in a matrix with two columns. The coded matrix is:

$$C = \begin{bmatrix} 25 & 80 \\ 21 & 102 \\ 7 & 24 \\ 20 & 100 \\ 1 & 30 \\ 30 & 62 \end{bmatrix}$$

To decode the matrix we multiply it on the right by the inverse of the encoding matrix to get:

$$CA^{-1} = \begin{bmatrix} 25 & 80 \\ 21 & 102 \\ 7 & 24 \\ 20 & 100 \\ 1 & 30 \\ 30 & 62 \end{bmatrix} \begin{bmatrix} 1 & -1 \\ 0 & \frac{1}{2} \end{bmatrix} = \begin{bmatrix} 25 & 15 \\ 21 & 30 \\ 7 & 5 \\ 20 & 30 \\ 1 & 14 \\ 0 & 1 \end{bmatrix}$$

Next we decode the matrix using the key from the text book.

$$25 \quad 15 \mid 21 \quad 30 \mid 7 \quad 5 \mid 20 \quad 30 \mid 1 \quad 14 \mid 30 \quad 1$$
$$\text{Y} \quad \text{O} \mid \text{U} \quad \_ \mid \text{G} \quad \text{E} \mid \text{T} \quad \_ \mid \text{A} \quad \text{N} \mid \_ \quad \text{A}$$

The message reads "You get an A".

19. The augmented matrix for the system is:

$$\begin{bmatrix} 3 & 0 & -1 & 0 & \mid & 0 \\ 0 & 2 & 1 & 0 & \mid & 0 \\ 1 & 1 & 1 & 1 & \mid & 6 \\ 2 & 3 & 5 & -10 & \mid & 15 \end{bmatrix}$$

We pivot on the three in row 1, column 1 to get the matrix:

$$\begin{bmatrix} 1 & 0 & \frac{-1}{3} & 0 & \mid & 0 \\ 0 & 2 & 1 & 0 & \mid & 0 \\ 0 & 1 & \frac{4}{3} & 1 & \mid & 6 \\ 0 & 3 & \frac{17}{3} & -10 & \mid & 15 \end{bmatrix}$$

Next we pivot on the two in row 2, column 2. This pivot results in the matrix at the top of the next column.

$$\begin{bmatrix} 1 & 0 & \frac{-1}{3} & 0 & \mid & 0 \\ 0 & 1 & \frac{1}{2} & 0 & \mid & 0 \\ 0 & 0 & \frac{5}{6} & 1 & \mid & 6 \\ 0 & 0 & \frac{25}{6} & -10 & \mid & 15 \end{bmatrix}$$

Next we pivot on the $\frac{5}{6}$ in row 3, column 3. Pivoting on this element results in the matrix:

$$\begin{bmatrix} 1 & 0 & 0 & \frac{2}{5} & \mid & \frac{12}{5} \\ 0 & 1 & 0 & \frac{-3}{5} & \mid & \frac{-18}{5} \\ 0 & 0 & 1 & \frac{6}{5} & \mid & \frac{36}{5} \\ 0 & 0 & 0 & -15 & \mid & -15 \end{bmatrix}$$

Finally we pivot on the -15 in row 4, column 4. Pivoting on this element results in the tableau:

$$\begin{bmatrix} 1 & 0 & 0 & 0 & \mid & 2 \\ 0 & 1 & 0 & 0 & \mid & -3 \\ 0 & 0 & 1 & 0 & \mid & 6 \\ 0 & 0 & 0 & 1 & \mid & 1 \end{bmatrix}$$

The solution to the system is:

$$x = 2, y = -3, z = 6, w = 1.$$

Chapter 5 Sample Test Answers

1. The constraints are:

$$2x + y \leq 8$$
$$x + 5y \leq 10$$
$$x \geq 0;\ y \geq 0$$

2. Pivot on the two in row 1, column 1 to obtain:

$$\begin{array}{cccc} x & y & s_1 & s_2 \end{array}$$
$$\begin{bmatrix} 1 & \frac{1}{2} & \frac{1}{2} & 0 & \mid & 4 \\ 0 & \frac{9}{2} & \frac{-1}{2} & 1 & \mid & 6 \end{bmatrix}$$

3. No, due to the fact that there is a negative in the row below the horizontal bar, a maximum for the objective function as not been reached.

4. At this stage, the value of f is 35. The point that corresponds with this function value is:

$$x = 2, y = 0, s_1 = 8, s_2 = 4, s_3 = 0$$

5. At this stage in the pivoting process, the first constraint has a surplus of eight $\left(s_1 = 8\right)$, the second constraint has a surplus of four $\left(s_2 = 4\right)$, and the third constraint has a surplus of zero $\left(s_3 = 0\right)$.

6. The feasible region is shown below:

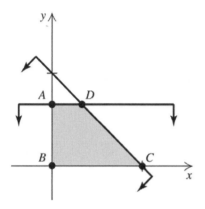

The values of the variables at each corner point are

Point	x	y	f	s_1	s_2
A	0	4	4	2	0
B	0	0	0	6	4
C	6	0	12	0	4
D	2	4	8	0	0

7. The initial tableau is:

$$
\begin{array}{ccccc}
x & y & s_1 & s_2 & f
\end{array}
$$
$$
\left[
\begin{array}{ccccc|c}
1 & 1 & 1 & 0 & 0 & 6 \\
0 & 1 & 0 & 1 & 0 & 4 \\
-2 & -1 & 0 & 0 & 1 & 0
\end{array}
\right]
$$

Pivoting on the one in row 1, column 1 results in the tableau:

$$
\begin{array}{ccccc}
x & y & s_1 & s_2 & f
\end{array}
$$
$$
\left[
\begin{array}{ccccc|c}
1 & 1 & 1 & 0 & 0 & 6 \\
0 & 1 & 0 & 1 & 0 & 4 \\
0 & 1 & 2 & 0 & 1 & 12
\end{array}
\right]
$$

The tableau at the bottom of the previous column indicates that the optimal solution has been found. The basic feasible solution that this tableau represents is:

$$x = 6, y = 0, f = 12, s_1 = 0, s_2 = 4$$

8. The dual problem is:

Maximize $\qquad d = 3u + 4v + w$
subject to

$$u + v + w \le 2$$
$$u + 2v \quad\ \le 3$$
$$u \ge 0, v \ge 0, w \ge 0$$

9. The initial tableau for the dual problem is:

$$
\begin{array}{cccccc}
u & v & w & s_1 & s_2 & d
\end{array}
$$
$$
\left[
\begin{array}{cccccc|c}
1 & 1 & 1 & 1 & 0 & 0 & 2 \\
1 & 2 & 0 & 0 & 1 & 0 & 3 \\
-3 & -4 & -1 & 0 & 0 & 1 & 0
\end{array}
\right]
$$

The simplex method indicates that we should pivot on the two in row 2, column 2. Pivoting results in the tableau:

$$
\begin{array}{cccccc}
u & v & w & s_1 & s_2 & d
\end{array}
$$
$$
\left[
\begin{array}{cccccc|c}
\frac{1}{2} & 0 & 1 & 1 & \frac{-1}{2} & 0 & \frac{1}{2} \\
\frac{1}{2} & 1 & 0 & 0 & \frac{1}{2} & 0 & \frac{3}{2} \\
-1 & 0 & -1 & 0 & 2 & 1 & 6
\end{array}
\right]
$$

Next we can pivot on the one in row 1, column 3, or the $\frac{1}{2}$ in row 1, column 1. We choose to pivot on the element in row 1, column 1. Pivoting results in the tableau:

$$
\begin{array}{cccccc}
u & v & w & s_1 & s_2 & d
\end{array}
$$
$$
\left[
\begin{array}{cccccc|c}
1 & 0 & 2 & 2 & -1 & 0 & 1 \\
0 & 1 & -1 & -1 & 1 & 0 & 1 \\
0 & 0 & 1 & 2 & 1 & 1 & 7
\end{array}
\right]
$$

The absence of negatives in the last row indicates that we have reached a solution to the dual problem. The maximum value of the objective function for the dual problem is $d = 7$. This value occurs when $u = 1, v = 1, w = 0$. Therefore, by The Duality Theorem, the solution to the primal problem can be stated. The minimum value of f is seven. The minimum occurs when $x = 2, y = 1$.

10. The feasible region is shown below:

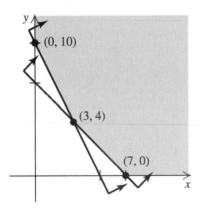

11. We must first rewrite the minimization problem as an appropriate maximization problem. The problem is:

Maximize $\qquad g = -f = -3x - y$

subject to

$$-x - y \le -7$$
$$-2x - y \le -10$$
$$x \ge 0, y \ge 0$$

The initial tableau is:

$$\begin{array}{cccc} x & y & s_1 & s_2 & g \end{array}$$
$$\left[\begin{array}{ccccc|c} -1 & -1 & 1 & 0 & 0 & -7 \\ -2 & -1 & 0 & 1 & 0 & -10 \\ \hline 3 & 1 & 0 & 0 & 1 & 0 \end{array}\right]$$

Crown's rules indicate that we should pivot on the negative one in row 2, column 2. (note we could have also pivoted on the negative one in row 1, column 1)

Pivoting on this element results in the tableau:

$$\begin{array}{cccc} x & y & s_1 & s_2 & g \end{array}$$
$$\left[\begin{array}{ccccc|c} 1 & 0 & 1 & -1 & 0 & 3 \\ 2 & 1 & 0 & -1 & 0 & 10 \\ \hline 1 & 0 & 0 & 1 & 1 & -10 \end{array}\right]$$

Thus the maximum for g is -10, so the minimum for f is 10. This occurs when $x = 0, y = 10$.

12. The chart below contains the necessary information.

Point	$f = 3x + y$
$(0,10)$	$f = 3(0) + (10) = 10$
$(3,4)$	$f = 3(3) + (4) = 13$
$(7,0)$	$f = 3(7) + (0) = 21$

Once again, we see that the minimum is $f = 10$ when $x = 0, y = 10$.

13. Rewrite the second constraint by multiplying by negative one. The initial tableau is:

$$\begin{array}{ccccccc} x & y & z & s_1 & s_2 & s_3 & f \end{array}$$
$$\left[\begin{array}{ccccccc|c} 1 & 1 & 2 & 1 & 0 & 0 & 0 & 6 \\ 0 & -2 & -1 & 0 & 1 & 0 & 0 & -4 \\ 2 & 0 & 1 & 0 & 0 & 1 & 0 & 4 \\ \hline -2 & -3 & -5 & 0 & 0 & 0 & 1 & 0 \end{array}\right]$$

Crown's rules indicate that we should pivot on the one in row 1, column 2. Pivoting on this element results in the tableau:

$$\begin{array}{ccccccc} x & y & z & s_1 & s_2 & s_3 & f \end{array}$$
$$\left[\begin{array}{ccccccc|c} 1 & 1 & 2 & 1 & 0 & 0 & 0 & 6 \\ 2 & 0 & 3 & 2 & 1 & 0 & 0 & 8 \\ 2 & 0 & 1 & 0 & 0 & 1 & 0 & 4 \\ \hline 1 & 0 & 1 & 3 & 0 & 0 & 1 & 18 \end{array}\right]$$

The absence of negative values in the last row indicates that we have reached the optimal solution. The solution is:

$$f = 18, \text{ when } x = 0, y = 6, z = 0.$$

14. Let x represent the number of professors, and y represent the number of lecturers. Let cost be in thousands of dollars. The linear programming problem is:

Minimize $\qquad C = 40x + 15y$

subject to

$$x + y \le 50$$
$$x + y \ge 35$$
$$x \ge 25$$
$$y \le 15$$
$$x \ge 0, y \ge 0$$

We will need to rewrite the problem on the previous page as the following maximization problem:

Maximize $\qquad g = -C = -40x - 15y$

subject to

$$x + y \le 50$$
$$-x - y \le -35$$
$$-x \quad \le -25$$
$$y \le 15$$
$$x \ge 0, y \ge 0$$

The initial simplex tableau is:

$$
\begin{array}{ccccccc}
x & y & s_1 & s_2 & s_3 & s_4 & g \\
\end{array}
$$

$$
\left[
\begin{array}{ccccccc|c}
1 & 1 & 1 & 0 & 0 & 0 & 0 & 50 \\
-1 & -1 & 0 & 1 & 0 & 0 & 0 & -35 \\
-1 & 0 & 0 & 0 & 1 & 0 & 0 & -25 \\
0 & 1 & 0 & 0 & 0 & 1 & 0 & 15 \\
\hline
40 & 15 & 0 & 0 & 0 & 0 & 1 & 0 \\
\end{array}
\right]
$$

Crown's rules indicate that we may pivot on the one in row 1, column 1, or the one in row 4, column 2. We choose the one in row 1, column 1 Pivoting on this element results in the tableau:

$$
\begin{array}{ccccccc}
x & y & s_1 & s_2 & s_3 & s_4 & g \\
\end{array}
$$

$$
\left[
\begin{array}{ccccccc|c}
1 & 1 & 1 & 0 & 0 & 0 & 0 & 50 \\
0 & 0 & 1 & 1 & 0 & 0 & 0 & 15 \\
0 & 1 & 1 & 0 & 1 & 0 & 0 & 25 \\
0 & 1 & 0 & 0 & 0 & 1 & 0 & 15 \\
\hline
0 & -25 & -40 & 0 & 0 & 0 & 1 & -2000 \\
\end{array}
\right]
$$

Now we use the simplex method to pivot on the one in row 2, column 3 of the matrix. Pivoting results in the tableau:

$$
\begin{array}{ccccccc}
x & y & s_1 & s_2 & s_3 & s_4 & g \\
\end{array}
$$

$$
\left[
\begin{array}{ccccccc|c}
1 & 1 & 0 & -1 & 0 & 0 & 0 & 35 \\
0 & 0 & 1 & 1 & 0 & 0 & 0 & 15 \\
0 & 1 & 0 & -1 & 1 & 0 & 0 & 10 \\
0 & 1 & 0 & 0 & 0 & 1 & 0 & 15 \\
\hline
0 & -25 & 0 & 40 & 0 & 0 & 1 & -1400 \\
\end{array}
\right]
$$

Next we will pivot on the one in row 3, column 2. Pivoting results in the tableau at the top of the next column.

$$
\begin{array}{ccccccc}
x & y & s_1 & s_2 & s_3 & s_4 & g \\
\end{array}
$$

$$
\left[
\begin{array}{ccccccc|c}
1 & 0 & 0 & 0 & -1 & 0 & 0 & 25 \\
0 & 0 & 1 & 1 & 0 & 0 & 0 & 15 \\
0 & 1 & 0 & -1 & 1 & 0 & 0 & 10 \\
0 & 0 & 0 & 1 & -1 & 1 & 0 & 5 \\
\hline
0 & 0 & 0 & 15 & 25 & 0 & 1 & -1150 \\
\end{array}
\right]
$$

Thus the optimal value of g is -1150. Therefore the minimum cost is \$1,150,000 with 25 professors and 10 lecturers.

Chapter 6
Mathematics of Finance

Exercises Section 6.1

1. We have the following information:

$$P = 2000, r = 0.075, m = 12, n = mt = 48$$

The compound interest formula gives us:

$$A = 2000\left(1 + \frac{0.075}{12}\right)^{48} = \$2697.20 .$$

3. We have the following information:

$$P = 3000, r = 0.06, m = 4, n = mt = 48$$

The compound interest formula gives us:

$$A = 3000\left(1 + \frac{0.06}{4}\right)^{48} = \$6130.43 .$$

5. We have the following information:

$$P = 6000, r = 0.12, t = 1$$

The simple interest formula gives us:

$$A = 6000\left(1 + (0.12)(1)\right) = \$6720.00 .$$

7. We have the following information for each part:

$$P = 800, r = 0.07, t = 12$$

a. The simple interest formula gives us:

$$A = 800\left(1 + (0.07)(12)\right) = \$1472.00 .$$

b. Compounding quarterly means: $m = 4, mt = 48$
The compound interest formula gives us:

$$A = 800\left(1 + \frac{0.07}{4}\right)^{48} = \$1839.68 .$$

c. Compounding monthly means: $m = 12, mt = 144$
The compound interest formula gives us:

$$A = 800\left(1 + \frac{0.07}{12}\right)^{144} = \$1848.58 .$$

9. We have the following information for each part:

$$P = 1, r = 0.12, t = 10$$

a. Compounding semiannually means: $m = 2, mt = 20$

The compound interest formula gives us:

$$A = 1\left(1 + \frac{0.12}{2}\right)^{20} = \$3.21 .$$

b. Compounding quarterly means: $m = 4, mt = 40$
The compound interest formula gives us:

$$A = 1\left(1 + \frac{0.12}{4}\right)^{40} = \$3.26 .$$

c. Compounding daily means: $m = 365, mt = 3650$
The compound interest formula gives us:

$$A = 1\left(1 + \frac{0.12}{365}\right)^{3650} = \$3.32 .$$

11. We have the following table:

Years	Simple Interest	Compounded Quarterly
1	$ 5500	$ 5519.06
2	$ 6000	$ 6092.01
3	$ 6500	$ 6724.44
4	$ 7000	$ 7422.53
5	$ 7500	$ 8193.08

The graph is shown below:

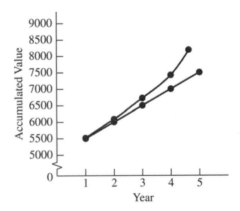

13. The information given in the problem is:

$$A = 20,000, P = 8000, m = 1, t = 10$$

The solution is continued on the next page.

Substituting the information on the previous page into the compound interest formula we get:

$$20,000 = 8000\left(1+\tfrac{r}{1}\right)^{10}$$

Solve the equation for r.

$$\tfrac{20,000}{8000} = \left(1+r\right)^{10}$$

$$1+r = \left(2.5\right)^{0.1} = 1.096$$

$$r = 0.096$$

Thus, investing $8000 at an interest rate of 9.6% compounded annually will yield $20,000 after ten years.

15. The information given in the problem is:

$$A = 5000, P = 1250, m = 4, t = 12$$

Substituting into the compound interest formula we get:

$$5000 = 1250\left(1+\tfrac{r}{4}\right)^{48}$$

Solve the equation for r.

$$\tfrac{5000}{1250} = \left(1+\tfrac{r}{4}\right)^{48}$$

$$1+\tfrac{r}{4} = \left(4\right)^{\frac{1}{48}} = 1.029$$

$$\tfrac{r}{4} = 0.0293$$

$$r = 0.117$$

Thus, investing $1250 at an interest rate of 11.7% compounded quarterly will yield $5000 after 12 years.

17. The information given in the problem is:

$$A = 7500, P = 2500, m = 2, t = 6$$

Substituting into the compound interest formula we get:

$$7500 = 2500\left(1+\tfrac{r}{2}\right)^{12}$$

Solve the equation for r.

$$\tfrac{7500}{2500} = \left(1+\tfrac{r}{2}\right)^{12}$$

$$1+\tfrac{r}{2} = \left(3\right)^{\frac{1}{12}} = 1.0959$$

$$\tfrac{r}{2} = 0.0959$$

$$r = 0.192$$

The solution is stated at the top of the next column.

Thus, investing $2500 at an interest rate of 19.2% compounded semiannually will yield $7500 after 6 years.

19. The information given in the problem is:

$$A = 3500, P = 1200, r = 0.08, m = 2$$

Substituting into the compound interest formula we get:

$$3500 = 1200\left(1+\tfrac{0.08}{2}\right)^{n}$$

Solve the equation for n.

$$\tfrac{3500}{1200} = \left(1+0.04\right)^{n}$$

$$\ln\left(1.04\right)^{n} = \ln\left(\tfrac{3500}{1200}\right)$$

$$n\ln\left(1.04\right) = \ln\left(\tfrac{3500}{1200}\right)$$

$$n = \frac{\ln\left(\tfrac{3500}{1200}\right)}{\ln\left(1.04\right)} = 27.29$$

Thus, it will take 28 compounding periods or 14 years to achieve the conditions set forth in the problem.

21. The information given in the problem is:

$$A = 6000, P = 1000, r = 0.12, m = 4$$

Substituting into the compound interest formula we get the equation:

$$6000 = 1000\left(1+\tfrac{0.12}{4}\right)^{n}$$

Solve the equation for n.

$$6 = \left(1+0.03\right)^{n}$$

$$\ln\left(1.03\right)^{n} = \ln\left(6\right)$$

$$n\ln\left(1.03\right) = \ln\left(6\right)$$

$$n = \frac{\ln\left(6\right)}{\ln\left(1.03\right)} = 60.62$$

Thus, it will take 61 compounding periods or 15.25 years to achieve the conditions set forth in the problem.

23. The information given in the problem is:

$$A = 50,000, P = 6000, r = 0.10, m = 1$$

Substituting into the compound interest formula we get:

$$50,000 = 6000\left(1 + \frac{0.10}{1}\right)^n$$

We solve for n:

$$\frac{50,000}{6000} = (1 + 0.10)^n$$

$$\ln(1.10)^n = \ln\left(\frac{50,000}{6000}\right)$$

$$n \ln(1.10) = \ln\left(\frac{50,000}{6000}\right)$$

$$n = \frac{\ln\left(\frac{50,000}{6000}\right)}{\ln(1.10)} = 22.25$$

Thus, it will take 23 compounding periods which is 23 years to achieve the conditions set forth in the problem.

25. The information given in the problem is:

$$A = 5000, r = 0.06, m = 1, t = 10$$

Substituting into the compound interest formula we get:

$$5000 = P\left(1 + \frac{0.06}{1}\right)^{10}$$

Solve the equation for P.

$$5000 = P(1 + 0.06)^{10}$$

$$P = \frac{5000}{(1.06)^{10}} = 2791.97$$

Thus, a principal of $ 2791.97 must be invested to achieve the stated conditions set forth in the problem.

27. The information given in the problem is:

$$A = 30,000, r = 0.12, m = 4, t = 10$$

Substituting into the compound interest formula we get:

$$30,000 = P\left(1 + \frac{0.12}{4}\right)^{40}$$

Solve the equation for P as shown at the top of the next column.

$$30,000 = P(1 + 0.03)^{40}$$

$$P = \frac{30,000}{(1.03)^{40}} = 9196.71$$

Thus, a principal of $ 9196.71 must be invested to achieve the stated conditions set forth in the problem.

29. The information given in the problem is:

$$A = 3000, r = 0.075, m = 12, t = 10$$

Substituting into the compound interest formula we get:

$$3000 = P\left(1 + \frac{0.075}{12}\right)^{120}$$

Solve the equation for P.

$$3000 = P(1 + 0.00625)^{120}$$

$$P = \frac{3000}{(1.00625)^{120}} = 1420.41$$

Thus, a principal of $ 1420.41 must be invested to achieve the stated conditions set forth in the problem.

31. The information given in the problem is:

$$P = 5000, r = 0.07, m = 2, t = 10$$

Substituting into the compound interest formula we get the equation:

$$A = 5000\left(1 + \frac{0.07}{2}\right)^{20} = 9948.94$$

The value of the investment will be $ 9948.94 ten years from now.

33. The information given in the problem is:

$$A = 20,000, r = 0.08, m = 4, t = 12$$

Substituting into the compound interest formula we get:

$$20,000 = P\left(1 + \frac{0.08}{4}\right)^{48}$$

The solution is continued on the next page.

Solving the equation on the previous page for P we get:

$$20,000 = P(1+0.02)^{48}$$

$$P = \frac{20,000}{(1.02)^{48}} = 7730.75$$

Dr. Bishop will need to invest $ 7,730.75 now in order to have saved $ 20,000 dollars in 12 years.

35. The information given in the problem is:

$$A = 20,000, P = 8000, m = 2, t = 12$$

Substituting into the compound interest formula we get:

$$20,000 = 8000\left(1+\tfrac{r}{2}\right)^{24}$$

Solving for r we get:

$$\frac{20,000}{8000} = \left(1+\tfrac{r}{2}\right)^{24}$$

$$1+\tfrac{r}{2} = (2.5)^{\frac{1}{24}} = 1.0389$$

$$\tfrac{r}{2} = 0.0389$$

$$r = 0.0778$$

The Tanners will need to invest at an interest rate of 7.78% compounded semiannually to have enough money for their child's college education in 12 years.

37. The information given in the problem is:

$$r = 0.08, m = 4$$

To triple the principle, we need $A = 3P$

Substituting into the compound interest formula we get the equation:

$$3P = P\left(1+\tfrac{0.08}{4}\right)^{4t}$$

Solving for t we get:

$$\tfrac{3P}{P} = (1+0.02)^{4t}$$

$$\ln(1.02)^{4t} = \ln(3)$$

$$4t\ln(1.02) = \ln(3)$$

$$t = \frac{\ln(3)}{4\ln(1.02)} = 13.87.$$

It will take 14 years for the principal to triple in value.

39. The information given in the problem is:

$$A = 8000, P = 5000, r = 0.075, m = 2$$

Substituting into the compound interest formula we get:

$$8000 = 5000\left(1+\tfrac{0.075}{2}\right)^{2t}$$

We solve the equation for t:

$$\tfrac{8000}{5000} = (1+0.0375)^{2t}$$

$$\ln(1.0375)^{2t} = \ln(1.6)$$

$$2t\ln(1.0375) = \ln(1.6)$$

$$t = \frac{\ln(1.6)}{2\ln(1.0375)} = 6.38.$$

Since interest is calculated at the end of the period, it will take 6.5 years compounding semiannually for the principal to reach $8000.

41. Since the reserve requirements are 15%, the bank will pay interest on the available reserves of

$$(\$1200)(0.85) = \$1020 \text{ in your account.}$$

The interest earned on an account is given by the formula $I = Pi$.

Since $P = 1020$, and $i = \tfrac{r}{12} = \tfrac{0.045}{12} = .00375$, the interest earned during the period is:

$$I = 1020(0.00375) = 3.825$$

The interest was earned this month was $ 3.83.

43. Using the compounded interest formula:

$$C = 50\left(1+\tfrac{0.04}{1}\right)^{5} = 50(1.04)^{5} = 60.83$$

The cost of $ 50 worth of goods will be $ 60.83 five years from now.

45. If the purchasing power decreases by 5% each year, then after n years, the purchasing power will be $(0.95)^n$. To find out how long it will take for the purchasing power to be cut in half, we set:

$$(0.95)^n = \tfrac{1}{2}$$

and solve for n.

$$\ln(0.95)^n = \ln\left(\tfrac{1}{2}\right)$$
$$n \ln(0.95) = \ln\left(\tfrac{1}{2}\right)$$

$$n = \frac{\ln\left(\tfrac{1}{2}\right)}{\ln(0.95)} = 13.51$$

It will take about 13.5 years for the purchasing power of the dollar to be cut in half.

47. If the rate of inflation is 5% per year, and they can live comfortably now on \$ 25,000, in order to maintain the same standard of living ten years from now, they will need:

$$A = 25,000\left(1 + \tfrac{0.05}{1}\right)^{10} = 40,722.37$$

Thus, they will need \$ 40,722.37 ten years from now to have the same standard of living as they do today.

49. If the rate of inflation is 5% per year, and a new car costs \$20,000 today. The cost of a new car eight years from now will be:

$$A = 20,000\left(1 + \tfrac{0.05}{1}\right)^8 = 29,549.11$$

Thus, the median value of a new car will be \$29,549.11 eight years from now.

51. If each floor costs 20% more than the floor immediately below it, the second floor will cost $250,000(1.2)$, the third floor will cost $250,000(1.2)^2$, etc.

a. Using the pattern described above for the fourth floor we have:

$$250,000(1.2)^3 = 432,000$$

The fourth floor will cost \$ 432,000.

b. Using the pattern described above for the eighth floor we have:

$$250,000(1.2)^7 = 895,795.20$$

The fourth floor will cost \$ 895,795.20

c. The total cost of the building will be the sum of the cost of the eight floors.

$$C = 250,000\sum_{i=0}^{7}(1.2)^i$$

$$C = 250,000\left(\frac{(1.2)^8 - 1}{1.2 - 1}\right)$$

$$C = \$4,124,771.20$$

The entire cost of the building will be \$4,124,771.20.

Exercises Section 6.2

1. The future value of the ordinary annuity will be:

$$A = 500\left[\frac{(1+0.025)^{30} - 1}{0.025}\right] = \$21,951.35 \,.$$

3. The future value of the ordinary annuity will be:

$$A = 700\left[\frac{(1+0.01)^{20} - 1}{0.01}\right] = \$15,413.30 \,.$$

5. The future value of the ordinary annuity will be:

$$A = 8000\left[\frac{(1+0.05)^{20} - 1}{0.05}\right] = \$264,527.63 \,.$$

7. Using the information given in the problem, the amount that should be deposited per period to reach \$100,000 is:

$$p = \frac{100,000(.025)}{(1+.025)^{20} - 1} = \$3914.71$$

9. Using the information given in the problem, the amount that should be deposited per period to reach $100,000 is:

$$p = \frac{100,000(.02)}{(1+.02)^{16}-1} = \$5365.01$$

11. Using the information given in the problem, the amount that should be deposited per period to reach $100,000 is:

$$p = \frac{100,000(.03)}{(1+.03)^{22}-1} = \$3274.74$$

13. The information in the problem is:

$$p = 500, r = 0.08, m = 1, n = mt = 16$$

The future value of the annuity will be:

$$A = 500\left[\frac{(1+0.08)^{16}-1}{0.08}\right] = \$15,162.14.$$

The value of the account will be $15,162.14 on Nannette's daughter's 16th birthday.

15. The information given in the problem is:

$$A = 20,000, r = 0.07, m = 4$$
$$i = \frac{r}{m} = 0.175, n = mt = 64$$

The amount deposited into the account each quarter is given by:

$$p = \frac{20,000\left(\frac{.07}{4}\right)}{\left(1+\frac{.07}{4}\right)^{64}-1} = 171.96$$

Elena should deposit $ 171.96 at the end of each quarter for the next 16 years in order to have $20,000 accumulated for her son's college expenses.

17. The information in the problem is:

$$p = 2000, r = 0.05, m = 1, n = mt = 25$$

The future value of the annuity will be:

$$A = 2000\left[\frac{(1+0.05)^{25}-1}{0.05}\right] = \$95,454.20.$$

The value of Eduardo's account will be $ 95,454.20 upon his retirement.

19. Compute the value of each account.

a. The information for this account is:

$$p = 2000, r = 0.07, m = 1, n = mt = 20.$$

The future value of the annuity will be:

$$A = 2000\left[\frac{(1+0.07)^{20}-1}{0.07}\right] = \$81,990.98$$

b. The information for this account is:

$$p = 1800, r = 0.065, m = 4, n = mt = 80.$$

The future value of the annuity will be:

$$A = 1800\left[\frac{\left(1+\frac{0.065}{4}\right)^{80}-1}{\frac{0.065}{4}}\right] = \$291,450.92$$

Obviously there is more money in the second account; however, this account requires more savings to be put into the annuity. The annuity in part (a) requires the individual to deposit $40,000 into the account and therefore earned $41,990.98 over the 20 years. The annuity in part (b) requires the individual to deposit 144,000 into the account and therefore earned $ 147,450.92 over the 20 years.

Therefore the annuity in part (b) earned more money over the 20 years.

21. The information that applies to all four parts of the problem is:

$$p = 500, r = 0.06, m = 4$$

The only variable that will change is the number of periods the account is active.

a. For this situation, $n = mt = (4)(35) = 140$. The future value of the annuity is shown at the top of the next page.

Using the information on the previous page the future value of the annuity is:

$$A = 500 \left[\frac{\left(1 + \frac{0.06}{4}\right)^{140} - 1}{\frac{0.06}{4}} \right] = \$234,660.41$$

The value of the retirement account will be $234,660.41 at the end of 35 years.

b. For this situation, $n = mt = (4)(30) = 120$. The future value of the annuity is:

$$A = 500 \left[\frac{\left(1 + \frac{0.06}{4}\right)^{120} - 1}{\frac{0.06}{4}} \right] = \$165,644.10$$

The value of the retirement account will be $165,644.10 at the end of 30 years.

c. For this situation, $n = mt = (4)(25) = 100$. The future value of the annuity is:

$$A = 500 \left[\frac{\left(1 + \frac{0.06}{4}\right)^{100} - 1}{\frac{0.06}{4}} \right] = \$114,401.52$$

The value of the retirement account will be $114,401.52 at the end of 25 years.

d. For this situation, $n = mt = (4)(20) = 80$. The future value of the annuity is:

$$A = 500 \left[\frac{\left(1 + \frac{0.06}{4}\right)^{80} - 1}{\frac{0.06}{4}} \right] = \$76,355.43$$

The value of the retirement account will be $76,355.43 at the end of 20 years.

23. We want to determine the present value of an annuity paying $500 a month with $r = 0.07, m = 12, n = mt = 240$.

Using the present value formula we get

$$P = 500 \left[\frac{1 - \left(1 + \frac{0.07}{12}\right)^{-240}}{\frac{0.07}{12}} \right] = \$64,491.25$$

Katy will need to have saved $64,491.25 in order to withdraw $500 each month for the next 20 years.

25. We want to determine the present value of an annuity paying $800 a month with $r = 0.06, m = 12, n = mt = 8$. Using the present value formula:

$$P = 800 \left[\frac{1 - \left(1 + \frac{0.06}{12}\right)^{-8}}{\frac{0.06}{12}} \right] = \$6258.37$$

The Meek brothers will need to have saved $6258.37 in order to be able to withdraw $800 each month for the next 8 months.

27. We want to determine the present value of an annuity paying $5000 a year with $r = 0.10, m = 1, n = mt = 10$.

Using the present value formula:

$$P = 5000 \left[\frac{1 - \left(1 + \frac{0.10}{1}\right)^{-10}}{\frac{0.10}{1}} \right] = \$30,722.84$$

The present value of your lottery is $30,722.84.

29. We want to determine the present value of an annuity paying $65 per month with $r = 0.09, m = 12, n = mt = 36$.

Using the present value formula:

$$P = 65 \left[\frac{1 - \left(1 + \frac{0.09}{12}\right)^{-36}}{\frac{0.09}{12}} \right] = \$2044.04$$

The company should expect to receive $2044.04 for its loan.

31. The information needed to compute the value of the annuity is:

$$p = 100, r = 0.0525, m = 12, n = mt = 36.$$

We compute the future value of the annuity at the top of the next page.

Using the information on the previous page, the future value of the annuity is:

$$A = 100 \left[\frac{\left(1 + \frac{0.0525}{12}\right)^{36} - 1}{\frac{0.0525}{12}} \right] = \$3889.80$$

Now, Bronwyn deposits the $3889.80 into a certificate of deposit paying 7.5% interest compounded quarterly for 2.5 years or 10 periods. Using the compounded interest formula for a single payment problem results in the following:

$$A = 3889.80 \left(1 + \frac{0.075}{4}\right)^{10} = 4683.86 .$$

Next, Bronwyn deposits the $4683.86 into a 90 day certificate of deposit paying 6.8% annually. Thus, $r = 0.068, m = 365, n = mt = 90$. Once again using the compound interest formula we see the final amount in the account is:

$$A = 4683.86 \left(1 + \frac{0.068}{365}\right)^{90} = 4763.05 .$$

When the certificate of deposit matures, Bronwyn will have accumulated $4763.05.

Exercises Section 6.3

1. The information in the problem is: $r = 0.065, m = 4$. The effective rate of interest formula tells us:

$$APR = \left(1 + \frac{0.065}{4}\right)^4 - 1 = 0.0666$$

The annual percentage rate is 6.66%.

The annual percentage rate is 8.30%.

3. The information in the problem is: $r = 0.04, m = 2$. The effective rate of interest formula tells us:

$$APR = \left(1 + \frac{0.04}{2}\right)^2 - 1 = 0.0404$$

The annual percentage rate is 4.04%.

5. The information in the problem is: $r = 0.16, m = 12$. The effective rate of interest formula tells us:

$$APR = \left(1 + \frac{0.16}{12}\right)^{12} - 1 = 0.1723$$

The annual percentage rate is 17.23%.

7. The information in the problem is: $r = 0.09, m = 365$. The effective rate of interest formula tells us:

$$APR = \left(1 + \frac{0.09}{365}\right)^{365} - 1 = 0.0942$$

The annual percentage rate is 9.42%.

9. Compute the effective rates of interest for each of the three accounts.

a. $r = 0.05, m = 2$. The effective rate of interest formula tells us:

$$APR = \left(1 + \frac{0.05}{2}\right)^2 - 1 = 0.0506$$

The annual percentage rate is 5.06%

b. $r = 0.048, m = 4$. The effective rate of interest formula tells us:

$$APR = \left(1 + \frac{0.048}{4}\right)^4 - 1 = 0.0489$$

The annual percentage rate is 4.89%.

c. $r = 0.046, m = 12$. The effective rate of interest formula tells us:

$$APR = \left(1 + \frac{0.046}{12}\right)^{12} - 1 = 0.0470$$

The annual percentage rate is 4.70%

Plan (a) will offer the highest interest per year, followed by plan (b), and then plan (c).

11. Compute the effective rates of interest for each of the three accounts.

a. $r = 0.063, m = 12$. The effective rate of interest formula tells us:

$$APR = \left(1 + \frac{0.063}{12}\right)^{12} - 1 = 0.0649$$

The annual percentage rate is 6.49%

b. $r = 0.065, m = 4$. The effective rate of interest formula tells us:

$$APR = \left(1 + \frac{0.065}{4}\right)^{4} - 1 = 0.0666$$

The annual percentage rate is 6.66%.

c. $r = 0.068, m = 1$. The effective rate of interest formula tells us:

$$APR = \left(1 + \frac{0.068}{1}\right)^{1} - 1 = 0.068$$

The annual percentage rate is 6.80%

Plan (c) will offer the highest interest per year, followed by plan (b), and then plan (a).

13. Compute the effective rates of interest for each of the three accounts.

a. $r = 0.08, m = 1$. The effective rate of interest formula tells us:

$$APR = \left(1 + \frac{0.08}{2}\right)^{2} - 1 = 0.0816$$

The annual percentage rate is 8.16%

b. $r = 0.082, m = 1$. The effective rate of interest formula tells us:

$$APR = \left(1 + \frac{0.082}{1}\right)^{1} - 1 = 0.082$$

The annual percentage rate is 8.20%.

c. $r = 0.06, m = 12$. The effective rate of interest formula tells us:

$$APR = \left(1 + \frac{0.06}{12}\right)^{12} - 1 = 0.0617$$

The annual percentage rate is 6.17%

Plan (b) will offer the highest interest per year, followed by plan (a), and then plan (c).

15. The problem gives the period interest rate $i = 0.01$. The number of periods in one year will be $m = 12$. The effective rate of interest formula can now be applied.

$$APR = \left(1 + 0.01\right)^{12} - 1 = 0.1268.$$

The annual percentage rate the credit card company charges is 12.68%.

17. The number of monthly payments in six years is $n = mt = (12)(6) = 72$. Monthly interest is $i = 0.03$, and $P = 20,000$. The amount of the payments needed to retire the loan given is:

$$p = \frac{(0.03)(20,000)}{1 - (1 + 0.03)^{-72}} = 681.08.$$

The monthly payments on the loan will be $681.08.

19. The number of monthly payments in 20 years is $n = mt = (12)(20) = 240$. Monthly interest is $i = 0.02$, and $P = 40,000$ The amount of the payments needed to retire the loan given is:

$$p = \frac{(0.02)(40,000)}{1 - (1 + 0.02)^{-240}} = 806.96.$$

The monthly payments on the loan will be $806.96.

21. The number of monthly payments is given to be $n = 60$. Annual interest is $r = 0.119$, and $m = 12$, so monthly interest is $i = \frac{0.119}{12}$. The amount borrowed is $P = 12,000$. The amount of the payments needed to retire the loan given is calculated at the top of the next page.

Using the information from the previous page, the payments needed to retire the loan are:

$$p = \frac{\left(\frac{0.119}{12}\right)(12,000)}{1-\left(1+\frac{0.119}{12}\right)^{-60}} = 266.33.$$

The monthly payments on the loan will be $266.33.

23. The number of quarterly payments in 12 years is $n = mt = (4)(12) = 48$. The annual interest is $r = 0.10$, and $m = 4$, so the quarterly interest is $i = \frac{0.10}{4}$. The amount borrowed is $P = 22,000$. The amount of the payments needed to retire the loan given is:

$$p = \frac{\left(\frac{0.10}{4}\right)(22,000)}{1-\left(1+\frac{0.10}{4}\right)^{-48}} = 792.13.$$

The quarterly payments on the loan will be $792.13

25. The information in the problem gives us:

$$P = 9643.68, r = 0.095, n = \frac{35}{365}.$$

The interest that will be accumulated over 35 days is given by the simple interest formula $I = Prt$. Plugging the information into this formula we get:

$$I = 9643.68(0.095)\left(\frac{35}{365}\right) = 87.85.$$

Thus, $87.85 of you next payment will go towards the interest.

27. The information given in the problem is

$$P = 12,500, r = 0.098, m = 12, n = 48.$$ The approximate size of the monthly payment is:

$$p = \frac{\left(\frac{0.098}{12}\right)(12,500)}{1-\left(1+\frac{0.098}{12}\right)^{-48}} = 315.83$$

The payment each month will be $315.83.

29. The information given in the problem is

$$P = 800, r = 0.079, m = 12, n = 48.$$

The approximate size of the monthly payment is stated at the top of the next column.

$$p = \frac{\left(\frac{0.079}{12}\right)(800)}{1-\left(1+\frac{0.079}{12}\right)^{-48}} = 19.49$$

The payment each month will be $19.49.

31. The information given in the problem is:

$$P = 20,000, r = 0.11, m = 1, n = 8.$$

The approximate size of the annual payment is:

$$p = \frac{\left(\frac{0.11}{1}\right)(20,000)}{1-\left(1+\frac{0.11}{1}\right)^{-8}} = 3886.41$$

The payment each year will be $3886.41.

33.
a. Since George finances 90% of the value of the boat $P = (.90)9200 = 8280$. To determine the monthly payments, note that the simple interest rate can be thought of as the A.P.R. therefore, we have:

$$r = 0.085, m = 12, n = 72.$$

The monthly payments on the loan are:

$$p = \frac{\left(\frac{0.085}{12}\right)(8280)}{1-\left(1+\frac{0.085}{12}\right)^{-72}} = 147.21.$$

George will pay $147.21 each month to pay off the loan.

b. To determine the unpaid balance after 24 timely payments have been made, first observe that 48 payments have yet to be paid. The unpaid balance will be the present value of an annuity whose payments are $147.21 earning 8.5% interest compounded monthly for 48 months. Thus, the unpaid balance on the loan is:

$$A = 147.21\left[\frac{1-\left(1+\frac{0.085}{12}\right)^{-48}}{\left(\frac{0.085}{12}\right)}\right] = 5972.42$$

The unpaid balance on the boat after two years is $5972.42.

(Notice, if you actually computed the monthly compound interest rate from the A.P.R. formula, you would see that $r = 0.0819$ compounded monthly, this gives a periodic payment of $145.94, the present value of an annuity whose payments are $145.94 per month compounded monthly at 8.19% is $5956.20. The error from using the A.P.R. is not substantial enough to worry about.)

35.

a. Since Diane financed 70% of the value of the farm, $P = (.70)95,000 = 66,500$. To determine the quarterly payments, we have:

$r = 0.10, m = 4, n = mt = 40$.

The quarterly payments on the loan are:

$$p = \frac{\left(\frac{0.10}{4}\right)(66,500)}{1 - \left(1 + \frac{0.10}{4}\right)^{-40}} = 2649.11.$$

Diane will pay $2649.11 each quarter to pay off the loan in ten years.

b. To determine the unpaid balance after 20 timely payments have been made, first observe that 20 payments have yet to be paid. The unpaid balance will be the present value of an annuity whose payments are $2649.11 earning 10% interest compounded quarterly for 5 years. Thus, the unpaid balance on the loan is:

$$A = 2649.11 \left[\frac{1 - \left(1 + \frac{0.10}{4}\right)^{-20}}{\left(\frac{0.10}{4}\right)} \right] = 41,297.41$$

Diane's unpaid balance on the farm after six years is $41,297.41.

In problems 37–41 the information for each problem is:

$$P = .9(130,000) = 117,000, r = 0.095, m = 12.$$

Only the time period is changing in each problem.

37.

a. Calculating the number of periods we have $n = mt = 12 \cdot 25 = 300$. Thus, the monthly payments are given by:

$$p = \frac{\left(\frac{0.095}{12}\right)(117,000)}{1 - \left(1 + \frac{0.095}{12}\right)^{-300}} = 1022.23$$

The payments on the loan are $1022.23 each month.

b. To find out the total amount paid to the finance company, multiply the amount of each payment by the number payments: $300(1022.23) = 306,669$.

Thus, $306,669 will be paid to the finance company over 25 years.

39.

a. Calculating the number of periods we have $n = mt = 12 \cdot 35 = 420$. Thus, the monthly payments are given by:

$$p = \frac{\left(\frac{0.095}{12}\right)(117,000)}{1 - \left(1 + \frac{0.095}{12}\right)^{-420}} = 961.29$$

The payments on the loan are $961.29 each month.

b. To find out the total amount paid to the finance company, multiply the amount of each payment by the number payments:

$$420(961.29) = 403,741.80.$$

Thus, $403,741.80 will be paid to the finance company over 35 years.

41.

a. Calculating the number of periods we have $n = mt = 12 \cdot 45 = 540$. Thus, the monthly payments are given by:

$$p = \frac{\left(\frac{0.095}{12}\right)(117,000)}{1 - \left(1 + \frac{0.095}{12}\right)^{-540}} = 939.54$$

The payments on the loan are $939.54 each month.

b. To find out the total amount paid to the finance company, multiply the amount of each payment by the number payments: $540(939.54) = 507,351.60$.

Thus, $507,351.60 will be paid to the finance company over 45 years.

For problems 43 – 49, the following information will remain the same.

$$P = \$125,000, m = 12, n = 180$$

43. We must first calculate the payment. Using the information given in the problem and an interest rate of $r = 0.0925$, the payment will be:

$$p = \frac{\left(\frac{0.0925}{12}\right)(125,000)}{1 - \left(1 + \frac{0.0925}{12}\right)^{-180}} = 1286.49.$$

Using an excel worksheet, the amortization table for the first 12 periods can be seen below:

Payment Number	Amount	Interest	Applied to Principal	Unpaid Balance		
Principal				$125,000.00		
1	$ 1,286.49	$ 963.54	$ 322.95	$ 124,677.05	r =	0.0925
2	$ 1,286.49	$ 961.05	$ 325.44	$ 124,351.61	m=	12
3	$ 1,286.49	$ 958.54	$ 327.95	$ 124,023.67	n=	180
4	$ 1,286.49	$ 956.02	$ 330.47	$ 123,693.19	I =	0.007708
5	$ 1,286.49	$ 953.47	$ 333.02	$ 123,360.17	Pmt	1286.49
6	$ 1,286.49	$ 950.90	$ 335.59	$ 123,024.58		
7	$ 1,286.49	$ 948.31	$ 338.18	$ 122,686.40		
8	$ 1,286.49	$ 945.71	$ 340.78	$ 122,345.62		
9	$ 1,286.49	$ 943.08	$ 343.41	$ 122,002.21		
10	$ 1,286.49	$ 940.43	$ 346.06	$ 121,656.16		
11	$ 1,286.49	$ 937.77	$ 348.72	$ 121,307.43		
12	$ 1,286.49	$ 935.08	$ 351.41	$ 120,956.02		

45. We must first calculate the payment. Using the information given in the problem and an interest rate of $r = 0.0795$, the payment will be:

$$p = \frac{\left(\frac{0.0795}{12}\right)(125,000)}{1 - \left(1 + \frac{0.0795}{12}\right)^{-180}} = 1190.96.$$

Using an excel worksheet, the amortization table for the first 12 periods can be seen on the next page.

Payment Number	Amount		Interest		Applied to Principal		Unpaid Balance		
Principal							$125,000.00		
1	$	1,190.96	$	828.13	$	362.83	$	124,637.17	r =	0.0795
2	$	1,190.96	$	825.72	$	365.24	$	124,271.93	m=	12
3	$	1,190.96	$	823.30	$	367.66	$	123,904.27	n=	180
4	$	1,190.96	$	820.87	$	370.09	$	123,534.17	I =	0.006625
5	$	1,190.96	$	818.41	$	372.55	$	123,161.63	Pmt	1190.96
6	$	1,190.96	$	815.95	$	375.01	$	122,786.61		
7	$	1,190.96	$	813.46	$	377.50	$	122,409.12		
8	$	1,190.96	$	810.96	$	380.00	$	122,029.12		
9	$	1,190.96	$	808.44	$	382.52	$	121,646.60		
10	$	1,190.96	$	805.91	$	385.05	$	121,261.55		
11	$	1,190.96	$	803.36	$	387.60	$	120,873.95		
12	$	1,190.96	$	800.79	$	390.17	$	120,483.78		

47. We must first calculate the payment. Using the information given in the problem and an interest rate of $r = 0.0820$, the payment will be:

$$p = \frac{\left(\frac{0.0820}{12}\right)(125,000)}{1-\left(1+\frac{0.0820}{12}\right)^{-180}} = 1209.04.$$

Using an excel worksheet, the amortization table for the first 12 periods can be seen below:

Payment Number	Amount		Interest		Applied to Principal		Unpaid Balance		
Principal							$125,000.00		
1	$	1,209.04	$	854.17	$	354.88	$	124,645.12	r =	0.082
2	$	1,209.04	$	851.74	$	357.30	$	124,287.82	m=	12
3	$	1,209.04	$	849.30	$	359.74	$	123,928.08	n=	180
4	$	1,209.04	$	846.84	$	362.20	$	123,565.88	I =	0.006833
5	$	1,209.04	$	844.37	$	364.68	$	123,201.21	Pmt	1209.042
6	$	1,209.04	$	841.87	$	367.17	$	122,834.04		
7	$	1,209.04	$	839.37	$	369.68	$	122,464.36		
8	$	1,209.04	$	836.84	$	372.20	$	122,092.16		
9	$	1,209.04	$	834.30	$	374.75	$	121,717.41		
10	$	1,209.04	$	831.74	$	377.31	$	121,340.11		
11	$	1,209.04	$	829.16	$	379.88	$	120,960.22		
12	$	1,209.04	$	826.56	$	382.48	$	120,577.74		

49. We must first calculate the payment. Using the information given in the problem and an interest rate of $r = 0.0475$, the payment will be:

$$p = \frac{\left(\frac{0.0475}{12}\right)(125,000)}{1-\left(1+\frac{0.0475}{12}\right)^{-180}} = 972.29.$$

Using an excel worksheet, the amortization table for the first 12 periods can be seen on the next page.

Payment Number	Amount	Interest	Applied to Principal	Unpaid Balance		
Principal				$125,000.00		
1	$ 972.29	$ 494.79	$ 477.50	$ 124,522.50	r =	0.0475
2	$ 972.29	$ 492.90	$ 479.39	$ 124,043.11	m=	12
3	$ 972.29	$ 491.00	$ 481.29	$ 123,561.83	n=	180
4	$ 972.29	$ 489.10	$ 483.19	$ 123,078.64	I =	0.003958
5	$ 972.29	$ 487.19	$ 485.10	$ 122,593.53	Pmt	972.2899
6	$ 972.29	$ 485.27	$ 487.02	$ 122,106.51		
7	$ 972.29	$ 483.34	$ 488.95	$ 121,617.56		
8	$ 972.29	$ 481.40	$ 490.89	$ 121,126.67		
9	$ 972.29	$ 479.46	$ 492.83	$ 120,633.84		
10	$ 972.29	$ 477.51	$ 494.78	$ 120,139.06		
11	$ 972.29	$ 475.55	$ 496.74	$ 119,642.32		
12	$ 972.29	$ 473.58	$ 498.71	$ 119,143.61		

Chapter 6 Summary Exercises

1. Using the simple interest formula with $P = 2000, r = 0.07, t = 3$ implies:

$$A = 2000\left(1 + 0.07(3)\right) = 2420$$

The accumulated value after three years earning simple interest is $2420.00.

3. Using the compound interest formula with $P = 2000, r = 0.07, t = 3, m = 4$ implies:

$$A = 2000\left(1 + \tfrac{0.07}{4}\right)^{12} = 2462.88$$

The accumulated value after three years earning quarterly compounded interest is $2462.88.

5. The information in the problem is:

$$A = 3000, P = 1500, r = 0.08, m = 4 .$$

Using the compound interest formula we have:

$$3000 = 1500\left(1 + \tfrac{0.08}{4}\right)^{n}$$

To find the number of periods needed, solve for n as shown at the top of the next column.

$$\tfrac{3000}{1500} = \left(1.02\right)^{n}$$

$$\ln\left(1.02\right)^{n} = \ln\left(2\right)$$

$$n\ln\left(1.02\right) = \ln\left(2\right)$$

$$n = \frac{\ln\left(2\right)}{\ln\left(1.02\right)} \approx 35$$

It will take 35 quarters, or eight years nine months for $1500 to accumulate to $3000.

7. The information in the problem is:

$$A = 5000, r = 0.09, m = 12, n = mt = 120 .$$

Using the compound interest formula we have:

$$5000 = P\left(1 + \tfrac{0.09}{12}\right)^{120}$$

Solving for P we get:

$$P = \frac{5000}{\left(1 + \tfrac{0.09}{12}\right)^{120}} = 2039.69$$

The principal needed to accumulate $5000 in 10 years at 9% interest compounded monthly is $2039.69.

9. The information in the problem is:

$A = 5000, r = 0.09, m = 1, n = mt = 10$.

Using the compound interest formula we have:

$5000 = P\left(1 + \frac{0.09}{1}\right)^{10}$

Solving for P we get:

$P = \dfrac{5000}{\left(1 + \frac{0.09}{1}\right)^{10}} = 2112.05$

The principal needed to accumulate $5000 in 10 years at 9% interest compounded monthly is $2112.05.

11. Using the future value of an ordinary annuity formula we get:

$A = 5000\left[\dfrac{(1 + 0.03)^{40} - 1}{(0.03)}\right] = 377,006.30$.

The future value of the annuity will be $377,006.30.

13. The information given in the problem is:

$A = 40,000, i = 0.04, n = 12$. To calculate the payments we use the formula:

$p = \dfrac{(0.04)(40,000)}{(1.04)^{12} - 1} = 2662.09$.

The payments to the sinking fund having a goal of $40,000 are $2662.09 per period for 12 periods.

15. The information in the problem is:
$r = 0.06, m = 4$. The effective rate of interest formula tells us:

$APR = \left(1 + \frac{0.06}{4}\right)^{4} - 1 = 0.0614$

The annual percentage rate is 6.14%.

17. The information in the problem is:
$r = 0.10, m = 2$. The effective rate of interest formula tells us:

$APR = \left(1 + \frac{0.10}{2}\right)^{2} - 1 = 0.1025$

The annual percentage rate is 10.25%.

19. The information given in the problem is:

$P = 14,000, r = 0.069, m = 12, n = mt = 36$.

The monthly payment will be:

$p = \dfrac{\left(\frac{0.069}{12}\right)(14,000)}{1 - \left(1 + \frac{0.069}{12}\right)^{-36}} = 431.64$.

The approximate monthly payment for the new car will be $431.64.

21. We compare the A.P.R of each rate by using the effective rate of interest formula for each interest rate.

a. The effective rate of interest is:

$APR = \left(1 + \frac{0.08}{1}\right)^{1} - 1 = 0.08$

The annual percentage rate is 8%.

b. The effective rate of interest is:

$APR = \left(1 + \frac{0.079}{2}\right)^{2} - 1 = 0.0806$

The annual percentage rate is 8.06%.

c. The effective rate of interest is:

$APR = \left(1 + \frac{0.078}{4}\right)^{4} - 1 = 0.0803$

The annual percentage rate is 8.03%.

d. The effective rate of interest is:

$APR = \left(1 + \frac{0.077}{12}\right)^{12} - 1 = 0.0798$

The annual percentage rate is 7.98%.

e. The effective rate of interest is:

$$APR = \left(1 + \frac{0.076}{365}\right)^{365} - 1 = 0.0790$$

The annual percentage rate is 7.90%.

f. The effective rate of interest is:

$$APR = e^{0.075} - 1 = 0.0779$$

The annual percentage rate is 7.79%.

Comparing all of the effective rates of interest, the best rate is the rate given by investment **b.**

Cumulative Review

23. To find the inverse of the matrix, consider the augmented matrix:

$$\begin{bmatrix} 1 & 0 & 3 & | & 1 & 0 & 0 \\ 2 & 4 & 6 & | & 0 & 1 & 0 \\ 0 & -2 & 1 & | & 0 & 0 & 1 \end{bmatrix}$$

Pivot on the one in row 1, column 1 to get:

$$\begin{bmatrix} 1 & 0 & 3 & | & 1 & 0 & 0 \\ 0 & 4 & 0 & | & -2 & 1 & 0 \\ 0 & -2 & 1 & | & 0 & 0 & 1 \end{bmatrix}$$

Next pivot on the four in row 2, column 2 to get:

$$\begin{bmatrix} 1 & 0 & 3 & | & 1 & 0 & 0 \\ 0 & 1 & 0 & | & \frac{-1}{2} & \frac{1}{4} & 0 \\ 0 & 0 & 1 & | & -1 & \frac{1}{2} & 1 \end{bmatrix}$$

Finally pivot on the one in row 3, column 3 to get:

$$\begin{bmatrix} 1 & 0 & 0 & | & 4 & \frac{-3}{2} & -3 \\ 0 & 1 & 0 & | & \frac{-1}{2} & \frac{1}{4} & 0 \\ 0 & 0 & 1 & | & -1 & \frac{1}{2} & 1 \end{bmatrix}$$

Therefore the inverse matrix is:

$$\begin{bmatrix} 4 & \frac{-3}{2} & -3 \\ \frac{-1}{2} & \frac{1}{4} & 0 \\ -1 & \frac{1}{2} & 1 \end{bmatrix}$$

25. Using the slope-intercept formula, the equation for the line with slope 3 and y-intercept 2 is given by:

$$y = 3x + 2.$$

Chapter 6 Sample Test Answers

1. The amount is given by:

$$A = 1200\left(1 + (0.06)5\right) = 1560$$

At the end of five years $1560 will have accumulated in the account.

2. The amount is given by:

$$A = 3000\left(1 + \frac{0.07}{2}\right)^{12} = 4533.21$$

At the end of six years $4533.21 will have accumulated in the account.

3. The amount is given by:

$$A = 8000\left(1 + \frac{0.09}{12}\right)^{8} = 8492.79$$

At the end of eight months $8492.79 will have accumulated in the account.

4. The amount is given by:

$$A = 5000\left(1 + (.10)\left(\frac{8}{12}\right)\right) = 5333.33$$

At the end of eight months $5333.33 will have accumulated in the account.

5. The compound interest formula gives us:

$$7000 = 2400\left(1 + \frac{0.08}{2}\right)^{n}.$$

Solving for n we get:

$$n = \frac{\ln\left(\frac{7000}{2400}\right)}{\ln\left(1 + \frac{0.08}{2}\right)} = 27.29.$$

Therefore we need 28 interest periods, or 14 years for $2400 to accumulate to $7000.

6. The compound interest formula gives us:

$$25000 = 3000\left(1 + \tfrac{0.10}{1}\right)^n.$$

Solving for n we get:

$$n = \frac{\ln\left(\frac{25000}{3000}\right)}{\ln\left(1 + \frac{0.10}{1}\right)} = 22.25.$$

Therefore we need 23 interest periods, or 23 years for $3000 to accumulate to $25,000.

7. The compound interest formula gives us:

$$20,000 = P\left(1 + \tfrac{0.12}{12}\right)^{84}.$$

Solving for P we get:

$$P = \frac{20,000}{\left(1 + \frac{0.12}{12}\right)^{84}} = 8670.31.$$

The required principal will be $8670.31.

8. The compound interest formula gives us:

$$30,000 = P\left(1 + \tfrac{0.065}{4}\right)^{48}.$$

Solving for P we get:

$$P = \frac{30,000}{\left(1 + \frac{0.065}{4}\right)^{48}} = 13,838.67.$$

The required principal will be $13,838.67.

9. The compound interest formula gives us:

$$2P = P\left(1 + \tfrac{0.08}{365}\right)^n.$$

Solving for n we get:

$$2 = \left(1 + \tfrac{0.08}{365}\right)^n$$

$$n = \frac{\ln(2)}{\ln\left(1 + \frac{0.08}{365}\right)} = 3162.83$$

It will take 3163 days or 8.67 years for the principal to double in value.

10. The future value of the annuity is given by:

$$A = 800\left[\frac{(1+0.05)^{20} - 1}{0.05}\right] = 26,452.76.$$

The value in the account after 20 periods is $26,452.76.

11. The future value of the annuity is given by:

$$A = 2000\left[\frac{(1+0.02)^{30} - 1}{0.02}\right] = 81,136.16.$$

The value in the account after 30 periods is $81,136.16.

12. The payments are calculated using the following formula:

$$p = \frac{(0.03)(200,000)}{(1+0.03)^{24} - 1} = 5809.48$$

Equal payments of $5809.48 are needed to save $200,000 in 24 periods at the period interest rate of 3%.

13. The payments are calculated using the following formula:

$$p = \frac{(0.04)(200,000)}{(1+0.04)^{20} - 1} = 6716.35$$

Equal payments of $6716.35 are needed to save $200,000 in 24 periods at the period interest rate of 3%.

14. The information in the problem is:

$$p = 120, r = 0.08, m = 12, n = 36.$$

The present value of the loan is:

$$P = 120\left[\frac{1 - \left(1 + \frac{0.08}{12}\right)^{-36}}{\frac{.08}{12}}\right] = 3829.42.$$

A fair price for the loan would be $3829.42.

15. The effective rate of interest is given by:

$$APR = \left(1 + \tfrac{0.07}{4}\right)^4 - 1 = 0.0719.$$

The annual percentage rate is 7.19%.

16. The effective rate of interest is given by:

$$APR = \left(1 + \tfrac{0.09}{365}\right)^{365} - 1 = 0.0942 \,.$$

The annual percentage rate is 9.42%.

17. The monthly payments are given by:

$$p = \frac{(0.02)(50{,}000)}{1 - (1 + 0.02)^{-180}} = 1029.14 \,.$$

The monthly payments will be $1029.14.

18. The number of payments is given by:

$$n = \frac{\ln\left(\dfrac{800}{800 - 70{,}000\left(\frac{0.94}{12}\right)}\right)}{\ln\left(1 + \frac{0.94}{12}\right)} = 148.22$$

There will be 149 payments, or 12 years and five months required to pay off the loan.

Chapter 7
Logic, Sets, and Counting Techniques

Exercises Section 7.1

1. The student is a female or a sophomore.

3. The student is not a female or the student is a sophomore.

5. If the student is a female, then she is a sophomore.

7. The hat is red or the belt is black.

9. The hat is not red or the belt is black.

11. If the hat is red, then the belt is black.

13. The disjunction is:

The card is not a five or the card is a diamond.

15. The disjunction is:

The student will not do her homework or her test grades will improve.

17. This is a straight forward sentence. The symbolic form is:

$p \wedge q$

19. "Ten percent of all cataract surgeries fail." is logically equivalent to "Ninety percent of all cataract surgeries are successful."; while "Twenty percent of all kidney transplants fail." is logically equivalent to "Eighty percent of all kidney transplants are successful.", so the symbolic form is:

$p \wedge q$

21. The symbolic form is:

$p \wedge q$

23. "At least one of the coins is not silver." is logically equivalent to "Not all of the coins are silver"; this is the negation of "All of the coins are silver." While, "None of the marbles are purple." is the negation of "Some of the marbles are purple." Therefore the symbolic form is:

$\sim p \vee \sim q$.

25. Either exactly four students are taking mathematics or exactly five students are taking mathematics.

27. Exactly zero shirts have flaws, or exactly one shirt has a flaw, or exactly two shirts have flaws or exactly three shirts have flaws.

29. The car is not red and it does not have a CD player.

31. The blouse is blue or the belt is not black.

33. Neither of the two marbles is red.

35. At most one of the three students graduated with honors.

37. All four marbles are green.

Exercises Section 7.2

1.
The truth table for the sentence can be completed as follows:

p	q	$\sim q$	$p \vee \sim q$
T	T	F	**T**
T	F	T	**T**
F	T	F	**F**
F	F	T	**T**

3. The truth table for the sentence can be completed as follows:

p	q	$\sim p$	$\sim q$	$\sim p \Rightarrow \sim q$
T	T	F	F	**T**
T	F	F	T	**T**
F	T	T	F	**F**
F	F	T	T	**T**

5. The truth table for the sentence can be completed as follows:

p	q	$\sim p$	$(\sim p \vee q)$	$\sim(\sim p \vee q)$
T	T	F	T	**F**
T	F	F	F	**T**
F	T	T	T	**F**
F	F	T	T	**F**

7. The truth tables for the two statements can be completed as follows:

p	q	$p \vee q$	$\sim(p \vee q)$
T	T	T	**F**
T	F	T	**F**
F	T	T	**F**
F	F	F	**T**

p	q	$\sim p$	$\sim q$	$\sim p \wedge \sim q$
T	T	F	F	**F**
T	F	F	T	**F**
F	T	T	F	**F**
F	F	T	T	**T**

The last column in each table shows that the two statements are logically equivalent.

9. The truth tables for the two statements can be completed as follows:

p	q	r	$q \vee r$	$p \wedge (q \vee r)$
T	T	T	T	**T**
T	T	F	T	**T**
T	F	T	T	**T**
T	F	F	F	**F**
F	T	T	T	**F**
F	T	F	T	**F**
F	F	T	T	**F**
F	F	F	F	**F**

p	q	r	$p \wedge q$	$p \wedge r$	$(p \wedge q) \vee (q \wedge r)$
T	T	T	T	T	**T**
T	T	F	T	F	**T**
T	F	T	F	T	**T**
T	F	F	F	F	**F**
F	T	T	F	F	**F**
F	T	F	F	F	**F**
F	F	T	F	F	**F**
F	F	F	F	F	**F**

The last column in each table shows that the two statements are logically equivalent.

11. The truth tables are shown below:

p	q	$p \wedge q$
T	T	**T**
T	F	**F**
F	T	**F**
F	F	**F**

p	q	$\sim p$	$(q \wedge p)$	$\sim(q \wedge p)$	$\sim p \Rightarrow \sim(q \wedge p)$
T	T	F	T	F	**T**
T	F	F	F	T	**T**
F	T	T	F	T	**T**
F	F	T	F	T	**T**

The last column in the truth tables shows that the two sentences are not logically equivalent.

13. The truth tables are shown below:

p	q	$\sim p$	$\sim p \vee q$	$q \wedge (\sim p \vee q)$
T	T	F	T	**T**
T	F	F	F	**F**
F	T	T	T	**T**
F	F	T	T	**F**

p	q	$\sim p$	$(\sim p \vee q)$	$\sim(\sim p \vee q)$
T	T	F	T	**F**
T	F	F	F	**T**
F	T	T	T	**F**
F	F	T	T	**F**

The last column in the truth tables shows that the two sentences are not logically equivalent.

15. The Earth is bigger than the moon, so this sentence is false.

17. $2 + 3 = 5$ is true. We only need one of the sentences to be true for the disjunction to be true. The sentence is true.

19. The sentence "Some college students drink coffee" is true; however, "All college students own automobiles" is false. For the conjunction to be true, we need both sentences to be true. Therefore the sentence is false.

21. The disjunction is true because there have been at least two male presidents.

23. By DeMorgan's laws of logic:

$$\sim(p \wedge q) \equiv \sim p \vee \sim q$$

Therefore the sentence will read:

The die does not show an odd number or the die shows a five.

This could read:

The die shows an even number or the die shows a five.

25. By DeMorgan's laws of logic:

$$\sim(p \vee \sim q) \equiv \sim p \wedge \sim(\sim q) \equiv \sim p \wedge q$$

Therefore the sentence will read:

The die does not show an odd number and the die does not show a five.

This could read:

The die shows and even number and the die does not show a five.

27. By DeMorgan's laws of logic:

$$\sim(\sim p \vee \sim q) \equiv \sim(\sim p) \wedge \sim(\sim q) \equiv p \wedge q$$

Therefore the sentence will read:

The die shows an odd number and the die does not show a five.

Exercises Section 7.3

1. $A' = \{0, 4, 5, 6, 7\}$

3. $\varnothing' = \{0, 1, 2, 3, 4, 5, 6, 7\}$

5. $A' \cap B = \{5\}$

7. $(A \cup C)' = \{0, 4, 5\}$

9. $A - B' = \{2, 3\}$

11. $B' = \{a, b, e, f\}$

13. $A \cap B = \{c\}$

15. $U' = \varnothing$

17. $B' \cup C = \{a, b, e, f\}$

19. $A - C = \{a, b, c\}$

21. Since A consists of regions I and II and B' consists of regions I and IV. We will shade the regions I, II, and IV.

The Venn diagram for $A \cup B'$ is displayed below:

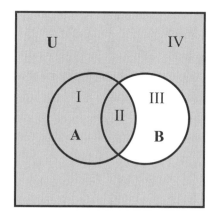

23. Since A' consists of regions III and IV, and B' consists of regions I and IV $A' \cap B'$ consists of region IV.

The Venn diagram for $A' \cap B'$ is:

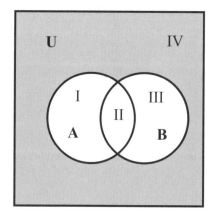

25. The Venn diagram corresponding to $(A \cap B) - C$ is:

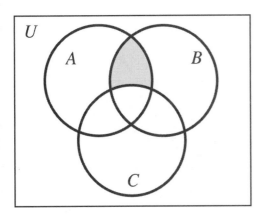

27. The Venn diagram corresponding to $A' \cap B$, if $A \subseteq B$ is:

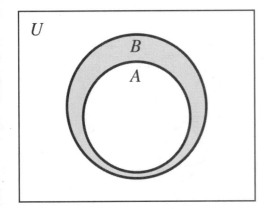

29. True

31. True

33. True.

35. The number is odd and the number is divisible by three.

37. The number is odd or the number is not divisible by three.

39. The number is odd and the number is not divisible by three.

41. The person is at most 40 years of age.

Alternatively we could state:

The person is 40 years of age or younger.

43. The person wears glasses and is over 40 years of age.

45. The person wears glasses and the person is not over 40 years of age.

Alternatively we could state the sentence as follows:

The person wears glasses and the person is at most 40 years of age.

47.
a.

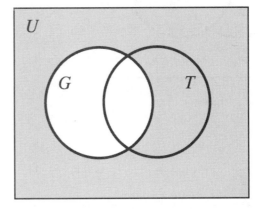

b. The symbolic expression identifying this set is:

G'.

49.
a.

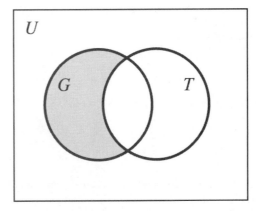

b. The symbolic expression identifying this set is:

$G \cap T'$ or $G - T$.

51.
a.

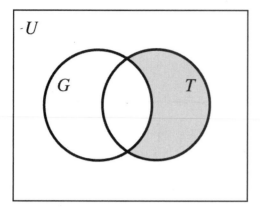

b. The symbolic expression identifying this set is:

$G' \cap T$ or $T - G$.

53.
a.

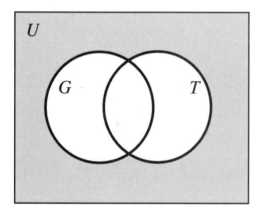

b. The symbolic expression identifying this set is:

$G' \cap T'$.

55.
a. The set of finite math students who are currently enrolled in English and history but not geology is shown at the top of the next column.

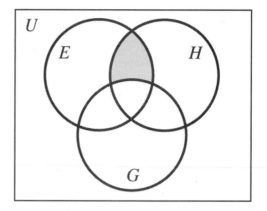

b. The symbolic expression identifying this set is:

$E \cap H \cap G'$.

57.
a. The set of finite math students who are enrolled in geology or English, but not history is:

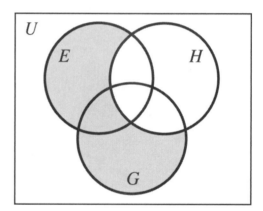

b. The symbolic expression identifying this set is:

$(E \cup G) \cap H'$ or $(E \cup G) - H$.

59. Since the universe consists of the set of finite math students, the set of students that are enrolled in exactly one of these four courses, must be the set of students who are enrolled only in the finite math course. This set is:

a.

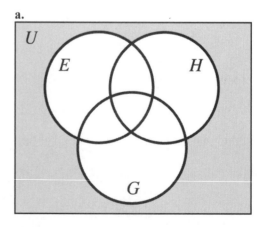

b. The symbolic expression identifying the set displayed on the previous page is:

$$\left(E \cup H \cup G\right)'.$$

61.

a. Since the universe consists of the set of finite math students, the set of students that are enrolled in at least one of these four courses is the universe since everyone is enrolled in the finite math course. The set of these students is:

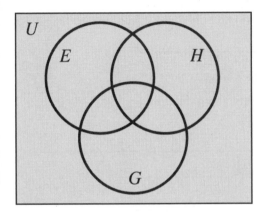

b. The symbolic expression identifying this set is U.

Exercises Section 7.4

1. In the same manner as the text, we label the various regions.

Set	Region Labels
A'	5,6,7,8
B	3,4,5,6
C	2,3,6,7
$A' \cup B$	3,4,5,6,7,8
$\left(A' \cup B\right) \cap C$	3,6,7

We shade the regions 3, 6, and 7 to get the appropriate diagram. The diagram is shown at the top of the next column.

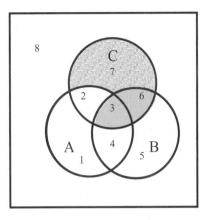

3. In the same manner as the text, we label the various regions.

Set	Region Labels
A	1,2,3,4
B	3,4,5,6
$A \cup B$	1,2,3,4,5,6
$\left(A \cup B\right)'$	7,8
C'	1,4,5,8
$\left(A \cup B\right)' \cap C'$	8

We shade the region 8 to get the appropriate diagram.

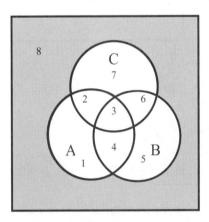

5. In the same manner as the text, we label the various regions. Notice that since A and B are disjoint sets, the Venn diagram shows the two sets without any overlap.

Set	Region Labels
A	1
B'	1,3
$A \cap B'$	1
$\left(A \cap B'\right)'$	2,3

We shade the regions 2 and 3 to get the appropriate diagram. The diagram is displayed at the top of the next page.

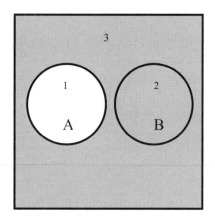

7. In the same manner as the text, we label the various regions.

Set	Region Labels
A	1,2,3,4
B	3,4,5,6
C'	1,4,5,8
$A \cap B$	3,4
$(A \cap B) \cup C'$	1,3,4,5,8

We shade the regions 1, 3, 4, 5, and 8 to get the appropriate diagram.

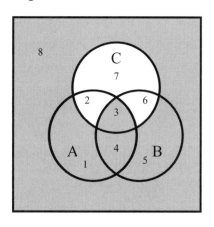

9. In the same manner as the text, we label the various regions.

Set	Region Labels
A	1,2,3,4
B	3,4,5,6
C	2,3,6,7
C'	1,4,5,8
$(A-C)$	1,4
$(B \cap C')$	4,5
$(A-C) \cup (B \cap C')$	1,4,5

We shade the regions 1, 4, and 5 to get the appropriate diagram.

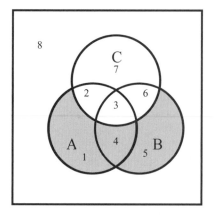

11. Labeling the various regions we get;

Set	Region Labels
A	1,2
B	2,3
$(A \cup B)$	1,2,3
$(A \cup B)'$	4

We shade the region 4 to get the appropriate diagram for $(A \cup B)'$.

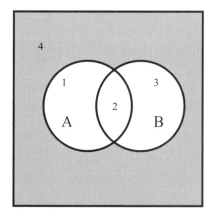

Likewise for $A' \cap B'$

Set	Region Labels
A'	3,4
B'	1,4
$A' \cap B'$	4

We shade the region 4 to get the appropriate diagram for $A' \cap B'$. The diagram is shown at the top of the next page.

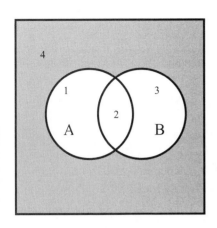

The two Venn diagrams show that $(A \cup B)' = A' \cap B'$.

13. Labeling the various regions we get;

Set	Region Labels
A'	5,6,7,8
B	3,4,5,6
$B \cup C$	2,3,4,5,6,7
$A' \cap (B \cup C)$	5,6,7
$[A' \cap (B \cup C)] - B$	7

We shade the region 7 to get the appropriate diagram for $[A' \cap (B \cup C)] - B$.

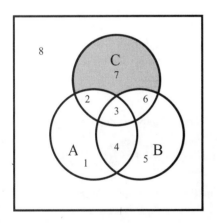

Likewise for $(A \cup B)' \cap C$

Set	Region Labels
$A \cup B$	1,2,3,4,5,6
$(A \cup B)'$	7,8
$(A \cup B)' \cap C$	7

We shade the region 7 to get the appropriate diagram for $(A \cup B)' \cap C$.

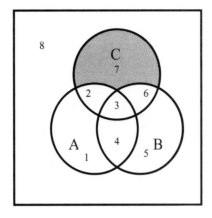

The two Venn diagrams show

$$[A' \cap (B \cup C)] - B = (A \cup B)' \cap C.$$

15. The Venn diagram for $A \cap B'$ is:

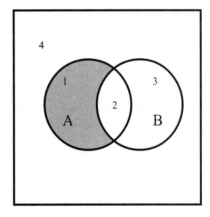

The Venn diagram for $A - B$ is:

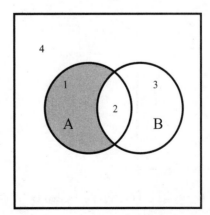

Therefore, $A \cap B' = A - B$

17. The Venn diagram for $(A \cup B)'$ is:

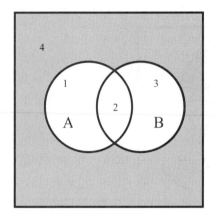

The Venn diagram for $A' \cap B$ is:

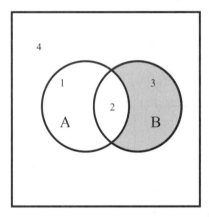

Therefore, $(A \cup B)' \neq A' \cap B$.

19. The Venn diagram for $C - (A \cup B)$ is:

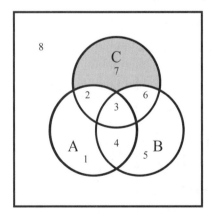

The Venn diagram for $(C \cap A) \cup (C \cap B)$ is shown at the top of the next column.

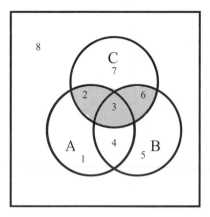

Therefore, $C - (A \cup B) \neq (C \cap A) \cup (C \cap B)$.

21. Using the Inclusion-Exclusion principle the number of elements in the union is:

$$n(A \cup B) = 8 + 12 - 3 = 17.$$

23. The Inclusion-Exclusion principle states:

$$n(A \cup B) = n(A) + n(B) - n(A \cap B).$$

Substituting the information given in the problem results in the following:

$$8 = n(A) + 5 - 2.$$

Solve for $n(A)$.
$$8 = n(A) - 3$$
$$n(A) = 5.$$

25. The Inclusion-Exclusion principle states:

$$n(A \cup B) = n(A) + n(B) - n(A \cap B).$$

Substituting the information given in the problem results in the following:

$$12 = 7 + 9 - n(A \cap B).$$

Solve for $n(A \cap B)$.
$$12 = 16 - n(A \cap B)$$
$$n(A \cap B) = 4.$$

27. Let F represent the set of those who read French and G represent the set of those who read German.

a. To construct the Venn diagram, insert the two who read both languages in the intersection of the two sets. This leaves ten people who read only French, and three people who read only German. We put those numbers in their respective regions of F and G. This leaves 15 people who do not read French or German. We place that number in the region outside the two sets. The Venn diagram that represents this situation is:

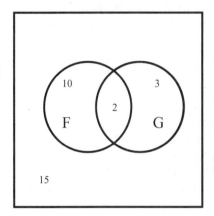

b. $n(F - G) = 10$

c. $n(F \cup G) = 15$

d. $n(F \cup G) = 15$

e. $n(F \cap G') \cup n(G \cap F') = 13$

f. $n\left((F \cup G)'\right) = 15$

g. The Inclusion-Exclusion principle states:

$$n(F \cup G) = n(F) + n(G) - n(F \cap G)$$

Therefore, replacing the symbols with the appropriate numerical value results in the following:

$$n(F \cup G) = 12 + 5 - 2 = 15.$$

29. Let $Q1$ represent the set of those who own a Chevrolet and $Q2$ represent the set of those who own at least two cars.

To construct the Venn diagram, insert the 18 who answered yes to both questions in the intersection of the two sets. Since a total 25 people answered yes to question one, this leaves seven people who answered yes to question one and no to question two. Since a total 52 people answered yes to question two, this leaves 34 people who answered yes to question two and no to question one. We put those numbers in their respective regions of $Q1$ and $Q2$. This leaves 21 people who answered no to both questions. We place that number in the region outside the two sets. The Venn diagram that represents this situation is:

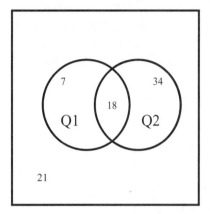

a. $n(Q1 \cup Q2) = 59$

59 people answered "yes" to at least one of these questions.

b. $n\left((Q1 - Q2) \cup (Q2 - Q1)\right) = 41$

41 people answered "yes" to exactly one of these questions.

c. $n\left((Q1 \cup Q2)'\right) = 21$

21 people answered "no" to both questions.

d. $n(Q1 \cap Q2') = 7$

Of the people that own only one car, seven own a Chevrolet.

31. Let $Q1$ represent the set of those who own a color television set and $Q2$ represent the set of those who own at least two radios.

To construct the Venn diagram, notice there are 120 people who answered "yes" to question one or question two, and 110 answered yes to question one, while 90 answered yes to question two.

The solution is continued on the next page.

We can find the number of people who answered yes to both questions by the Inclusion-Exclusion principle. Applying this principle we see that

$$120 = 110 + 90 - n(Q1 \cap Q2).$$

Solving this equation we see that 80 people answered yes to both questions. Since a total 110 people answered "yes" to question one, this leaves 30 people who answered "yes" to question one and "no" to question two. Since a total 90 people answered "yes" to question two, this leaves 10 people who answered "yes" to question two and "no" to question one. We put those numbers in their respective regions of $Q1$ and $Q2$. This leaves 30 people who answered "no" to both questions. We place that number in the region outside the two sets. The Venn diagram that represents this situation is:

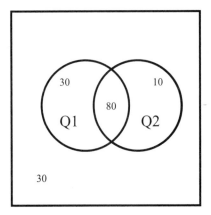

a. $n(Q1 \cap Q2) = 80$

Eighty people answered "yes" to both of these questions.

b. $n\left((Q1 \cup Q2)'\right) = 30$

Thirty people answered "no" to both questions.

c. $n(Q1 \cap Q2') = 30$

Thirty people own a color television but own less than two radios.

d. $n(Q1' \cap Q2) = 10$. Since 150 people were asked the questions, the percentage of respondents that own at least two radios but no color television is:

$\frac{10}{150} = 0.0667$.

So 6.67% of the respondents own at least two radios, but do not own a color television set.

33. Let B represent the set of those who purchased books and J represent the set of those who purchased jewelry and let F represent the set of those who purchased film.

To construct the Venn diagram, notice there are a total of 3 people who purchased all three items. We place this number in the intersection of the two sets. Since a total six people had purchased film and jewelry this leaves three people who purchased film and jewelry but not books. We place this number in the region of the diagram that represents $F \cap J \cap B'$. Since eight people purchased both books and jewelry this leaves five people that purchased books and jewelry but not film. We place this number in the region of the diagram that represents $F' \cap J \cap B$. Since 15 people had purchased both books and film this leaves 12 people who purchased books and film, but not jewelry. We place this number in the region of the diagram that represents $F \cap J' \cap B$. Twenty people purchased books, and we already have 20 people in the B set of the diagram. This means that nobody purchased only books, so the region that represents $F' \cap J' \cap B$ has zero elements. A total of 45 people purchased film, and we have already counted 18 people in the set F, this leaves 27 people who purchased only film. This number will be placed in the region that represents $F \cap J' \cap B'$. A total of 38 people purchased jewelry, and we have already counted 11 people in the set J, this leaves 27 people who purchased only jewelry. This number will be placed in the region that represents $F' \cap J \cap B'$. This accounts for a total of 77 of the 90 customers, so there must be 13 customers that did not purchase any of the three items. This number will be placed in the region that represents $F' \cap J' \cap B'$. The Venn diagram that represents this situation is:

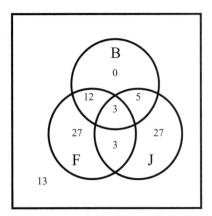

a. $n(B \cup F \cap J') = 39$

Thirty nine customers purchased books or film but not jewelry.

The solution is continued on the next page.

b. $n\left(B \cap F \cap J'\right) = 12$

Twelve customers purchased books and film, but not jewelry.

c. $n\left(J \cap \left(B \cup F\right)'\right) = 27$

Twenty-seven customers purchased Jewelry, but not books or film.

d. $n\left(B \cup F \cup J\right) = 77$.

Seventy-seven customers purchased at least one of these three items.

e. We figure out exactly how many customers purchased only books, only jewelry and only film, and then sum those numbers. The customers that purchased only one item are

$$n\left(B - \left(J \cup F\right)\right) = 0$$
$$n\left(J - \left(B \cup F\right)\right) = 27$$
$$n\left(F - \left(B \cup J\right)\right) = 27$$

Therefore the total number of customers that purchased exactly one item is 54.

35. Let $Q1, Q2$, and $Q3$ represent the set of those people who answered yes to questions 1, 2 and 3 respectively. Twelve people answered "yes" to all three questions. This number represents the region $Q1 \cap Q2 \cap Q3$. Since a total of 15 people answered "yes" to questions 1 and 2, this leaves 3 people that answered "yes" to questions 1 and 2, but "no" to question 3. This set represents the region $Q1 \cap Q2 \cap Q3'$. Since a total of 42 people answered "yes" to questions 1 and 3, this leaves 30 people who answered "yes" to questions 1 and 3 but "no" to question 2. This set represents the region $Q1 \cap Q2' \cap Q3$. Since a total of 40 people answered "yes" to questions 2 and 3, this leaves 28 people who answered "yes" to questions 2 and 3 but "no" to question 1. This set represents the region $Q1' \cap Q2 \cap Q3$. Since a total of 63 answered "yes" to question one, and we already have 45 of those people accounted in the set, that means that 18 people answered "yes" to question 1 only. This set represents the region $Q1 \cap \left(Q2 \cup Q3\right)'$ Since a total of 82 people answered "yes" to question three and we have already accounted for 70 people in the set, this leaves 12 people who answered "yes" to question 3 only.

This set represents the region $Q3 \cap \left(Q1 \cup Q2\right)'$ We also know that 27 people answered "no" to all three questions. Therefore 27 people are in the region $Q1' \cap Q2' \cap Q3'$. This accounts for a total of 130 out of 150 residents surveyed. This means that 20 people answered "yes" to question 2 only. This set represents the region

$Q2 \cap \left(Q1 \cup Q3\right)'$. The Venn diagram for this problem is:

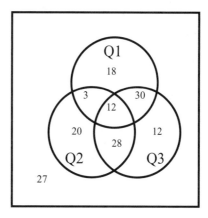

a. $n\left(Q2 \cup Q3\right) = 105$

One hundred and five people answered "yes" to question 2 or question 3.

b. $n\left(Q1 \cup Q2 \cup Q3\right) = 123$

One hundred and twenty-three residents answered "yes" to at least one of the questions.

c. The number of people that answered "yes" to exactly one of the three questions is given by the sum of the following sets.

$$n\left(Q1 - \left(Q2 \cup Q3\right)\right) = 18$$
$$n\left(Q2 - \left(Q1 \cup Q3\right)\right) = 20$$
$$n\left(Q3 - \left(Q1 \cup Q2\right)\right) = 12$$

Fifty residents answered "yes" to exactly one of the three questions.

d. $n\left(Q1 \cap \left(Q2 \cup Q3\right)'\right) = 18$

Eighteen residents favored the East-West expressway but did not favor the other two projects.

37. Let T represent the set of monkeys who could be taught to ride a tricycle, C represent the set of monkeys who could be taught to do chin-ups, and B represent the set of monkeys who could be taught to shoot a basketball.

We start with the 2 monkeys that could be taught to do all three activities in the intersection of all three sets $T \cap B \cap C$. Since four monkeys could be taught to ride a tricycle and shoot a basketball, this leaves two monkeys in the set $T \cap B \cap C'$. Since three monkeys could be taught to do chin-ups and shoot a basketball, this leaves one monkey in the set $T' \cap B \cap C$. Since five monkeys could be taught to ride a tricycle and do chin-ups, this leaves 3 monkeys in the set $T \cap B' \cap C$. Since there are nine monkeys in set T and seven of those monkeys have already been accounted for, then there must be two monkeys that could only ride a tricycle. They represent the set $T \cap B' \cap C'$. Since 8 monkeys could be taught to shoot a basketball, and five of those monkeys have already been accounted for, then there must be 3 monkeys that could only be taught to shoot a basketball. They represent the set $T' \cap B \cap C'$. Five monkeys could not be taught any of the three activities. They represent the set $T' \cap B' \cap C'$. So far we have accounted for 18 of the 20 monkeys. This means that there must be 2 monkeys that could only be taught to do chin-ups. They represent the set $T' \cap B' \cap C$. The Venn diagram for this information is shown below:

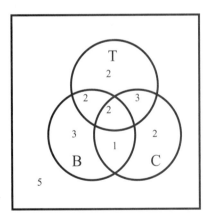

a. $n\left(C \cap T'\right) = 3$

Three monkeys could be taught how to do chin-ups but not how to ride a tricycle.

b. The number of monkeys that could be taught to do at least two of these activities is given by the number of monkeys that can do exactly two of the three activities plus the number of monkeys that could do all three of the activities. We need to add the following cases at the top of the next column.

$$n\left(\left(T \cap B\right) - C\right) = 2$$
$$n\left(\left(T \cap C\right) - B\right) = 3$$
$$n\left(\left(B \cap C\right) - T\right) = 1$$
$$n\left(T \cap B \cap C\right) = 2$$

Eight monkeys could be taught to do at least two of the activities.

c. $n\left(\left(T \cup B\right)'\right) = 7$

Seven monkeys could not be taught how to ride a tricycle or shoot a basketball.

39. The student should not be believed. His report that 50 people bought cinnamon, 40 people bought pepper, and 5 people bought both, means that $50 + 40 - 5 = 85$ bought cinnamon or pepper. This is inconsistent with the report that only 60 people bought a spice of some kind.

41.
a. Using the table, there is a total of 70 males in the town. There is also a total of 5 unemployed females, so the number of people who are male or unemployed is $70 + 5 = 75$.

b. Using the table, we see that there are 40 employed females in the town.

Exercises Section 7.5

1. There are six possible outcomes for rolling a single die..

3. Assuming fractional scores are not possible, there are 11 possible scores on a ten-point quiz. (Remember zero is a possibility)

5. There are 16 possible outcomes. They are:
HHHH, HHHT, HHTH, HTHH, HHTT, HTHT, HTTH, HTTT, THHH, THHT, THTH, TTHH, TTTH, TTHT, THTH, TTTT.

7. There are 12 possible two-digit numbers where zero cannot be the first digit and repetition is allowed. They are: 10, 11, 12, 13, 20, 21, 22, 23, 30, 31, 32, 33.

9. There are three possible ways that exactly two letters are placed in the wrong envelope. To count this let letters 'a', 'b', 'c' correspond to envelopes A,B,C. Exactly two letters are placed in the wrong envelope is the same as saying exactly one letter is placed in the correct envelope. The three ways this can happen are letter 'a' placed in envelope 'A', letter 'b' placed in envelope 'C' and letter 'c' placed in envelope 'B'. or, letter 'b' placed in envelope 'B', letter 'a' placed in envelope 'C', and letter 'c' placed in envelope 'A', or letter 'c' placed in envelope 'C', letter 'a' placed in envelope 'B' and letter 'b' placed in envelope 'A'.

Simplified we see the three possibilities as:

'a' in 'A', 'b' in 'C', 'c' in 'B'
'b' in 'B', 'a' in 'C', 'c' in 'A'
'c' in 'C', 'a' in 'B', 'b' in 'A'

11. Let $S1$, $S2$, and $S3$ represent the three varieties of soybeans. Let $F1$ and $F2$ represent the two levels of fertilizer. Let $I1$ and $I2$ represent the two types of insecticides. The possible combinations are:

$S1, F1, I1$;	$S2, F1, I1$;	$S3, F1, I1$
$S1, F1, I2$;	$S2, F1, I2$;	$S3, F1, I2$
$S1, F2, I1$;	$S2, F2, I1$;	$S3, F2, I1$
$S1, F2, I2$;	$S2, F2, I2$;	$S3, F2, I2$

Therefore, twelve plots will be needed to test every combination.

13.
a. If there are no restrictions there are ten possible digits for each position in the social security number. Therefore the total number of possible social security numbers is:

$$10 \cdot 10 \cdot 10 \cdot 10 \cdot 10 \cdot 10 \cdot 10 \cdot 10 \cdot 10$$
$$= 1,000,000,000.$$

b. If no number can begin with a zero, there are nine possible digits for the first position and then ten possible digits for the remaining eight. The total number of possible social security numbers is:

$$9 \cdot 10 \cdot 10 \cdot 10 \cdot 10 \cdot 10 \cdot 10 \cdot 10 \cdot 10$$
$$= 900,000,000.$$

c. If neither of the first two digits can be zero, there are nine possible digits for the first two positions and ten possible digits for the remaining seven positions. The total number of possible social security numbers is:

$$9 \cdot 9 \cdot 10 \cdot 10 \cdot 10 \cdot 10 \cdot 10 \cdot 10 \cdot 10$$
$$= 810,000,000.$$

15.
a. If there are no restrictions, there are
$$8 \cdot 7 \cdot 6 \cdot 5 \cdot 4 \cdot 3 \cdot 2 \cdot 1 = 40,320 \text{ different ways of seating}$$
the eight children in a row.

b. If boys and girls must be seated alternately, there are zero ways of seating the children. There are not enough girls to seat between the boys.

c. If one of the girls must be seated on both end seats, then there are $3 \cdot 2 = 6$ ways to arrange the girls in the end seats, and then there are no restrictions in arranging the other children. There are
$$3 \cdot 2 \cdot 6 \cdot 5 \cdot 4 \cdot 3 \cdot 2 \cdot 1 = 4320 \text{ different ways of seating the}$$
children with girls occupying the end seats.

17.
a. If there are no restrictions, there are
$$8 \cdot 7 \cdot 6 \cdot 5 \cdot 4 \cdot 3 \cdot 2 \cdot 1 = 40,320 \text{ different ways of arranging}$$
eight books on the shelf.

b. If *Of Mice and Men* must be displayed on the left end, then there are $1 \cdot 7 \cdot 6 \cdot 5 \cdot 4 \cdot 3 \cdot 2 \cdot 1 = 5040$ different ways of arranging the books on the shelf.

19.
a. If there are no restrictions, we have 10 choices for each digit in the number. There are
$$10 \cdot 10 \cdot 10 \cdot 10 = 10,000 \text{ different four-digit numbers.}$$

b. If zero can not be the first digit, and no digit can be repeated then there are $9 \cdot 9 \cdot 8 \cdot 7 = 4536$ possible four-digit numbers that meet these requirements.

c. The first digit must be even, so there are four possible choices for the first digit (assuming that 0 can not be the first digit), the number must be odd so there are 5 possible choices for the last digit, and none of the choices for the middle two digits can duplicate the first or the last digit. That leaves 8 choices for the second digit, and 7 choices for the third digit. There are $4 \cdot 8 \cdot 7 \cdot 5 = 1120$ different possible four-digit numbers that satisfy the requirements in the problem.

21. If a coin is tossed twice, there are $2 \cdot 2 = 4$ possible outcomes. If it is tossed five times there are $2 \cdot 2 \cdot 2 \cdot 2 \cdot 2 = 32$ possible outcomes. If it is tossed seven times there are $2^7 = 128$ possible outcomes.

23. Provided that we allow for repetition then we know that $10^x = 600$. Solving this exponential equation for x. We see that $x = \log(600) \approx 2.77$. Therefore, the fewest number of digits that an employee I.D. number can have is three.

Similarly if the company had 12,000 employees we know that $x = \log(12,000) \approx 4.08$. So the shortest I.D. number would need to be five digits in order to have enough for everyone in the company.

25. Assuming that every question is answered, there are two choices for each question. Therefore there are $2 \cdot 2 \cdot 2 \cdot 2 \cdot 2 \cdot 2 \cdot 2 \cdot 2 \cdot 2 \cdot 2 = 1024$ different ways to answer a ten-question true-false quiz.

27. There are three choices for the material, and two choices for the color, so there are $3 \cdot 2 = 6$ different ways of selecting a bathtub.

29.
a. If there are no restrictions, we have 26 choices for each letter, and 10 choices for each number. Therefore there are $26 \cdot 26 \cdot 10 \cdot 10 \cdot 10 \cdot 10 = 6,760,000$ possible license plates.

b. If no letter can be repeated, we have 26 choices for the first letter and 25 choices for the second letter. Therefore, we have $26 \cdot 25 \cdot 10 \cdot 10 \cdot 10 \cdot 10 = 6,500,000$ possible license plates.

c. If the last digit must be a four, then there is only one choice for that digit. Since no digit can be repeated, and four is already chosen for the last digit, we have 9 choices for the first digit, 8 choices for the second digit and 7 choices for the third digit. Since there are no restrictions on the letters, we have 26 choices for each letter, and. Therefore, there are $26 \cdot 26 \cdot 9 \cdot 8 \cdot 7 \cdot 1 = 340,704$ possible license plates.

31.
a. Since the call signal must begin with a K or a W, we have two choices for the first letter. The remaining letters are chosen without restriction, so there are $2 \cdot 26 \cdot 26 \cdot 26 = 35,152$ different possible four-letter call signals. We also have the choice of using three digits after the K or the W. There are $2 \cdot 10 \cdot 10 \cdot 10 = 2000$ different possible combinations using three digits. Therefore there are $35,152 + 2000 = 37,152$ different possibilities for the radio station call signals.

b. We still have the same number of four-letter call signals as part (a). The restriction that no digit can be repeated means that we have $2 \cdot 10 \cdot 9 \cdot 8 = 1440$ different call signals that use three non-repeated digits. The total number of call signals that meet these requirements is $35,152 + 1440 = 36,592$.

c. Now we have the same number of one letter three digit call signals. The restriction that no letter can be repeated means that we have $2 \cdot 25 \cdot 24 \cdot 23 = 27,600$ possible four-letter call signals. Therefore, there are $27,600 + 1440 = 29,040$ radio station call signals that meet these requirements.

33. Since each runner is a different person, we do not allow for repetition. This means that there are 30 different runners who can finish first, 29 different runners who can finish second, 28 different runners who can finish third and 27 different runners who can finish first. Therefore, there are $30 \cdot 29 \cdot 28 \cdot 27 = 657,720$ different ways the runners can place first through fourth.

35. Since each member on the team is unique, we do not allow for repetition. Therefore, the team as 40 different choices for captain and 39 different choices for co-captain. Thus, there are $40 \cdot 39 = 1560$ different possibilities of choosing a captain and co-captain.

37. Since each person is unique, we do not allow for repetition. Therefore there are $8 \cdot 7 \cdot 6 \cdot 5 = 1680$ possible ways to seat four of eight people into four chairs.

39.
a. There are $1 \cdot 2 \cdot 2 = 4$ possible outcomes that have a head on the first toss, and $2 \cdot 2 \cdot 1 = 4$ possible outcomes that have a head on the last toss. However, this counts the arrangements *HHH* and *HTH* twice. The inclusion-exclusion principle tells us that we need to subtract the intersection out of the final total. So the number of possible outcomes which have a head on the first toss or the last toss is $4 + 4 - 2 = 6$.

b. There are $1 \cdot 2 \cdot 2 = 4$ possible outcomes that have a head on the first toss, and $2 \cdot 1 \cdot 2 = 4$ possible outcomes that have a tail on the second toss. However, this counts the arrangements *HTH* and *HTT* twice. The inclusion-exclusion principle tells us that we need to subtract the intersection out of the final total. So the number of possible outcomes which have a head on the first toss or a tail on the second toss is $4 + 4 - 2 = 6$

41.

a. The restrictions mean that there is only one choice for the first and last digit. There are no restrictions for the middle two digits, so there are $1 \cdot 10 \cdot 10 \cdot 1 = 100$ possible I.D. numbers that fit these restrictions.

b. If we put the restriction that the I.D. number must begin with a zero *or* end with an eight, then we have many more possibilities. There are $1 \cdot 10 \cdot 10 \cdot 10 = 1000$ possible I.D. numbers that begin with zero. Likewise, there are $10 \cdot 10 \cdot 10 \cdot 1 = 1000$ possible I.D. numbers that end with eight. The inclusion-exclusion principle tells us that we must subtract out the intersection of these two groups. Therefore, there are $1000 + 1000 - 100 = 1900$ possible four digit I.D. numbers that begin with zero or end in eight.

43.

a. Without replacement, we have eight choices for the first selection, seven choices for the second selection and six choices for the third selection. Therefore, there are $8 \cdot 7 \cdot 6 = 336$ possible samples that can be selected without replacement, and where order is important.

b. If the sampling is done with replacement, there are eight ways of selecting each computer in the sample. Therefore, there are $8 \cdot 8 \cdot 8 = 512$ possible samples that can be selected with replacement if the order of the selection is important.

Exercises Section 7.6

1. $6! = 6 \cdot 5 \cdot 4 \cdot 3 \cdot 2 \cdot 1 = 720$

3. $8! = 8 \cdot 7 \cdot 6 \cdot 5 \cdot 4 \cdot 3 \cdot 2 \cdot 1 = 40,320$

5. $\dfrac{8!}{4!} = \dfrac{8 \cdot 7 \cdot 6 \cdot 5 \cdot 4 \cdot 3 \cdot 2 \cdot 1}{4 \cdot 3 \cdot 2 \cdot 1} = 8 \cdot 7 \cdot 6 \cdot 5 = 1680$

7. $\dfrac{12!}{7!5!} = 792$

9. $\dfrac{52!}{47!5!} = 2,598,960$

11. $(2!)! = (2 \cdot 1)! = 2! = 2$

13. $P(8,4) = \dfrac{8!}{(8-4)!} = \dfrac{8!}{4!} = 1680$

15. $P(6,1) = \dfrac{6!}{(6-1)!} = \dfrac{6!}{5!} = 6$

17. $P(7,7) = \dfrac{7!}{(7-7)!} = \dfrac{7!}{0!} = \dfrac{7!}{1} = 5040$

19. $P(10,0) = \dfrac{10!}{(10-0)!} = \dfrac{10!}{10!} = 1$

21. $P(4,3) = \dfrac{4!}{(4-3)!} = \dfrac{4!}{1!} = 24$

23. $P(0,0) = \dfrac{0!}{(0-0)!} = \dfrac{0!}{0!} = \dfrac{1}{1} = 1$

25.

a. The problem can be stated in permutation notation as $P(8,4)$.

b. $P(8,4) = \dfrac{8!}{(8-4)!} = \dfrac{8!}{4!} = 1680$. There are 1680 ways to arrange four of eight books on a shelf.

27.

a. The problem can be stated in permutation notation as $P(7,7)$.

b. $P(7,7) = \dfrac{7!}{(7-7)!} = \dfrac{7!}{0!} = 5040$. There are 5040 ways to arrange seven monkeys in a row.

29. $P(5,5) = \dfrac{5!}{(5-5)!} = \dfrac{5!}{0!} = 120$. There are 120 different ways to arrange five candidates on a ballot.

31. $P(8,3) = \dfrac{8!}{(8-3)!} = \dfrac{8!}{5!} = 336$. There are 336 possible ways the board of directors can elect the officers.

33. $P(5,5) = \dfrac{5!}{(5-5)!} = \dfrac{5!}{0!} = 120$. The five pieces of music can be arranged 120 different ways in the program.

35. $P(4,4) = \dfrac{4!}{(4-4)!} = \dfrac{4!}{0!} = 24$. There are 24 different ways to arrange the letters in the word *page*.

37. We notice that the two *l*'s and the two *e*'s in college are indistinguishable. Therefore the number of distinguishable permutations can be given by

$$\frac{P(7,7)}{2!2!} = \frac{\frac{7!}{(7-7!)}}{4} = \frac{7!}{4} = 1260 \,.$$

There are 1260 distinguishable permutations of the word *college*.

39. The way to think about this problem is to arrange all 12 students in a line. The first seven students will be the first group, and the last five students will be the second group. There are $12!$ different ways to arrange the 12 students in a line. However, since we are forming a group of seven and a group of five, any different ordering of the first seven or the second five would be the same group. There are $7!$ such orderings of the group of seven, and $5!$ such orderings of the group of five. Therefore, the number of ways that the 12 students can be divided into a group of seven and a group of five is:

$$\frac{12!}{7!5!} = 792 \,.$$

41. If each song within a category is indistinguishable, then there are $\dfrac{13!}{6!4!3!} = 60,060$ different ways to arrange her concert.

43. There are 10 balls total to be arranged in a line. Dividing out the indistinguishable arrangement we see that there are $\dfrac{10!}{5!3!2!} = 2520$ distinguishable arrangements of the blue, green and red balls in a line.

45. There are $P(7,2)$ different ways of constructing the two-digit I.D. numbers and $P(7,3)$ different ways of constructing the three-digit I.D. numbers. Because no two-digit I.D. number is the same as a three-digit I.D. number, we simply need to add the two groups together. Therefore, there are $P(7,2) + P(7,3) = 42 + 210 = 252$ two-digit or three-digit I.D. numbers.

47. We assume repetition of initials is possible. In other words someone's first name can begin with the same letter as their last name. With this in mind, there are $26 \cdot 26 = 676$ possible arrangements of 2 letter initials. Even if 676 employees all had different initials, there are 74 employees that would be forced to have the same initials as another employee in the firm.

Exercises Section 7.7

1. $C(6,2) = \dfrac{6!}{2!(6-2)!} = \dfrac{6!}{2!4!} = 15$

3. $C(9,3) = \dfrac{9!}{3!(9-3)!} = \dfrac{9!}{3!6!} = 84$

5. $C(15,2) = \dfrac{15!}{2!(15-2)!} = \dfrac{15!}{2!13!} = 105$

7. $C(48,4) = \dfrac{48!}{4!(48-4)!} = \dfrac{48!}{4!44!} = 194,580$

9. $C(14,7) = \dfrac{14!}{7!(14-7)!} = \dfrac{14!}{7!7!} = 3432$

11.
a.

$$C(5,2) = \frac{5!}{2!(5-2)!} = \frac{5!}{2!3!} = 10$$

$$C(5,3) = \frac{5!}{3!(5-3)!} = \frac{5!}{3!2!} = 10$$

b. $C(9,7) = \dfrac{9!}{7!(9-7)!} = \dfrac{9!}{7!2!} = 36$

$$C(9,2) = \frac{9!}{2!(9-2)!} = \frac{9!}{2!7!} = 36$$

c.

$$C(8,5) = \frac{8!}{5!(8-5)!} = \frac{8!}{5!3!} = 56$$

$$C(8,3) = \frac{8!}{3!(8-3)!} = \frac{8!}{3!5!} = 56$$

d. The rule that we see from the above examples is

For all permissible values of n and r :

$$C(n,r) = C(n,n-r) \,.$$

13. The set of possible values is $r = \{0,1,2,3,4,5,6\}$.
The appropriate combinations are:

$$C(6,0) = \frac{6!}{0!(6-0)!} = \frac{6!}{0!6!} = 1$$

$$C(6,1) = \frac{6!}{1!(6-1)!} = \frac{6!}{1!5!} = 6$$

$$C(6,2) = \frac{6!}{2!(6-2)!} = \frac{6!}{2!4!} = 15$$

$$C(6,3) = \frac{6!}{3!(6-3)!} = \frac{6!}{3!3!} = 20$$

$$C(6,4) = \frac{6!}{4!(6-4)!} = \frac{6!}{4!2!} = 15$$

$$C(6,5) = \frac{6!}{5!(6-5)!} = \frac{6!}{5!1!} = 6$$

$$C(6,6) = \frac{6!}{6!(6-6)!} = \frac{6!}{6!0!} = 1$$

15. A different line is formed by every pair of points. So we want to know how many combinations are there of five points taken two at a time. This is calculated by:

$$C(5,2) = \frac{5!}{2!(5-2)!} = \frac{5!}{2!3!} = 10 .$$

There are 10 possible lines that can be determined from five points in a plane, when no three points lie on the same line.

17. $C(52,3) = \dfrac{52!}{3!(52-3)!} = \dfrac{52!}{3!49!} = 22,100$.

There are 22,100 possible three-card hands.

19. The question wants us to count how many different ways we can select 6 working modems from a group of 15 working modems. The combination is:

$$C(15,6) = \frac{15!}{6!(15-6)!} = \frac{15!}{6!9!} = 5005 .$$

There are 5005 possible samples of six in which all six modems are in working order.

21.
a. If all of the disks are to have no bad sectors, then the three must be selected from the 15 good disks. Therefore there are $C(15,3) = 455$ different samples of three which have no bad sectors.

b. If all of the disks are to have bad sectors, then the three must be selected from the five bad disks. Therefore there are $C(5,3) = 10$ different samples of three, all of which have bad sectors.

c. If exactly two do not have bad sectors, then that means exactly one must have must have a bad sector. There are $C(15,2) = 105$ different ways of selecting exactly two disks with no bad sectors and $C(5,1) = 5$ different ways of selecting one disk that has a bad sector. Therefore there are $105 \cdot 5 = 525$ different combinations of 3 disks, two of which have no bad sectors and one that has a bad sector.

23.
a. If the mayor must be included in the five person committee, then there is only one way of selecting the mayor's position. The remaining four people can be selected from the seven city officials. There are $C(7,4) = 35$ ways of filling the other four positions. Therefore there are 35 different ways of selecting a five person committee, if one of the positions is occupied by the mayor.

b. If the mayor is not on the committee, then the members must be chosen from the seven city officials. There are $C(7,5) = 21$ ways of selecting five people from the group of seven.

c. If there are no restrictions, then we are selecting a group of five officials from a total of eight. There are $C(8,5) = 56$ ways of selecting a five person committee from a group of eight people.

25.
a. Since Data Fix sent five members to the conference, in order for all four members of the committee to be employees of Data Fix, we must select the committee from those five representatives. There are $C(5,4) = 5$ ways of selecting a committee of four from the five Data Fix representatives.

b. If none of the representatives are to be from Data Fix, that means that the committee must be chosen from the ten representatives that are not employed by Data Fix. There are $C(10,4) = 210$ different ways of selecting a committee of four where none of the members represent Data Fix.

c. Since all of the members must be from Data Mix or Data Six, This also means that none of the members must be from Data Fix. This is identical to the problem posed in part (b). Therefore there are $C(10,4) = 210$ different ways of selecting a committee of four whose members represent Data Mix or Data Six.

d. If all of the members are to come from Data Mix, or all of the members are to come from Data Six, we will need to find out how many ways it is possible to select a committee from each group. Since both groups sent five representatives, there are $C(5,4) = 5$ different ways of selecting a committee of four so that all the members represent one company. Therefore there are $5 + 5 = 10$ different ways of selecting a committee of four so that all of the committee members represent Data Mix or all of the committee members represent Data Six.

27.

a. There are $C(13,8) = 1287$ different ways of selecting an eight card hand made up entirely of spades.

b. Since there are $C(13,8) = 1287$ different ways of selecting an eight card hand made up from one suit, and there are four suits, there are $4C(13,8) = 4 \cdot 1287 = 5148$ different ways to select an eight card hand made up of cards from one suit.

c. There are $C(13,5) = 1287$ ways of selecting five clubs from the deck, and there are $C(39,3) = 9139$ ways of selecting the other three cards. Therefore there are $1287 \cdot 9139 = 11,761,893$ different ways of selecting an eight card hand where exactly five of the cards are clubs.

d. There are $C(13,8) = 1287$ different ways of selecting an eight card hand that consists of one suit. Since we want the eight card hand to be made up only of hearts or only of spades there are $2C(13,8) = 2 \cdot 1287 = 2574$ different ways of selecting an eight card hand that is made up either entirely of hearts or entirely of spades.

29.

a. If none of the committee members are to be conservative, this means that all three must be liberal. Since there are five liberals to choose from, there are $C(5,3) = 10$ different ways of selecting a committee of three liberals from the group.

b. If exactly one is to be conservative, that means exactly two must be liberal as well. There are $C(3,1) = 3$ ways of selecting a conservative and there are $C(5,2) = 10$ ways of selecting a liberal, so there are $3 \cdot 10 = 30$ different ways of constructing a committee consisting of exactly one conservative.

c. If at least two members of the committee are liberal, then this means that either exactly two are to be liberals and one is to be a conservative, or all three members of the committee are to be liberal. Since we know from part (a) and part (b) that there are 10 different ways of selecting a committee consisting of liberals and 30 different ways of selecting a committee consisting of two liberals and one conservative, we must have $30 + 10 = 40$ different ways of selecting a committee with at least two liberals on it.

31.

a. If all are to be black cards, then we have 26 cards from which to select the four-card hand. Therefore there are $C(26,4) = 14,950$ different possible combinations of four-card hands consisting only of black cards.

b. Since there are only four cards that have the number eight on them, then there are $C(4,4) = 1$ way of selecting four eights.

c. If at least two of the cards are to be hearts, then we wish to count all the different four-card hands that either have exactly two hearts and two non-hearts, exactly three hearts and one non-heart, or all four hearts. There are $C(13,2)C(39,2) = 78 \cdot 741 = 57,798$ different ways of selecting exactly two hearts and two non-hearts. There are $C(13,3)C(39,1) = 286 \bullet 39 = 11,154$ different ways of selecting exactly three hearts and one non-heart. There are $C(13,4) = 715$ different ways of selecting four hearts from the deck. Therefore there are $57,798 + 11,154 + 715 = 69,667$ different ways of selecting a four-card hand where at least two of the cards must be hearts.

33.

a. There are $C(6,4) = 15$ ways of selecting four balls all of which are blue and there are $C(7,4) = 35$ ways of selecting four balls all of which are purple. Therefore there are $15 + 35 = 50$ ways of selecting four balls that are either all blue or all purple.

b. If exactly two of the balls are to be blue, then exactly two of the balls are to be a color other than blue. There are $C(6,2) = 15$ ways of choosing two balls which are blue. The remaining two balls must be select from the five red balls or seven purple balls. Therefore there are $C(12,2) = 66$ ways of selecting two balls that are not blue. Therefore there are $15 \cdot 66 = 990$ ways of selecting four balls, where exactly two of the balls are blue.

c. If exactly one of the balls is to be purple, then the other three balls must be a color other than purple. There are $C(7,1) = 7$ ways of selecting one purple ball from the group. There are $C(11,3) = 165$ ways of selecting two balls that are not purple. Therefore there are $7 \cdot 165 = 1155$ different ways of selecting four balls, exactly one of which is purple.

35. There are $C(11,4) = 330$ ways of choosing the four positions to place the F's, hence there are 330 distinct arrangements of four F's and 11 S's in a row.

37.
a. If all of the light bulbs are to be good, then we must choose the three from the group of 94 good light bulbs. Therefore there are $C(94,3) = 134,044$ ways of selecting three good light bulbs.

b. If all of the bulbs are to be defective, then we must choose the three bulbs from the group of six defective bulbs. There are $C(6,3) = 20$ ways to select three defective bulbs from the group.

c. If exactly two are defective then exactly one must be good. There are $C(6,2) = 15$ ways of selecting two defective bulbs. There is $C(94,1) = 94$ ways of selecting one good bulb. Therefore, there are $15 \cdot 94 = 1410$ ways of selecting exactly two defective bulbs.

39.
a. There are $C(3,2) = 3$ ways of selecting two red marbles. There are $C(4,2) = 6$ ways of selecting two green marbles. There are $C(6,1) = 6$ ways of selecting one purple marble. Therefore, there are $3 \cdot 6 \cdot 6 = 108$ ways of selecting five marbles where exactly two of the marbles are red, two of the marbles are green, and one of the marbles is purple.

b. Since there are three red marbles and six purple marbles, there are nine marbles that are red or purple. Therefore, there are $C(9,5) = 126$ ways of selecting five marbles all of which are red or purple.

c. Since there are only three red marbles, there are zero ways to select five red marbles from the group. There are $C(6,5) = 6$ ways of selecting five purple marbles. Therefore there are six ways of selecting a group of five marbles if all are red or all are purple.

d. If at least four of the marbles are to be purple, then we want to count the number of ways we can draw exactly four purple marbles and one non-purple marble or the number of ways we can draw five purple marbles. There are $C(6,4)C(7,1) = 15 \cdot 7 = 105$ ways of drawing exactly four purple marbles and one non-purple marble. There are $C(6,5) = 6$ ways of drawing five purple marbles. Therefore there are $105 + 6 = 111$ ways of drawing five marbles where at least four of the marbles are purple.

41.
a. There is $C(3,3) = 1$ way of choosing three suits from the size 38 suits, and there is $C(3,3) = 1$ of choosing three suits from the size 40 suits. There is no way to select three suits from the shipment of two size 42 suits. Therefore there are two ways to select three suits that are the same size from the shipment.

b. There are $C(3,2)C(5,1) = 3 \cdot 5 = 15$ ways of selecting exactly two size 38 suits.

There are $C(3,2)C(5,1) = 3 \cdot 5 = 15$ ways of selecting exactly two size 40 suits.

There are $C(2,2)C(6,1) = 1 \cdot 6$ different ways of selecting exactly two size 42 suits.

Therefore there are $15 + 15 + 6 = 36$ ways of selecting three suits where exactly two of the suits are the same size.

43. Combination locks should actually be called permutation locks, because the order in which the numbers are selected is important. Mathematically speaking, a combination is an arrangement in which the order of the selection does not matter, whereas a permutation is an arrangement in which the order of the selection does matter.

Chapter 7 Summary Exercises

1.

p	q	$q \wedge p$	$p \vee (q \wedge p)$
T	T	T	T
T	F	F	T
F	T	F	F
F	F	F	F

p	q	$q \vee p$	$(q \vee p) \wedge q$
T	T	T	T
T	F	T	F
F	T	T	T
F	F	F	F

The tables show that the statements are not equivalent.

3. The current machine does break-down and we do not buy a new machine.

5. $B \cap C = \{5, 6\}$

7. $B' = \{1, 2, 4\}$

9. $(C')' = C = \{2, 4, 5, 6\}$

11. $A \cap B = \{\ \} = \varnothing$

13. $(C \cap A) \cup B = \{2, 3, 5, 6\}$

15. There are two possible outcomes for each toss of the coin, therefore there are $2^7 = 128$ possible outcomes when a coin is tossed seven times.

17. $P(96, 5) = \dfrac{96!}{(96 - 5)!} = \dfrac{96!}{91!} = 7,334,887,680$

19. $C(96, 91) = \dfrac{96!}{91!(96 - 91)!} = \dfrac{96!}{91!5!} = 61,124,064$

21. $C(47, 28) = \dfrac{47!}{28!(47 - 28!)} =$

$C(47, 28) = \dfrac{47!}{28!19!} = 1,362,649,145$

23. "Not all Fords are red and blue" is one way of negating this sentence. We could also say "At least one Ford is not red and is not blue."

25. Let W, WW, and R represent those who bought white, whole wheat, and rye bread during the past week. We draw a blank Venn diagram with three intersecting sets labeled with these three letters. To fill in the diagram, we start with the fact that three shoppers bought none of the above types of bread, so we place a 3 in the region outside the three sets. The number of people who bought white or whole wheat is given by :

$$n(W \cup WW) = n(W) + n(WW) - n(W \cap WW)$$
$$n(W \cup WW) = 43 + 39 - 19 = 63$$

The number of people who bought rye bread only is:

$$n(R) = n(U) - n(W \cup WW) - n(W' \cap WW' \cap R')$$
$$n(R) = 75 - 63 - 3 = 9$$

We place a nine in the region that represents the people who only bought rye bread.

Similarly, $n(W \cup R) = 43 + 16 - 5 = 54$, so the number of people who bought whole wheat bread only is given by the equation:

$$n(WW) = n(U) - n(W \cup R) - n(W' \cap WW' \cap R')$$
$$n(WW) = 75 - 54 - 3 = 18.$$

We place an 18 in the region that represents the people who only bought whole wheat bread.

Finally, $n(WW \cap R) = 39 + 16 - 6 = 49$, so the number of people who bought white bread only is given by:

$$n(W) = n(U) - n(WW \cup R) - n(W' \cap WW' \cap R')$$
$$n(W) = 75 - 49 - 3 = 23.$$

We place a 23 in the region that represents the people who only bought white bread.

Now, since 43 people bought white bread, 18 bought whole wheat bread only, nine bought rye bread only, and three bought none of the three types of bread, this accounts for 73 of the 75 customers. Thus, two people must have bought whole wheat bread and rye bread, but not white bread.

The solution is continued at the top of the next page.

Since 39 people bought whole wheat bread, 23 bought white bread alone, nine bought rye bread alone and three bought none of the three types of bread, this accounts for 74 of the 75 customers. Thus one person must have bought white bread and rye bread, but not whole wheat bread.

Since 16 people bought rye bread, 18 bought only whole wheat bread, 23 bought only white bread, and three people bought none of the three types of bread, this accounts for 60 of the 75 customers. This means that 15 people bought white and whole wheat, but did not buy rye.

Now we can total all of the regions together. We see that we have accounted for 71 customers; therefore four customers must have bought all three types of bread. This is represented by the four in the intersection of all three sets of the Venn diagram shown below.

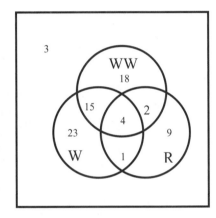

27. Since the president of the club must go, there is only one way to fill the first position. The remaining four delegates must be chosen from the 16 members who are not officers. Since order selected doesn't matter there are

$C(16,4) = 1820$ different ways of choosing the other four delegates. Thus, there are 1820 different delegations of five that could be selected to visit the museum if the president must be one of the delegates and none of the other officers can go.

29. $C(83,61) = 6.915656429 \times 10^{19}$

$$= 69,156,564,288,396,165,360$$

Cumulative Review

31. Two associated points with this cost function are: $(3,3.70)$ and $(10,8.25)$. Thus we can find the slope of the linear cost function. The slope

is: $m = \dfrac{8.25-3.70}{10-3} = 0.65$.

Now choosing one of the points (we chose the first one) we place the information in the point-slope formula as shown.

$$y - 3.70 = 0.65(x-3)$$
$$y - 3.70 = 0.65x - 1.95$$
$$y = 0.65x + 1.75$$

The linear cost function is given by the equation $y = 0.65x + 1.75$, where x is the number of miles traveled in the cab, and y is the cost of the fare in dollars.

33.

$$AB = \begin{bmatrix} -1 & 5 \\ -15 & 11 \end{bmatrix}$$

$$BA = \begin{bmatrix} 2 & -6 & 0 \\ 7 & 9 & -5 \\ 1 & 3 & -1 \end{bmatrix}$$

Chapter 7 Sample Test Answers

1. The first ball is not red or the second ball is blue.

2. The first ball drawn is not red and the second ball is not blue.

3. The first ball drawn is red and the second ball drawn is blue.

4. If the first ball drawn is red, then the second ball drawn is blue.

5. False. $\{2\}$ is an element, but 2 is not.

6. False. $\{\{2\}\}$ would be the subset. $\{2\}$ is simply an element of C.

7. True. The empty set is a subset of any set.

8. False 2 is not a set.

9. True. The empty set is an element as well as a subset of C.

10. False. 2 is and element of A, but not of C. $\{2\}$ is an element of C, but not of A.

11. True. Some students do drink tea, so the disjunction is valid.

12. $\left(A \cup C\right)' = \{4\}$

13. $A \cap C' = \{2\}$

14. $A - B = \{\varnothing, 3\}$

15. The indicated Venn diagram is:

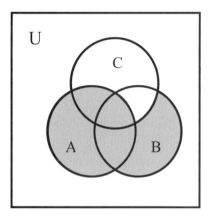

16 and 17.

Let B,L,D represent the sets of students who ate breakfast, lunch and dinner respectively. The Venn diagram for the problem is:

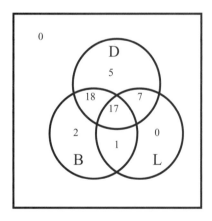

16. There were zero students who only ate lunch at the university dining hall.

17. 47 students ate dinner at the dining hall.

18. Since zero can not be the first digit, there are nine choices for the first digit, and ten choices for the remaining two digits. Therefore, there are $9 \cdot 10 \cdot 10 = 900$ different three-digit numbers that do not begin with zero.

19. There are $C\left(13,3\right) = 286$ ways in which we can draw three hearts. There are $C\left(39,2\right) = 741$ ways in which we can draw two cards that are not hearts. Therefore there are $286 \cdot 741 = 211,926$ different five card hands with exactly three hearts.

20. There are two *n*'s and three *a*'s that are not distinguishable. Therefore there are

$\dfrac{6!}{2!3!} = 60$ distinguishable arrangements of the word *banana*.

21. There are $C\left(9,4\right) = 126$ different groups of five and four.

22. There are $P\left(2,2\right) = 2$ ways of assigning which car Juanita and Kareem will drive. Then there are $P\left(7,7\right) = 5040$ different ways of seating the remaining seven passengers. (remember the seat the passenger sits in is important) Therefore there are $2 \cdot 5040 = 10,080$ different seating arrangements for the vehicles if two people are designated drivers.

The other way of looking at this problem is if the seat that the person sits in does not matter, just the car that the passenger rides in. In other words it does not matter if you sit in the front of the van, or in one of the back seats. This is analogous to asking how many groups of five and four are there if Juanita and Kareem are designated drivers. There are still 2 ways of assigning which car Juanita and Kareem will drive. There are $C\left(7,4\right) = 35$ ways of assigning the other four people to ride in the passenger van. Therefore there are $2 \cdot 35 = 70$ different groups of five and four people if Juanita drives on vehicle and Kareem drives the other.

23. There are $2 \cdot 3 \cdot 6 = 36$ possible outfits (assuming you can not wear two different shoes with one outfit).

24. Since the couples can't be split up, we can think of the problem as sitting 4 people into four seats. There are $P\left(4,4\right) = 24$ ways of doing this. There are $2^4 = 16$ ways of permuting the four couples among themselves. Therefore, there are $16 \cdot 24 = 384$ different ways in which the four couples can be seated in a row if each couple is to be seated together.

25. There are a total of $2^{10} = 1024$ different ways of answering the test. There is only one way to answer the test such that every answer is wrong. So there are $1024 - 1 = 1023$ different ways of answering a ten-question true-false exam so that at least one answer is correct.

Chapter 8
Basic Concepts of Probability

Exercises Section 8.1

1.
a. The sample space is: $S = \{A, B, C, D\}$

b. Since there are four equally likely outcomes we should assign a probability of $\frac{1}{4}$ to each outcome.

c. Let the event E represent the result the outcome is not a vowel or $E = \{B, C, D\}$. The probability of E is $\frac{3}{4}$.

3. Let M represent a male child and F represent a female child. Therefore FM would mean that the couple had a female child first, and then a male child second.

a. The sample space for this experiment is
$S = \{FF, FM, MF, MM\}$

b. Since the sociologist assumed equally likely possibilities, we assign a probability of $\frac{1}{4}$ to each possible outcome.

c. Let G represent the event "the family has at least one girl". $G = \{FF, FM, MF\}$. Therefore the probability of G is $\frac{3}{4}$.

d. Let E represent the event "The family has a child of each sex". $E = \{FM, MF\}$. Therefore the probability of E is $\frac{2}{4} = \frac{1}{2}$.

5. Let M represent a male child and F represent a female child. Therefore FMM would mean that the couple had a female child first, a male child second, and a male child third.

a. The sample space is $S =$
$\{MMM, MMF, MFM, MFF, FMM, FMF, FFM, FFF\}$

b. If we assume that each outcome is equally likely, we should assign a probability of $\frac{1}{8}$ to each outcome.

c. Let G1 represent the event "at least one girl". The only outcome in the sample space that is not in G1 is the outcome MMM. Therefore, there are seven outcomes in G1. The probability of G1 occurring is $\frac{7}{8}$.

d. Let G2 represent the event "exactly two girls". $G2 = \{MFF, FMF, FFM\}$. Therefore the probability of G2 occurring is $\frac{3}{8}$.

7. Let T represent Tigist, N represent Noora, and R represent Ran. We write, for example, TN for the outcome Tigist is elected president and Noora is elected secretary.

a. The sample space is:
$S = \{TN, TR, NT, NR, RT, RN\}$

b. Since each outcome is equally likely, we assign a probability of $\frac{1}{6}$ to each outcome.

c. Let E represent the event "Tigist will hold one of the offices". $E = \{TN, TR, NT, RT\}$. Therefore, the probability that E will occur is $\frac{4}{6} = \frac{2}{3}$.

d. Let G represent the event "Noora will be elected president". $G = \{NT, NR\}$. The probability that T will occur is $\frac{2}{6} = \frac{1}{3}$.

9.
a. The sample space is $S \{1, 2, 3, 4\}$.

b. Since there are four equally likely outcomes, we assign a probability of $\frac{1}{4}$ to each outcome.

c. Let E represent the event "an even number is rolled". $E = \{2, 4\}$. The probability of E occurring is $\frac{2}{4} = \frac{1}{2}$.

d. Let G represent the event "a number greater than three is rolled". $G = \{4\}$. The probability of G occurring is $\frac{1}{4}$.

11.
a. The sample space is
$S = \{H1, H2, H3, H4, T1, T2, T3, T4\}$.

b. There are eight possible outcomes which are equally likely. Therefore, we should assign a probability of $\frac{1}{8}$ to each outcome.

c. Let E represent the event "the coin turns up heads, and an even number of pips show on the bottom face of the die". $E = \{H2, H4\}$. The probability of E occurring is $\frac{2}{8} = \frac{1}{4}$.

d. Let G represent the even "the die shows a number greater than three on the bottom". $G = \{H4, T4\}$. The probability that G will occur is $\frac{2}{8} = \frac{1}{4}$.

13.

a. The sample space is: $S = \{1, 2, 3, 4, 5, 6, 7, 8, 9\}$.

b. Since there are nine equally likely outcomes, a probability of $\frac{1}{9}$ should be assigned to each outcome.

c. Let G represent the event "a number greater than seven is selected". $G = \{8, 9\}$. The probability that G will occur is $\frac{2}{9}$.

d. Let E represent the event "an odd number is selected". $E = \{1, 3, 5, 7, 9\}$. The probability that E will occur is $\frac{5}{9}$.

15. We write, for example, 12 to indicate a one drawn first and a two drawn second.

a. The sample space is $S = \{12, 13, 21, 23, 31, 32\}$

b. Since there are six equally likely outcomes, we assign a probability of $\frac{1}{6}$ to each outcome.

c. Let E represent the event "the sum of the two numbers drawn is five". $E = \{23, 32\}$. The probability that E will occur is $\frac{2}{6} = \frac{1}{3}$.

17. The sample space consists of 36 possible outcomes. $S = \{11, 12, 13, 14, ..., 46, 56, 66\}$. Each outcome is assigned a probability of $\frac{1}{36}$.

a. Let E represent the event "a sum of seven is rolled". $E = \{16, 25, 34, 43, 52, 61\}$. The probability that E will occur is $\frac{6}{36} = \frac{1}{6}$.

b. Let G represent the event "a sum of eleven is rolled". $G = \{56, 65\}$. The probability that G will occur is $\frac{2}{36} = \frac{1}{18}$.

c. Let F represent the event "a sum greater than eight is rolled". $F = \{36, 45, 46, 54, 55, 56, 63, 64, 65, 66\}$. The probability that F will occur is $\frac{10}{36} = \frac{5}{18}$.

19. There are 36 possible outcomes, so we assign a probability of $\frac{1}{36}$ to each outcome.

a. There are zero ways of rolling a one, so the probability of rolling a one is $\frac{0}{36} = 0$.

b. Let E represent the event "a sum of two is rolled". $E = \{11\}$. Therefore the probability of E occurring is $\frac{1}{36}$.

c. Let F represent the event "the red die shows two". If we let the red die represent the first number in the outcome, $F = \{21, 22, 23, 24, 25, 26\}$. The probability that F will occur is $\frac{6}{36} = \frac{1}{6}$.

21. There are $6 \times 6 \times 6 = 216$ possible outcomes for this experiment.

a. There is no way to get a sum of two when one rolls three dice. Therefore the probability of rolling a sum less than two is $\frac{0}{216} = 0$.

b. Let E represent the event "rolling a sum of four". $E = \{112, 121, 211\}$. The probability of E occurring is $\frac{3}{216} = \frac{1}{72}$.

c. Let F represent the event "rolling a sum of seven". $F = \{115, 124, 133, 142, 151, 214, 223, 232, 241, 313, 322, 331, 412, 421, 511\}$.

There are 15 outcomes in F, so the probability that F will occur is $\frac{15}{216} = \frac{5}{72}$.

23.
a. There are 52 equally likely outcomes in the sample space.

b. There are 13 clubs, so the probability of drawing a club is $\frac{13}{52} = \frac{1}{4}$.

c. There are four kings, so the probability of drawing a king is $\frac{4}{52} = \frac{1}{13}$.

25.
a. There are $C(52, 5) = 2,598,960$ possible outcomes.

There are $C(13, 5) = 1287$ possible hands that have five clubs, so the probability of drawing five clubs is

$$\frac{1287}{2,598,960} \approx 0.0004952.$$

b. There are $4 \times 1287 = 5148$ different hands with five cards from the same suit. So the probability of drawing five cards all from the same suit is $\dfrac{5148}{2,598,960} \approx 0.00198$.

c. There are $C(39,5) = 575,757$ different ways of drawing five cards where none of the cards are hearts. The probability of drawing a five-card hand with no hearts is $\dfrac{575,757}{2,598,960} \approx 0.2215$.

27.

a. There are $C(52,6) = 20,358,520$ possible outcomes. There are $C(12,6) = 924$ possible hands that have six face cards, so the probability of drawing six face cards is $\dfrac{924}{20,358,520} \approx 0.00004539$.

b. Since there are only four kings in a deck of cards, there is no possible way of drawing six kings. Therefore the probability of drawing six cards, all of which are kings is zero.

c. There are $C(48,6) = 12,271,512$ different ways of drawing six cards without drawing a king. The probability of drawing a six-card hand with no kings is $\dfrac{12,271,512}{20,358,520} \approx 0.6028$.

29.

a. There are $C(52,5) = 2,598,960$ possible outcomes. There are $C(13,2) = 78$ ways of drawing two spades, and there are $C(39,3) = 9139$ ways of drawing the other three cards. Therefore the probability of drawing a five-card hand with exactly two spades is

$$\frac{C(13,2) \cdot C(39,3)}{C(52,5)} =$$

$$\frac{78 \cdot 9139}{2,598,960} = \frac{712,842}{2,598,960} \approx 0.2743.$$

b. There are $C(4,2) = 6$ ways of drawing two queens, and there are $C(48,3) = 17,296$ ways of drawing the other three cards. Therefore the probability of drawing a five-card hand with exactly two queens is calculated at the top of the next column.

$$\frac{C(4,2) \cdot C(48,3)}{C(52,5)} =$$

$$\frac{6 \cdot 17,296}{2,598,960} = \frac{103,776}{2,598,960} \approx 0.03993.$$

c. There are $C(39,3) = 9139$ ways of drawing three cards that are not a heart, and there are $C(13,2) = 78$ ways of drawing the other two cards which must be hearts. Therefore the probability of drawing a five-card hand with exactly three cards that are not hearts is the same as the probability of drawing a five-card hand with exactly two cards that are hearts. The probability is

$$\frac{C(13,2) \cdot C(39,3)}{C(52,5)} =$$

$$\frac{78 \cdot 9139}{2,598,960} = \frac{712,842}{2,598,960} \approx 0.2743.$$

31.

a. There are $C(200,10)$ different samples. There are $C(190,10)$ possible samples that do not have a flaw. Therefore the probability of selecting a sample of ten shirts that do not have a flaw is $\dfrac{C(190,10)}{C(200,10)} \approx 0.591$.

b. There are $C(190,8)$ ways of selecting eight shirts without a flaw and $C(10,2)$ ways of selecting two shirts with a flaw. The probability of selecting a sample of ten shirts, exactly two of which have a flaw is

$$\frac{C(190,8) \cdot C(10,2)}{C(200,10)} \approx 0.0727.$$

c. There are $C(190,5)$ ways of selecting five shirts without a flaw and $C(10,5)$ ways of selecting two shirts with a flaw. The probability of selecting a sample of ten shirts, exactly five of which have a flaw is

$$\frac{C(190,5) \cdot C(10,5)}{C(200,10)} \approx 0.000022.$$

33.

a. There are $P(5,5)$ different arrangements of the five

children. There are $P(3,1)$ ways of seating a boy in the

middle seat, and $P(4,4)$ ways of seating the other four

children around the boy in the middle seat. Therefore the
probability that a boy will be seated in the middle seat is

$$\frac{P(3,1)\,P(4,4)}{P(5,5)} = \frac{3\cdot4\cdot3\cdot2\cdot1}{5!} = 0.6 .$$

b. There are $P(2,2)$ different ways of seating the girls on

each end and there are $P(3,3)$ different ways of arranging
the boys in the middle. Therefore the probability that the
girls will be seated on both ends is

$$\frac{P(2,2)\,P(3,3)}{P(5,5)} = \frac{2\cdot1\cdot3\cdot2\cdot1}{5!} = 0.1 .$$

c. In order for boys and girls to be seated alternatively, a
boy must be seated on both ends. There are three ways to
select a boy for the first seat, two ways to select a girl for
the second seat, two ways to select a boy for the third seat,
one way to select a girl for the fourth seat, and one way to
select a boy for the last seat. The probability that the boys

and girls will be alternately seated is $\dfrac{3\cdot2\cdot2\cdot1\cdot1}{5!} = 0.1 .$

Note, for this problem we assumed that each boy and girl
was distinguishable. The analysis would not change if we
assumed that the boys and girls were indistinguishable and
used combinations instead of permutations.

35.

a. There are $26 \times 26 \times 26 \times 10 \times 10 \times 10$ different
possible license plates.

There are $1 \times 26 \times 26 \times 10 \times 10 \times 1$ the license plate must
start with an A and end with an 8.

The probability of having a license plate that starts with and
A and ends with an 8 is

$$\frac{1 \times 26 \times 26 \times 10 \times 10 \times 1}{26 \times 26 \times 26 \times 10 \times 10 \times 10} = \frac{1}{260} \approx 0.0038 .$$

b. There are $26 \times 25 \times 24 \times 10 \times 9 \times 8$ possible license
plates that have no letter or digit repeated. So the probability
of having a license plate that has no letter or digit repeated

is $\dfrac{26 \times 25 \times 24 \times 10 \times 9 \times 8}{26 \times 26 \times 26 \times 10 \times 10 \times 10} = \dfrac{108}{169} \approx 0.639 .$

c. There are $25 \times 25 \times 25 \times 9 \times 9 \times 9$ license plates that do
not contain the letter Z or the number 0. Therefore the
probability of selecting a license plate that does not contain
the letter Z or the digit zero is

$$\frac{25 \times 25 \times 25 \times 9 \times 9 \times 9}{26 \times 26 \times 26 \times 10 \times 10 \times 10} \approx 0.648 .$$

37.

a. There are $C(80,6)$ different samples of light bulbs.

There are $C(60,6)$ different samples that contain no
defects. The probability of selecting a sample of six light

bulbs with no defects is $\dfrac{C(60,6)}{C(80,6)} \approx 0.166 .$

b. There are $C(20,3)$ ways of selecting three defective

bulbs and $C(60,3)$ ways of selecting 3 non-defective
bulbs. The probability of selecting a sample of six light
bulbs that contain exactly three defective bulbs is

$$\frac{C(20,3)\cdot C(60,3)}{C(80,6)} \approx 0.130 .$$

c. There are $C(20,6)$ ways of selecting six defective

bulbs. The probability of selecting a sample of six light

bulbs that are all defective is $\dfrac{C(20,6)}{C(80,6)} \approx 0.000129 .$

39.

a. There are $10 \times 10 \times 10 \times 10 = 10,000$ different
samples of spark plugs if we allow replacement. There are
$7 \times 7 \times 7 \times 7$ different samples that contain no defects. The
probability of selecting a sample of four spark plugs with no

defects is $\dfrac{7^4}{10^4} \approx 0.2401 .$

b. There are $3 \times 3 \times 10 \times 10$ ordered samples of four spark
plugs where the first two spark plugs are defective. The
probability of testing four plugs where the first two are

defective is $\dfrac{3 \times 3 \times 10 \times 10}{10,000} = 0.09 .$

c. There are $3 \times 3 \times 3 \times 3$ ordered samples of four spark
plugs where all of the plugs are defective. The probability

of testing four defective plugs is $\dfrac{3 \times 3 \times 3 \times 3}{10,000} = 0.0081 .$

41.

a. There are $P(8,3)$ different ordered samples of three balls without replacement. There are $P(5,3)$ different ordered samples of three balls in which all of the balls are green. Therefore the probability of selecting three green balls without replacement is $\dfrac{P(5,3)}{P(8,3)} \approx 0.179$.

b. There are $P(3,3)$ different ordered samples of three red balls. Therefore the probability of drawing an ordered sample of three red balls without replacement is

$$\frac{P(3,3)}{P(8,3)} \approx 0.0179 \,.$$

c. There are $P(3,1)$ different ways to select the first ball to be red. There are $P(5,2)$ different ways to select the remaining two balls to be green. The probability of drawing three balls without replacement so that the first ball is red and the last two balls are green is

$$\frac{P(3,1) \cdot P(5,2)}{P(8,3)} \approx 0.179 \,.$$

43.

a. There are $C(10,4)$ different samples of four balls that can be drawn without replacement and without regard to order. There are $C(6,4)$ different ways that you can draw four green balls. The probability of drawing four green balls is $\dfrac{C(6,4)}{C(10,4)} = \dfrac{1}{14} \approx 0.0714$.

b. There are $C(4,2)$ ways of selecting two red balls and $C(6,2)$ ways of selecting two green balls. The probability of selecting a sample of four balls, exactly two of which are red is $\dfrac{C(4,2) \cdot C(6,2)}{C(10,4)} = \dfrac{3}{7} \approx 0.4286$.

c. There is $C(4,4) = 1$ way of selecting four red balls. The probability of selecting a sample of four balls such that all four are red is $\dfrac{C(4,4)}{C(10,4)} \approx 0.00476$.

45.

a. There are $C(10,5)$ different ways of selecting a committee of five from ten managers assuming that order is not important. There are $C(6,5)$ different ways of selecting five women to be on the committee. The probability that the committee is made up entirely of women is $\dfrac{C(6,5)}{C(10,5)} = \dfrac{1}{42} \approx 0.0238$.

b. There are $C(4,2)$ ways of selecting two men to be on the committee and $C(6,3)$ ways of selecting the remaining three committee members from the pool of women. The probability that exactly two men will serve on the committee is $\dfrac{C(4,2) \cdot C(6,3)}{C(10,5)} = \dfrac{10}{21} \approx 0.476$.

c. There are $C(6,3)$ ways of selecting two women to be on the committee and $C(4,2)$ ways of selecting the remaining two committee members from the pool of men. The probability that exactly three women will serve on the committee is $\dfrac{C(6,3) \cdot C(4,2)}{C(10,5)} = \dfrac{10}{21} \approx 0.476$.

Exercises Section 8.2

1. $P(d) = 1 - P(a) - P(b) - P(c) =$
$1 - \frac{1}{3} - \frac{1}{3} - \frac{1}{6} = \frac{1}{6}$.

Therefore, $P(d) = \frac{1}{6}$.

3. We know that $P(E) + P(a) = 1$.

So $P(a) = 1 - P(E) = 1 - \frac{1}{2} = \frac{1}{2}$

Therefore, $P(a) = \frac{1}{2}$.

5. The assignment is not valid because:

$$P(a) + P(b) + P(c) + P(d) =$$
$$\tfrac{1}{3} + \tfrac{1}{3} + \tfrac{1}{3} + \tfrac{1}{3} = \tfrac{4}{3} > 1.$$

7. The assignment is not valid because:

$$P(a)+P(b)+P(c)+P(d)=$$
$$\tfrac{1}{6}+\tfrac{1}{6}+\tfrac{1}{12}+\tfrac{1}{12}=\tfrac{1}{2}<1.$$

9.
a. Assuming the arrow does not have the possibility of landing on a dividing line, the sample space is:
$$S=\{A,B,C\}.$$

The probability of each outcome is:
$$P(A)=\tfrac{1}{4};\ P(B)=\tfrac{1}{4};\ P(C)=\tfrac{1}{2}.$$

b. $P(A\cup C)=P(A)+P(C)=\tfrac{1}{4}+\tfrac{1}{2}=\tfrac{3}{4}.$

c. $P(\sim A)=1-P(A)=1-\tfrac{1}{4}=\tfrac{3}{4}.$

11. The sample space is $S=\{s,f\}$.
The probability assignment is $P(s)=\tfrac{1}{6};\ P(f)=\tfrac{5}{6}.$

13.
a. The sample space for this experiment is: $S=\{r,g,b\}.$
The probability assignment is
$$P(r)=\tfrac{6}{15};\ P(g)=\tfrac{2}{15};\ P(b)=\tfrac{7}{15}.$$

b. $P(r\cup g)=P(r)+P(g)=\tfrac{6}{15}+\tfrac{2}{15}=\tfrac{8}{15}.$

The probability that the ball will be red or green is $\tfrac{8}{15}.$

c. $P(\sim r)=1-P(r)=1-\tfrac{6}{15}=\tfrac{9}{15}$

The probability that the ball will not be red is $\tfrac{9}{15}.$

15. The sample space for this experiment is $S=\{s,f\}$

Where $s=\{7,8\};f=\{2,3,4,5,6,9,10,11,12\}.$

$$P(s)=P(7\cup8)=P(7)+P(8)=\tfrac{6}{36}+\tfrac{5}{36}=\tfrac{11}{36}$$
$$P(f)=1-P(s)=1-\tfrac{11}{36}=\tfrac{25}{36}.$$

Therefore the probability assignment is

$$P(s)=\tfrac{11}{36};\ P(f)=\tfrac{25}{36}.$$

17. The sample space for this experiment is $S=\{s,f\}$

Where $s=$ "a heart is drawn" and $f=$ "a card other than a heart is drawn".

Since there are 13 hearts in deck of cards, the probability assignment is:

$$P(s)=\tfrac{13}{52}=\tfrac{1}{4}.$$
$$P(f)=1-P(s)=1-\tfrac{1}{4}=\tfrac{3}{4}.$$

19. The sample space for this experiment is $S=\{s,f\}$

Where $s=\{HH\}$ and $f=\{HT,TH,TT\}.$

The probability assignment is

$$P(s)=\tfrac{1}{4}.$$
$$P(f)=1-P(s)=1-\tfrac{1}{4}=\tfrac{3}{4}.$$

21.
a. The sample space is $S=\{0,1,2,3\}.$

Where the simple events in each outcome are:
$$0=\{TTT\}$$
$$1=\{HTT,THT,TTH\}$$
$$2=\{HHT,HTH,THH\}$$
$$3=\{HHH\}$$

So the probability assignment is:

$$P(0)=\tfrac{1}{8};\ P(1)=P(2)=\tfrac{3}{8};\ P(3)=\tfrac{1}{8}$$

b. The probability that at least two heads turn up is the same as the probably that two heads turn up or three heads turn up. The probability of this happening is:

$$P(2\cup3)=P(2)+P(3)=\tfrac{3}{8}+\tfrac{1}{8}=\tfrac{1}{2}.$$

The probability of at least two heads turning up is $\tfrac{1}{2}$

c. The probability that two or fewer heads will turn up is the same as the probability that exactly three heads will not turn up. The probability of this happening is:
$$P(\sim3)=1-P(3)=1-\tfrac{1}{8}=\tfrac{7}{8}.$$

The probability of two or fewer heads turning up is $\tfrac{7}{8}.$

d.

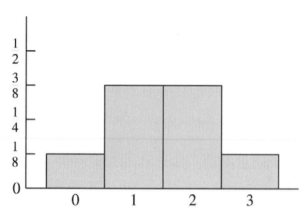

23. The probability that the selected customer purchased a large car is 0.076.

25. The probability that the selected customer purchases a mid-size or small car is $0.527 + 0.232 = 0.759$.

27.
a. The sample space for the experiment is:
$$S = \{1, 2, 3, 4, 5, 6\}.$$

Since each outcome is equally likely, the probability for each outcome is $\frac{1}{6}$.

b. The probability that the red die shows at least four pips on the top face is the same as the probability showing exactly four pips or exactly five pips or exactly six pips. The probability is:

$$P(4 \cup 5 \cup 6) = P(4) + P(5) + P(6) =$$
$$\frac{1}{6} + \frac{1}{6} + \frac{1}{6} = \frac{3}{6} = \frac{1}{2}.$$

The probability of at least four pips showing on the top face is $\frac{1}{2}$.

c. The probability of showing at most two pips on the top face is the same as showing exactly one pip on the top face or exactly two pips on the top face. The probability is:

$$P(1 \cup 2) = P(1) + P(2) = \frac{1}{6} + \frac{1}{6} = \frac{2}{6} = \frac{1}{3}$$

The probability of at most two pips showing on the top face is $\frac{1}{3}$.

d.

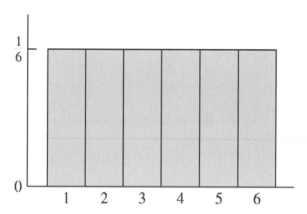

29.
a. The sample space is $S = \{0, 1, 2\}$.

There are $\dfrac{P(5,5)}{3!2!} = 10$ distinguishable arrangements of three boys and two girls. Of these ten arrangement only one (BBBGG) has two girls in the last two seats. Six of these arrangements (BBGBG, BGBBG, GBBBG, BBGGB, BGBGB,GBBGB) have exactly one girl in the last two seats. The remaining three arrangements (BGGBB, GBGBB, GGBBB) have no girls in the last two seats.

So the probability assignment is:

$$P(2) = \tfrac{1}{10}; \ P(1) = \tfrac{6}{10} = \tfrac{3}{5}; \ P(0) = \tfrac{3}{10}$$

b. The probability that at least one boy is seated in the last two seats is the same as exactly one girl sits in the last two chairs or no girl sits in the last two chairs. The probability of this happening is:

$$P(0 \cup 1) = P(0) + P(1) = \tfrac{3}{10} + \tfrac{6}{10} = \tfrac{9}{10}.$$

The probability of at least one boy sitting in the last two seats is $\tfrac{9}{10}$

c.

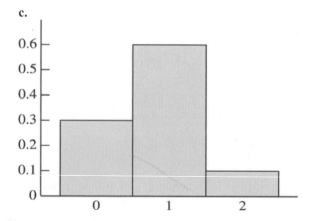

31.

a. We are counting the number of 2's in the last two digit positions of a three-digit I.D. number. There can be no 2's in the last two positions, one 2 in one of the last two positions or both of the last two positions could be 2. The sample space for this experiment is $S = \{0, 1, 2\}$

There are $10 \times 10 \times 10 = 1000$ possible I.D. numbers. There are $10 \times 9 \times 9 = 810$ different I.D. numbers that do not have a two in the last two positions. Therefore

$$P(0) = \frac{810}{1000} = 0.81$$

There are $10 \times 1 \times 9 = 90$ ways of having one two in the second to last position while not having a two in the last position and $10 \times 9 \times 1 = 90$ ways of having a two in the last position while not having a two in the second to last position. Therefore there are $90 + 90 = 180$ different I.D. numbers that have exactly one zero appearing in the last two digit positions. Therefore, $P(1) = \frac{180}{1000} = 0.18$.

There are $10 \times 1 \times 1 = 10$ different I.D. numbers that have twos in both of the last positions. Therefore,

$$P(2) = \frac{10}{1000} = 0.01.$$

The probability assignment is:

$$P(0) = 0.81;\ P(1) = 0.18;\ P(2) = 0.01.$$

b. The probability that at most one two appears in the last two positions is $P(0) + P(1) = 0.81 + 0.18 = 0.99$.

c.

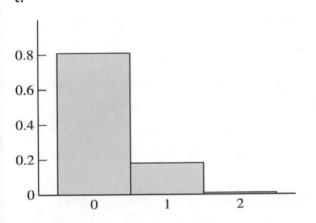

33.

a. The sample space for this experiment is $S = \{0, 1, 2\}$.

There are $C(52, 2) = 1326$ different two-card hands.

There are $C(48, 2) = 1128$ two-card hands that do not contain a king. Therefore, $P(0) = \frac{1128}{1326} \approx 0.8507$

There are $C(4, 4) \cdot C(48, 1) = 4 \cdot 48 = 192$ two-card hands that contain exactly one king. Therefore,

$$P(1) = \frac{192}{1326} \approx 0.1448 .$$

There are $C(4, 2) = 6$ two-card hands that contain two kings. Therefore, $P(2) = \frac{6}{1326} \approx 0.0045$.

The probability assignment is:

$$P(0) \approx 0.8507;\ P(1) \approx 0.1448;\ P(2) \approx 0.0045 .$$

b. The probability that at least one king will be dealt is

$$P(1) + P(2) \approx 0.1448 + 0.0045 \approx 0.1493 .$$

c.

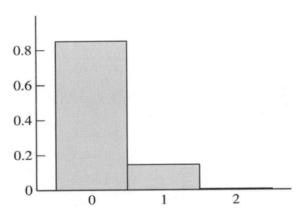

35.

a. The sample space for this experiment is $S = \{0, 1, 2\}$

There are $C(10, 2) = 45$ possible samples.

There are $C(7, 2) = 21$ samples that contain no defective computers. Therefore, $P(0) = \frac{21}{45} \approx 0.4667$.

The solution is continued on the next page.

There are $C(7,1) \cdot C(3,1) = 7 \cdot 3 = 21$ samples that contain exactly one defective computer. Therefore,

$$P(1) = \frac{21}{45} \approx 0.4667 \,.$$

There are $C(3,2) = 3$ samples that contain two defective computers. Therefore, $P(2) = \frac{3}{45} \approx 0.0667$.

The probability distribution is:

$$P(0) \approx 0.4667; \ P(1) \approx 0.4667; \ P(2) \approx 0.0667$$

b. The probability that at least one computer will be defective is

$$P(1) + P(2) \approx 0.4667 + 0.0667 \approx 0.5334 \,.$$

c.

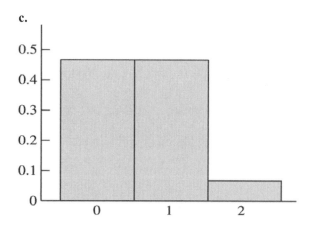

37.
a. The frequency distribution for this experiment is:

Number of Radios Owned	Relative Frequency
3	$\frac{5}{20} = 0.25$
4	$\frac{8}{20} = 0.40$
5	$\frac{7}{20} = 0.35$

b.

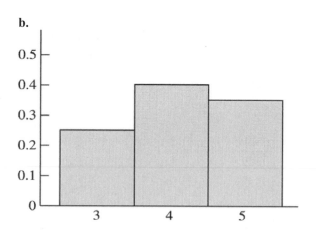

c. The probability that a household will own at least four radios is $P(4) + P(5) = 0.40 + 0.35 = 0.75$.

39.
a. The frequency distribution for this experiment is:

Number of Beans in a Pod	Relative Frequency
2	$\frac{2}{18} \approx 0.11$
3	$\frac{5}{18} \approx 0.28$
4	$\frac{8}{18} \approx 0.44$
5	$\frac{3}{18} \approx 0.17$

b.

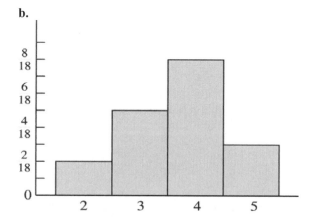

c. The probability that a pod will have 3 or more beans is
$P(3) + P(4) + P(5) = \frac{5}{18} + \frac{8}{18} + \frac{3}{18} = \frac{16}{18} = \frac{8}{9}$.

41.
a. The empirical probability distribution for this information is:

Number of Heads	Assigned Probability
0	$\frac{290}{1000} = 0.29$
1	$\frac{510}{1000} = 0.51$
2	$\frac{200}{1000} = 0.20$

b. Using the table on the previous page, the probability that at least one head will occur in two tosses of the coin is:

$$P(x \geq 1) = P(1) + P(2) = 0.51 + 0.20 = 0.71$$

43. We need to assume that all of the economists are equally well-informed, so the predictions are equally valid.

a. The empirical probability distribution for this information is:

Prediction	Assigned Probability
Recession (R)	$\frac{64}{120} \approx 0.533$
Slow Growth (SG)	$\frac{42}{120} \approx 0.350$
Higher Growth (HG)	$\frac{14}{120} = 0.117$

b. The empirical probability that there will not be a recession within the next year is:

$$P(\sim R) = P(SG \cup HG) =$$
$$P(SG) + P(HG) \approx 0.350 + 0.117 \approx 0.467.$$

45.
a. The empirical probability distribution for this information is:

Test Result	Assigned Probability
Usable (U)	0.92
Unusable (N)	0.08

b. The empirical probability that a chip selected at random is usable is:

$$P(U) = 0.92.$$

47.
a. Sixteen of the 450 people were in the 30-39 age group and renewed their lease three times, so the empirical probability of a person in this age group renewing their lease three times is $\frac{16}{450} \approx 0.036$.

b. In the age group 50-59, 46 people renewed their lease twice, 29 people renewed their lease three times, and 35 people renewed their lease four times, so $46 + 29 + 35 = 110$ people in this age group renewed their lease at least twice. The empirical probability of a person in this age group renewing their lease at least twice is $\frac{110}{450} \approx 0.244$.

c. From part (b) we know that 110 people renewed their least at least twice. We also know that 56 people total renewed their lease one time. Therefore a total of $110 + 56 = 166$ renewed their lease one time, or are in the 50-59 age group.

The empirical probability that a person is in the 50-59 or renewed their lease one time is $\frac{166}{450} \approx 0.369$.

d. There are $8 + 20 + 28 + 46 + 42 = 144$ people who renewed their lease twice, so the empirical probability of a person renewing their lease twice is $\frac{144}{450} \approx 0.32$.

49.
a. There are $100 + 82 = 182$ billing statements that are correct. The empirical probability that a statement selected at random is correct is: $\frac{182}{250} = 0.728$.

b. There are $82 + 20 + 8 = 110$ billing statements that are long-haul routes. The empirical probability that a statement selected at random is a long-haul route is: $\frac{110}{250} = 0.44$

c. There are 30 short-haul routes that were over billed. So the empirical probability that a statement selected at random was a short-haul route and over billed is $\frac{30}{250} = 0.12$.

d. There are 110 long-haul route statements, eight of those long-haul statements were under billed. 18 statements total were under billed. The number of statements that were long-haul routes or under billed is $110 + 18 - 8 = 120$. The empirical probability that a statement selected at random was a long-haul route or under billed is $\frac{120}{250} = 0.48$.

51.
a. There are four freshmen who made three or more trips to the library each week. The empirical probability that a student selected at random will be a freshman who made three or more trips to the library each week is $\frac{4}{200} = 0.02$.

b. There are eight sophomores who did not visit the library in a week, and four juniors who did not visit the library in a week. Therefore there are $8 + 4 = 12$ students who are sophomores or juniors that did not visit the library in a week. The empirical probability that a student selected at random will be a sophomore or a junior who did not visit the library once in a week is $\frac{12}{200} = 0.06$.

c. There are 20 juniors who visited the library twice a week, so the empirical probability that a student selected at random will be a junior who visited the library twice in a week is $\frac{20}{200} = 0.10$.

d. A total of 47 students visited the library three times each week. There are 88 students in the sample that were seniors. Of those 88 students, 22 visited the library 3 times a week. Therefore $47 + 88 - 22 = 113$ students are seniors or visited the library three times each week. The empirical probability that a student selected at random will be a senior or visited the library three times each week is $\frac{113}{200} = 0.565$.

53.

a. The probability that the card will be a king is $P(K) = \frac{4}{52}$. The probability that the card will not be a king is $P(\sim K) = \frac{48}{52}$. So the odds for the card being a king are:

$$\frac{P(K)}{P(\sim K)} = \frac{\frac{4}{52}}{\frac{48}{52}} = \frac{4}{48} = \frac{1}{12}$$

The odds for the card being a king are 1 to 12.

b. The probability that the card will be a spade is $P(S) = \frac{13}{52}$. The probability that the card will not be a spade is $P(\sim S) = \frac{39}{52}$. So the odds for the card being a spade are:

$$\frac{P(S)}{P(\sim S)} = \frac{\frac{13}{52}}{\frac{39}{52}} = \frac{13}{39} = \frac{1}{3}$$

The odds for the card being a spade are 1 to 3.

c. The probability that the card will be red is $P(R) = \frac{26}{52}$. The probability that the card will not be red is $P(\sim R) = \frac{26}{52}$. So the odds for the card being red are:

$$\frac{P(R)}{P(\sim R)} = \frac{\frac{26}{52}}{\frac{26}{52}} = \frac{26}{26} = \frac{1}{1}$$

The odds for the card being red are 1 to 1.

d. The probability that the card will be a club is $P(C) = \frac{13}{52}$. The probability that the card will not be a club is $P(\sim C) = \frac{39}{52}$. So the odds against the card being a club are:

$$\frac{P(\sim C)}{P(C)} = \frac{\frac{39}{52}}{\frac{13}{52}} = \frac{39}{13} = \frac{3}{1}$$

The odds against the card being a club are 3 to 1.

55.

a. The probability that the three-card will contain all spades is $P(S) = \frac{C(13,3)}{C(52,3)} = \frac{286}{22,100}$. The probability that three-card hand will not contain all spades is $P(\sim S) = 1 - P(S) = \frac{21,814}{22,100}$. So the odds for the three-card hand contain only spades are:

$$\frac{P(S)}{P(\sim S)} = \frac{\frac{286}{22,100}}{\frac{21,814}{22,100}} = \frac{286}{21,814} = \frac{11}{839}$$

The odds for the three-card hand to be all spades are 11 to 839.

b. The probability that the three-card will contain all kings is $P(K) = \frac{C(4,3)}{C(52,3)} = \frac{4}{22,100}$. The probability that the three-card hand will not contain all kings is $P(\sim K) = 1 - P(K) = \frac{22,096}{22,100}$. So the odds for the three-card hand to contain all kings are:

$$\frac{P(K)}{P(\sim K)} = \frac{\frac{4}{22,100}}{\frac{22,096}{22,100}} = \frac{4}{22,096} = \frac{1}{5524}$$

The odds for the card being a spade are 1 to 5524.

c. The probability that the three-card will contain all jacks is $P(J) = \frac{C(4,3)}{C(52,3)} = \frac{4}{22,100}$. The probability that the three-card hand will not contain all kings is $P(\sim J) = 1 - P(J) = \frac{22,096}{22,100}$. So the odds against the three-card hand containing all jacks are:

$$\frac{P(\sim K)}{P(K)} = \frac{\frac{22,096}{22,100}}{\frac{5}{22,100}} = \frac{22,096}{4} = \frac{5524}{1}$$

The odds against the three-card hand containing all jacks are 5524 to 1.

57.

a. The probability of a household owning two chainsaws is $P(2) = \frac{4}{20}$. The probability of a household not owning two chainsaws is $P(\sim 2) = \frac{16}{20}$. The odds for a household owning two chainsaws are:

$$\frac{P(2)}{P(\sim 2)} = \frac{\frac{4}{20}}{\frac{16}{20}} = \frac{4}{16} = \frac{1}{4}.$$

The odds for a household owning two chainsaws are 1 to 4.

b. The probability of a household owning at least one chainsaw is $P(x \geq 1) = \frac{14}{20}$. The probability of a household not owning at least one chainsaw is $P(x < 1) = \frac{6}{20}$. The odds for a household owning at least one chainsaw are:

$$\frac{P(x \geq 1)}{P(x < 1)} = \frac{\frac{14}{20}}{\frac{6}{20}} = \frac{14}{6} = \frac{7}{3}.$$

The odds for a household owning at least one chainsaw are 7 to 3.

c. The probability of a household owning two chainsaws is $P(2) = \frac{4}{20}$. The probability of a household not owning two chainsaws is $P(\sim 2) = \frac{16}{20}$. The odds against a household owning two chainsaws are:

$$\frac{P(\sim 2)}{P(2)} = \frac{\frac{16}{20}}{\frac{4}{20}} = \frac{16}{4} = \frac{4}{1}.$$

The odds for a household owning two chainsaws are 4 to 1.

59. If the odds of winning a free double cheeseburger are 1 to 100, the probability of winning a double cheeseburger is

$$P(W) = \frac{1}{100 + 1} = \frac{1}{101} \approx 0.009901.$$

Exercises Section 8.3

1.

a. The underlying sample space is $S = \{H, T\}$ with each outcome having a probability of $\frac{1}{2}$. The expected value of the game is $\left(\frac{1}{2}\right)(1) + \left(\frac{1}{2}\right)(-2) = -0.5$.

The expected earnings form the game is a loss of 50 cents.

b. The expected result from 1500 games is
$1500 \times (-0.5) = -750$. Therefore you should expect to lose \$750 if you play the game 1500 times.

3.

a. The sample space for this experiment is $S = \{R, G\}$. Where $P(R) = \frac{5}{11}$ and $P(G) = \frac{6}{11}$. The expected movement in this game is $\left(\frac{5}{11}\right)(4) + \left(\frac{6}{11}\right)(-3) = \frac{2}{11}$. You should expect to move forward $\frac{2}{11}$ of a space each turn.

b. In 44 turns, you should expect to move
$44\left(\frac{2}{11}\right) = 8$ spaces forward from your present position.

5. The sample space for this experiment is
$S = \{0, 1, 2\}$ Where, $P(0) = \frac{1}{4}$, $P(1) = \frac{1}{2}$, $P(2) = \frac{1}{4}$.

The expected number of boys is
$\left(\frac{1}{4}\right)(0) + \left(\frac{1}{2}\right)(1) + \left(\frac{1}{4}\right)(2) = 1$.

A two-child family should expect one boy.

7. The sample space for the experiment is
$S = \{4, 6, 9, 18\}$. Each outcome has a probability of $\frac{1}{4}$.

The expected value of the experiment is
$\left(\frac{1}{4}\right)(4) + \left(\frac{1}{4}\right)(6) + \left(\frac{1}{4}\right)(9) + \left(\frac{1}{4}\right)(18) = \frac{37}{4}$.

9. The sample space for the experiment is $S = \{R, G\}$.

The probability of drawing a red is $P(R) = \frac{6}{14} = \frac{3}{7}$. The probability of drawing a green is $P(G) = \frac{8}{14} = \frac{4}{7}$. The value of a red ball is $\$10 - \$5 = \$5$. The value of a green ball is $-\$5$.

The expected value of the game is:
$\left(\frac{3}{7}\right)(5) + \left(\frac{4}{7}\right)(-5) = \frac{-5}{7} \approx -0.71$.

The expected value of the game is to lose 71 cents.

11. The sample space for the experiment is $E = \{S, \sim S\}$.
The probability of drawing a spade is $P(S) = \frac{13}{52} = \frac{1}{4}$. The probability of not drawing a spade is $P(\sim S) = \frac{39}{52} = \frac{3}{4}$.
The value of drawing a spade is $\$5$. The value of not drawing a spade is $-\$2$.

The expected value of the game is:
$\left(\frac{1}{4}\right)(5) + \left(\frac{3}{4}\right)(-2) = \frac{-1}{4} \approx -0.25$.

You should expect to lose 25 cents if you play the game.

13. The sample space for the game is $S = \{W, L\}$, where
$W = \{6, 7, or 8\}$ and $L = \{2, 3, 4, 5, 9, 10, 11, 12\}$.

The probability of winning is
$P(W) = P(6, 7, or 8) = \frac{5}{36} + \frac{1}{6} + \frac{5}{36} = \frac{16}{36} = \frac{4}{9}$

The solution is continued on the next page.

The probability of losing is
$$P(L) = 1 - P(W) = 1 - \tfrac{4}{9} = \tfrac{5}{9}.$$

The value of winning is $\$5 - \$2 = \$3$, the value of losing is $-\$2$.

The expected value of the game is
$$\left(\tfrac{4}{9}\right)(3) + \left(\tfrac{5}{9}\right)(-2) = \tfrac{2}{9} \approx 0.22.$$

You should expect to win 22 cents when you play this game.

15. The sample space for this game is $S = \{W, L\}$, where $W = \{HH, TT\}$, and $L = \{HT, TH\}$. The probability of each outcome is $\tfrac{1}{2}$. So the expected value of matching pennies is $\left(\tfrac{1}{2}\right)(1) + \left(\tfrac{1}{2}\right)(-1) = 0$.

You should expect to break even when playing this game.

17. The probability that you will win is the probability that your number is among the 20 selected which is $\tfrac{20}{80} = \tfrac{1}{4}$. The value of winning is $\$3 - \$1 = \$2$. The value of losing is $-\$1$. The expected value of this carnival game is
$$\left(\tfrac{1}{4}\right)(2) + \left(\tfrac{3}{4}\right)(-1) = \tfrac{-1}{4} = -0.25.$$

You should expect to lose 25 cents when you play the game.

19. The sample space is $S = \{T, NT\}$, where T stands for theft, and NT stands for no theft. The probabilities of which are $P(T) = 0.2$ and $P(NT) = 0.8$. The premium of the insurance is $12 so if the C.D. player is stolen, the value of the insurance policy to Avery is $\$300 - \$12 = \$288$. If the C.D. player is not stolen, the value of the insurance policy is $-\$12$. Therefore, based on the agents estimate, the expected value is $(0.2)(288) + (0.8)(-12) = 48.00$.

The expected return on the insurance policy is $48 per year.

21.

a. The sample space for this situation is $S = \{L, D\}$ where L means the policyholder will survive for one year, and D means the policyholder will die within one year. The probabilities of each outcome in the sample space are $P(L) = 0.99$ and $P(D) = 0.01$. If the policy holder survives for one year, the insurance company will receive the premium of $\$1100$.

However, if the policyholder dies within a year, the insurance company will have to pay out the value of the policy less the premium. Therefore the insurance company will receive $\$1100 - \$100,000 = -\$98,900$.

The insurance companies expected earnings for each policy is $(0.99)(1100) + (0.01)(-98,900) = 100$.

The insurance company expects to earn $100 for each policy.

b. If the company sells 2000 of these policies they could expect $2000 \times 100 = 200,000$ dollars in income from this source.

23. The sample space for this situation is $S = \{L, D\}$ where L means the client will survive for one year, and D means the policy holder will die within one year. The probabilities of each outcome in the sample space are $P(L) = 0.92$ and $P(D) = 0.08$. If the insurance company charges x dollars for each premium, then it expects to earn x if the policyholder survives one year, and it expects to earn $x - 50,000$ if the policyholder dies within the year.

In order for insurance company expects to earn $500 for each policy the following equation must be satisfied:

$$E(X) = 500 = (0.92)x + (0.08)(x - 50,000)$$

Solving this equation for x we get:

$$500 = 0.92x + 0.08x - 4000$$
$$x = 4500.$$

The insurance company should charge a premium of $4500 if they wish to earn an average of $500 on each policy.

25. The probability of guessing the correct answer for a true-false question is $\tfrac{1}{2}$. The probability of guessing the correct answer for a five answer multiple choice question is $\tfrac{1}{5}$. The expected number of correct responses on the exam is $E(X) = \left(\tfrac{1}{2}\right)(4) + \left(\tfrac{1}{5}\right)(8) = 3.6$.

The student should expect to get 3.6 questions correct on the exam.

27. The probability of winning is $\frac{1}{38}$, the probability of losing is $\frac{37}{38}$. The value of winning is 35 and the value of losing is $-\$1$. Therefore the expected value of Roulette is

$$E(X) = \left(\tfrac{1}{38}\right)(35) + \left(\tfrac{37}{38}\right)(-1) \approx -0.053 .$$

The player should expect to lose 5.3 cents for each bet.

29. The probability distribution for the random variable X is:

		Campus	Business District
	x_i	p_i	p_i
Morning	$5	0.20	0.40
Afternoon	$10	0.30	0.50
Evening	$3	0.50	0.10

The expected income from each copy job at the campus location is:

$$E(X) = (0.20)(5) + (0.30)(10) + (0.50)(3) = 5.5$$

The expected income from each copy job at the business district location is:

$$E(X) = (0.40)(5) + (0.50)(10) + (0.10)(3) = 7.3$$

Since Antonio expects 2000 customers at either location, the copy shop would be more profitable at the business district location.

31. Using the frequency distribution table we construct an empirical probability distribution table for the random variable X

x_i	p_i
0	$\frac{5}{60} = \frac{1}{12}$
1	$\frac{7}{60}$
2	$\frac{8}{60} = \frac{2}{15}$
3	$\frac{10}{60} = \frac{1}{6}$
4	$\frac{20}{60} = \frac{1}{3}$
5	$\frac{6}{60} = \frac{1}{10}$
6	$\frac{2}{60} = \frac{1}{30}$
7	$\frac{2}{60} = \frac{1}{30}$

The solution is continued at the top of the next column.

The expected number of years of college attended is:

$$E(X) = \left(\tfrac{1}{12}\right)(0) + \left(\tfrac{7}{60}\right)(1) + \left(\tfrac{2}{15}\right)(2) + \left(\tfrac{1}{6}\right)(3) + \left(\tfrac{1}{3}\right)(4) + \left(\tfrac{1}{10}\right)(5) + \left(\tfrac{1}{30}\right)(6) + \left(\tfrac{1}{30}\right)(7) =$$

$$E(X) = 3.15$$

We expect the average person to have attended 3.5 years of college.

33. Assuming that Garrett did not offer odds to his sister, the value of the bet is 5 if Garrett wins and $-\$5$ if Garrett loses. Since the odds for his team winning are 89 to 10, the probability of his team winning is $\frac{89}{89+10} = \frac{89}{99}$. The probability of his team losing is $1 - \frac{89}{99} = \frac{10}{99}$. Therefore, Garrett's expected winnings are:

$$E(X) = \left(\tfrac{89}{99}\right)(5) + \left(\tfrac{10}{99}\right)(-5) = 3.99 .$$

Garrett expects to win $3.99 from his sister.

35. If the odds of winning the $500 dollars are 1 to 62,586, the probability of winning the $500 dollars is $\frac{1}{1+62,586} \approx 0.000016$. Thus the probability of losing is $1 - 0.000016 = .999984$. If you win, you will earn $497. If you lose, you will lose $3. Therefore the expected value of your lottery ticket is:

$$E(X) = (0.000016)(497) + (0.999984)(-3)$$
$$E(X) = -2.99 .$$

You should expect to lose $2.99 from the purchase of the lottery ticket.

37. There are three possible outcomes: zero red balls, one red ball, or two red balls. There are $C(8,2) = 28$ different samples of two balls. There are $C(3,2) = 3$ samples that contain two red balls. There are $C(5,1) \cdot C(3,1) = 15$ samples that contain exactly one red ball. There are $C(5,2) = 10$ samples where both balls are black.

a. The probability distribution table for the random variable X is shown on the next page.

The probability distribution table is:

x_i	p_i
0	$\frac{10}{28}$
1	$\frac{15}{28}$
2	$\frac{3}{28}$

b.

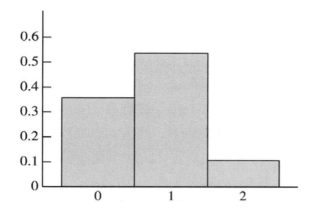

c. The expected number of red balls drawn is:

$$E(X) = \left(\tfrac{10}{28}\right)(0) + \left(\tfrac{15}{28}\right)(1) + \left(\tfrac{3}{28}\right)(2) = \tfrac{3}{4}.$$

39. Let the random variable X represent the number of kings. There are three outcomes: zero kings, one king, or two kings. The probabilities associated with the random variable are:

$$P(x = 0) = \frac{C(48,2)}{C(52,2)} = \frac{188}{221}$$

$$P(x = 1) = \frac{C(48,1) \cdot C(4,1)}{C(52,2)} = \frac{32}{221}$$

$$P(x = 2) = \frac{C(4,2)}{C(52,2)} = \frac{1}{221}.$$

The expected value of X is:

$$E(X) = \left(\tfrac{188}{221}\right)(0) + \left(\tfrac{32}{221}\right)(1) + \left(\tfrac{1}{221}\right)(2) = \tfrac{34}{221}$$

The expected number of kings drawn in a two-card hand is $\frac{34}{221}$.

41. The probability distribution table for the random variable X, where x_i is the number of red marbles drawn, is:

x_i	p_i
0	$\frac{C(3,2)}{C(8,2)} = \frac{3}{28}$
1	$\frac{C(3,1) \cdot C(5,1)}{C(8,2)} = \frac{15}{28}$
2	$\frac{C(5,2)}{C(8,2)} = \frac{10}{28}$

The expected value of the random variable is:

$$E(X) = \left(\tfrac{3}{28}\right)(0) + \left(\tfrac{15}{28}\right)(1) + \left(\tfrac{10}{28}\right)(2) = 1.25.$$

We should expect to draw 1.25 red balls from the box.

43. The probability distribution table for the random variable X, where x_i is the number of red balls drawn, is:

x_i	p_i
0	$\frac{C(11,2)}{C(12,2)} = \frac{5}{6}$
1	$\frac{C(1,1) \cdot C(11,1)}{C(12,2)} = \frac{1}{6}$
2	0

The expected value of the random variable is:

$$E(X) = \left(\tfrac{5}{6}\right)(0) + \left(\tfrac{1}{6}\right)(1) + (0)(2) = \tfrac{1}{6}.$$

We should expect to draw 0.167 red balls from the box.

Likewise, for green balls, we construct the probability distribution:

x_i	p_i
0	$\frac{C(10,2)}{C(12,2)} = \frac{45}{66}$
1	$\frac{C(2,1) \cdot C(10,1)}{C(12,2)} = \frac{10}{33}$
2	$\frac{C(2,2)}{C(12,2)} = \frac{1}{66}$

The expected value of the random variable is:

$$E(X) = \left(\tfrac{45}{66}\right)(0) + \left(\tfrac{10}{33}\right)(1) + \left(\tfrac{1}{66}\right)(2) = \tfrac{1}{3}.$$

Therefore the expected number of green balls in our sample of two is $\frac{1}{3}$.

45. The probability distribution for the random variable X, where x_i is the number of hearts drawn in a two-card hand is:

x_i	p_i
0	$\frac{C(39,2)}{C(52,2)} = \frac{741}{1326} = \frac{19}{34}$
1	$\frac{C(39,1) \cdot C(13,1)}{C(52,2)} = \frac{507}{1326} = \frac{13}{34}$
2	$\frac{C(13,2)}{C(52,2)} = \frac{78}{1326} = \frac{1}{17}$

The expected number of spades is:

$$E(X) = \left(\tfrac{19}{34}\right)(0) + \left(\tfrac{13}{34}\right)(1) + \left(\tfrac{1}{17}\right)(2) = \tfrac{1}{2}.$$

The expected number of hearts in a two-card hand is $\frac{1}{2}$.

47. There are six possible outcomes:
$$S = \{(2,4);(2,9);(4,2);(4,9);(9,2);(9,4)\}$$

The possible values for the random variable X are:

$$X = \{18, 33, 39\}.$$

The probability distribution for the random variable X is:

S	x_i	p_i
$(2,4);(4,2)$	18	$\frac{1}{3}$
$(2,9);(9,2)$	33	$\frac{1}{3}$
$(4,9);(9,4)$	39	$\frac{1}{3}$

The expected value of the random variable X is:

$$E(X) = \left(\tfrac{1}{3}\right)(18) + \left(\tfrac{1}{3}\right)(33) + \left(\tfrac{1}{3}\right)(39) = 30.$$

The expected value of the experiment is 30.

49. The random variable X contains the values
$X = \{2,3,4,5,6,7,8,9,10,11,12\}$.

The probability distribution for the random variable X is displayed at the top of the next column.

S	x_i	p_i
$(1,1)$	2	$\frac{1}{36}$
$(1,2);(2,1)$	3	$\frac{2}{36} = \frac{1}{18}$
$(1,3);(2,2);(3,1)$	4	$\frac{3}{36} = \frac{1}{12}$
$(1,4);(2,3);(3,2),(4,1)$	5	$\frac{4}{36} = \frac{1}{9}$
$(1,5);(2,4);(3,3);(4,2);(5,1)$	6	$\frac{5}{36}$
$(1,6);(2,5);(3,4);(4,3);(5,2);(6,1)$	7	$\frac{6}{36} = \frac{1}{6}$
$(2,6);(3,5);(4,4);(5,3);(6,2)$	8	$\frac{5}{36}$
$(3,6);(4,5);(5,4);(6,3)$	9	$\frac{4}{36} = \frac{1}{9}$
$(4,6);(5,5);(6,4)$	10	$\frac{3}{36} = \frac{1}{12}$
$(5.6);(6,5)$	11	$\frac{2}{36} = \frac{1}{18}$
$(6,6)$	12	$\frac{1}{36}$

The expected value of X is:

$$E(X) = \left(\tfrac{1}{36}\right)(2) + \left(\tfrac{1}{18}\right)(3) + \left(\tfrac{1}{12}\right)(4) + \left(\tfrac{1}{9}\right)(5) + $$
$$\left(\tfrac{5}{36}\right)(6) + \left(\tfrac{1}{6}\right)(7) + \left(\tfrac{5}{36}\right)(8) + \left(\tfrac{1}{9}\right)(9) + $$
$$\left(\tfrac{1}{12}\right)(10) + \left(\tfrac{1}{18}\right)(11) + \left(\tfrac{1}{36}\right)(12) = 7$$

$$E(X) = 7.$$

The expected sum of the pips showing on the top face of a pair of dice is seven.

Chapter 8 Summary Exercises

1. The number of outcomes is $52 \times 6 = 312$.

3. There are eight ways to draw a queen or a king from a standard deck. There are 52 ways of drawing a single card. The probability of drawing a queen or a king from a standard deck is $\frac{8}{52} = \frac{2}{13}$. There are 44 ways of not drawing a king or a queen from a standard deck. The probability of not drawing a king or a queen is $\frac{44}{52} = \frac{11}{13}$. The odds for drawing a queen or a king are $\dfrac{\frac{2}{13}}{\frac{11}{13}} = \dfrac{2}{11}$.

The odd in favor of drawing a queen or a king are 2 to 11.

5. There are $C(29,5) = 118,755$ different 5 person groups that can be formed. There are $C(17,2) \cdot C(12,3) = 29,920$ ways of forming a committee with two women and three men. The probability that the committee will have two women and three men on it is $\frac{29,920}{118,755} \approx 0.2519$.

7. The sample space for this experiment is $S = \{Success, Failure\}$, where

$Success = \{6,7,8,9,10,11,12\}$ and

$Failure = \{2,3,4,5\}$.

9. The probability distribution for the random variable X, where x_i is the number of girls in a three-child family, is displayed below:

S	x_i	p_i
BBB	0	$\frac{1}{8}$
BBG;BGB;GBB	1	$\frac{3}{8}$
BGG;GBG;GGB	2	$\frac{3}{8}$
GGG	3	$\frac{1}{8}$

The expected value for this random variable is:

$$E(X) = \left(\tfrac{1}{8}\right)(0) + \left(\tfrac{3}{8}\right)(1) + \left(\tfrac{3}{8}\right)(2) + \left(\tfrac{1}{8}\right)(3) = \tfrac{3}{2}.$$

A family with three children should expect 1.5 girls.

11. The expected number of DVD's can only be found if we assume that the people in the category of five or more DVD's per month only rent five. In other words, the expected value that we compute will be the lower bound for the actual expected value. Finding the empirical probabilities for each group, we see that the expected value is:

$$E(X) = \left(\tfrac{5}{50}\right)(0) + \left(\tfrac{7}{50}\right)(1) + \left(\tfrac{6}{50}\right)(2) + \left(\tfrac{8}{50}\right)(3) +$$
$$\left(\tfrac{11}{50}\right)(4) + \left(\tfrac{13}{50}\right)(5) = 3.04$$

The expected number of DVD's rented each month is 3.04.

13. The expected value to the company is:

$$E(X) = (0.98)(1800) + (0.02)(-48,200)$$
$$E(X) = 1764 - 964 = 800.$$

The company expects to earn \$800 on each policy.

15. If $P(E) = 1$, then the event is a certain to happen. Examples of such an event are "Drawing a card from a single deck of cards and the event E is 'the card is red or black' " or "A single marble is drawn from a bag containing six red marbles and the event E is 'a red marble is drawn' " (Note: we assume a finite sample space in this explanation)

17. Each outcome in an experiment is assigned a value by the random variable, and that value is weighted by the probability of the outcome occurring. The expected value of a random variable is the long run average value observed during the repeated experiment.

19. There are $C(52,5)$ ways of drawing a five-card hand with out replacement. In order to draw a hand that contains the king of spades, exactly one other king, and exactly two queens, we observe the following. There is only one way to draw the king of spades. There are $C(3,1)$ to draw exactly one other king. There are $C(4,2)$ ways to draw exactly two queens. Finally there are $C(44,1)$ ways to draw the final card in such a way that it is not a king or a queen. Therefore the probability of drawing the hand described is:

$$\frac{C(3,1) \cdot C(4,2) \cdot C(44,1)}{C(52,5)}$$

In other words answer B.

21. The expected value for each box of bagels is calculated as follows:

For the first 150 boxes the expected value per box is:

$$E(X) = (0.7)(4) + (0.3)(-1) = 2.5$$

So the expected value is $150 \times 2.5 = 375$

For the additional 200 boxes the distributor purchases the expected value is:

$$E(X) = (0.3)(4) + (0.7)(-1) = 0.5$$

Therefore, if the distributor decides to by 200 boxes, his expected value is $150 \times 2.5 + 200 \times 0.5 = 475$ or \$475 dollars.

Cumulative Review

23.

$$A^2 - 3A = \begin{bmatrix} -2 & 0 & 9 \\ 4 & 4 & -8 \\ -3 & 0 & 13 \end{bmatrix} - \begin{bmatrix} 3 & 0 & -9 \\ 0 & 6 & 12 \\ 3 & 0 & -12 \end{bmatrix}$$

$$= \begin{bmatrix} -5 & 0 & 18 \\ 4 & -2 & -20 \\ -6 & 0 & 25 \end{bmatrix}$$

25. Computing the probability for distribution table for the random variable X, where x_i is the number of girls in the four person family, we see:

x_i	p_i
0	$\frac{1}{16}$
1	$\frac{4}{16}$
2	$\frac{6}{16}$
3	$\frac{4}{16}$
4	$\frac{1}{16}$

The probability of having four girls is $\frac{1}{16}$. The probability that one will be a girl and three will be a boy is $\frac{4}{16}$. Also the probability that three will be girls and one will be a boy is $\frac{4}{16}$. Therefore the probability of having three children of one sex and one of the other is $\frac{8}{16}$. The probability of having two girls and two boys is $\frac{6}{16}$.

Therefore the most likely outcome in a family of four children is to have three of one sex and one of the other.

Chapter 8 Sample Test Answers

1. If we assume that the marbles are distinguishable and are drawn without replacement, then there are $C(8,3) = 56$ equally likely outcomes in the sample space.

2. There are $4! = 24$ seating arrangements. There are $2 \cdot 1 \cdot 2 \cdot 1 = 4$ arrangements with 2 boys on the left and 2 girls on the right. There are also 4 arrangements with two girls on the left and two boys on the right. The probability that the boys will sit next to each other and the girls will sit next to each other is $\frac{8}{24} = \frac{1}{3}$

3. There are $2^4 = 16$ possible outcomes. There are $C(4,2) = 6$ outcomes that have two heads and two tails. Therefore the probability of getting two heads and two tails on four tosses is $\frac{6}{16}$. The probability of not getting two heads and two tails on four tosses is $\frac{10}{16}$. The odds in favor of getting two heads and two tails on four tosses is

$$\frac{\frac{6}{16}}{\frac{10}{16}} = \frac{6}{10} = \frac{3}{5} \text{ or 3 to 5.}$$

4. The odds against E are 5 to 9 so the odds for E are 9 to 5. Therefore the probability of E is $P(E) = \frac{9}{9+5} = \frac{9}{14}$.

5. There are 36 possible outcomes. Five outcomes result in the sum of six. The probability that a sum of six will be rolled is $\frac{5}{36}$.

6. The probability that both dice will show four pips on the top side is $\frac{1}{36}$.

7. The probability that the green die will not show two pips on the top side is $\frac{5}{6}$.

8. The probability that none of the cards will be aces is:
$$\frac{C(48,3)}{C(52,3)} = \frac{4324}{5525} \approx 0.7826.$$

9. The probability that exactly two of the cards will be diamonds is: $\frac{C(13,2) \cdot C(39,1)}{C(52,3)} = \frac{117}{850} \approx 0.1376.$

10. The probability that all three cards will be from the same suit is: $\frac{4 \cdot C(13,3)}{C(52,3)} = \frac{22}{425} \approx 0.0518.$

11. The probability that exactly two calculators will be defective is: $\frac{C(3,2) \cdot C(5,1)}{C(8,3)} = \frac{15}{56} \approx 0.2679.$

12. The probability that zero calculators are defective is:
$$\frac{C(5,3)}{C(8,3)} = \frac{10}{56}.$$

Therefore the probability that at least one calculator is defective is:
$$1 - \frac{10}{56} = \frac{46}{56}.$$

13. The probability distribution histogram is:

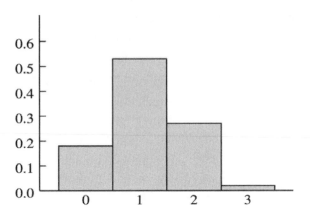

14. There are $2^5 = 32$ possible combinations of boys and girls for a five-child family. There are $C(5,2) = 10$ combinations that have exactly two girls. Therefore the probability that a five-child family will have exactly two girls is: $\frac{10}{32} = 0.3125$.

15. There are $C(5,2) = 10$ ways of having two boys, $C(5,1) = 5$ ways of having one boy, and $C(5,0) = 1$ way of having zero boys. This means that there are $10 + 5 + 1 = 16$ combinations of children that have at most two boys. The probability that a five-child family will have at most two boys is: $\frac{16}{32} = 0.5$.

16. The expected number of girls is:

$$E(X) = \left(\tfrac{1}{32}\right)(0) + \left(\tfrac{5}{32}\right)(1) + \left(\tfrac{10}{32}\right)(2) +$$
$$\left(\tfrac{5}{32}\right)(4) + \left(\tfrac{1}{32}\right)(5) = \tfrac{80}{36}$$

$$E(X) = \tfrac{80}{36} = 2.5.$$

A family of five children should expect 2.5 girls.

17. There are $C(52,2) = 1326$ different two-card hands.

There are $C(4,2) = 6$ different two-card hands with two jacks. The probability of drawing a hand with two jacks is $\frac{6}{1326} = \frac{1}{221}$.

There are $C(4,1) \cdot C(48,1) = 192$ different two-card hands that contain one jack. The probability of drawing a two-card hand that has one jack is $\frac{192}{1326} = \frac{32}{221}$.

There are $C(48,2) = 1128$ different two-card hands that have no jacks. The probability of drawing a two-card hand with no jacks is $\frac{1128}{1326} = \frac{188}{221}$.

Thus the expected number of jacks in a two-card hand is:

$$\left(\tfrac{188}{221}\right)(0) + \left(\tfrac{32}{221}\right)(1) + \left(\tfrac{1}{221}\right)(2) = \tfrac{34}{221} = \tfrac{2}{13}.$$

18. If a student guesses at every question, the probability of getting a true-false question correct is $\frac{1}{2}$, and the probability of getting a five-answer multiple-choice question correct is $\frac{1}{5}$. The expected number of correct responses on an exam that has 60 true-false questions and 40 multiple-choice questions is $\left(\tfrac{1}{2}\right)(60) + \left(\tfrac{1}{5}\right)(40) = 38$.

The student should expect to answer 38 questions out of 100 correctly.

Chapter 9
Additional Topics in Probability

c. $P(A' \cup B) = 0.5$

Exercises Section 9.1

3. We construct the Venn diagram as follows:

1. We construct the Venn diagram as follows:

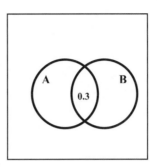

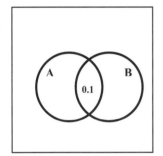

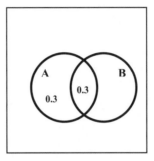

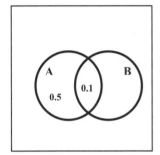

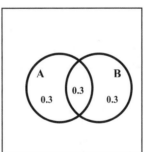

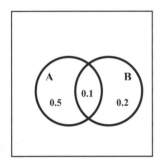

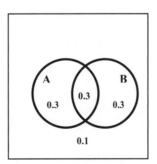

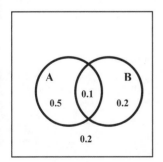

We can now state the desired probabilities:

a. $P(A \cup B) = 0.9$

b. $P(A' \cap B) = 0.3$

c. $P(A \cup B') = 0.7$

We can now state the desired probabilities:

a. $P(A \cup B) = 0.8$

b. $P(A \cap B') = 0.5$

5. We construct the Venn diagram as follows: Since
$P(A \cup B) = 0.8$ and $P(B) = 0.7$,
then $P(A - B) = 0.8 - 0.7 = 0.1$. Since $P(A) = 0.4$ it
follows that $P(A \cap B) = 0.4 - 0.1 = 0.3$

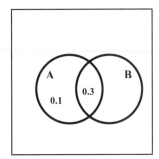

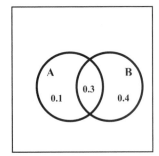

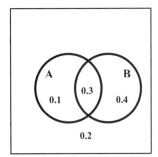

We can now state the desired probabilities:

a. $P(A \cap B) = 0.3$

b. $P(A' \cap B') = 0.2$

c. $P(A' \cup B') = 0.7$

7. We construct the Venn diagram as follows:

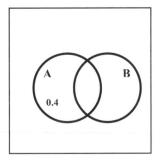

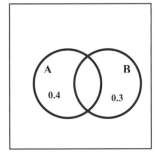

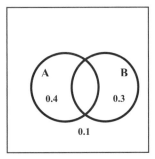

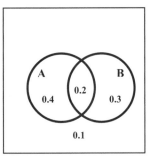

We can now state of the desired probabilities:

a. $P(A) = 0.6$

b. $P(A \cup B) = 0.9$

c. $P\left((A \cap B)' \right) = 0.8$

9. Let N represent the set of people who get their news from newspapers and T represent the set of people who get their news from television. We construct the Venn diagram as follows:

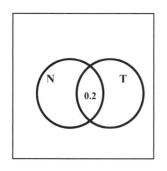

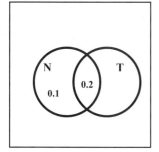

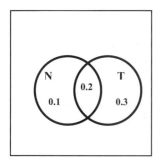

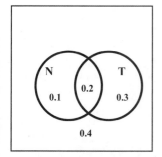

The probability that a person will get their news through a newspaper or a television is $0.1 + 0.2 + 0.3 = 0.6$.

11. Let F represent the students who ate in a fast-food place and C represent the students who ate in the cafeteria. We construct the Venn diagram as:

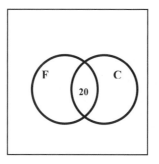

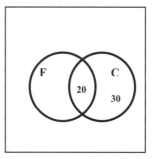

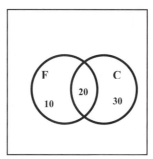

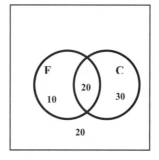

The probability that a student ate in fast-food place is $\frac{30}{80} = \frac{3}{8} = 0.375$.

13. Let B represent the households that own a bicycle, S represent the households that own a skateboard and BB represent the households that own a baseball bat. We construct the Venn diagram on the following page.

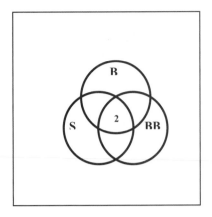

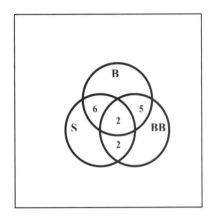

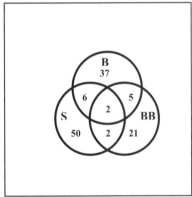

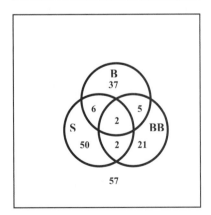

The solution is continued at the top of the next column.

a. There are 86 households that own a skateboard or a baseball bat, so the probability that a household will own a skateboard or a baseball bad is $\frac{86}{180} = \frac{43}{90} \approx 0.4778$.

b. There are $50 + 6 + 37 = 93$ households that have a bicycle or a skateboard but do not own a baseball bat. The probability that a household will own a bicycle or a skateboard but not a baseball bat is $\frac{93}{180} = \frac{31}{60} \approx 0.5167$.

c. There are $6 + 5 + 2 = 13$ households that own exactly two of these items. The probability that a household will own exactly two of these items is $\frac{13}{180} \approx 0.0722$.

15.
a. Yes, it is possible. Since the events are mutually exclusive, if $P(B) = 0.3$ then the situation would be valid.

b. No, it is not possible for $P(A \cap B)$ to be greater than $P(A)$.

c. It is not possible for $P(A \cap B)$ to be greater than $P(B)$.

17. There are $90 + 50 + 15 = 155$ people in the town that are either employed or female. Therefore the probability that a person selected at random in this town is either employed or female is $\frac{155}{165} = \frac{31}{33} \approx 0.9394$

19. There are $90 + 50 = 140$ people in the town who are employed. Therefore the probability that person selected at random is employed is $\frac{140}{165} = \frac{28}{33} \approx 0.8485$.

21. There were $4542 + 125 + 856 = 5523$ people who were female or killed unintentionally in 1997. Therefore the probability that a person who was killed by a firearm was female or killed unintentionally is $\frac{5523}{32166} \approx 0.1717$.

23. There were $26,643 + 856 = 27,497$ males whose death involved a firearm in 1997. Therefore the probability that a male was involved in a fatal shooting is $\frac{27,497}{32,166} \approx 0.8548$.

25. There were $10 + 1 = 11$ shells that had three or more peanuts. The probability that a shell has three or more peanuts in it is $\frac{11}{20} = 0.55$.

27.

a. There are eight cards that are a 5 or an 8. The probability that card drawn from a standard deck is either a 5 or an 8 is $\frac{8}{52} = \frac{2}{13}$.

b. The events are not mutually exclusive because of the jack of clubs. Therefore, there are $13 + 4 - 1 = 16$ cards that are either a jack or a club. The probability that a card drawn from a standard deck will be a jack or a club is $\frac{16}{52} = \frac{4}{13}$.

c. There is only one card that is a jack and a club. The probability that a card drawn from a standard deck will be the jack of clubs is $\frac{1}{52}$.

d. As in part b, there are 16 cards that are an Ace or a spade, so the probability is $\frac{16}{52} = \frac{4}{13}$.

29.

a. Since each of the children has an equal probability of seating in any seat, the probability that a boy will occupy the far left end seat is $\frac{4}{7}$.

b. There are $4 \bullet P(6,6) = 2880$ ways that a boy can occupy the left end seat. There are $3 \bullet P(6,6) = 2160$ ways that a girl can occupy the right end seat. There are $3 \bullet 4 \bullet P(5,5) = 1440$ ways that a girl can occupy the right end seat and a boy can occupy the left end seat. By the inclusion-exclusion principle there are $2880 + 2160 - 1440 = 3600$ different ways a boy can occupy the left end seat or a girl can occupy the right end seat. Since there are $P(7,7) = 5040$ different ways to arrange the children, the probability that a boy will occupy the left end seat or a girl will occupy the right end seat is $\frac{3600}{5040} = \frac{5}{7}$.

c. The reasoning and the computation are identical to part b. The probability is $\frac{3600}{5040} = \frac{5}{7}$.

31.

a. Since each of the balls has an equally likely chance of being drawn first, the probability that a red ball will be on the left end of the row is $\frac{2}{5}$.

b. There is only one way that the balls can alternate colors $GRGRG$. There is also only one way that the first two balls on the left end are red $RRGGG$. Since there are $C(5,3) = 10$ ways of arranging the balls, the probability that the balls will alternated color or the first two balls on the left end will be red is $\frac{2}{10} = \frac{1}{5}$.

33.

a.

$$P(at\ least\ 1\ head) = P(1\ head) + P(2\ heads)$$

$$P(at\ least\ 1\ head) = \frac{1}{2} + \frac{1}{4} = \frac{3}{4}$$

b.

$$P(at\ least\ 1\ head) = 1 - P(0\ heads)$$

$$P(at\ least\ 1\ head) = 1 - \frac{1}{4} = \frac{3}{4}$$

35.

a.

$$P(at\ least\ a\ 3) = P(3) + P(4) + P(5) + P(6)$$
$$P(7) + P(8) + P(9) + P(10) + P(11) + P(12)$$

$$P(at\ least\ a\ 3) = \frac{2}{36} + \frac{3}{36} + \frac{4}{36} + \frac{5}{36} + \frac{6}{36} + \frac{5}{36} +$$
$$\frac{4}{36} + \frac{3}{36} + \frac{2}{36} + \frac{1}{36} =$$

$$P(at\ least\ a\ 3) = \frac{35}{36}.$$

b.

$$P(at\ least\ a\ 3) = 1 - P(2)$$

$$P(at\ least\ a\ 3) = 1 - \frac{1}{36} = \frac{35}{36}.$$

37. There are $C(52,3) = 22,100$ different three-card hands. There are $C(48,3) = 17,296$ different three-card hands with no kings (and three non-kings). There are $C(4,1) \bullet C(48,2) = 4512$ different three-card hands with exactly one king (and two non-kings). There are $C(4,2) \bullet C(48,1) = 288$ different three-card hands with exactly two kings (and one non-king). There are $C(4,3) = 4$ different three-card hands with exactly three kings. We are now ready to compute the probability of being dealt at least one king.

a.

$$P(at\ least\ 1\ K) = P(1) + P(2) + P(3)$$
$$P(at\ least\ 1\ K) = \frac{4512}{22,100} + \frac{288}{22,100} + \frac{4}{22,100} = \frac{4804}{22,100}$$
$$P(at\ least\ 1\ K) = \frac{4804}{22,100} = \frac{2101}{5525}.$$

The solution is continued on the next page.

b.

$$P(at\ least\ 1K) = 1 - P(0)$$

$$P(at\ least\ 1K) = 1 - \frac{17,296}{22,100} = \frac{4804}{22,100} = \frac{1201}{5525}.$$

39. Each slip of paper has a probability of $\frac{1}{6}$ of being drawn. The probability of drawing a number greater than two, which is the same as drawing at least a three is:

a.

$$P(at\ least\ 3) = P(3) + P(4) + P(5) + P(6)$$

$$P(at\ least\ 3) = \frac{1}{6} + \frac{1}{6} + \frac{1}{6} + \frac{1}{6} = \frac{4}{6} = \frac{2}{3}.$$

b.

$$P(at\ least\ 3) = 1 - P(1) - P(2)$$

$$P(at\ least\ 3) = 1 - \frac{1}{6} - \frac{1}{6} = \frac{4}{6} = \frac{2}{3}.$$

41.
a. There is one way to roll a sum of two, two ways to roll a sum of three, three ways to roll a sum of four. So there are six total ways to roll a number less than five. There are 36 total outcomes. The probability that a roll has a sum less than five is $P(x < 5) = \frac{6}{36} = \frac{1}{6}$.

b. From part (a) we know that there are six ways to roll a sum less than five, and there are also three ways to roll a sum greater than ten. So the probability that a roll has a sum of less than five is $\frac{1}{6}$. The probability that a roll has a sum that is greater than ten is $\frac{3}{36} = \frac{1}{12}$. The probability that a roll has a sum that is less than five or greater than ten is $P(x < 5\ or\ x > 10) = \frac{1}{6} + \frac{1}{12} = \frac{3}{12} = \frac{1}{4}$.

c. The probability that the sum is greater than five or less than eight is one. It stands to reason that if a number is not greater than five, it must be less than eight, and if the number is not less than eight, it must be greater than five.

d. The probability that a sum greater than two is rolled is the same as finding the probability that a two isn't rolled.

$$P(x > 2) = 1 - P(2) = 1 - \frac{1}{36} = \frac{35}{36}.$$

e. The probability that a sum greater than three is rolled is the same as finding the probability that a two or a three is not rolled. $P(x > 3) = P\left((2 \cup 3)'\right) = 1 - P(2 \cup 3)$

$$P(x > 3) = 1 - \left(\frac{1}{36} + \frac{2}{36}\right) = 1 - \frac{1}{12} = \frac{11}{12}.$$

f. There are no ways a sum of one can be rolled with a pair of standard dice. Therefore the probability that a sum of 1 is rolled is zero.

43.
a. There are 36 equally likely outcomes in the sample space. There is one $\{(1,1)\}$ outcome where the sum is two and there are four $\{(3,6),(4,5),(5,4),(6,3)\}$ outcomes where the sum is nine. The probability that a two or a nine will be rolled is

$$P(2 \cup 9) = P(2) + P(9) = \frac{1}{36} + \frac{4}{36} = \frac{5}{36}.$$

b. The probability that the sum on the top faces is greater than five is $\frac{26}{36} = \frac{13}{18}$. Since seven is a sum greater than five, it is included in this case. Therefore the probability that the sum is greater than five or is a seven is $\frac{13}{18}$

c. There are six ways that the red die can show four pips on the top face. There are 21 ways that the sum of the dice can be greater than six. There are four ways that the red die can show four pips on the top face and the sum of the pips on the dice is greater than six. By the inclusion-exclusion principle there are $21 + 6 - 4 = 23$ ways that the red die can show four pips on the top face or the sum of the dice can be six. The probability that the red die will show four pips on the top face or the sum of the dice will be six is $\frac{23}{36}$.

45.
a. There are $C(52,3)$ possible three-card hands. There are $C(39,3)$ three-card hands that do not contain any clubs. The probability that the three-card hand will contain at least one club is $1 - \frac{C(39,3)}{C(52,3)} = \frac{997}{1700}$.

b. There are $C(4,3)$ different three-card hands that contain all kings. Likewise, there are $C(4,3)$ different three-card hands that contain all queens. The probability that a three-card hand will contain all kings or will contain all queens is $\frac{C(4,3)}{C(52,3)} + \frac{C(4,3)}{C(52,3)} = \frac{8}{22,100} \approx 3.62 \times 10^{-4}$.

c. There are $C(12,3)$ different three-card hands that contain all face cards. The four ways that a three-card hand contains all kings is a subset of the group that contains all face cards, so there are $C(12,3)$ different three-card hands that contain all face cards or all kings. The probability that a three-card hand will contain all face cards or all kings is $\frac{C(12,3)}{C(52,3)} = \frac{220}{22,100} = \frac{11}{1105} \approx 0.00995$.

47.

a. There are $C(9,4) = 126$ possible committees. There are $C(6,4) = 15$ committees with zero sophomores. There are $C(6,3) \cdot C(3,1) = 60$ committees with exactly one sophomore on it. The probability that there will be at least two sophomores on the committee is:

$$P(s \geq 2) = 1 - P(0) - P(1) = 1 - \tfrac{15}{126} - \tfrac{60}{126} = \tfrac{17}{42}.$$

b. As seen in part (a), the probability that the sophomores will have at least one representative on the committee is:

$$P(s \geq 1) = 1 - P(0) = 1 - \tfrac{15}{126} = \tfrac{37}{42}$$

c. The probability that the sophomores will not have any representatives on the committee is $P(0) = \tfrac{15}{126} = \tfrac{5}{42}$.

49. There are $C(150,5)$ different ways of selecting five children from the school. There are $C(120,5)$ ways of selecting five children who do not have vision problems. Therefore, the probability that at least one child among the five selected will have vision problems is

$$1 - \tfrac{C(120,5)}{C(150,5)} \approx 0.67785.$$

51.

a. Since it is equally likely that the first child is a boy or a girl, the probability that the first child is a boy is $\tfrac{1}{2}$.

b. There are eight equally likely combinations of boys and girls in a three-child family. There are six $\{BBB, BBG, BGB, BGG, GBB, GGB\}$ different ways in which the first child is a boy or the third child is a boy. The probability that the first child is a boy or the third child is a boy is $\tfrac{6}{8} = \tfrac{3}{4}$.

c. There are two $\{BBB, BGB\}$ different combinations where the first child is a boy and the third child is a boy. The probability that the first child is a boy and the third child is a boy is $\tfrac{2}{8} = \tfrac{1}{4}$.

d. There is only one combination of children that has no girls. Therefore the probability that the family will have at least one girl is $1 - P(0) = 1 - \tfrac{1}{8} = \tfrac{7}{8}$.

53. There are only four outcomes, they are:
$$\{(3B, 0W); (2B, 1W); (1B, 2W); (0B, 3W)\}.$$

Each of these outcomes results in a matching pair, so the probability that a matching pair of socks (in color) will be chosen is one.

Exercises Section 9.2

1.

a. The sample space is $S = \{1, 2, 3, 4, 5, 6\}$

b. There are three possible outcomes in the reduced sample space $S^* = \{1, 3, 5\}$. The probability that three pips will show on the top face of the die, given an odd number was rolled is $P(3 \mid odd) = \tfrac{1}{3}$.

c. $P(3 \mid odd) = \dfrac{\frac{1}{6}}{\frac{3}{6}} = \tfrac{1}{3}$

3.

a. The sample space is $S = \{1, 2, 3, 4, 5, 6\}$

b. There are three possible outcomes in the reduced sample space $S^* = \{1, 3, 5\}$. Since two is not a possible outcome given that an odd number was rolled, the probability that two was rolled is $P(2 \mid odd) = 0$.

c. $P(2 \mid odd) = \dfrac{0}{\frac{3}{6}} = 0$

5.

a. The sample space is:
$$S = \left\{ \begin{array}{l} HHH, HHT, HTH, HTT, \\ THH, THT, TTH, TTT \end{array} \right\}$$

b. Since we were given that at least one head occurred, there are seven possible outcomes. There are three outcomes that have exactly two tails, namely $S^* = \{HTT, THT, TTH\}$. The conditional probability that exactly two tails appeared, given that at least one head appeared is $\tfrac{3}{7}$.

c. $P(exactly\ 2T \mid at\ least\ 1H) = \dfrac{\frac{3}{8}}{\frac{7}{8}} = \tfrac{3}{7}$

7.
a. The sample space is:

$$S = \left\{ \begin{array}{l} HHH, HHT, HTH, HTT, \\ THH, THT, TTH, TTT \end{array} \right\}$$

b. Given that exactly one head occurred, there are three possible outcomes in the reduced sample space $S^* = \{HTT, THT, TTH\}$. There is only one of the three possible outcomes in which the second toss resulted in a head. Therefore the conditional probability that a head resulted on the second toss given the condition that exactly one head appeared is $\frac{1}{3}$.

c. $P\left(2^{nd}\, H \,\middle|\, exactly\, 1H \right) = \dfrac{\frac{1}{8}}{\frac{3}{8}} = \frac{1}{3}$

9. The reduced sample space has 51 cards, including four jacks. The probability that the second card was a jack, given that the first card was not a jack is $\frac{4}{51}$.

11. The reduced sample space has 51 cards. Since an ace is not a face card, there are still four aces to choose from. The conditional probability that the second card is an ace, given that the first card was a face card is $\frac{4}{51}$.

13. The reduced sample space has 51 cards. If we know that the first card was the nine of diamonds, then there are three nines left to choose from. The conditional probability that the second card was a nine, given that the first card was a nine and a diamond is $\frac{3}{51}$.

15. Since there is replacement, the sample space for each draw is 52 cards, and all of the cards are available to be drawn. The deck has no memory, so the probability that the second card is a heart, given that the first card and the third card were hearts is $\frac{13}{52} = \frac{1}{4}$.

17.
a. Since there is replacement, the sample space for each draw contains 14 marbles and all the marbles are available for each draw. The probability that the second marble was green given that the first marble was green is $\frac{9}{14}$.

b. Since there is replacement, the sample space for each draw contains 14 marbles and all the marbles are available for each draw. The probability that the second marble was red given that the first marble was red is $\frac{5}{14}$.

c. Since there is replacement, the sample space for each draw contains 14 marbles and all the marbles are available for each draw. The probability that the second marble was red given that the first marble was green is $\frac{5}{14}$.

19. Since 70 TV sets with 28-inch screens were in inventory in November, and of those, 30 TV sets were NuView brand, the probability that a TV set was NuView brand given that it was a 28-inch screen is $\frac{30}{70} = \frac{3}{7}$.

21. There are 52 VuTech brand TV's in inventory. of those 52 TV's, 12 of them have 36-inch screens. The conditional probability that a TV will have a 36-inch screen given that it is a VuTech brand is $\frac{12}{52} = \frac{3}{13}$.

23. There are a total of 90 TV's in inventory. Of those 90 TV's, 38 are NuView brand. The probability that at TV in inventory will be a NuView brand is $\frac{38}{90} = \frac{19}{45}$.

25. There are 40 in the class and 20 are freshmen so the probability that a student selected at random is a freshman is $P\left(F \right) = \frac{20}{40} = \frac{1}{2}$.

27. There are 20 freshmen in the class. Five students made an A in the class. Two students were freshman and made an A. The probability that a student selected at random is a freshman or made an A is

$$P\left(F \cup A \right) = P\left(F \right) + P\left(A \right) - P\left(F \cap A \right)$$
$$P\left(F \cup A \right) = \frac{20}{40} + \frac{5}{40} - \frac{2}{40} = \frac{23}{40}.$$

29. There are 20 sophomores in the class. Of those 20 sophomores, eight made a C. So the conditional probability that a student selected at random made a C, given that they are a student is $P\left(C \,\middle|\, S \right) = \frac{8}{20} = \frac{2}{5}$.

31. There are 12 females who placed ads. Of those 12 females, four communicated with a lover while two were looking for a mate. The conditional probability that the add was placed for the purpose of communicating with a lover or looking for a mate, given that the ad was placed by a female is $P\left(C \cup L \,\middle|\, F \right) = \frac{6}{12} = \frac{1}{2}$.

33. The sample space has 25 possibilities. Seven ads were placed to communicate with a lover and eight ads were placed looking for a mate. The probability that an ad was placed to communicate with a lover or to seek a mate is

$$P\left(C \cup L \right) = \frac{15}{25} = \frac{3}{5}.$$

35. The sample space is the set of 44 Democrats. The conditional probability that a person selected at random will answer "yes" given that they were a Democrat is

$$P(Yes|D) = \frac{20}{44} = \frac{5}{11}.$$

37. The sample space is the 25 adults that answered "No". Of the 25 people that answered "No", 17 of them were not Republicans. The conditional probability that a person will not be a Republican, given that they answered "No" is

$$P(\sim R|No) = \frac{17}{25}.$$

39. Of the 100 adults that were surveyed, 18 of them were Libertarians. The probability that a person selected at random will be a Libertarian is $P(L) = \frac{18}{100} = \frac{9}{50}.$

41. There are 36 outcomes in the sample space. There are two ways to roll a three, and four ways to roll a nine. The probability that a three or a nine is rolled is

$$P(3 \cup 9) = P(3) + P(9) = \frac{2}{36} + \frac{4}{36} = \frac{6}{36} = \frac{1}{6}.$$

43. Since we are given that a sum of less than five is rolled the reduced sample space has six outcomes. Of those six outcomes, there are two ways to roll a three and 3 ways to roll a four. The conditional probability that a three or a four is rolled, given that a sum of less than five is rolled is

$$P(s = 3 \cup s = 4|s < 5) = \frac{5}{6}.$$

45. There are six outcomes given that the green die shows four pips. Of those six outcomes only the outcome where the red die shows three pips to go with the four pips on the green die will give a sum of seven. So the conditional probability that the sum will be seven give the green die shows four pips is $P(s = 7|4G) = \frac{1}{6}.$

47.
a. Since we are drawing without replacement, there are only 12 possible outcomes. Four of these 12 outcomes result in a sum of five. The outcomes are $\{14, 23, 32, 41\}$. The probability that the two numbers drawn will sum to five is

$$P(s = 5) = \frac{4}{12} = \frac{1}{3}.$$

b. The reduced sample space given that the sum of the numbers is four is $\{13, 31\}$. One of these two outcomes has the property that the second number drawn was a three. Therefore the conditional probability that the second number drawn was a three, given that the sum of the two numbers was four is $P(3 on 2^{nd}|s = 4) = \frac{1}{2}.$

c. The sum of the two numbers cannot be nine. Therefore the probability that the sum of the two numbers is nine, given that the first number drawn is a two is

$$P(s = 9|2 on 1^{st}) = 0.$$

d. The reduced sample space is $\{41, 42, 43, 31, 32, 34\}$.

Exactly two of these outcomes $\{41, 32\}$ result in a sum of five. The conditional probability that the sum of the two numbers is a five, given that the first number was a four or a three is $P(s = 5|4 on 1^{st} \cup 3 on 1^{st}) = \frac{2}{6} = \frac{1}{3}.$

49. The reduced sample space contains seven outcomes. Three of these outcomes have two boys. The conditional probability that two of the children are boys, given that at least one of the children is a girl is

$$P(2 \, boys|at \, least \, 1 \, girl) = \frac{3}{7}.$$

For questions 51–53, we construct a Venn diagram for this situation as follows:

Let T represent the percentage of farmers that bought a new tractor and let C represent the percentage of farmers that bought a new car.

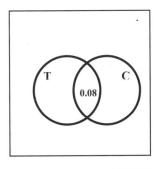

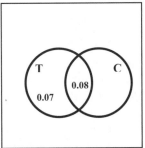

The construction is continued on the following page.

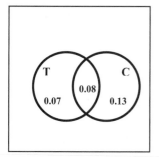

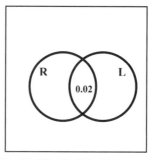

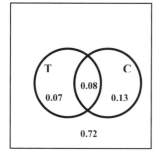

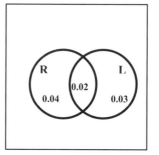

51. The conditional probability that the farmer bought a new tractor, given that she bought a new car is

$$P(T|C) = \frac{P(T \cap C)}{P(C)} = \frac{0.08}{0.21} = \frac{8}{21} \approx 0.381.$$

53. The conditional probability that a farmer bought exactly one of these items given that she bought at least one is

$$P((T \cap C') \cup (T' \cap C) | (T \cup C)) =$$
$$P((T \cap C')|(T \cup C)) + P((T' \cap C)|(T \cup C)) =$$
$$\frac{P((T \cap C') \cap (T \cup C))}{P(T \cap C)} + \frac{P((T' \cap C) \cap (T \cup C))}{P(T \cap C)} =$$
$$\frac{0.07}{0.28} + \frac{0.13}{0.28} = \frac{0.20}{0.28} \approx 0.714$$

Therefore the conditional probability that a farmer bought exactly one of these two items, given that she bought at least one of these two items is approximately 0.714.

55. Let R represent the percentage of people that reported a rise in blood pressure and let L represent the percentage of people that reported loss of sleep. Construct the Venn diagram at the top of the next column.

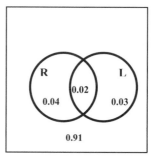

The conditional probability that a person taking this drug will experience loss of sleep, given that they reported a rise in blood pressure is

$$P(L|R) = \frac{P(L \cap R)}{P(R)} = \frac{0.02}{0.06} = \frac{1}{3} \approx 0.333.$$

57. $P(A|B) = \frac{P(A \cap B)}{P(B)} = \frac{0.3}{0.7} = \frac{3}{7}.$

59. Since $P(A|B) = \frac{P(A \cap B)}{P(B)}$

We can solve for $P(A \cap B)$ to get:

$$P(A \cap B) = P(B)P(A|B)$$

The solution is continued on the next page.

Substituting the information on the previous page into the equation we get:

$$P(A \cap B) = P(B)P(A|B) = \left(\tfrac{5}{7}\right)\left(\tfrac{4}{5}\right) = \tfrac{4}{7}.$$

61. $P(A \cup B) = P(A) + P(B) - P(A \cap B)$

Solving for $P(A \cap B)$ we get:

$$P(A \cap B) = P(A) + P(B) - P(A \cup B)$$

Therefore:

$$P(A \cap B) = 0.4 + 0.6 - 0.8 = 0.2.$$

Now we can find:

$$P(A|B) = \frac{P(A \cap B)}{P(B)}$$

$$P(A|B) = \frac{0.2}{0.6} = \frac{1}{3}.$$

63. Let D represent the event that the refrigerator selected is defective, and let G represent the event that the refrigerator selected is not defective. The tree diagram for the experiment is:

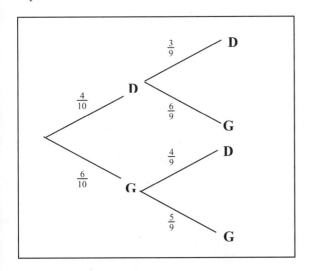

65. The tree diagram for the experiment is:

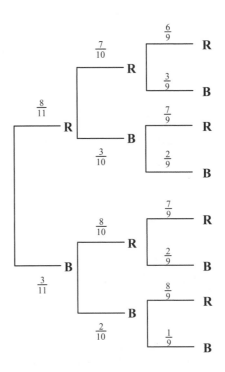

67. Let S represent the event that a spade was drawn and let N represent the event that a spade was not drawn. The tree diagram for the experiment is:

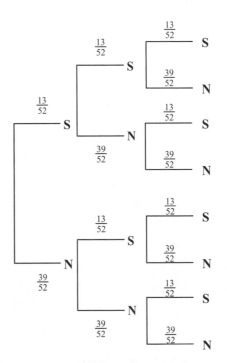

69. There are $26^4 = 456,976$ possible four-letter code words. There are $25^3 = 15,625$ different code words where the only A occupies the first spot. Since the single A could occupy any of the four spots in the code word, there are $4 \cdot 25^3 = 62,500$ possible code words that have exactly one A. Therefore the probability that a code word has exactly one A is $\frac{62,500}{456,976} \approx 0.1368$.

71. We know that there are 26^4 possible four-letter code words. There are $25^4 = 390,625$ possible code words that do not contain the letter A. Therefore there are $26^4 - 25^4 = 66,351$ four-letter code words with at least one A.

There are $25^3 = 15,625$ different code words where the only A occupies the first spot. Since the single A could occupy any of the four spots in the code word, there are $4 \cdot 25^3 = 62,500$ possible code words that have exactly one A.

Therefore the conditional probability that a four-letter code word will contain exactly one A given that it has at least one A in it is $\frac{62,500}{66,351} \approx 0.9420$.

Exercises Section 9.3

1. The reduced sample space has 51 cards, 12 of those cards are hearts. The probability that the second card is a heart, given the first card is a heart is

$$P\left(H \ on \ 2^{nd} \middle| H \ on \ 1^{st}\right) = \frac{12}{51} = \frac{4}{17}.$$

3. The probability that the first card is a spade is $\frac{13}{52} = \frac{1}{4}$. The probability that the second card is not a spade, given that the first card is a spade is $\frac{39}{51} = \frac{13}{17}$. The probability that the first card is a spade and the second card is not a spade is $\frac{1}{4} \times \frac{13}{17} = \frac{13}{68}$.

5. The probability that the first card is an ace is $\frac{4}{52} = \frac{1}{13}$. The probability that the second card is an ace, given the first card is an ace is $\frac{3}{51} = \frac{1}{17}$. The probability that both cards are aces is $\frac{1}{13} \times \frac{1}{17} = \frac{1}{221}$.

7. The probability that the first bulb is good is $P\left(1^{st} \ good\right) = \frac{16}{20} = \frac{4}{5}$. The probability that the second bulb is bad given the first bulb is good is $P\left(2^{nd} \ bad \middle| 1^{st} \ good\right) = \frac{4}{19}$. The probability that the third bulb is bad given the first bulb is good and the second bulb is bad is $P\left(3^{rd} \ bad \middle| 1^{st} \ good \cap 2^{nd} \ bad\right) = \frac{3}{18} = \frac{1}{6}$.

The probability that the first bulb is good and the last two bulbs are bad is $P\left(GBB\right) = \left(\frac{4}{5}\right)\left(\frac{4}{19}\right)\left(\frac{1}{6}\right) = \frac{8}{285}$.

9. The probability that the first bulb is bad is $P\left(1^{st} \ bad\right) = \frac{4}{20} = \frac{1}{5}$. The probability that the second bulb is bad given the first bulb is bad is $P\left(2^{nd} \ bad \middle| 1^{st} \ bad\right) = \frac{3}{19}$. The probability that the third bulb is bad given the first two bulbs are bad is $P\left(3^{rd} \ bad \middle| 1^{st} \ two \ bad\right) = \frac{2}{18} = \frac{1}{9}$.

The probability that all three bulbs are bad is:
$$P\left(BBB\right) = \left(\frac{1}{5}\right)\left(\frac{3}{19}\right)\left(\frac{1}{9}\right) = \frac{1}{285}.$$

11. The probability that the first three tosses are heads and the last two tosses are tails is $\left(\frac{1}{2}\right)\left(\frac{1}{2}\right)\left(\frac{1}{2}\right)\left(\frac{1}{2}\right)\left(\frac{1}{2}\right) = \frac{1}{32}$.

13. The probability of getting a question correct is $\frac{1}{4}$. Since the events are independent, the probability of getting all five questions correct is $\left(\frac{1}{4}\right)\left(\frac{1}{4}\right)\left(\frac{1}{4}\right)\left(\frac{1}{4}\right)\left(\frac{1}{4}\right) = \frac{1}{1024}$.

15. The probability of getting a question correct is $\frac{1}{4}$. The probability of getting a question incorrect is $\frac{3}{4}$. Since the events are independent, the probability of getting the first question right, the second question wrong, the third question right, the fourth question wrong, and the fifth question right is $\left(\frac{1}{4}\right)\left(\frac{3}{4}\right)\left(\frac{1}{4}\right)\left(\frac{3}{4}\right)\left(\frac{1}{4}\right) = \frac{9}{1024}$.

17. The probability that the player will get a hit is 0.245 for each at bat. The probability that the player will not hit safely is 0.755. There are three ways that a player can get exactly two hits in the next three at bats. Each of those three ways carries the probability of $\left(0.245\right)^2\left(0.755\right) \approx 0.0453$. Therefore the probability that a player will get exactly two hits in the next three at bats is $3\left(0.0453\right) \approx 0.1360$.

19. For independent events it is known that
$P(A \cap B) = P(A)P(B)$.

a. Substituting in the known values we get:
$0.4 = P(A) \cdot 0.6$
$P(A) = \frac{0.4}{0.6} = \frac{2}{3}$.

b. If $P(B') = 0.7$ then
$P(B) = 1 - P(B') = 1 - 0.7 = 0.3$.

Substituting in the proper values we get:

$0.1 = P(A) \cdot 0.3$
$P(A) = \frac{0.1}{0.3} = \frac{1}{3}$.

21.

a. $P\left((A \cap B)'\right) = 0.6$, therefore

$P(A \cap B) = 1 - 0.6 = 0.4$. If $P(A') = 0.5$, then
$P(A) = 1 - 0.5 = 0.5$.

Now substituting the proper values we get:

$P(A \cap B) = P(A)P(B)$
$0.4 = 0.5 \cdot P(B)$
$P(B) = \frac{0.4}{0.5} = \frac{4}{5}$.

b. $P(A') = 0.4$, so $P(A) = 1 - 0.4 = 0.6$. Since A and B are independent, it follows that

$P(A \cap B) = P(A)P(B) = (0.6)(0.5) = 0.3$.

Therefore,
$P(A \cup B) = P(A) + P(B) - P(A \cap B)$
$P(A \cup B) = 0.6 + 0.5 - 0.3 = 0.8$.

23. Reading the probabilities from the tree we see:

a. $P(R|N) = 0.4$

b. $P(M \cap S) = P(M)P(S|M) =$
$\qquad (0.1)(0.3) = 0.03$.

c. $P(N \cap R) = P(N)P(R|N) =$
$\qquad (0.3)(0.4) = 0.12$

d. $P(R) = P(M \cap R) + P(N \cap R) + P(Q \cap R)$
$P(R) = (0.1)(0.7) + (0.3)(0.4) + (0.6)(0.8)$
$P(R) = 0.67$.

e. $P(S) = 1 - P(R) = 1 - 0.67 = 0.33$.

25.
a. The probability that the first marble is red is $\frac{4}{9}$. If the first marble is red, this leaves eight marbles, three red and five green. The probability that the second marble is green, given the first marble is red is $\frac{5}{8}$. The probability that the third marble is green, given the first marble is red and the second marble is green is $\frac{4}{7}$. The probability that the first marble is red and the last two marbles are green is

$\left(\frac{4}{9}\right)\left(\frac{5}{8}\right)\left(\frac{4}{7}\right) = \frac{10}{63}$.

b. In a manner similar to part a, the probability that the first marble is red, the second marble is green and the third marble is red is $\left(\frac{4}{9}\right)\left(\frac{5}{8}\right)\left(\frac{3}{7}\right) = \frac{5}{42}$.

c. In a manner similar to part a, the probability that all three marbles are red is $\left(\frac{4}{9}\right)\left(\frac{3}{8}\right)\left(\frac{2}{7}\right) = \frac{1}{21}$.

27.
a. The probability that the first ball is red is $\frac{5}{12}$. Since we replace the first ball drawn back into the box, there are still 12 balls left from which to choose. The probability that the second ball is purple is $\frac{4}{12} = \frac{1}{3}$. Likewise the probability that the third ball is purple is $\frac{4}{12} = \frac{1}{3}$.
The probability that the first ball is red, and the last two are purple is $\left(\frac{5}{12}\right)\left(\frac{1}{3}\right)\left(\frac{1}{3}\right) = \frac{5}{108}$.

b. In a technique similar to the one used in part a, the probability that the first ball is red, the second ball is green, and the third ball is purple is $\left(\frac{5}{12}\right)\left(\frac{3}{12}\right)\left(\frac{4}{12}\right) = \frac{5}{144}$.

c. In the same manner, the probability that all of the balls are green is $\left(\frac{3}{12}\right)\left(\frac{3}{12}\right)\left(\frac{3}{12}\right) = \frac{1}{64}$.

29. Each roll is independent of the previous roll. Therefore the probability of any particular number, on any particular roll is $\frac{1}{6}$. It stands to reason that the probability of not rolling a particular number on a given roll is $\frac{5}{6}$.

a. The probability of rolling a two is $\frac{1}{6}$. The probability of not rolling a two is $\frac{5}{6}$. Therefore, the probability of rolling the sequence {two, two, not a two, not a two} on four rolls is $\left(\frac{1}{6}\right)\left(\frac{1}{6}\right)\left(\frac{5}{6}\right)\left(\frac{5}{6}\right) = \frac{25}{1296}$.

b. Similarly to part a, the probability of rolling the sequence {three, five, not a five, two} on four rolls is $\left(\frac{1}{6}\right)\left(\frac{1}{6}\right)\left(\frac{5}{6}\right)\left(\frac{1}{6}\right) = \frac{5}{1296}$.

c. The probability of rolling a number greater than four is equal to the probability of rolling a five or a six which is $\frac{1}{6} + \frac{1}{6} = \frac{2}{6} = \frac{1}{3}$. It stands to reason, the probability of rolling a number less than five is $\frac{2}{3}$. Therefore, the probability of rolling the sequence {a number greater than four, a three, a number less than five, a four} on four sequential rolls is $\left(\frac{1}{3}\right)\left(\frac{1}{6}\right)\left(\frac{2}{3}\right)\left(\frac{1}{6}\right) = \frac{1}{162}$.

31.
a. The two rolls are independent. The probability of rolling a sum of six on the first roll is $\frac{5}{36}$. The probability of rolling a sum of nine on the second roll is $\frac{4}{36} = \frac{1}{9}$. The probability of rolling a sum of six on the first roll and a sum of nine on the second roll is $\left(\frac{5}{36}\right)\left(\frac{1}{9}\right) = \frac{5}{324}$.

b. The probability of rolling a sum of seven on either roll is $\frac{6}{36} = \frac{1}{6}$. Therefore, the probability of rolling a sum of seven on both rolls is $\left(\frac{1}{6}\right)\left(\frac{1}{6}\right) = \frac{1}{36}$.

c. The probability of rolling a sum of five is $\frac{4}{36} = \frac{1}{9}$. The probability of not rolling a sum of five is $1 - \frac{1}{9} = \frac{8}{9}$. Therefore, the probability of rolling a sum of five on the first roll and not rolling a sum of five on the second roll is $\left(\frac{1}{9}\right)\left(\frac{8}{9}\right) = \frac{8}{81}$.

33. Since the events are independent of each other, the probability that person A will get the flu and that person B will get the flu is:
$$P(A \cap B) = P(A)P(B)$$
$$P(A \cap B) = (0.03)(0.03) = 0.0009$$

The probability that both people will get the flu is 0.0009.

35. Since the events are independent, the probability that person A will get the flu or person B will get the flu, or both will get the flu is:
$$P(A \cup B) = P(A) + P(B) - P(A \cap B)$$
$$P(A \cup B) = 0.03 + 0.03 - 0.0009 = 0.0591.$$

The probability that at least one of the two people will get the flu is 0.0591.

37. The probability that the first is a junior is $\frac{4}{9}$. The probability that the second is a junior given the first is a junior is $\frac{3}{8}$. The probability that the third is a junior given that the first two are juniors is $\frac{2}{7}$. Thus, the probability that all three committee members are juniors is
$\left(\frac{4}{9}\right)\left(\frac{3}{8}\right)\left(\frac{2}{7}\right) = \frac{1}{21}$.

39. Let A be the event Maria's mom is alive ten years from now, and let B be the event Magda's mom is alive ten years from now. The events are independent so:
$$P(A \cap B) = P(A)P(B)$$
$$P(A \cap B) = (0.91)(0.91) = 0.8281.$$

The probability that both moms will be alive ten years from now is 0.8281.

41. Let A be the event Maria's mom is alive ten years from now, and let B be the event Magda's mom is alive ten years from now. The events are independent. Since
$P(A) = P(B) = 0.91$, it must be the case that
$P(A') = P(B') = 1 - 0.91 = 0.09$. The probability that neither of the mothers are alive ten years from now is:
$$P(A' \cap B') = P(A')P(B')$$
$$P(A' \cap B') = (0.09)(0.09) = 0.0081.$$

The probability that neither of the mothers are alive ten years from now is 0.0081.

43. Since the events are independent, the probability that the first two components will fail and the third one will not fail is $(0.002)(0.002)(0.998) = 0.000003992$.

45. If two parts are installed, the probability that at least one will function properly is $1 - (0.002)^2 = 0.999996$.
Therefore, the installation of two components will be enough to guarantee at least one of the components will function properly with a probability of 0.99999.

47. The probability that both will fail is $(0.04)(0.10) = 0.004$. The probability that at least one computer will work is the same as the probability that not both of the computers will fail. The probability that at least one of the computers will function properly is $1 - 0.004 = 0.996$.

49. The probability that all of the tosses will result in heads turning up is $\left(\frac{1}{2}\right)^7 = \frac{1}{128}$.

51. Each toss of the coin is independent from the previous toss. The fact that the first four tosses turned up tails gives us no new information. The probability that the last three tosses will be heads, given that the first four were tails is $\left(\frac{1}{2}\right)^3 = \frac{1}{8}$.

53. The probability that at least one of the bulbs will still work after 800 hours is the same as the probability that not all of the bulbs will burn out in less than 800 hours. The probability of all the bulbs burning out in less than 800 hours is $(0.01)^4$. Therefore the probability that at least one of the bulbs will still be burning after 800 hours is $1 - (0.01)^4 = 0.99999999$.

55. There are $C(12,3) = 220$ different samples, but only $C(3,3) = 1$ sample in which all three are defective. The probability that all three TV sets are defective is $\frac{1}{220}$.

57. There are $C(9,3) = 84$ samples in which all three TV's are not defective. This means that there are $220 - 84 = 136$ samples in which at least one TV is defective. The probability that at least one TV in the sample of three is defective is $\frac{136}{220} = \frac{34}{55}$.

59. Assuming the gender of each child is independent, and the likelihood of having a boy and a girl are equal, the probability of having three girls is $\left(\frac{1}{2}\right)^3 = \frac{1}{8}$.

61. Assuming that the population of voters is large enough to consider each individual result independent, the probability that all ten of the voters contacted are Republicans is $(0.25)^{10} = 9.5367 \times 10^{-7}$.

63. Assume that the population is large enough to consider each event independent. The probability that at least one is a Republican is the same as the probability that not all of the voters contacted are non Republicans. Therefore, the probability that at least one of the voters contacted is a Republican is $1 - (0.75)^{10} \approx 0.9437$.

65. The probability that at least one will be wearing a seatbelt is equivalent to the negation that all three will not be wearing a seatbelt. Thus, the probability that at least one will be wearing their seatbelt is $1 - (0.4)^3 = 0.936$.

67. The probability that someone will get through all three security systems is $(0.02)^3 = 0.000008$.

69. Assuming the population of fruit flies is large enough so that each sample is independent, the probability that all three flies selected will have red eyes is $(0.10)^3 = 0.001$.

71. If the order in which the flies were selected did not matter, the probability that exactly one would have red eyes is $(0.10)(0.90)(0.90)$. However, there are three ways in which this could occur, so the probability that exactly one fly will have red eyes is $3(0.10)(0.90)(0.90) = 0.243$.

Exercises Section 9.4

1. $P(A \cap D) = P(A)P(D|A) =$
$$(0.2)(0.6) = 0.12.$$

3. The conditional probability is:

$$P(E|B) = 0.9$$

5. This problem requires the use of Bayes's theorem.

$$P(A|D) = \frac{P(A \cap D)}{P(D)} =$$

$$\frac{P(A)P(D|A)}{P(A)P(D|A) + P(B)P(D|B)} =$$

$$\frac{(0.2)(0.6)}{(0.2)(0.6) + (0.8)(0.1)} = 0.6.$$

7. This problem requires the use of Bayes's theorem.

$$P(A|E) = \frac{P(A \cap E)}{P(E)} =$$

$$\frac{P(A)P(E|A)}{P(A)P(E|A) + P(B)P(E|B) + P(C)P(E|C)} =$$

$$\frac{(0.2)(0.2)}{(0.2)(0.2) + (0.5)(0.3) + (0.3)(0.6)} = \frac{4}{37}$$

$$P(A|E) = \frac{4}{37} \approx 0.1081$$

9. This problem requires the use of Bayes's theorem.

$$P(C|D) = \frac{P(C \cap D)}{P(D)} =$$

$$\frac{P(C)P(D|C)}{P(A)P(D|A) + P(B)P(D|B) + P(C)P(D|C)} =$$

$$\frac{(0.3)(0.4)}{(0.2)(0.8) + (0.5)(0.7) + (0.3)(0.4)} = \frac{4}{21}$$

$$P(C|D) = \frac{4}{21} \approx 0.1905.$$

11. This problem requires the use of Bayes's theorem.

$$P(B|E) = \frac{P(B \cap E)}{P(E)} =$$

$$\frac{P(B)P(E|B)}{P(A)P(E|A) + P(B)P(E|B) + P(C)P(E|C)} =$$

$$\frac{(0.5)(0.3)}{(0.2)(0.2) + (0.5)(0.3) + (0.3)(0.6)} = \frac{15}{37}$$

$$P(B|E) = \frac{15}{37} \approx 0.4054.$$

The tree diagram for question 13 is shown below:

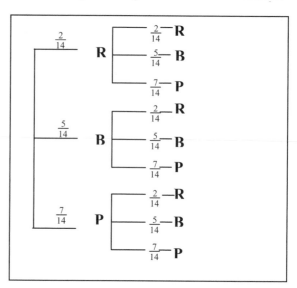

13. The probability that the second ball is blue, given that the first ball is purple is $P(2^{nd}\, isB | 1^{st}\, isP) = \frac{5}{14}$.

The tree diagram for questions 15-17 is shown below:

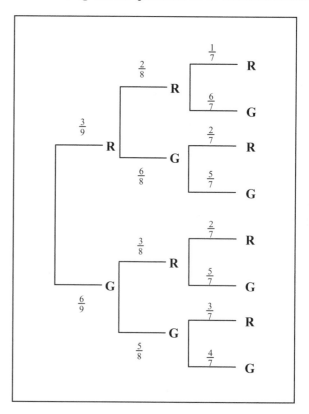

15. The probability that all three balls are red is
$P(RRR) = \left(\frac{3}{9}\right)\left(\frac{2}{8}\right)\left(\frac{1}{7}\right) = \frac{1}{84} \approx 0.0119$.

17. Using the tree diagram on the previous page, the probability that the third ball is green is

$$P\left(3^{rd} \text{ is } G\right) = \left(\tfrac{3}{9}\right)\left(\tfrac{2}{8}\right)\left(\tfrac{6}{7}\right) + \left(\tfrac{3}{9}\right)\left(\tfrac{6}{8}\right)\left(\tfrac{5}{7}\right) +$$
$$\left(\tfrac{6}{9}\right)\left(\tfrac{3}{8}\right)\left(\tfrac{5}{7}\right) + \left(\tfrac{6}{9}\right)\left(\tfrac{5}{8}\right)\left(\tfrac{4}{7}\right) = \tfrac{2}{3}$$
$$P\left(3^{rd} \text{ is } G\right) = \tfrac{2}{3} \approx 0.6667 .$$

19. The probability that the first number is a three and the second number is a one is

$$P\left(1^{st} \text{ is } 3 \cap 2^{nd} \text{ is } 1\right) = \left(\tfrac{1}{3}\right)\left(\tfrac{1}{3}\right) = \tfrac{1}{9} .$$

The tree diagram for questions 21-23 is shown below:

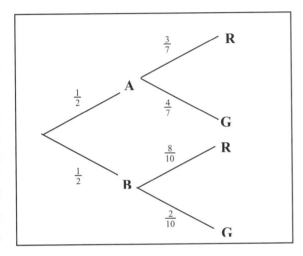

21. The probability that the ball is red, given that it came from box A is $P\left(R|A\right) = \tfrac{3}{7} \approx 0.4286$.

23. The probability that the ball is red is

$$P(R) = P(A \cap R) + P(B \cap R) =$$
$$P(R) = \left(\tfrac{1}{2}\right)\left(\tfrac{3}{7}\right) + \left(\tfrac{1}{2}\right)\left(\tfrac{8}{10}\right) = \tfrac{43}{70} \approx 0.6143 .$$

The tree diagram for questions 25-27 is shown below:

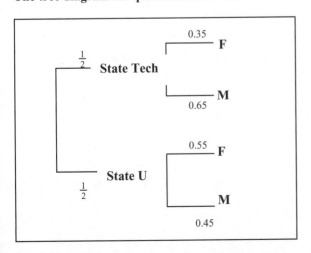

25. The probability that a student chosen at random will be female is: $P(F) = \left(\tfrac{1}{2}\right)(0.35) + \left(\tfrac{1}{2}\right)(0.55) = 0.45$.

27. The probability that the student comes from State Tech given that the student is female is:

$$P\left(ST|F\right) = \frac{P\left(ST \cap F\right)}{P\left(F\right)} =$$
$$\frac{\left(\tfrac{1}{2}\right)(0.35)}{\left(\tfrac{1}{2}\right)(0.35) + \left(\tfrac{1}{2}\right)(0.55)} = \frac{7}{18} .$$

The tree diagram for questions 29-31 is shown below:

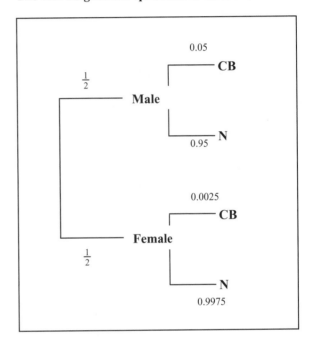

29. The probability that a person selected at random is color blind, given that the person is male is:

$$P\left(CB|M\right) = 0.05 .$$

31. The probability that a person selected at random is male, given that the person is color blind is:

$$P\left(M|CB\right) = \frac{P\left(M \cap CB\right)}{P\left(CB\right)} =$$
$$\frac{\left(\tfrac{1}{2}\right)(0.05)}{\left(\tfrac{1}{2}\right)(0.05) + \left(\tfrac{1}{2}\right)(0.0025)} = \frac{20}{21} \approx 0.9524 .$$

The tree diagram for question 33 is shown below:

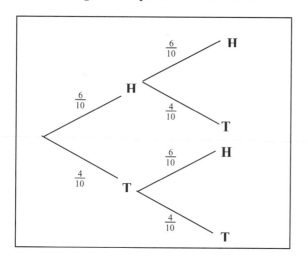

33. The probability that the second toss comes up heads is

$$P\left(H \, on \, 2^{nd}\right) = P\left(H \cap H\right) + P\left(T \cap H\right)$$

$$P\left(H \, on \, 2^{nd}\right) = \left(\tfrac{6}{10}\right)\left(\tfrac{6}{10}\right) + \left(\tfrac{4}{10}\right)\left(\tfrac{6}{10}\right) = \tfrac{3}{5} = 0.6$$

The tree diagram for questions 35-37 is shown below:

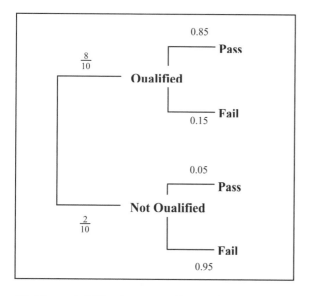

35. The probability that an applicant selected at random is qualified and will pass the test is:

$$P\left(Q \cap pass\right) = \left(0.8\right)\left(0.85\right) = 0.68.$$

The probability that the event will not happen is $1 - 0.68 = 0.32$. So the odds of the event happening are

$$\frac{P\left(Q \cap pass\right)}{1 - P\left(Q \cap pass\right)} = \frac{0.68}{1 - 0.68} = \frac{17}{8}.$$

The odds that an applicant will be qualified and pass the test are 17 to 8.

37. The probability that an applicant is unqualified, given that the applicant passed the test is:

$$P\left(NQ\big|\,pass\right) = \frac{P\left(NQ \cap pass\right)}{P\left(pass\right)}$$

$$P\left(NQ\big|\,pass\right) = \frac{\left(0.2\right)\left(0.05\right)}{\left(0.8\right)\left(0.85\right) + \left(0.2\right)\left(0.05\right)} =$$

$$P\left(NQ\big|\,pass\right) = \frac{1}{69} \approx 0.01449.$$

The tree diagram for question 39 is shown below:

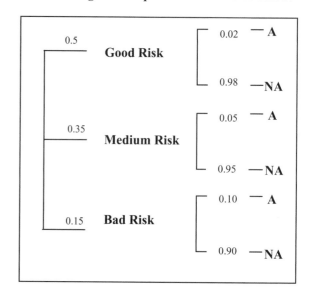

39. The probability that an insured driver will have an accident next year is:

$$P\left(A\right) = P\left(GR \cap A\right) + P\left(MR \cap A\right) + P\left(BR \cap A\right)$$
$$P\left(A\right) = \left(0.5\right)\left(0.02\right) + \left(0.35\right)\left(0.05\right) + \left(0.15\right)\left(0.1\right)$$
$$P\left(A\right) = 0.0425.$$

The probability is 0.0425 that an insured driver will have an accident during the next year.

The tree diagram for questions 41-43 is displayed on the following page.

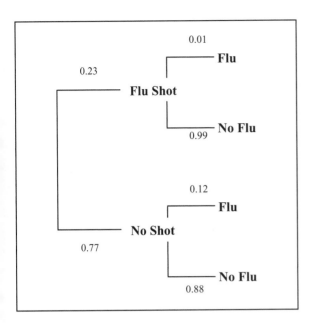

41. The probability that a person did not get the flu, given that she had received the flu shot is:

$$P(No\ Flu|Shot) = 0.99.$$

43. The probability that he did not receive the shot, given that he had the flu is:

$$P(No\ Shot|Flu) = \frac{P(No\ Shot \cap Flu)}{P(Flu)} =$$

$$\frac{(0.77)(0.12)}{(0.23)(0.01)+(0.77)(0.12)} = \frac{924}{947}$$

$$P(No\ Shot|Flu) = \frac{924}{947} \approx 0.9757.$$

The tree diagram for question 45 is shown below:

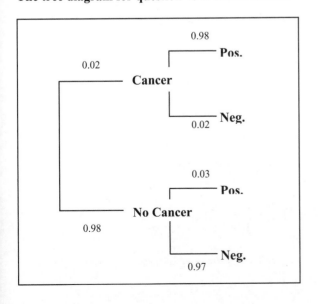

45. The probability that the person has this type of cancer given that the test indicates she has this type of cancer is:

$$P(C|P) = \frac{P(C \cap P)}{P(P)} =$$

$$\frac{(0.02)(0.98)}{(0.02)(0.98)+(0.98)(0.03)} = \frac{2}{5}$$

$$P(C|P) = \frac{2}{5} = 0.4.$$

The tree diagram for questions 47-49 is shown below:

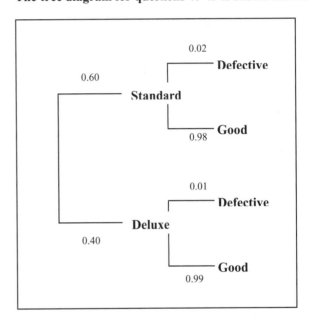

47. The probability that a PDA selected at random is a deluxe model and has a defect is:

$$P(Deluxe \cap Defect) = (0.40)(0.01) = 0.004.$$

49. The probability that a PDA selected at random is a deluxe model given that it has a defect is:

$$P(Deluxe|Defect) = \frac{P(Deluxe \cap Defect)}{P(Defect)} =$$

$$\frac{(0.4)(0.01)}{(0.6)(0.02)+(0.4)(0.01)} = \frac{1}{4}$$

$$P(Deluxe|Defect) = \frac{1}{4} = 0.25.$$

The tree diagram for questions 51-53 is displayed below:

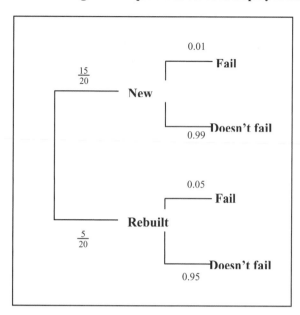

51. The probability that a fuel pump will be rebuilt and will fail within a year is:

$$P(R \cap F) = \left(\tfrac{5}{20}\right)(0.05) = 0.0125 .$$

On average a rebuilt fuel pump will fail within one year 1.25% of the time.

53. The probability that a fuel pump is rebuilt, given that it failed within one year is:

$$P(R|F) = \frac{P(R \cap F)}{P(F)} =$$

$$\frac{\left(\tfrac{5}{20}\right)(0.05)}{\left(\tfrac{15}{20}\right)(0.01) + \left(\tfrac{5}{20}\right)(0.05)} = \frac{5}{8}$$

$$P(R|F) = \frac{5}{8} = 0.625 .$$

Given that the fuel pump failed within one year, the probability that it was rebuilt is 0.625.

The tree diagram for questions 55-59 is shown at the top of the next column.

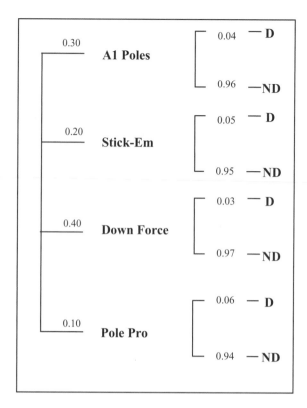

55. The probability that a ski pole will have a defect is:

$$P(D) = (0.3)(0.04) + (0.2)(0.05) +$$
$$(0.4)(0.03) + (0.1)(0.06) = 0.04 .$$

On average, 4% of ski poles will have a defect in a given shipment.

57. The probability that a pole came from Down Force and does not have a defect is:

$$P(DF \cap ND) = (0.4)(0.97) = 0.388 .$$

On average, 38.8% of the poles come from Down Force and do not have defects.

59. The probability that a pole came from Stick-Em, given that the pole had a defect is:

$$P(SE|D) = \frac{P(SE \cap D)}{P(D)} =$$

$$\frac{(0.2)(0.05)}{(0.3)(0.04) + (0.2)(0.05) + (0.4)(0.03) + (0.1)(0.06)}$$

$$P(SE|D) = \frac{1}{4} = 0.25 .$$

On average, 25% of the defective poles came from Stick-Em.

The tree diagram for question 61 is shown below:

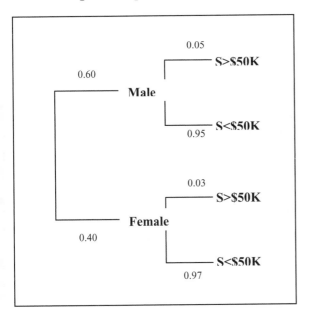

61. The probability that the employee is a man, given that the salary exceeds $50,000 is:

$$P\left(M|S>50K\right)=\frac{P\left(M\cap S>50K\right)}{P\left(S>50K\right)}$$

$$\frac{\left(0.6\right)\left(0.05\right)}{\left(0.6\right)\left(0.05\right)+\left(0.4\right)\left(0.03\right)}=\frac{5}{7}$$

$$P\left(M|S>50K\right)=\frac{5}{7}\approx 0.7143.$$

Given the employee's salary exceeds $50,000, the probability that the employee is a man is 0.7143.

63. The tree diagram for the problem is shown below:

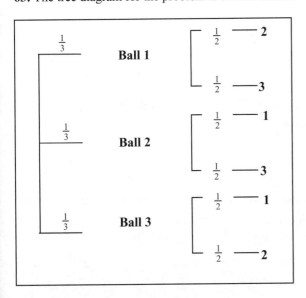

The probability that the second ball is labeled with the number two is summing all of the paths in which you can get a two on the second draw. In other words, the probability of getting a two on the second draw is the probability of drawing one then two plus the probability of drawing three then two.

$$\left(\tfrac{1}{3}\right)\left(\tfrac{1}{2}\right)+\left(\tfrac{1}{3}\right)\left(\tfrac{1}{2}\right)=\tfrac{2}{6}=\tfrac{1}{3}.$$

Exercises Section 9.5

1. The probability of getting four heads and two tails is:

$$C\left(6,4\right)\cdot\left(0.55\right)^{4}\left(0.45\right)^{2}\approx 0.2780.$$

3. The probability of getting exactly five heads is:

$$C\left(6,5\right)\cdot\left(0.55\right)^{5}\left(0.45\right)^{1}\approx 0.13589.$$

The probability of getting exactly six heads is:

$$C\left(6,6\right)\cdot\left(0.55\right)^{6}\left(0.45\right)^{0}\approx 0.02768.$$

The probability of getting at least five heads is the probability of getting exactly five heads plus the probability of getting exactly six heads. This probability is $0.13589+0.027681=0.16357$.

5. The probability of getting a hit is 0.335 and the probability of not getting a hit is 0.665. The probability that he will get exactly no hits in his next five at bats is:

$$C\left(5,0\right)\cdot\left(0.335\right)^{0}\left(0.665\right)^{5}\approx 0.13005.$$

7. The probability that the bulb will be defective is 0.01. The probability that the bulb will not be defective is 0.99. The probability that exactly one bulb out of six will be defective is:

$$C\left(6,1\right)\cdot\left(0.01\right)^{1}\left(0.99\right)^{5}\approx 0.05706$$

9. The probability that the bulb will be defective is 0.01. The probability that the bulb will not be defective is 0.99. The probability that all six bulbs will be good (which is the same as saying none of the bulbs are defective) is:

$$C\left(6,0\right)\cdot\left(0.01\right)^{0}\left(0.99\right)^{6}\approx 0.94148.$$

11. The probability that a voter will be a Republican is 0.38. The probability that a voter will not be a Republican is 0.62. The probability that exactly eight of the people surveyed will not be Republicans and thus exactly two will be Republicans is:

$$C\left(10,2\right)\cdot\left(0.38\right)^{2}\left(0.62\right)^{8}\approx 0.14188.$$

13. The probability that the cataract surgery is successful is 0.98. The probability that the cataract surgery is not successful is 0.02. The probability that exactly six out of the eight surgeries are successful is:

$$C(8,6) \cdot (0.98)^6 (0.02)^2 \approx 0.009921.$$

On average, 0.9921% of the time exactly six of the eight surgeries will be successful.

15. The probability that the cataract surgery is successful is 0.98. The probability that the cataract surgery is not successful is 0.02.

The probability that all eight surgeries are successful is:

$$C(8,8) \cdot (0.98)^8 (0.02)^8 \approx 0.8508.$$

85.1% of the time all eight surgeries will be successful.

17. The probability that multiple births will occur is 0.20. The probability that multiple births will not occur is 0.80. The probability that none of the 12 women will have multiple births is: $C(12,0) \cdot (0.2)^0 (0.8)^{12} \approx 0.06872$.

6.82% of the time, none of the twelve women will have multiple births.

19. The probability that a person selected at random is in favor of the bond issue is 0.56.The probability that a person selected at random is not in favor of the bond issue is 0.44. The probability that among five voters, exactly four will favor the bond issue is:

$$C(5,4) \cdot (0.56)^4 (0.44)^1 \approx 0.2164.$$

21. The probability that a person selected at random is in favor of the bond issue is 0.56. The probability that a person selected at random is not in favor of the bond issue is 0.44. The probability that among five voters, all five will favor the bond issue is:

$$C(5,5) \cdot (0.56)^5 (0.44)^0 \approx 0.0551.$$

23. The probability that a child will be a girl is $\frac{1}{2}$. The probability that the child will not be a girl is $\frac{1}{2}$. The probability that an eight child family will have exactly four boys and exactly four girls is:

$$C(8,4) \cdot \left(\frac{1}{2}\right)^4 \left(\frac{1}{2}\right)^4 = \frac{35}{128} \approx 0.27344.$$

25. The probability that a sum of seven is rolled is $\frac{6}{36} = \frac{1}{6}$.

The probability that a sum of seven is not rolled is $\frac{5}{6}$. The probability that a sum of seven is rolled all four times is:

$$C(4,4) \cdot \left(\frac{1}{6}\right)^4 \left(\frac{5}{6}\right)^0 = \frac{1}{1296} \approx 0.000772.$$

27. The probability that a sum of seven is rolled is $\frac{6}{36} = \frac{1}{6}$.

The probability that a sum of seven is not rolled is $\frac{5}{6}$. The probability that at least one seven is rolled is equivalent to the probability that no sevens are rolled is:

$$C(4,0) \cdot \left(\frac{1}{6}\right)^0 \left(\frac{5}{6}\right)^4 = \frac{625}{1296} \approx 0.48225.$$

29. The probability that a participant prefers Diet Raspberry is 0.62. The probability that a participant prefers Diet Banana is 0.38. The probability that all ten participants will prefer Diet Raspberry is:

$$C(10,10) \cdot (0.62)^{10} (0.38)^0 \approx 0.008393.$$

31. The probability that a participant prefers Diet Raspberry is 0.62. The probability that a participant prefers Diet Banana is 0.38. The probability that at least one of the ten participants will prefer Diet Banana is equivalent to the probability that not all of the ten will prefer Diet Raspberry. This probability is:

$$1 - C(10,10) \cdot (0.62)^{10} (0.38)^0 \approx 0.991607.$$

33. The probability that a student will guess at the correct answer is $\frac{1}{4}$. The probability that a student will guess an incorrect answer is $\frac{3}{4}$.

The probability that a student will get exactly eight out of ten questions correct is:

$$C(10,8) \cdot \left(\frac{1}{4}\right)^8 \left(\frac{3}{4}\right)^2 \approx 0.00038624.$$

The probability that a student will get exactly nine out of ten questions correct is:

$$C(10,9) \cdot \left(\frac{1}{4}\right)^9 \left(\frac{3}{4}\right)^1 \approx 0.00002861.$$

The probability that a student will get exactly ten out of ten questions correct is:

$$C(10,10) \cdot \left(\frac{1}{4}\right)^{10} \left(\frac{3}{4}\right)^0 \approx 0.0000009537.$$

Therefore, the probability that a student will get at least eight questions out of ten correct on the exam is:
$0.0003862 + 0.00002861 + 0.0000009537 = 0.00041577$.

35. The probability that a student will guess at the correct answer is $\frac{1}{2}$. The probability that a student will guess an incorrect answer is $\frac{1}{2}$. The probability that a student will get exactly five out of ten questions correct is:

$$C(10,5) \cdot \left(\frac{1}{2}\right)^5 \left(\frac{1}{2}\right)^5 = \frac{63}{256} \approx 0.2461.$$

37. The probability that a student will guess at the correct answer is $\frac{1}{2}$. The probability that a student will guess an incorrect answer is $\frac{1}{2}$.

The probability that none are answered correctly is:

$C(10,0) \cdot \left(\frac{1}{2}\right)^0 \left(\frac{1}{2}\right)^{10} = \frac{1}{1024} \approx 0.0009766$.

The probability that exactly one question is answered correctly is:

$C(10,1) \cdot \left(\frac{1}{2}\right)^1 \left(\frac{1}{2}\right)^9 = \frac{10}{1024} \approx 0.009766$.

The probability that at least two questions are answered correctly is:

$1 - \frac{1}{1024} - \frac{10}{1024} = \frac{1013}{1024} \approx 0.9893$.

39. The probability that a club is drawn on any particular draw is $\frac{1}{4}$ because we are replacing the cards. The probability that a club is not drawn on any particular draw is $\frac{3}{4}$. The probability that all five of the cards drawn are clubs is: $C(5,5) \cdot \left(\frac{1}{4}\right)^5 \left(\frac{3}{4}\right)^0 \approx 0.0009766$.

41. The probability that a mirror will be broken is 0.01. The probability that a mirror will not be broken is 0.99. The probability that none of the twelve mirrors randomly selected will be broken is:

$C(12,0) \cdot (0.01)^0 (0.99)^{12} \approx 0.88638$.

43. The probability that a seed will germinate is 0.90. The probability that a seed will not germinate is 0.10. The probability that exactly nine out of ten seeds will germinate is:

$C(10,9) \cdot (0.90)^9 (0.10)^1 \approx 0.38742$.

45. The probability that a seed will germinate is 0.90. The probability that a seed will not germinate is 0.10.

The probability that exactly seven out of ten seeds will germinate is:

$C(10,7) \cdot (0.90)^7 (0.10)^3 \approx 0.05740$.

The probability that exactly eight out of ten of the seeds will germinate is:

$C(10,8) \cdot (0.90)^8 (0.10)^2 \approx 0.19371$.

The probability that exactly nine out of ten seeds will germinate is:

$C(10,9) \cdot (0.90)^9 (0.10)^1 \approx 0.38742$.

The probability that all ten of the seeds will germinate is:

$C(10,10) \cdot (0.90)^{10} (0.10)^0 \approx 0.34868$.

The probability that at least seven of the ten seeds will germinate is:

$0.05740 + 0.1937 + 0.38742 + 0.34868 = 0.9872$.

47. The probability that a driver in Burlington is wearing a seatbelt is 0.48. The probability that a driver is not wearing a seatbelt is 0.52. If ten motorists are stopped the probability that exactly eight out of ten drivers will be wearing a seatbelt is:

$C(10,8) \cdot (0.48)^8 (0.52)^2 \approx 0.03429$.

49. The probability that a message is delivered on time is 0.92. The probability that a message is not delivered on time is 0.08. If the firm sends four separate envelopes, the probability that none of the four will be delivered on time is:

$C(4,0) \cdot (0.92)^0 (0.08)^4 \approx 0.00004096$.

51.

a. The binomial distribution table for calls on five customers is shown below. The number of successes is in the left hand column, and the probability of that outcome is in the right hand column.

Number of successes	Probability
0	$C(5,0) \cdot (0.6)^0 (0.4)^5 \approx 0.01024$
1	$C(5,1) \cdot (0.6)^1 (0.4)^4 \approx 0.0768$
2	$C(5,2) \cdot (0.6)^2 (0.4)^3 \approx 0.2304$
3	$C(5,3) \cdot (0.6)^3 (0.4)^2 \approx 0.3456$
4	$C(5,4) \cdot (0.6)^4 (0.4)^1 \approx 0.2592$
5	$C(5,5) \cdot (0.6)^5 (0.4)^0 \approx 0.07776$

b. Using the above table the expected number of successes is:

$0(0.01024) + 1(0.0768) + 2(0.2304) + 3(0.3456) + 4(0.2592) + 5(0.07776) = 3.0000$

On the other hand, the expected number of successes from one customer is 0.6, so the expected number from five customers is $5(0.6) = 3.0$.

c. If the salesman calls on 200 customers, the expected number of successes is $200(0.6) = 120$.

If the salesman calls on 800 customers, the expected number of successes is $800(0.6) = 480$.

53.
a. The binomial distribution table for the number of field goals made when the kicker attempts three is shown below. The number of successful field goal tries is in the left hand column, and the probability of that outcome is in the right hand column.

Number of successes	Probability
0	$C(3,0) \cdot (0.95)^0 (0.05)^3 \approx 0.00013$
1	$C(3,1) \cdot (0.95)^1 (0.05)^2 \approx 0.00713$
2	$C(3,2) \cdot (0.95)^2 (0.05)^1 \approx 0.1354$
3	$C(3,3) \cdot (0.95)^3 (0.05)^0 \approx 0.8574$

b. Using the above table the expected number of field goals made when three field goals are attempted from 30 yards is:

$$0(0.00013) + 1(0.00713) + 2(0.1354) +$$
$$3(0.8574) = 2.85$$

On the other hand, each field goal from 30 yards is expected to be made with a probability of 0.95, so the expected number of made field goals when three are attempted from 30 yards is $3(0.95) = 2.85$.

c. If 50 field goals are attempted from 30 yards, the expected number of field goals made is $50(0.95) = 47.5$.

If 200 field goals are attempted from 30 yards, the expected number of field goals made is $200(0.95) = 190$.

55.
a. The probability that a phone call on Kwasi Caller ID box is for Amma is 0.65. Therefore, $10(0.65) = 6.5$ of the next ten phone calls are expected to be for Amma.

b. The probability that the call is for Amma is 0.65. The probability that the call is not for Amma is 0.35.

The probability that exactly eight of the next ten phone calls are for Amma is:

$$C(10,8) \cdot (0.65)^8 (0.35)^2 \approx 0.17565.$$

The probability that exactly nine of the next ten phone calls are for Amma is:

$$C(10,9) \cdot (0.65)^9 (0.35)^1 \approx 0.07249.$$

The probability that all of the next ten phone calls are for Amma is:

$$C(10,10) \cdot (0.65)^{10} (0.35)^0 \approx 0.01346.$$

Thus, the probability that at least eight of the next ten phone calls will be for Amma is:

$$0.17565 + 0.07249 + 0.01346 = 0.2616$$

c. Since the probability that a phone call will be for someone other than Amma is 0.35. Kwasi should expect $100(0.35) = 35$ of the next 100 phone calls to be for someone other than Amma.

57. Since the order does not matter, there are $C(48,1) \cdot C(2,2) = 48$ different samples in which two of the computers are defective and one is not. There are $C(50,3) = 19,600$ total possible samples of three computers. The probability that exactly two computers will be defective and one will not be defective in a sample of three is:

$$\frac{48}{19,600} \approx 0.00245.$$

59. The probability that a student will guess correctly is $\frac{1}{3}$.

The probability that a student will guess incorrectly is $\frac{2}{3}$.

The probability that exactly three of the questions are answered correctly is:

$$C(5,3) \cdot \left(\frac{1}{3}\right)^3 \left(\frac{2}{3}\right)^2 = \frac{40}{243}.$$

The probability that exactly four questions are answered correctly is:

$$C(5,4) \cdot \left(\frac{1}{3}\right)^4 \left(\frac{2}{3}\right)^1 = \frac{10}{243}.$$

The probability that all five questions are answered correctly is:

$$C(5,5) \cdot \left(\frac{1}{3}\right)^5 \left(\frac{2}{3}\right)^0 = \frac{1}{243}.$$

Therefore, the probability that at least three of the five questions are answered correctly is:

$$\frac{40}{243} + \frac{10}{243} + \frac{1}{243} = \frac{51}{243} \approx 0.2099.$$

61. Consider the tree diagram:

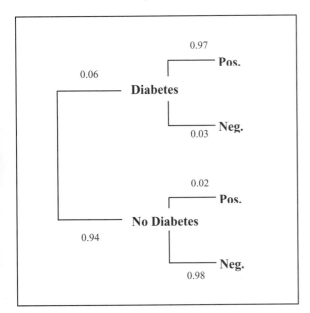

The probability that she has diabetes given that she tested positive is:

$$P(D|P) = \frac{P(D \cap P)}{P(P)} =$$

$$\frac{(0.06)(0.97)}{(0.06)(0.97)+(0.94)(0.02)} = \frac{291}{385}$$

$$P(D|P) = \frac{291}{385} \approx 0.7558.$$

63. The probability that both will live 20 more years is $\left(\frac{1}{5}\right)\left(\frac{4}{9}\right) = \left(\frac{4}{45}\right) \approx 0.0889$.

The probability that neither will be alive 20 years from now is $\left(\frac{4}{5}\right)\left(\frac{5}{9}\right) = \frac{4}{9}$.

So the probability that at least one of the two will be alive 20 years from now is $1 - \frac{4}{9} = \frac{5}{9} \approx 0.5556$.

65. There are $C(4,1) \cdot C(4,1)$ different combinations consisting of an Ace and a queen. There are $C(52,2)$ different two card combinations. The probability that one of the cards is an Ace and one of the cards is a queen is $\frac{C(4,1) \cdot C(4,1)}{C(52,2)} = \frac{16}{1326} \approx 0.01207$.

67. There are $6! = 720$ different ways to seat the children. There are $3 \times 3 \times 2 \times 2 \times 1 \times 1 = 36$ different ways of seating the children alternately such that a boy is in the first chair on the left. There are also 36 ways of seating the children alternately such that a girl is in the first chair on the left. Therefore there are 72 different ways of seating the children such that the boys and girls alternate seats. The probability that the boys and girls will be in alternate seats is $\frac{72}{720} = \frac{1}{10}$.

69. Since we are selecting with replacement, the draws are independent of each other. The probability that a given draw will be a club is $\frac{13}{52} = \frac{1}{4}$. Therefore, the probability that all three cards drawn are clubs is $\left(\frac{1}{4}\right)^3 = \frac{1}{64} \approx 0.0156$.

71. We have the formula:

$$P(A \cup B) = P(A) + P(B) - P(A \cap B)$$

Substituting in the known values we get:

$$0.8 = 0.6 + 0.5 - P(A \cap B)$$
Therefore,

$$P(A \cap B) = 0.6 + 0.5 - 0.8 = 0.3$$

73. Consider the Venn diagram where M and E represent those who failed math and English respectively.

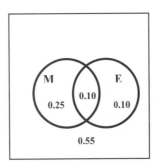

The probability that a freshman failed exactly one of the two classes is $0.25 + 0.1 = 0.35$.

75. We assume that the digits can be repeated, therefore the events are independent. The probability that the last two digits are both even is $\left(\frac{1}{2}\right)\left(\frac{1}{2}\right) = \frac{1}{4}$.

Chapter 9 Summary Exercises

1.. $P(W|B) = 0.2$.

3. $P(W) = (0.4)(0.4) + (0.5)(0.2) + (0.1)(0.7)$

$P(W) = 0.33$.

5. $P(B \cup Y) = P(B) + P(Y) - P(B \cap Y)$

The needed probabilities are:

$P(B) = 0.5$

$P(Y) = (0.4)(0.6) + (0.5)(0.8) + (0.1)(0.3)$

$P(Y) = 0.67$

$P(B \cap Y) = (0.5)(0.8) = 0.40$.

Therefore,

$P(B \cup Y) = 0.5 + 0.67 - 0.4 = 0.77$.

7. If A and B are mutually exclusive, then

$P(B) = P(A \cup B) - P(A)$

$P(B) = 0.8 - 0.5 = 0.3$.

Therefore,

$P(B') = 1 - P(B) = 1 - 0.3 = 0.7$.

9. The probability formula for the union of two events states:

$P(A \cup B) = P(A) + P(B) - P(A \cap B)$

Since the events A and B are independent, we know

$P(A \cap B) = P(A) P(B)$.

Substituting this equation into the formula for the union we get:

$P(A \cup B) = P(A) + P(B) - P(A) P(B)$

Solving the equation for the $P(B)$ we get:

$P(B) = \dfrac{P(A \cup B) - P(A)}{1 - P(A)}$

Substituting in the correct probabilities we get:

$P(B) = \dfrac{0.8 - 0.5}{1 - 0.5} = \dfrac{0.3}{0.5} = 0.6$

11. We use the formula:

$P(A \cup B) = P(A) + P(B) - P(A \cap B)$

Thus,

$P(A \cap B) = P(A) + P(B) - P(A \cup B)$

$P(A \cap B) = 0.5 + 0.7 - 0.8 = 0.4$.

By Bayes's Formula

$P(A \mid B) = \dfrac{P(A \cap B)}{P(B)} = \dfrac{0.4}{0.7} = \dfrac{4}{7} \approx 0.5714$.

13. Because C and D are mutually exclusive, we know $P(C \cup D) = P(C) + P(D) = \frac{3}{7} + \frac{4}{7} = 1$.

Thus, we can calculate $P(E)$ in the following manner:

$P(E) = P(E \cap C) + P(E \cap D)$

$P(E) = P(C) P(E \mid C) + P(D) P(E \mid D)$

$P(E) = \left(\frac{3}{7}\right)\left(\frac{2}{3}\right) + \left(\frac{4}{7}\right)\left(\frac{3}{4}\right) = \frac{2}{7} + \frac{3}{7} = \frac{5}{7}$.

15. We use the formula:

$P(D \mid E) = \dfrac{P(D \cap E)}{P(E)}$.

To get the needed information we calculate the following:

$P(D \cap E) = P(D) P(E \mid D) = \left(\frac{4}{7}\right)\left(\frac{3}{4}\right) = \frac{3}{7}$.

$P(E) = P(C) P(E \mid C) + P(D) P(E \mid D)$

$P(E) = \left(\frac{3}{7}\right)\left(\frac{2}{3}\right) + \left(\frac{4}{7}\right)\left(\frac{3}{4}\right) = \frac{2}{7} + \frac{3}{7} = \frac{5}{7}$

Therefore,

$P(D \mid E) = \dfrac{P(D \cap E)}{P(E)} = \dfrac{\frac{3}{7}}{\frac{5}{7}} = \dfrac{3}{5} = 0.6$.

17. If her success rate on first serves is 0.6, then her failure rate on first serves is 0.4. If her success rate on second serves is 0.9, then her failure rate on second serves is 0.1. The probability that the first serve is good is 0.6. The probability that the first serve was a fault, and the second serve is good is $(0.4)(0.9) = 0.36$. Therefore the probability that one of the two serves is good is $0.6 + 0.36 = 0.96$.

The probability that the first serve is good, given one serve is good is $\frac{0.6}{0.96} = 0.625$.

19. It seems reasonable to assume that the probability of traffic below will be independent of an avalanche occurring. So the probability of traffic being on the road, given an avalanche was triggered naturally should be the probability of traffic being on the road or $\frac{1}{25} = 0.04$.

21. If a skier is on the slope, the probability of an avalanche occurring is $\frac{1}{40} = 0.025$.

23. The probability of a completion on a given pass is 0.58. The probability of an incompletion on a given pass is 0.42. The probability that he completes exactly five passes is:

$$C(10,5) \cdot (0.58)^5 (0.42)^5 \approx 0.2162.$$

25. The probability that he completes more passes than not is equivalent to the probability that he completes at least six passes.

The probability that he completes exactly six passes is:

$$C(10,6) \cdot (0.58)^6 (0.42)^4 \approx 0.24876$$

The probability that he completes exactly seven passes is:

$$C(10,7) \cdot (0.58)^7 (0.42)^3 \approx 0.19630$$

The probability that he completes exactly eight passes is:

$$C(10,8) \cdot (0.58)^8 (0.42)^2 \approx 0.10166$$

The probability that he completes exactly six passes is:

$$C(10,9) \cdot (0.58)^9 (0.42)^1 \approx 0.03120$$

The probability that he completes exactly ten passes is:

$$C(10,10) \cdot (0.58)^{10} (0.42)^0 \approx 0.004308.$$

The probability that he completes more passes than not is:

$$0.24876 + 0.19630 + 0.10166 +$$
$$0.03120 + 0.00431 = 0.58223$$

27. $P(S) = P(S \cap T) + P(S \cap T')$

Thus,

$$P(S) = \tfrac{1}{10} + \tfrac{1}{5} = \tfrac{3}{10}.$$

Now, since $P(S)P(T) = P(S \cap T)$

$$P(T) = \frac{P(S \cap T)}{P(S)} = \frac{\frac{1}{10}}{\frac{3}{10}} = \frac{1}{3}.$$

We can now find

$$P(T') = 1 - \tfrac{1}{3} = \tfrac{2}{3}$$
$$P(S') = 1 - \tfrac{3}{10} = \tfrac{7}{10}.$$

$$P\left((S \cup T)' \right) = P(S' \cap T') = \left(\tfrac{2}{3}\right)\left(\tfrac{7}{10}\right) = \tfrac{7}{15}.$$

Therefore, the correct answer is 'c'.

29. The theoretical distribution would be:

SUM	FREQ	SUM	FREQ
2	10	8	50
3	20	9	40
4	30	1	30
5	40	11	20
6	50	12	10
7	60		

However, running empirical probabilities will probably not match the theoretical distribution exactly. The pattern should be similar with values of 6, 7, 8 occurring more frequently than 2, 3, 11, or 12.

Cumulative Review

31. The augmented matrix is:

$$\begin{bmatrix} 3 & -1 & 2 & | & 19 \\ -1 & 3 & 10 & | & -1 \\ 2 & 4 & 5 & | & 7 \end{bmatrix}$$

We perform Gauss-Jordan elimination to reduce the matrix to the following:

$$\begin{bmatrix} 1 & 0 & 0 & | & 5 \\ 0 & 1 & 0 & | & -2 \\ 0 & 0 & 1 & | & 1 \end{bmatrix}$$

Therefore, the solution is:

$$x = 5; \; y = -2; \; z = 1.$$

33. The matrix product is:

$$AB = \begin{bmatrix} 1 & 0 & 3 & 2 \\ 2 & -5 & 1 & 0 \end{bmatrix} \begin{bmatrix} 3 & -1 & 0 \\ -2 & 1 & 2 \\ 1 & 0 & -1 \\ 2 & -1 & 0 \end{bmatrix} =$$

$$AB = \begin{bmatrix} 10 & -3 & -3 \\ 17 & -7 & -11 \end{bmatrix}$$

Chapter 9 Sample Test Answers

1. The probability is $\left(\frac{4}{7}\right)\left(\frac{3}{6}\right) = \frac{2}{7}$.

3. Since there are only three boys, if the girls and the boys are to be seated alternately a girl must be seated in the right hand seat. So the probability is the same as that of a girl being seated in the right hand seat, which is $\frac{4}{7}$.

5. The probability that a finite math student is a female is $\frac{140}{260} = \frac{7}{13}$.

7. The probability that a finite math student is a female and a freshman is $\frac{86}{260} = \frac{43}{130}$.

9. There are 150 total freshmen. Of those 150 freshmen, 86 of the students are female. The conditional probability that a finite math student is a female, given that the student is a freshman is $\frac{86}{150} = \frac{43}{75}$.

11. The probability that a finite math student is not a male or not a freshman is $\frac{196}{260} = \frac{49}{65}$.

13. The probability that the card will be a 9 given that it is a 10 is zero.

15. The probability is $(0.002)(0.2) = 0.0004$.

17. There are $6^3 = 216$ possible outcomes. There are $5^3 = 125$ possible outcomes where no six is rolled. Therefore there are $216 - 125 = 91$ outcomes with at least one six. The probability that at least one six will be rolled is $\frac{91}{216}$.

For question 19 consider the Venn diagram.

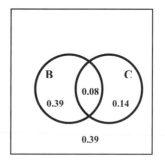

19. Given that the person did not buy crackers, the probability that a person bought bread is $\frac{0.39}{0.39 + 0.39} = 0.5$

21. Given that the third child is a girl, the sample space is reduced to:

$$\{BBG, GBG, BGG, GGG\}$$

In this reduced samples space two outcome have exactly two girls. Therefore, the probability that exactly two children are girls given that the third child is a girl is $\frac{2}{4} = \frac{1}{2} = 0.5$.

For question 23 consider the tree diagram.
Note: L is the set of students who like crab legs and N is the set of student who do not like crab legs.

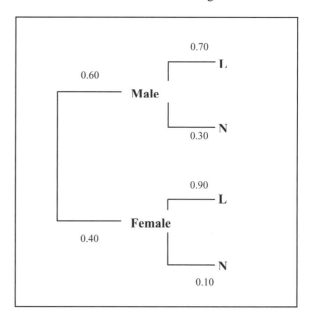

23. The probability that a member is female and likes eating crab for dinner is $(0.4)(0.9) = 0.36$

The probability that a member likes eating crab for dinner is

$$(0.6)(0.7) + (0.4)(0.9) = 0.78.$$

The solution is continued on the following page.

The probability that a member is female given that the member likes to eat crab legs for dinner is:

$$P\left(F|L\right) = \frac{P\left(F \cap L\right)}{P\left(L\right)} = \frac{0.36}{0.78} = \frac{36}{78} = \frac{6}{13}.$$

25. The probability that exactly 18 will sprout is:

$$C\left(20,18\right) \cdot \left(0.9\right)^{18}\left(0.1\right)^{2} \approx 0.28518.$$

The probability that exactly 19 will sprout is:

$$C\left(20,19\right) \cdot \left(0.9\right)^{19}\left(0.1\right)^{1} \approx 0.27017.$$

The probability that exactly 20 will sprout is:

$$C\left(20,20\right) \cdot \left(0.9\right)^{20}\left(0.1\right)^{0} \approx 0.12158.$$

Therefore the probability that at least 18 will sprout is:

$$0.28518 + 0.27017 + 0.12158 = 0.67693.$$

Chapter 10
Statistics

Exercises Section 10.1

1. The expanded frequency table is:

Class	Frequency	Relative Frequency	Cumulative Frequency
0 up to 3	4	0.114	4
3 up to 6	8	0.229	12
6 up to 9	12	0.343	24
9 up to 12	7	0.2	31
12 up to 15	4	0.114	35

The frequency histogram is:

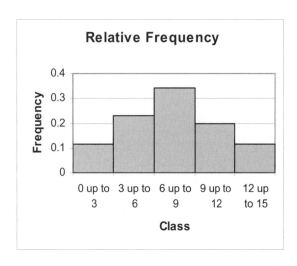

The relative frequency histogram is:

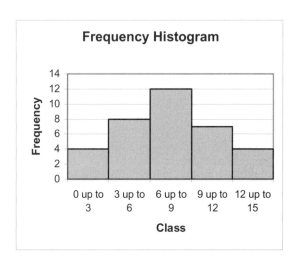

The cumulative frequency histogram is:

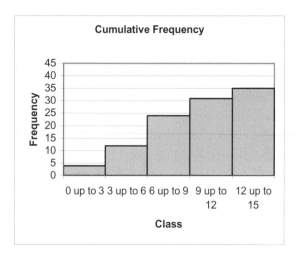

The data seems to be normally distributed. Maybe skewed slightly to the right.

3. The expanded frequency table is:

Class	Frequency	Relative Frequency	Cumulative Frequency
3	9	0.3	9
4	12	0.4	21
5	4	0.133333	25
6	3	0.1	28
7	2	0.066667	30

The frequency histogram is:

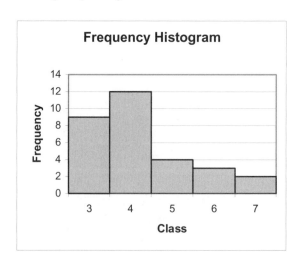

The relative frequency histogram is displayed at the top of the next page.

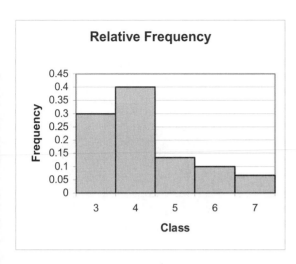

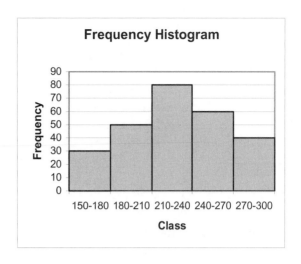

The cumulative frequency histogram is:

The relative frequency histogram is:

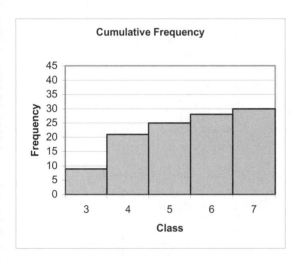

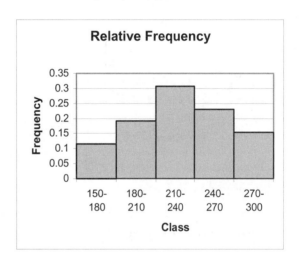

The data appears to be skewed to the right.

The cumulative frequency histogram is:

5. The expanded frequency table is:

Class	Frequency	Relative Frequency	Cumulative Frequency
150-180	30	0.115385	30
180-210	50	0.192308	80
210-240	80	0.307692	160
240-270	60	0.230769	220
270-300	40	0.153846	260

The frequency histogram is displayed at the top of the next column.

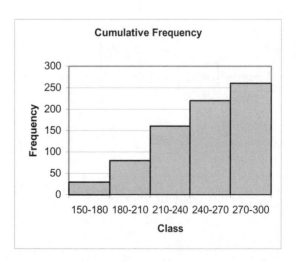

The data appears to be normal, maybe slightly skewed to the left.

7. The expanded frequency table is:

Class	Frequency	Relative Frequency	Cumulative Frequency
0	2	0.1	2
1	3	0.15	5
2	8	0.4	13
3	5	0.25	18
4	2	0.1	20

The frequency histogram is:

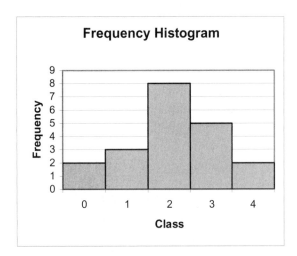

The relative frequency histogram is:

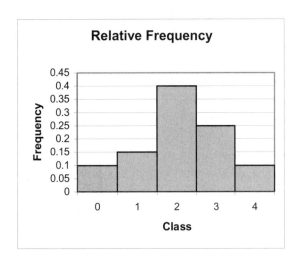

The cumulative frequency histogram is shown at the top of the next column.

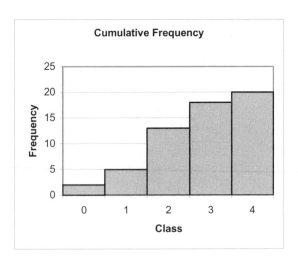

The data appears to be normally distributed, maybe slightly skewed left.

9. There are $6 + 5 + 3 = 14$ people who have spent three or more minutes waiting in line.

11. Forty-eight of the 62 people observed spent less than three minutes waiting in line. Since $\frac{48}{62} \approx 0.774$, approximately 77.4% of the people observed spent less than three minutes in line.

13. The logical classes of equal width would be six classes of width one starting at the value ten hours. using this, the frequency tables for the data are:

Class	Frequency	Relative Frequency	Cumulative Frequency
10-11	2	0.04	2
11-12	4	0.08	6
12-13	11	0.22	17
13-14	16	0.32	33
14-15	11	0.22	44
15-16	6	0.12	50

15. Adding up the values in the relative frequency table from problem 13 for the classes 12 and higher we get: $0.22 + 0.32 + 0.22 + 0.12 = 0.88$. This means 88% of the batteries have a life of 12 or more hours.

17. The frequency histogram for the data on bean pods is:

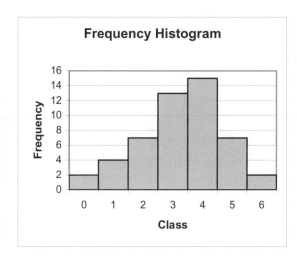

The relative frequency histogram for the data on bean pods is:

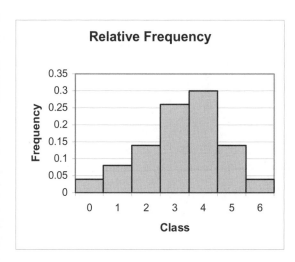

19. The relative frequency histogram indicates that 48% of the pods have four or more beans in them. this obtained by adding the relative frequencies for classes 4 beans per pod, 5 beans per pod, and 6 beans per pod.

21. The probability that a pod will have three or more beans in it is: $\frac{37}{50} = 0.74$. The probability that a pod will not have three or more beans in it is: $\frac{13}{50} = 0.26$. Therefore the odds that a pod will have three or more beans in it are: $\dfrac{\frac{37}{50}}{\frac{13}{50}} = \dfrac{37}{13}$, or 37 to 13. (Note, this is slightly less than 3 to 1.)

23. The frequency histogram for the number of can openers in a household is:

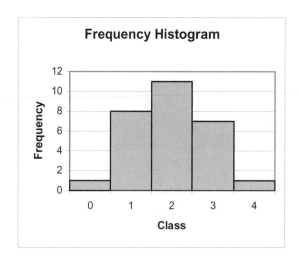

The relative frequency histogram for the above data is:

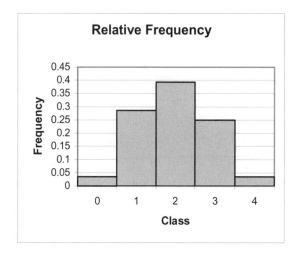

The data appears to be normally distributed.

25. The data indicates that 11 out of the 28 households questioned own exactly two can openers. Therefore, $\frac{11}{28} \approx 0.393$ or 39.3% of the households questioned own exactly two can openers.

27. The probability that a home will have three or more can openers is $\frac{8}{28}$. The probability that a household will not have three or more can openers is: $1 - \frac{8}{28} = \frac{20}{28}$. The odds for a household owning three or more can openers are determined by:

$$\dfrac{\frac{8}{28}}{\frac{20}{28}} = \dfrac{8}{20}$$

Therefore, the odds for a household owning 3 or more can openers are 9 to 19. (Note: this slightly less than 2 to 5.)

29. The frequency histogram for the data is:

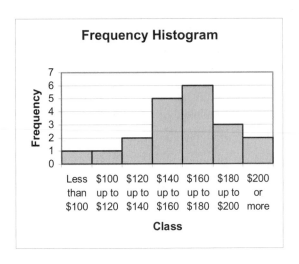

The relative frequency histogram for the data is:

31. The frequency distribution and relative frequency distribution tables are:

Class	Frequency	Relative Frequency
1 to 2 min.	4	0.133333
2 to 3 min.	5	0.166667
3 to 4 min.	6	0.2
4 to 5 min.	5	0.166667
5 to 6 min.	3	0.1
6 to 7 min.	4	0.133333
7 to 8 min.	3	0.1

33. The cumulative frequency histogram is displayed at the top of the next column.

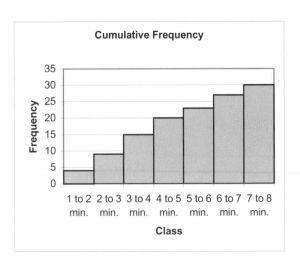

35. The frequency distribution and relative-frequency distribution tables for the data are:

Class	Frequency	Relative Frequency
2 to 3 months	2	0.083333
4 to 5 months	2	0.083333
6 to 7 months	2	0.083333
8 to 9 months	6	0.25
10 to 11 months	4	0.166667
12 to 13 months	4	0.166667
14 to 15 months	4	0.166667

37. The probability that the cab company will keep a care 11 months or longer is $\frac{9}{24} = \frac{3}{8}$.

39. To make the salaries appear as low as possible, we first observe that many of the salaries end in 25 or 75. The highest weekly salary is $525, and the lowest salary is $175. If we choose to group the salaries in $100 intervals ending in $225, $325, $425, and $525, and decide to include the right hand endpoint in our interval, most of the salaries in each grouping will be in the highest end, making the salaries appear lower than they actually are. The frequency table is displayed on the following page.

Class	Frequency
$126 to $225	3
$226 to $325	6
$326 to $425	6
$426 to $525	5

The frequency histogram for the above data is:

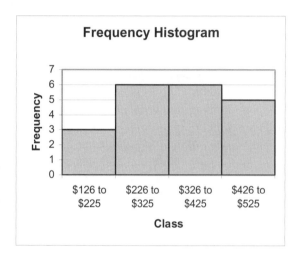

41.
a. The cumulative relative frequency histogram for these heights is:

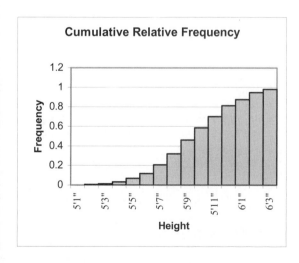

b. In order to calculate the relative frequency for each height, we simply subtract the cumulative relative frequency of the height in question by the cumulative relative frequency of the proceeding height. For example, to find the relative frequency for the height of those males who are 5'2" we simply calculate $0.005 - 0.001 = 0.004$.

The relative frequencies are shown in the expanded table:

Height	Cumulative Relative Frequency	Relative Frequency
5'1"	0.001	0.001
5'2"	0.005	0.004
5'3"	0.013	0.008
5'4"	0.034	0.021
5'5"	0.069	0.035
5'6"	0.117	0.048
5'7"	0.208	0.091
5'8"	0.32	0.112
5'9"	0.463	0.143
5'10"	0.587	0.124
5'11"	0.701	0.114
6'0"	0.812	0.111
6'1"	0.874	0.062
6'2"	0.947	0.073
6'3"	0.979	0.032

c. The relative frequency histogram for these heights is:

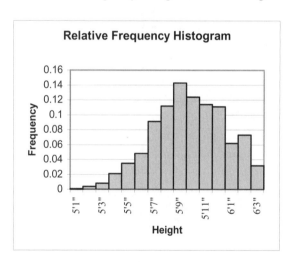

43.

a. The relative frequency histogram for the data is:

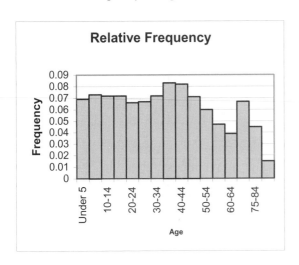

b. The cumulative relative frequency for each class interval is shown in the table below:

Class	Relative Frequency	Cumulative Relative Frequency
Under 5	0.069	0.069
5-9	0.073	0.142
10-14	0.072	0.214
15-19	0.072	0.286
20-24	0.066	0.352
25-29	0.067	0.419
30-34	0.072	0.491
35-39	0.083	0.574
40-44	0.082	0.656
45-49	0.071	0.727
50-54	0.06	0.787
55-59	0.047	0.834
60-64	0.039	0.873
65-74	0.067	0.94
75-84	0.045	0.985
85 and over	0.015	1

The cumulative relative frequency distribution is shown at the top of the next column.

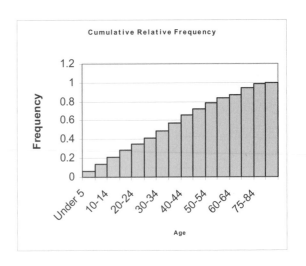

45-51. Answers will vary depending on who is surveyed.

Exercises Section 10.2

1.

a. There is no mode.

b. The mean of the population is:

$$\mu = \frac{6+9+18+26+43}{5} = \frac{102}{5} = 20.4 \ .$$

c. The median is 18.

3.

a. The mode is 102 (102 appears twice).

b. The mean of the sample is:

$$\bar{x} = \frac{96+102+102+112+173+225}{6} = \frac{810}{6} = 135 \ .$$

c. There are six scores, the second highest score is 102, and the third highest score is 112. Therefore the median is $\frac{102+112}{2} = \frac{214}{2} = 107$.

5.

a. There is no mode, each value appears only once.

b. The mean of the population is:

$$\mu = \frac{-12-8-6-5-3+0+3+5+6+12}{11} = \frac{0}{11} = 0 \ .$$

c. There are 11 scores. The sixth highest is zero . Therefore the median is zero.

7. The median is $22,000. The median best describes the data. The one extreme value of $60,000 distorts the mean of $31,000, and there is no mode.

9. The median is 20. The mean is 28.75. There is no mode. The mean is overstates the typical number of Girl Scout cookies sold per scout, because of the one girl who sold 80 boxes. The median best describes the data.

11. The median is 7. The mean is 6.73. The mode is 7. The vast majority of the show sizes are 7. This might be of interest to the shoe manufacturer. The mode is of more interest than usual in this data set.

13. The median is 20. The mean is 22. The population is bimodal 19 and 20 appear three times each. In this case the median best describes the data. The mean is distorted by the two older students.

15. The weighted mean is:

$$\left(\tfrac{1}{2}\right)(5) + \left(\tfrac{1}{2}\right)(7) = 6.$$

17. The weighted mean is:

$$\left(\tfrac{1}{2}\right)(2) + \left(\tfrac{1}{4}\right)(3) + \left(\tfrac{1}{8}\right)(4) + \left(\tfrac{1}{8}\right)(5) = \tfrac{23}{8}.$$

19. The weighted mean is:

$$\left(\tfrac{5}{8}\right)(-4) + \left(\tfrac{1}{4}\right)(-2) + \left(\tfrac{1}{8}\right)(0) = -3.$$

21. The mean is the "average" being used. The mean is distorted by the two large donations of $250 and $500. The rest of the donations were all below $50, with the mode of $25 and a median of $30. Either the mode or the median would give a more accurate picture of the "average" gift that was donated.

23. The median for the sample is $\frac{360+370}{2} = 365$.

The median best describes the sample, because the mean is too high due to the high score of 990.

25. The mean of this population is:

$$\frac{4+6+8+9+12+12}{5} = 8.5.$$

The median of this population is $\frac{8+9}{2} = 8.5$.

27. The mean price is $\frac{(700)(6)+(300)(3.5)}{1000} = \frac{5250}{1000} = 5.25$.

29. The mean of the two selling prices is:

$$\frac{6.00+3.50}{2} = \$4.75.$$

31. The mean purchase price per six-pack is:

$$\frac{(20)(1.79)+(15)(1.59)}{35} = \$1.70.$$

33. The mean of the two purchase prices is:

$$\frac{1.59+1.79}{2} = \$1.69.$$

35. There were a total of 90 treadmills purchased. The median purchase price will be the average of the 45th and 46th treadmill. Since both the 45th and the 46th treadmill were purchased for $150, the median purchase price per unit is $150.

37. The mean price per unit for which the calculators were sold is:

$$\frac{(25)(2000)+(35)(1500)+(80)(1000)+(95)(800)+(120)(500)}{5800} =$$

$$\frac{\$318,500}{5800} \approx \$54.91$$

39. The mean of the unit prices is:

$$\frac{25+35+80+95+120}{5} = \frac{355}{5} = \$71.00.$$

41. The mean quiz score is:

$$\frac{(2)(5)+(2)(6)+(9)(7)+(12)(8)+(7)(9)+(5)(10)}{37} = \frac{294}{37} = 7.95.$$

43. The mean number of automobiles per household is:

$$\frac{(1)(0)+(3)(1)+(12)(2)+(9)(3)+(2)(4)}{27} = \frac{62}{27} \approx 2.296.$$

45. There are 27 automobiles, so the median number of automobiles per household is the number of automobiles in the 14th highest household. If we list the households in order from lowest number of automobiles to highest, we see that the median number of automobiles per household is two.

47. Since 28 pods were opened, the median number of peas per pod will be the average number of peas in the pods that have the 14th highest total and the 15th highest total. Both of these pods contained 4 peas, so the median number of peas per pod is four.

49. If we assume that the amount in each account is the midpoint of the range the amount falls in, the total amount in the 30 accounts is:

$$(5)(\$25)+(16)(\$75)+(5)(\$150)+(4)(\$250)$$

Which totals to $\$3075$. Thus, the mean amount of the 30 accounts is:

$$\frac{\$3075}{30}=\$102.50.$$

51. If we assume the scores in each interval to be at the average, the sum of the 30 scores is:

$$(6)(14)+(12)(18)+(9)(22)+(3)(26)=576$$

Thus, the estimated mean of the 30 scores is:

$$\frac{576}{30}=19.2.$$

We can approximate the median by noticing the 15th and 16th score both lie in the 16-20 range. Once again assuming the scores represent the midpoint of the range it falls in, the approximate median score is 18.

53. Assuming that the average age of the persons in each age group is at the midpoint of the group, the sum of the ages is:

$$(8)(64.5)+(5)(74.5)+(8)(84.5)=1564.5.$$

Thus the approximate mean age on the sightseeing trip is:

$$\frac{1564.5}{21}\approx 74.5.$$

55. Since each one-hour test score has the weight of one test, and the final has the weight of two tests, the weighted mean is:

$$\frac{78+92+84+(2)(85)}{5}=84.8.$$

The mean class average is 84.8.

57. Let x be the score on the final needed to obtain a final average of 90. Using the weighted mean we have:

$$\frac{88+79+91+2x}{5}=90$$

Solve the equation for x as shown at the top of the next column.

$$\frac{258+2x}{5}=90$$
$$258+2x=450$$
$$2x=192$$
$$x=96$$

Therefore, you will have to make 96 out of 100 on the final to receive an 'A' in the course.

59. To find the frequency from the relative frequency in each class interval, simply multiply each relative frequency by the sample size. There will be some round off error when this is done. The results are shown in the table below:

Class	Relative Frequency	Frequency
$1 to $100	0.222	604
$101 to $200	0.125	340
$201 to $300	0.099	269
$301 to $400	0.08	218
$401 to $500	0.061	166
$501 to $600	0.047	128
$601 to $700	0.037	101
$701 to $999	0.061	166
$1000 or more	0.268	729

61. Since the sample size is 2719, the median household charitable contribution will be the 1359th highest contribution. Looking at the frequencies of each class interval, we see that the 1359th highest charitable contribution falls in the $301 to $400 dollar range. Assuming that the contributions are the midpoint of the range, the estimated median of U.S. household charitable contributions is 1998 is $350.5.

63. Create the cumulative relative frequency by summing the relative frequencies as shown in the table below:

Class	Relative Frequency	Weighted Mean	Cumulative Relative Frequency
Under 5	0.069	0.138	0.069
5-9	0.073	0.511	0.142
10-14	0.072	0.864	0.214
15-19	0.072	1.224	0.286
20-24	0.066	1.452	0.352
25-29	0.067	1.809	0.419
30-34	0.072	2.304	0.491
35-39	0.083	3.071	0.574
40-44	0.082	3.444	0.656
45-49	0.071	3.337	0.727
50-54	0.06	3.12	0.787
55-59	0.047	2.679	0.834
60-64	0.039	2.418	0.873
65-74	0.067	4.6565	0.94
75-84	0.045	3.5775	0.985
85 and over	0.015	1.275	1
Total	1	35.88	

Note: For the age group 85 and over, we used the age 85 as the average age.

63. Looking at the cumulative relative frequency, we notice that the cumulative relative frequency of 0.5 falls somewhere in the 35-39 age range. Therefore, we estimate the median of the population to be 37 years old.

65. Answers will vary based on the data collected.

Exercises Section 10.3

1.

a. To calculate the mean use $n = 5$ and $\sum x = 20$. Therefore,

$$\bar{x} = \frac{20}{5} = 4.$$

b.

x	$x - \bar{x}$	$(x - \bar{x})^2$
2	-2	4
3	-1	1
4	0	0
5	1	1
6	2	4
SUM 20	0	10

c. From the table, we see the sum of the deviations is zero.

d. From the table, we see the sum of the squares of the deviations is 10.

e. The sample variance is $\frac{10}{4} = 2.5$.

f. The sample standard deviation $\sqrt{2.5} \approx 1.5811$.

3.

a. To calculate the mean use $n = 10$ and $\sum x = 477$. Therefore,

$$\mu = \frac{477}{10} = 47.7.$$

b. The information is given in the table:

x	$x - \mu$	$(x - \mu)^2$
12	-35.7	1274.49
18	-29.7	882.09
20	-27.7	767.29
30	-17.7	313.29
32	-15.7	246.49
33	-14.7	216.09
80	32.3	1043.29
82	34.3	1176.49
84	36.3	1317.69
86	38.3	1466.89
SUM 477	0	8704.1

c. From the table, we see the sum of the deviations is zero.

d. From the table, we see the sum of the squares of the deviations is 8704.1.

e. The population variance is $\frac{8704.1}{10} = 870.41$.

f. The population standard deviation $\sqrt{870.41} \approx 29.502$.

5.

a. To calculate the mean use $n = 10$ and $\sum x = 531$. Therefore,

$$\bar{x} = \frac{531}{10} = 53.1.$$

b.

x	$x - \overline{x}$	$\left(x - \overline{x}\right)^2$
50	-3.1	9.61
51	-2.1	4.41
52	-1.1	1.21
52.5	-0.6	0.36
53	-0.1	0.01
53.5	0.4	0.16
54	0.9	0.81
54.5	1.4	1.96
55	1.9	3.61
55.5	2.4	5.76
SUM 531	0.0	27.9

c. From the table, we see the sum of the deviations is zero.

d. From the table, we see the sum of the squares of the deviations is 27.9.

e. The sample variance is $\frac{27.9}{9} = 3.1$.

f. The sample standard deviation $\sqrt{3.1} \approx 1.76068$.

7. Since the data varies more in sample 'b' then it does in sample 'a', sample 'b' will have the larger standard deviation.

9. The extreme value of 50 in sample 'a' causes the standard deviation in sample 'a' to be larger than the standard deviation in sample 'b'.

11.

a. To calculate the mean consider $n = 5$ and $\sum x = 48$

Therefore, the mean is:

$$\mu = \frac{48}{5} = 9.60$$

The population standard deviation is:

$$\sum \left(x - \mu\right)^2 = 15.3125$$
$$\sigma^2 = \frac{15.3125}{5} = 3.0625$$
$$\sigma = \sqrt{3.0625} = 1.75.$$

The population standard deviation is 1.75.

b. To calculate the mean consider:

$$n = 5 \text{ ,and} \sum x = 48$$

Therefore, the mean is:

$$\mu = \frac{48}{5} = 9.60$$

To calculate the population standard deviation:

$$\sum \left(x - \mu\right)^2 = 105.524$$
$$\sigma^2 = \frac{105.524}{5} = 21.1048$$
$$\sigma = \sqrt{21.1048} = 4.594.$$

The population standard deviation is 4.594.

13.

a. The mean of the sample is $1276.31. This is found by summing the values in the sample and dividing by 10. The standard deviation of the sample is $567.26. First find the variance of the sample by summing the squares of the deviations from the mean and dividing by nine. Then take the square root of the variance to find the sample standard deviation.

b. The range of claims that would be within one standard deviation of the mean is $709.05 up to $1845.57. Eight of the ten claims fall within this range.

15. The population mean is 79.375, and the population standard deviation is 8.2906. To calculate the standard deviation, first find the variance of the population by summing the squares of the deviations from the mean and dividing by 8. Then take the square root of the variance to find the population standard deviation.

b. The range of data that is within two standard deviations of the mean is 62.7937 to 95.956. All eight scores are in this range.

17.

a. The population mean is 76.25, and the population standard deviation is 6.476.

b. The range of data from 56.825 to 95.675 is within three standard deviations from the mean. All eight values fall within this range.

19.

a. The sample mean is 45.2.

b. The sample variance is 88.7. The units of the sample variance are degrees squared.

c. The sample standard deviation is 9.41807. The units of the standard deviation are degrees.

21.

a. The population mean is 315.6. The median is the third highest value or 345.

b. The population variance is 6494.64. The units of the population variance are expressed in tourist squared.

c. The population standard deviation is approximately 80.59. The units of the population standard deviation are expressed in tourists.

23.
a. The sample mean is 78.5. The median of the sample is the average of the fourth and fifth largest number. Therefore the median is $\frac{80+81}{2} = 80.5$.

b. The sample variance is 52.5. The units of the sample variance are expressed in blooms squared.

c. The sample standard deviation is 7.25. The units of the sample standard deviation are expressed in blooms.

25. We are given that the population mean is $\mu = 8$ and the population standard deviation is $\sigma = 1$. To last between six and ten years is to vary less than 2 years from the mean. The probability that this will happen is :

$$P\left((\mu - 2) \leq x \leq (\mu + 2)\right) \geq 1 - \left(\tfrac{1}{2}\right)^2$$
$$P\left((\mu - 2) \leq x \leq (\mu + 2)\right) \geq \tfrac{3}{4} = 0.75 .$$

Thus, the probability that a dishwasher will last between six and ten years is at least 0.75.

27. We are given that the population mean is $\mu = 35,000$ and the population standard deviation is $\sigma = 2500$. For a tire to last less than 30,000 miles or more than 40,000 mile is the same as the tire not lasting between 30,000 miles and 40,000 miles. The probability that a tire will last between 30,000 miles and 40,000 miles is:

$$P\left((\mu - 5000) \leq x \leq (\mu + 5000)\right) \geq 1 - \left(\tfrac{2500}{5000}\right)^2$$
$$P\left((\mu - 5000) \leq x \leq (\mu + 5000)\right) \geq \tfrac{3}{4} \approx 0.75 .$$

So the probability that a tire will last between 30,000 miles and 40,000 miles is at least 0.75. Therefore, the probability that a tire will last less than 30,000 miles or more than 40,000 miles is at most 0.25.

29. Let $k = 2\sigma$. The probability that an observation will lie within two standard deviations of the mean is given by:

$$P\left((\mu - k) \leq x \leq (\mu + k)\right) \geq 1 - \left(\tfrac{\sigma}{k}\right)^2$$
$$P\left((\mu - k) \leq x \leq (\mu + k)\right) \geq 1 - \left(\tfrac{\sigma}{2\sigma}\right)^2$$
$$P\left((\mu - k) \leq x \leq (\mu + k)\right) \geq 1 - \tfrac{1}{4} = \tfrac{3}{4} .$$

The probability that an observation will always lie within two standard deviations of the mean is at least 0.75.
31.
a. We assume that the scores in each range are concentrated at the middle of the range. Consider the following table:

Score	Frequency	
54.5	5	272.5
64.5	15	967.5
74.5	50	3725
84.5	26	2197
94.5	26	378
SUM	100	7540

The mean is approximately $\frac{7540}{100} = 75.4$.

b. Again assume the scores are concentrated at the middle of each range. We have the following table:

Score- x	Frequency	$(x - \mu)$	$f \cdot (x - \mu)^2$
54.5	5	-20.9	2184.05
64.5	15	-10.9	1782.15
74.5	50	-0.9	40.5
84.5	26	9.1	2153.06
94.5	26	19.1	1459.24
SUM	100	--	7619.00

So the population variance is approximately $\frac{7619.00}{100} = 76.19$ and the population standard deviation is approximately the square root of the population variance or $\sqrt{76.19} = 8.73$ points.

33. Assume that all the mileages in each range are at the midpoint of each range. If we make this assumption, we get the following table:

Miles per gallon - x	Frequency- f	$f \cdot x$
23	6	138
25	12	300
27	20	540
29	10	290
31	2	62
SUM	50	1330

The sample mean is approximately $\frac{1330}{50} = 26.6$ miles per gallon.

b. Again, we assume that the scores are concentrated at the midpoint of each range. Consider the table on the following page.

x	f	$(x-\mu)^2$	$f\cdot(x-\mu)^2$
23	6	12.96	77.76
25	12	2.56	30.72
27	20	0.16	3.20
29	10	5.76	57.60
31	2	19.36	38.72
SUM	50	--	208.00

The sample variance is approximately $\frac{208}{49} = 4.245$. The sample standard deviation is the square root of the sample variance, so the sample standard deviation is

$\sqrt{4.245} = 2.0603$.

35. We assume the data points in each range are concentrated at the midpoint of each range, and use 2000 to estimate the average value in the last interval. Consider the table below:

Contribution-x	Frequency-f	$f\cdot x$
50.5	604	30502
150.5	340	51170
250.5	269	67384.5
350.5	218	76409
450.5	166	74783
550.5	128	70464
650.5	101	65700.5
850	166	141100
2000	729	1458000
SUM	2721	2035513

The sample mean of this data is $\frac{2035513}{2721} = 748.08$

To find the sample variance consider the following table:

x	f	$(x-\mu)^2$	$f\cdot(x-\mu)^2$
50.5	604	486617.9	293917185.3
150.5	340	357101.9	121414631.2
250.5	269	486617.9	130900203.4
350.5	218	158069.9	34459228.7
450.5	166	88553.86	14699940.16
550.5	128	39037.86	4996845.619
650.5	101	9521.856	961707.4964
850	166	10387.69	1724355.942
2000	729	1567304	1142564387
SUM	2721	3203212	1745638485

Therefore the sample variance of the data is

$s^2 = \dfrac{1,745,638,485}{2720} = 641,778.85$.

Therefore, the sample standard deviation of the data is:

$s = \sqrt{s^2} = \sqrt{641,778.85} = 801.11$.

37. The expected earnings are:

$E(X) = \left(\frac{3}{8}\right)(5) + \left(\frac{5}{8}\right)(-2) = 0.625$.

On average you should expect to win $0.63 each time you play.

The variance is:

$\sigma^2 = \left(\frac{3}{8}\right)(5-0.625)^2 + \left(\frac{5}{8}\right)(-2-0.625)^2$
$\sigma^2 = 11.4843$

Therefore, the standard deviation from the expected earnings is:

$\sigma = \sqrt{\sigma^2} = \sqrt{11.4843} = 3.389$.

The standard deviation from the expected earnings is $3.39.

39. The expected value of a policy in thousands of dollars is:

$E(X) = (0.98)(1) + (0.02)(-49) = 0$.

On average the company expects to break even on each policy.

The variance is:

$\sigma^2 = (0.98)(1-0)^2 + (0.02)(-49-0)^2$
$\sigma^2 = 49$

Therefore, the standard deviation from the expected value is:

$\sigma = \sqrt{\sigma^2} = \sqrt{49} = 7$.

The standard deviation from the expected value is $7,000.

Exercises Section 10.4

1. Probability curve B has the larger standard deviation.

3. The graph of a normal curve with a mean of $\mu = 20$ and a standard deviation of $\sigma = 2$ changes from concave up to concave down at $x = 20 - 2 = 18$, and from concave down back to concave up at $x = 20 + 2 = 22$.

5.

a. The point on the x-axis lies at:

$$x = 20 + 2(1.5) = 23.$$

b. The point on the x-axis lies at:

$$x = 20 - 3(1.5) = 15.5.$$

c. The point on the x-axis lies at:

$$x = 20 - 1.6(1.5) = 17.6.$$

7. Since $\mu = 20$, and $\mu + \sigma = 23$, the area in question is one standard deviation above the mean. Therefore, 34% of the area under the curve lies between $x = 20$ and $x = 23$.

9. Since the area is symmetric about the mean and $\mu = 20$, 50% of the area is to the left of $x = 20$.

11. Since $\mu - 3\sigma = 11$, we are looking to find the area of the curve that is less than three standard deviations below the mean. We know that 49.85% of the area under the curve is between $x = 11$ and $x = 20$. Since 50% of the area is to the left of $x = 20$, this means that $0.50 - 0.4985 = 0.0015$, or 0.15% of the area under the curve lies to the left of $x = 11$.

13. Since $\mu - 2\sigma = 88$, we are looking to find the area of the curve that is less than two standard deviations below the mean. We know that 47.5% of the area under the curve lies within two standard deviations below the mean. Since 50% of the area is to the left of $x = 100$, this means that $0.50 - 0.475 = 0.025$, or 2.5% of the area under the curve lies to the left of $x = 88$.

15. Since $\mu = 100$ and $\mu + \sigma = 106$, The area of the curve that lies between $x = 100$ and $x = 106$ is 34%. Furthermore, we know that 50% of the area underneath the curve lies to the left of $x = 100$. Therefore, 84% of the curve is to the left of $x = 106$.

17. Since $\mu = 0$, $\mu - \sigma = -1$, and $\mu + \sigma = 1$, The area in question is one standard deviation below the mean and one standard deviation above the mean. Therefore, 68% of the area under the curve lies between $z = -1$ and $z = 1$.

19. Since $\mu = 0$, $\mu - \sigma = -1$, 34% of the area under the curve lies between $z = -1$ and $z = 0$. Because 50% of the curve lies to the right of $z = 0$, we know that $50\% + 34\% = 84\%$ of the area under the curve lies to the right of $z = -1$.

21. The table shows that the area to the left of $z = 2$ is 0.9773.

23. The table shows that the area to the left of $z = -2.3$ is 0.0107.

25. Remember, the area to the right of a z-value is equal to one minus the area to the left of the z-value. Thus, the table shows that the area to the right of $z = 0.03$ is $1 - 0.5120 = 0.4880$.

27. The table shows that the area to the left of $z = 0.67$ is 0.7484 and the area to the left of $z = -2.1$ is 0.0179. Therefore, the area between $z = -2.1$ and $z = 0.67$ is $0.7484 - 0.0179 = 0.7305$.

29. The area between $z = -6$ and $z = 0$ is the same as the area to the left of $z = 0$ (rounded to four decimals) Therefore, the area between $z = -6$ and $z = 0$ is 0.5000.

31.
a. $90 = \mu + 2\sigma$, so $x = 90$ corresponds to $z = 2$. Thus,
$$P(x > 90) = P(z > 2).$$

b. Use the fact that $P(z > 2) = 1 - P(z \leq 2)$. The table shows that $P(z \leq 2) = 0.9773$. Therefore,
$$P(z > 2) = 1 - 0.9773 = 0.0227.$$

33.
a. $62.8 = \mu - 17.2$, since $\sigma = 5$, $k\sigma = 17.2$ implies that $k = \frac{17.2}{5} = 3.44$ so
$$\mu - 17.2 = \mu - 3.44\sigma.$$ Thus $x = 62.8$ corresponds to $z = -3.44$. Thus, $P(x < 62.8) = P(z < -3.44)$.

b. Using a calculator, we find that
$$P(z < -3.44) = 0.0003.$$

35.
a. Since $\mu = 80$ and $\sigma = 5$, the z-score that corresponds to $x = 93.67$ is $z = \frac{93.67 - 80}{5} = 2.734$. Therefore,
$$P(x < 93.67) = P(z < 2.73).$$

b. The probability given by the table is

$$P(z < 2.73) = 0.9968.$$

37.
a. Since $\mu = 80$ and $\sigma = 5$, the z-score that corresponds

to $x = 76.5$ is $z = \frac{76.5-80}{5} = -0.7$. Therefore,

$$P(x > 76.5) = P(z > -0.7).$$

b. Use the fact that $P(z > -0.7) = 1 - P(z < -0.7)$.

From the table we get:

$$P(z < -0.7) = 0.2420$$

.
Therefore,

$$P(z > -0.7) = 1 - 0.2420 = 0.7580.$$

39. Since $\mu = 283\,gm$ and $\sigma = 4\,gm$, we know that two
standard deviations below the mean is
$\mu - 2\sigma = 283 - 2(4) = 275\,gm$. Therefore, at least 275
grams must be placed in each box so that the weight will not
be more than two standard deviations below the mean.

41. Crystal 193 points on the Dr. Schein's final which
corresponds to the z-score of $z = \frac{193-180}{20} = 0.65$. Xiou
scored 182 points on Dr. Feldman's final which corresponds
to the z-score of $z = \frac{182-165}{10} = 1.7$. If both professors
grade on a normal curve, Xiou will get the higher grade.

43.
a. The z-score corresponding to 55 months is
$z = \frac{55-52}{3} = 1$. The table shows that $P(z < 1) = 0.8413$.
Therefore, $P(z > 1) = 1 - 0.8413 = 0.1587$. We would
expect 15.87% of the batteries to last more than 55 months.

b. The z-score corresponding to 46 months is
$z = \frac{46-52}{3} = -2$. The table shows that

$$P(z < -2) = 0.0227.$$ Therefore, 2.27% of the batteries

will last less than 46 months.

c. Referring to part 'b', 2.27% of the batteries will last less
than 46 months. If the store sells 200 of these batteries, we
would expect $(200)(0.0227) = 4.54$ or around five of
the batteries to last less than 46 months.

d. The z-score corresponding to 48 months is
$z = \frac{48-52}{3} = -1.33$. The table shows that

$$P(z < -1.33) = 0.0918.$$ Therefore, 9.18% of the

batteries will last less than 48 months. It stands to reason
that the company can expect to have 9.18% of the batteries
returned under warranty.

45.
a. The z-score corresponding to 775 hours is
$z = \frac{775-750}{25} = 1$. The table shows that

$$P(z < 1) = 0.8413.$$ Thus, the probability that the bulbs

will last more than 775 hours is

$$P(z > 1) = 1 - 0.8413 = 0.1587.$$

In other words, 15.87% of the bulbs should last longer than
775 hours. Since a sample of 10,000 was selected,
$(10,000)(0.1587) = 1587$ bulbs should last longer than
775 hours.

b. The z-score corresponding to 700 hours is
$z = \frac{700-750}{25} = -2$. The table shows that

$$P(z < -2) = 0.0227.$$ Therefore, 2.27% of the bulbs will

last less than 700 hours. Since a sample of 10,000 was
selected, $(10,000)(0.0227) = 227$ of the bulbs should
last less than 700 hours.

c. The z-score corresponding to 700 hours is
$z = \frac{700-750}{25} = -2$. The table shows that

$$P(z < -2) = 0.0227..$$ The z-score corresponding to

800 hours is $z = \frac{800-750}{25} = 2$. The table shows that

$$P(z < 2) = 0.9773.$$ The probability that a bulb will last

between 700 and 800 hours is:

$$P(-2 < z < 2) = 0.9773 - 0.0227 = 0.9546$$

Therefore, 95.46% of the bulbs should last between 700 and
800 hours. Since a sample of 10,000 bulbs was selected,
9,546 of the bulbs should last between 700 and 800 hours.

47.
a. Converting all the values to seconds we see that the times
are normally distributed with a mean of $\mu = 270$ second
and a standard deviation of $\sigma = 105$ seconds. The
percentage of students that spend more than five minutes or
300 seconds complaining corresponds to the z-score
$z = \frac{300-270}{105} = 0.29$.

The solution is continued on the following page.

The table shows that $P(z < 0.29) = 0.6141$. Therefore, the probability of students that spend more than 300 seconds complaining is:

$P(z > 0.29) = 1 - 0.6141 = 0.3859$.

Therefore, 38.59% of the students spend more than five minutes complaining.

b. The z-score that corresponds to two minutes or 120 seconds is $z = \frac{120 - 270}{105} = -1.43$. The table shows that

$P(z < -1.43) = 0.0764$. Therefore, 7.64% of the students spend less than two minutes complaining.

49.

a. Converting all the values to seconds we see that the times are normally distributed with a mean of $\mu = 1110$ second and a standard deviation of $\sigma = 132$ seconds. The z-value that corresponds to 20 minutes or 1200 seconds is

$z = \frac{1200 - 1110}{132} = 0.68$.

The table shows that $P(z < 0.68) = 0.7518$. Therefore, the probability that it will take Sharokh longer than 20 minutes to get to class is:

$P(z > 0.68) = 1 - 0.7518 = 0.2482$.

b. To get to class on time, Sharokh will have to arrive in less than 15 minutes, or 900 seconds. The z-score that corresponds to 900 second is $z = \frac{900 - 1110}{132} = -1.59$. The table shows that $P(z < -1.59) = 0.0559$.

Therefore, Sharokh will make it to class on time 5.59% of the time.

51.

a. The z-score that corresponds to 14 ounces is $z = \frac{14 - 14.1}{0.2} = -0.5$. The table shows that

$P(z < -0.5) = 0.3085$. Therefore, the probability that bag of peanuts will weigh at least 14 ounces is

$P(z \geq -0.5) = 1 - 0.3085 = 0.6915$.

b. The z-score that corresponds to 13.8 ounces is $z = \frac{13.8 - 14.1}{0.2} = -1.5$. The table shows that

$P(z < -1.5) = 0.0668$.

The z-score that corresponds to 14.5 ounces is $z = \frac{14.5 - 14.1}{0.2} = 2$. The table shows that

$P(z < 2) = 0.9773$.

The probability that a bag of peanuts will weigh between 13.8 and 14.5 ounces is:

$P(-1.5 < z < 2) = 0.9773 - 0.0668$
$P(-1.5 < z < 2) = 0.9105$.

53.

a. The z-score that corresponds to $12,000 is $z = \frac{12,000 - 10,000}{2500} = 0.80$. The table shows that

$P(z < 0.80) = 0.7881$. Therefore, the probability that an account will have a balance of more than $12,000 is

$P(z \geq 0.80) = 1 - 0.7881 = 0.2119$.

Given that the bank has 100,000 accounts, there are $(100,000)(0.2119) = 21,190$ accounts that have balances of more than $12,000.

b. The z-score that corresponds to $5000 is $z = \frac{5000 - 10,000}{2500} = -2$. The table shows that

$P(z < -2) = 0.0227$. Therefore, the probability that an account will have a balance greater than $5000 is

$P(z \geq -2) = 1 - 0.0227 = 0.9773$.

Therefore, 97.73% of the accounts will have a balance more than $5000.

55.

a. Converting the distribution to months, we see that the mean of 48 years and 6 months is equivalent to the mean of 582 months. Likewise, the standard deviation of 7 years and 2 months is equivalent to the standard deviation of 86 months. The z-score that corresponds to 65 years or 780 months is $z = \frac{780 - 582}{86} = 2.30$. The table shows that

$P(z < 2.30) = 0.9893$. Therefore, the probability that a faculty member will be 65 years of age or older is

$P(z \geq 2.30) = 1 - 0.9893 = 0.0107$.

We conclude that approximately 1.1% of the faculty are eligible for the early retirement incentive program.

b. Using the conversions from part 'a'. The z-score that corresponds to 40 years of age or 480 months is

$z = \frac{480 - 582}{86} = -1.19$. The table shows that

$P(z < -1.19) = 0.1170$. Therefore, the probability that a faculty member is under 40 years of age is 0.1170.

57.

a. The z-score that corresponds to 23,000 miles is

$z = \frac{23{,}000 - 25{,}000}{1600} = -1.25$. The table shows that

$P(z < -1.25) = 0.1056$.

The z-score that corresponds to 28,000 miles is

$z = \frac{28{,}000 - 25{,}000}{1600} = 1.88$. The table shows that

$P(z < 1.88) = 0.9700$.

Therefore, the probability that a tire will last between 23,000 and 28,000 miles is:

$P(-1.25 < z < 1.88) = 0.9700 - 0.1056$

$P(-1.25 < z < 1.88) = 0.8644$

b. The z-score that corresponds to 22,500 miles is

$z = \frac{22{,}500 - 25{,}000}{1600} = -1.56$. The table shows that

$P(z < -1.56) = 0.0594$.

Therefore, 5.94% of the tires will not last more than 22,500 miles and will be returned under warranty.

59.

a. The z-score that corresponds to a seven pounds is

$z = \frac{7-10}{2} = -1.5$. The table shows that

$P(z < -1.5) = 0.0668$.

The z-score that corresponds to a 13 pounds is

$z = \frac{13-10}{2} = 1.5$. The table shows that

$P(z < 1.5) = 0.9332$.

Therefore, the probability that a domestic short hair cat weights between 7 and 13 pounds is

$P(-1.5 < z < 1.5) = 0.9332 - 0.0668$

$P(-1.5 < z < 1.5) = 0.8664$.

b. The z-score that corresponds to a 19 pounds is

$z = \frac{19-10}{2} = 4.5$. Calculating the probability with a

calculator application shows that $P(z < 4.5) \approx 1$.

Therefore, the probability that a cat will weigh at least as much as Ender is $P(z > 4.5) \approx 1 - 1 \approx 0$. The percentage of short hair domestic cats that weigh as much or more than Ender is approximately zero.

61. From the table we find that the requested z-score is $z = 1.53$.

63. From the table we find that 0.1 is between $z = -1.29$ and $z = -1.28$. Taking the average of the two scores, the requested z-score is $z = -1.285$

65. If the area to the right of z is 0.001, this means that the area to the left of z is $1 - 0.001 = 0.9990$. Using a calculator, we see that the requested z-score is $z \approx 3.09$

67. According to the table, the z-score that corresponds to the area less than 0.04 is $z = -1.75$. Since $z = \frac{x-\mu}{\sigma}$ and we know that the life of the refrigerators are normally distributed with a mean of 15 years and a standard deviation of 3 years, we get the equation:

$-1.75 = \frac{x-15}{3}$

Solving for x we get:

$(3)(-1.75) = x - 15$

$-5.25 = x - 15$

$9.75 = x$

Therefore, less than 4% of the refrigerators will fail under warranty if the company sets the warranty period to be 9.75 years or nine years and nine months.

69. According to the table, the z-score that corresponds to the area less than 0.05 is $z = -1.64$. Since $z = \frac{x-\mu}{\sigma}$ and we know that the life of the air conditioners are normally distributed with a mean of 10 years and a standard deviation of 2.5 years, we get the equation:

$-1.64 = \frac{x-10}{2.5}$

Solving for x we get:

$(2.5)(-1.64) = x - 10$

$-4.1 = x - 10$

$5.9 = x$

Therefore, less than 5% of the air conditioners will fail under warranty if the company sets the warranty period to be 5.9 years or just over five years and ten months.

Chapter 10 Summary Exercises

1. The mode is two, since that is value that appears most frequently.

3. The mean is $\frac{2+28+13+2+57}{5} = 20.40$.

5. For the standard normal curve, the area to the left of $z = -1.3$ is approximately 0.1515.

7. Using the table, the area to the left of $z = -2.4$ is 0.0082 and the area to the left of $z = 0.6$ is 0.7258. Therefore, the area between $z = -2.4$ and $z = 0.6$ is $0.7258 - 0.0082 = 0.7176$.

9. Once again assuming that all of the data in each range is concentrated at the midpoint: Consider the table:

x	Freq.	$x - \mu$	$f \cdot (x - \mu)^2$
7.5	2	-5.25	55.125
10.5	6	-2.25	30.375
13.5	8	0.75	4.5
16.5	3	3.75	42.1875
19.5	1	6.75	45.5625
SUM	20	3.75	177.75

The sample variance is:

$$s^2 = \frac{177.75}{19} = 9.36$$

The sample standard deviation is:

$$s = \sqrt{s^2} = \sqrt{9.36} = 3.06$$

11. The z-score that corresponds to 5.75 ounces is $z = \frac{5.75 - 6}{0.5} = -0.5$. The table shows that

$$P(z < -0.5) = 0.3085.$$

The z-score that corresponds to 6.25 ounces is $z = \frac{6.25 - 6}{0.5} = 0.5$. The table shows that

$$P(z < 0.5) = 0.6915.$$

Therefore, the probability that a randomly selected bag weighs between 5.75 and 6.25 ounces is

$$P(-0.5 < z < 0.5) = 0.6915 - 0.3085$$
$$P(-0.5 < z < 0.5) = 0.3830.$$

13. The expanded table is shown below:

Number of Repair Calls	Frequency	Relative Frequency
0	5	0.25
1	7	0.35
2	4	0.2
3	3	0.15
4	1	0.05

15. Summing the frequencies in the table from problem number 14 we see that $0.15 + 0.05 = 0.20$ or 20% of the families called a professional to make household repairs more than two times in the past year.

17. In order to find the population variance and standard deviation, look at the following table:

x	Freq.	$x - \mu$	$f \cdot (x - \mu)^2$
0	5	-1.4	9.8
1	7	-0.4	1.12
2	4	0.6	1.44
3	3	1.6	7.68
4	1	2.6	6.76
SUM	20	3	26.8

The population variance is:

$$\sigma^2 = \frac{26.8}{20} = 1.34.$$

The population standard deviation is:

$$\sigma = \sqrt{\sigma} = \sqrt{1.34} \approx 1.1576.$$

19. The actual mean of the data is 5.9625. The sum of the squares of the deviations from the mean is 50.03625. The population variance is $\frac{50.03625}{24} = 2.084844$, and the population standard deviation is

$$\sqrt{2.084844} = 1.499421$$

There are several ways in which we can break down the groups into different intervals. The way that we will be doing so is to take intervals of equal length starting, and ending at appropriate values.

a. To construct the tables for five intervals, we start with the minimum interval of 2.5 with equal lengths of 1.25.

The frequency table is shown at the top of the next page.

Class	Midpoint	Frequency	$f \cdot x$
2.5-3.75	3.12	3	9.36
3.75-5.0	4.37	5	21.85
5.0-6.25	5.62	3	16.86
6.25-7.5	6.87	10	68.7
7.5-8.75	8.12	3	24.36
SUM		24	141.13

Assuming that the data is concentrated at the midpoint of the interval, the mean is $\frac{141.13}{24} = 5.88$.

To calculate the standard deviation, consider the table:

x	Freq.	$x - \mu$	$f \cdot (x - \mu)^2$
3.12	3	-2.760416	22.8597
4.37	5	-1.510416	11.40679
5.62	3	-0.260416	0.203451
6.87	10	0.9895833	9.792752
8.12	3	2.2395833	15.0472
SUM	24	-1.302083	59.3099

The population variance is:

$$\sigma^2 = \frac{59.3099}{24} = 2.47.$$

The population standard deviation is:

$$\sigma = \sqrt{\sigma} = \sqrt{2.47} \approx 1.57.$$

b. For six intervals, we use the minimum value 3 and the maximum value 9. The frequency table looks like:

Class	Midpoint	Frequency	$f \cdot x$
3 to 4	3.49	3	10.47
4 to 5	4.49	5	22.45
5 to 6	5.49	2	10.98
6 to 7	6.49	8	51.92
7 to 8	7.49	4	29.96
8 to 9	8.49	2	16.98
SUM		24	142.76

Assuming that the data is concentrated at the midpoint of the interval, the mean is $\frac{142.76}{24} = 5.948$.

To calculate the standard deviation, consider the table at the top of the next column.

Midpoint	Frequency	$x - \mu$	$f \cdot (x - \mu)^2$
3.49	3	-2.458333	18.13021
4.49	5	-1.458333	10.63368
5.49	2	-0.458333	0.420139
6.49	8	0.5416667	2.347222
7.49	4	1.5416667	9.506944
8.49	2	2.5416667	12.92014
SUM	24	0.25	53.95833

The population variance is:

$$\sigma^2 = \frac{59.958}{24} = 2.248.$$

The population standard deviation is:

$$\sigma = \sqrt{\sigma} = \sqrt{2.248} \approx 1.499.$$

c. For seven intervals we use the length of 0.9 starting with 2.8.

The frequency table is shown below:

Class	Midpoint	Frequency	$f \cdot x$
2.8 to 3.7	3.25	3	9.75
3.7 to 4.6	4.15	3	12.45
4.6 to 5.5	5.05	2	10.1
5.5 to 6.4	5.95	3	17.85
6.4 to 7.3	6.85	9	61.65
7.3 to 8.2	7.75	3	23.25
8.2 to 9.1	8.65	1	8.65
SUM		24	143.7

Assuming that the data is concentrated at the midpoint of the interval, the mean is

$\frac{143.7}{24} = 5.9875$.

To calculate the standard deviation, consider the table:

x	Freq.	$x - \mu$	$f \cdot (x - \mu)^2$
3.25	3	-2.7375	22.48172
4.15	3	-1.8375	10.12922
5.05	2	-0.9375	1.757813
5.95	3	-0.0375	0.004219
6.85	9	0.8625	6.695156
7.75	3	1.7625	9.319219
8.65	1	2.6625	7.088906
SUM	24	-0.2625	57.47625

The population variance is:

$$\sigma^2 = \frac{57.47625}{24} = 2.395.$$

The solution is continued on the following page.

The population standard deviation is:

$$\sigma = \sqrt{\sigma} = \sqrt{2.395} \approx 1.548 \,.$$

d. For eight intervals, we start with a minimum value of 2.75 and use interval lengths of 0.75.

The frequency table is shown below:

Class	Midpoint	Frequency	$f \cdot x$
2.75 to 3.5	3.12	1	3.12
3.5 to 4.25	3.88	2	7.76
4.25 to 5	4.63	5	23.15
5 to 5.75	5.38	1	5.38
5.75 to 6.5	6.13	3	18.39
6.5 to 7.25	6.88	8	55.04
7.25 to 8	7.63	2	15.26
8 to 8.75	5.38	2	10.76
SUM		24	138.86

Assuming that the data is concentrated at the midpoint of the interval, the mean is
$\frac{138.86}{24} = 5.786$.

To calculate the standard deviation, consider the table:

x	Freq.	$x - \mu$	$f \cdot (x - \mu)^2$
3.12	1	-2.665833	7.106667
3.88	2	-1.905833	7.264401
4.63	5	-1.155833	6.679753
5.38	1	-0.405833	0.164701
6.13	3	0.3441667	0.355352
6.88	8	1.0941667	9.577606
7.63	2	1.8441667	6.801901
5.38	2	-0.405833	0.329401
SUM	24	-3.256666	38.27978

The population variance is:

$$\sigma^2 = \frac{38.2798}{24} = 1.595 \,.$$

The population standard deviation is:

$$\sigma = \sqrt{\sigma} = \sqrt{1.595} \approx 1.263 \,.$$

e. For nine intervals, we start with a minimum value of 3 and use interval lengths of 0.6.

The frequency table is displayed the top of the next column.

Class	Midpoint	Frequency	$f \cdot x$
3 to 3.6	3.25	1	3.25
3.6 to 4.2	3.85	2	7.7
4.2 to 4.8	4.45	5	22.25
4.8 to 5.4	5.05	0	0
5.4 to 6	5.65	2	11.3
6 to 6.6	6.25	3	18.75
6.6 to 7.2	6.85	7	47.95
7.2 to 8	7.45	2	14.9
8 to 8.6	8.25	2	16.5
SUM		24	142.6

Assuming that the data is concentrated at the midpoint of the interval, the mean is
$\frac{142.6}{24} = 5.942$.

To calculate the standard deviation, consider the table below:

x	Freq.	$x - \mu$	$f \cdot (x - \mu)^2$
3.25	1	-2.691667	7.245069
3.85	2	-2.091667	8.750139
4.45	5	-1.491667	11.12535
5.05	0	-0.891667	0
5.65	2	-0.291667	0.170139
6.25	3	0.3083333	0.285208
6.85	7	0.9083333	5.775486
7.45	2	1.5083333	4.550139
8.25	2	2.3083333	10.65681
SUM	24	-2.425	48.55833

The population variance is:

$$\sigma^2 = \frac{48.558}{24} = 2.023 \,.$$

The population standard deviation is:

$$\sigma = \sqrt{\sigma} = \sqrt{2.023} \approx 1.422 \,.$$

f. For ten intervals, we start with a minimum value of 2.6 and use interval lengths of 0.6.

The frequency table is displayed at the top of the following page.

Class	Midpoint	Frequency	$f \cdot x$
2.6 to 3.2	2.85	1	2.85
3.2 to 3.8	3.45	2	6.9
3.8 to 4.4	4.05	0	0
4.4 to 5.0	4.65	5	23.25
5.0 to 5.6	5.25	1	5.25
5.6 to 6.2	5.85	1	5.85
6.2 to 6.8	6.45	4	25.8
6.8 to 7.4	7.05	7	49.35
7.4 to 8.0	7.65	1	7.65
8.0 to 8.6	8.25	2	16.5
SUM		24	143.4

Assuming that the data is concentrated at the midpoint of the interval, the mean is

$\frac{143.4}{24} = 5.975$.

To calculate the standard deviation, consider the table:

x	Freq.	$x - \mu$	$f \cdot (x - \mu)^2$
2.85	1	-3.125	9.765625
3.45	2	-2.525	12.75125
4.05	0	-1.925	0
4.65	5	-1.325	8.778125
5.25	1	-0.725	0.525625
5.85	1	-0.125	0.015625
6.45	4	0.475	0.9025
7.05	7	1.075	8.089375
7.65	1	1.675	2.805625
8.25	2	2.275	10.35125
	24	-4.25	53.985

The population variance is:

$\sigma^2 = \frac{53.985}{24} = 2.2494$.

The population standard deviation is:

$\sigma = \sqrt{\sigma} = \sqrt{2.249} \approx 1.500$.

Cumulative Review

21. To solve the matrix equation, we use the matrix capabilities of a graphing calculator to find A^{-1}.

$$A^{-1} = \begin{bmatrix} -1 & -3 & 5 \\ \frac{1}{2} & 1 & \frac{-3}{2} \\ \frac{1}{2} & 2 & \frac{-5}{2} \end{bmatrix}$$

Multiplying both sides of the matrix equation on the left by A^{-1} we get:

$$\begin{bmatrix} x \\ y \\ z \end{bmatrix} = \begin{bmatrix} -1 & -3 & 5 \\ \frac{1}{2} & 1 & \frac{-3}{2} \\ \frac{1}{2} & 2 & \frac{-5}{2} \end{bmatrix} \begin{bmatrix} 4 \\ 1 \\ 0 \end{bmatrix} = \begin{bmatrix} -7 \\ 3 \\ 4 \end{bmatrix}$$

Therefore, the solution to the matrix equation is: $x = -7$; $y = 3$; $z = 4$.

23. The corner points are:

A. The intersection of the line $x + y = 10$ and $-2x + y = 1$ which is the point $A(3, 7)$.

B. The intersection of the line $x + y = 10$ and $-x + 2y = 0$ which is the point $B\left(\frac{20}{3}, \frac{10}{3}\right)$.

C. The intersection of the line $7x + 3y = 21$ and $-x + 2y = 0$ which is the point $C\left(\frac{42}{17}, \frac{21}{17}\right)$.

D. The intersection of the line $7x + 3y = 21$ and $-2x + y = 1$ which is the point $D\left(\frac{18}{13}, \frac{49}{13}\right)$.

Chapter 10 Sample Test Answers

1. The completed table is:

Number of Secretaries	Frequency	Relative Frequency
1	2	0.2
3	3	0.3
5	5	0.5

2. The mean is: $(0.2)(1) + (0.3)(3) + (0.5)(5) = 3.6$

3. The median is the average of three and five (the fifth and the sixth highest value), or $\frac{3+5}{2} = 4$.

4. There are 36 total secretaries in this group and $2 \cdot 1 + 3 \cdot 3 = 11$ of them work for a firm that employees three or fewer secretaries. Therefore, the probability that a secretary selected at random works for a firm that employees three or fewer secretaries is $\frac{11}{36} \approx 0.306$.

5. The median is the third highest value, or 2.

6. The mode is one, because one occurs more often then any other value.

7. The mean is $\frac{1+1+2+3+8}{5} = 3$.

8. The sample variance is:

$s^2 = 8.5$.

The sample standard deviation is:

$s = \sqrt{s^2} = \sqrt{8.5} \approx 2.915$

9. The range of values that lie within one standard deviation of the mean is $77.8 \le x \le 104.2$.

Five of the seven salaries lie within this range.

10. The probability distribution table is:

X	p
0	$\frac{C(13,0) \cdot C(39,3)}{C(52,3)} \approx 0.4135$
1	$\frac{C(13,1) \cdot C(39,2)}{C(52,3)} \approx 0.4359$
2	$\frac{C(13,2) \cdot C(39,1)}{C(52,3)} \approx 0.1377$
3	$\frac{C(13,3) \cdot C(39,0)}{C(52,3)} \approx 0.0129$

11.
The population mean is $\mu = 72.6$.

The median is 77.

The mode is 82.

If the data represents a population, the population standard deviation is: $\sigma = 19.835$.

If the data represents a sample, the sample standard deviation is $s = 20.908$

12. The expected profit is:

$$E(X) = (0.6)(25,000) + (0.4)(-12,000)$$

$$E(X) = 10,200.$$

The business manager expects to gain an additional $10,200 profit if she expands her business.

13. The z-score that corresponds to 1100 hours is $z = \frac{1100-1200}{160} = -0.63$. The table shows that

$P(z < -0.63) = 0.2643$. Therefore, the probability that a bulb will burn out in less than 1100 hours is 0.2643.

14. The top 10% correspond to a z-score of $z = 1.285$. Using the equation $1.285 = \frac{x-520}{75}$, we find that the minimum score to get accepted into the honors program is $x = 616.375$

15. The relative frequency histogram for the data shown is:

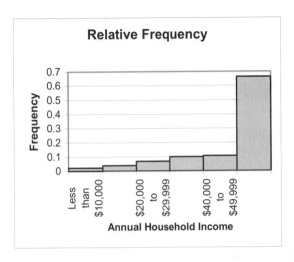

16. Assuming the data is concentrated around the midpoint of each range and using $75,000 to represent the range of the last interval. We estimate the mean household income among those with Internet access in the year 2000 was approximately $60,503.50.

17. Assuming the data is concentrated around the midpoint of each range and using $75,000 to represent the range of the last interval. We estimate the median household income among those with Internet access in the year 2000 was approximately $75,000.

18. Assuming the data is concentrated around the midpoint of each range and using $75,000 to represent the range of the last interval. We estimate the standard deviation of household income among those with Internet access in the year 2000 was approximately $21,594.33.

19. Summing up the relative frequencies, the probability that a household selected at random had an annual income of at least $20,000 is 0.9389.

Chapter 11
Markov Chains

Exercises Section 11.1

1. The transition diagram with all of the probabilities shown is:

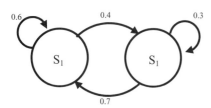

The probability of staying in state 1 is 0.6, so the probability of going from state 1 to state 2 is 0.4. The probability of going from state 2 to state one is 0.7, so the probability of staying in state 2 is 0.3.

The transition matrix is:

$$T = \begin{bmatrix} 0.6 & 0.4 \\ 0.7 & 0.3 \end{bmatrix}$$

3. The transition diagram with all of the probabilities shown is:

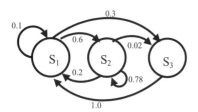

The probability of going from state 1 to state 2 is 0.6 and the probability of going from state 2 to state 3 is 0.3, so the probability of going from state 1 to state 1 is 0.1. The probability of going from state 2 to state 1 is 0.2, the probability of going from state 2 to state 3 is 0.02, so the probability of staying in state 2 is 0.78. The probability of going from state 3 to state 1 is 1. The probability of going from state 3 to state 2 or state 3 is zero.

The transition matrix is displayed at the top of the next column.

$$T = \begin{bmatrix} 0.1 & 0.6 & 0.3 \\ 0.2 & 0.78 & 0.02 \\ 1 & 0 & 0 \end{bmatrix}$$

5. The transition diagram with all of the probabilities shown is:

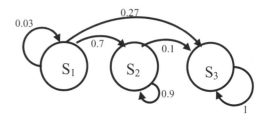

The transition matrix is:

$$T = \begin{bmatrix} 0.03 & 0.7 & 0.27 \\ 0 & 0.9 & 0.1 \\ 0 & 0 & 1 \end{bmatrix}$$

7. The transition diagram corresponding to the given matrix is:

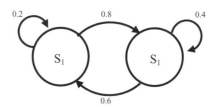

9. The transition diagram corresponding to the given matrix is:

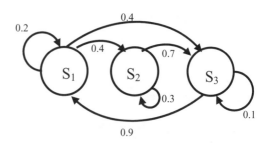

11. Let F represent the state of owning a foreign car and A represent the state of owning an American car. The transition diagram that models the problem is:

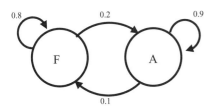

The transition matrix where state 1 is owning a foreign car and state 2 is owning an American car is:

$$T = \begin{bmatrix} 0.8 & 0.2 \\ 0.1 & 0.9 \end{bmatrix}$$

13. Let States 1 and 2 represent having the psychological characteristic and not having the psychological characteristic respectively. The transition diagram for this problem is:

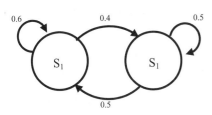

The transition matrix is:

$$T = \begin{bmatrix} 0.6 & 0.4 \\ 0.5 & 0.5 \end{bmatrix}$$

15. Let E represent state of subscribing to the Evening Star, let M represent the state of subscribing to the Morning Tribune, and let G represent the state of subscribing to the Gazette. The transition diagram for the problem is:

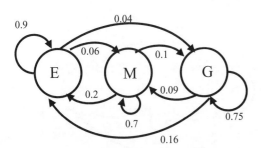

The transition matrix is:

$$T = \begin{bmatrix} 0.90 & 0.06 & 0.04 \\ 0.20 & 0.70 & 0.10 \\ 0.16 & 0.09 & 0.75 \end{bmatrix}$$

17. The sum of the entries in row two is 1.1, therefore this is not a transition matrix.

19. Since each entry is non-negative and each row sums to one, this is a transition matrix.

21.
a. The transition matrix for the Markov chain is:

$$T = \begin{bmatrix} 0.80 & 0.20 \\ 0.40 & 0.60 \end{bmatrix}$$

b. The tree diagram is:

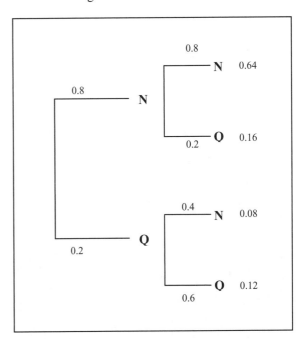

The probability that a person now using a NumberKrunch will have switched to a QuickDigit two years later is $0.16 + 0.12 = 0.28$.

c. The square of the transition matrix is:

$$T^2 = \begin{bmatrix} 0.72 & 0.28 \\ 0.56 & 0.44 \end{bmatrix}$$

The 0.28 in row 1, column 2 represents the probability that someone using a NumberKrunch will have switched to a QuickDigit 2 years later.

d. The 0.72 in row 1, column 1 of the matrix represents the probability that someone using a NumberKrunch will purchase a NumberKrunch 2 years later.

23.

a. Let state 1 represent the state of being aligned 'A' and let state two represent the state of being misaligned 'M'. The transition diagram for this Markov chain is:

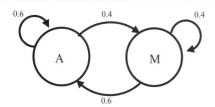

The transition matrix for the Markov chain is:

$$T = \begin{bmatrix} 0.6 & 0.4 \\ 0.6 & 0.4 \end{bmatrix}$$

b. Assuming that the probe is aligned now, the tree diagram is:

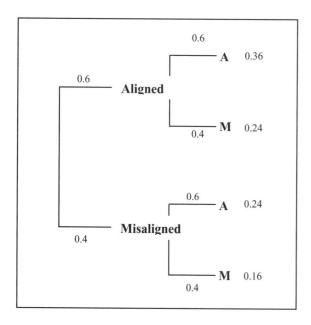

From the tree diagram, there is a probability of 0.36 that the probe will remain aligned for each of the next two minutes and a probability of 0.24 that it will go out of alignment and be corrected in the next two minutes. Thus, the probability that the probe will be aligned two minutes from now is $0.36 + 0.24 = 0.60$.

c. Knowing that it is currently aligned, the state vector is $\begin{bmatrix} 1 & 0 \end{bmatrix}$. The state vector one minute from now is:

$$\begin{bmatrix} 1 & 0 \end{bmatrix} \begin{bmatrix} 0.6 & 0.4 \\ 0.6 & 0.4 \end{bmatrix} = \begin{bmatrix} 0.6 & 0.4 \end{bmatrix}$$

and the state vector two minutes from now is:

$$\begin{bmatrix} 0.6 & 0.4 \end{bmatrix} \begin{bmatrix} 0.6 & 0.4 \\ 0.6 & 0.4 \end{bmatrix} = \begin{bmatrix} 0.6 & 0.4 \end{bmatrix}.$$

Thus, the probability that the probe will be aligned two minutes from now is 0.6.

d. If it is not aligned, we start with the state vector $\begin{bmatrix} 0 & 1 \end{bmatrix}$. The state vector one minute from now is:

$$\begin{bmatrix} 0 & 1 \end{bmatrix} \begin{bmatrix} 0.6 & 0.4 \\ 0.6 & 0.4 \end{bmatrix} = \begin{bmatrix} 0.6 & 0.4 \end{bmatrix}$$

and the state vector two minutes from now is:

$$\begin{bmatrix} 0.6 & 0.4 \end{bmatrix} \begin{bmatrix} 0.6 & 0.4 \\ 0.6 & 0.4 \end{bmatrix} = \begin{bmatrix} 0.6 & 0.4 \end{bmatrix}$$

and three minutes from now, the state vector will be:

$$\begin{bmatrix} 0.6 & 0.4 \end{bmatrix} \begin{bmatrix} 0.6 & 0.4 \\ 0.6 & 0.4 \end{bmatrix} = \begin{bmatrix} 0.6 & 0.4 \end{bmatrix}$$

So the probability that the probe will be misaligned three minutes from now is 0.4.

25. Let states 1, 2 and 3 represent preferences for 2-door, 4-door and station wagons respectively. The transition diagram is:

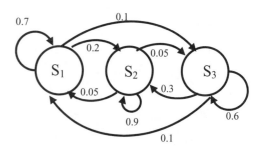

The transition matrix for the Markov chain on the previous page is:

$$T = \begin{bmatrix} 0.70 & 0.20 & 0.10 \\ 0.05 & 0.90 & 0.05 \\ 0.10 & 0.30 & 0.60 \end{bmatrix}$$

b. Consider the tree diagram:

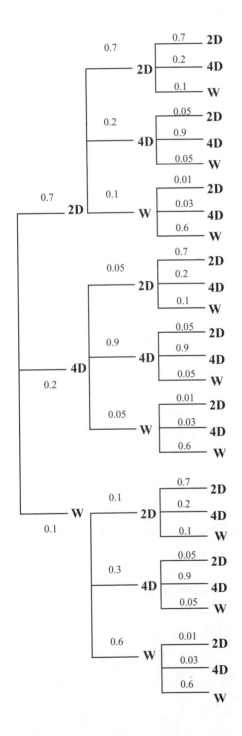

The probability that a person leasing a two-door sedan now will be leasing a station wagon three choices later is:

$$(0.7)(0.7)(0.1)+(0.7)(0.2)(0.05)+$$
$$(0.7)(0.1)(0.6)+(0.2)(0.05)(0.1)+$$
$$(0.2)(0.9)(0.05)+(0.2)(0.05)(0.6)+$$
$$(0.1)(0.1)(0.1)+(0.1)(0.3)(0.05)+$$
$$(0.1)(0.6)(0.6)=0.1525$$

The probability that a person leasing a two-door sedan now will be leasing a station wagon three choices later is 0.1525.

c. Assuming we start with a two-door sedan, the initial state vector is $\begin{bmatrix} 1 & 0 & 0 \end{bmatrix}$. The state vector after one choice is:

$$\begin{bmatrix} 1 & 0 & 0 \end{bmatrix} \begin{bmatrix} 0.70 & 0.20 & 0.10 \\ 0.05 & 0.90 & 0.05 \\ 0.10 & 0.30 & 0.60 \end{bmatrix} = \begin{bmatrix} 0.7 & 0.2 & 0.1 \end{bmatrix}.$$

The state vector after two choices is:

$$\begin{bmatrix} 0.7 & 0.2 & 0.1 \end{bmatrix} \begin{bmatrix} 0.70 & 0.20 & 0.10 \\ 0.05 & 0.90 & 0.05 \\ 0.10 & 0.30 & 0.60 \end{bmatrix} =$$
$$\begin{bmatrix} 0.51 & 0.35 & 0.14 \end{bmatrix}$$

The state vector after three choices is:

$$\begin{bmatrix} 0.51 & 0.35 & 0.14 \end{bmatrix} \begin{bmatrix} 0.70 & 0.20 & 0.10 \\ 0.05 & 0.90 & 0.05 \\ 0.10 & 0.30 & 0.60 \end{bmatrix} =$$
$$\begin{bmatrix} 0.3885 & 0.459 & 0.1525 \end{bmatrix}$$

The probability that a person leasing a two-door sedan now will be leasing a station wagon three choices later is 0.1525.

27. The following tree diagram illustrates the probabilities on Denita's next two free throws after she makes her first:

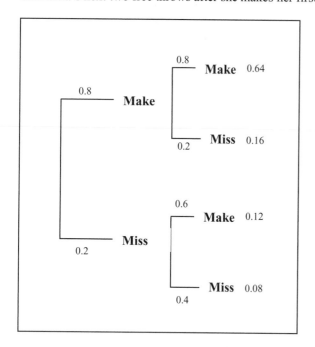

The probability that she will miss the 3rd free throw is the sum of the probabilities of the two branches with a miss on the 3rd throw, or $0.16 + 0.08 = 0.24$.

29. The first row in the transition matrix means that if you start in (or move to) state 1, you must remain in state 1 with a probability of one. In other words, state 1 is an absorbing state. This property does apply for all powers of P.

Exercises Section 11.2

1. $P_1 = P_0 T$

$$P_1 = \begin{bmatrix} 0.3 & 0.7 \end{bmatrix} \begin{bmatrix} 1 & 0 \\ 0.2 & 0.8 \end{bmatrix} = \begin{bmatrix} 0.44 & 0.56 \end{bmatrix}.$$

3. Since $P_2 = P_1 T$ and $P_1 = P_0 T$ then,

$$P_2 = P_0 TT = P_0 T^2.$$

$$P_2 = \begin{bmatrix} 0.25 & 0.75 \end{bmatrix} \left(\begin{bmatrix} 1.0 & 0.0 \\ 0.2 & 0.8 \end{bmatrix} \right)^2 = \begin{bmatrix} 0.52 & 0.48 \end{bmatrix}.$$

5.

$$P_1 = P_0 T$$

$$P_1 = \begin{bmatrix} 0.60 & 0.20 & 0.20 \end{bmatrix} \begin{bmatrix} 0.25 & 0.50 & 0.25 \\ 0.30 & 0.30 & 0.40 \\ 0.50 & 0.00 & 0.50 \end{bmatrix} =$$

$$P_1 = \begin{bmatrix} 0.31 & 0.36 & 0.33 \end{bmatrix}.$$

7. Since $P_2 = P_1 T$ and $P_1 = P_0 T$ then,

$$P_2 = P_0 TT = P_0 T^2.$$

$$P_2 = \begin{bmatrix} 0.70 & 0.15 & 0.15 \end{bmatrix} \left(\begin{bmatrix} 0.25 & 0.50 & 0.25 \\ 0.30 & 0.30 & 0.40 \\ 0.50 & 0.00 & 0.50 \end{bmatrix} \right)^2 =$$

$$P_2 = \begin{bmatrix} 0.34725 & 0.266 & 0.38675 \end{bmatrix}.$$

9. Since $P_3 = P_2 T$, $P_2 = P_1 T$ and $P_1 = P_0 T$ then,

$$P_3 = \left(P_1 T \right) T = \left(P_0 T \right) T^2 = P_0 T^3.$$

$$P_3 = \begin{bmatrix} 0.5 & 0.5 \end{bmatrix} \left(\begin{bmatrix} 1.0 & 0.0 \\ 0.2 & 0.8 \end{bmatrix} \right)^3 = \begin{bmatrix} 0.744 & 0.256 \end{bmatrix}.$$

11.
a. After one transition, the state vector is:

$$\begin{bmatrix} 0.7 & 0.3 \end{bmatrix} \begin{bmatrix} 0.6 & 0.4 \\ 0.2 & 0.8 \end{bmatrix} = \begin{bmatrix} 0.48 & 0.52 \end{bmatrix}.$$

b. After two transitions, the state vector is:

$$\begin{bmatrix} 0.48 & 0.52 \end{bmatrix} \begin{bmatrix} 0.6 & 0.4 \\ 0.2 & 0.8 \end{bmatrix} = \begin{bmatrix} 0.392 & 0.608 \end{bmatrix}.$$

13.
a. After one transition, the state vector is:

$$\begin{bmatrix} 0.1 & 0.4 & 0.5 \end{bmatrix} \begin{bmatrix} 0.3 & 0.2 & 0.5 \\ 0.4 & 0.0 & 0.6 \\ 0.1 & 0.8 & 0.1 \end{bmatrix} = \begin{bmatrix} 0.24 & 0.42 & 0.34 \end{bmatrix}.$$

b. After two transitions, the state vector is:

$$\begin{bmatrix} 0.24 & 0.42 & 0.34 \end{bmatrix} \begin{bmatrix} 0.3 & 0.2 & 0.5 \\ 0.4 & 0.0 & 0.6 \\ 0.1 & 0.8 & 0.1 \end{bmatrix} =$$

$$\begin{bmatrix} 0.274 & 0.320 & 0.406 \end{bmatrix}.$$

15. The transition matrix for this problem is:

$$T = \begin{array}{cc} & \begin{array}{cc} F & A \end{array} \\ \begin{array}{c} F \\ A \end{array} & \begin{bmatrix} 0.8 & 0.2 \\ 0.1 & 0.9 \end{bmatrix} \end{array}.$$

The initial state vector for the problem is:

$$P_0 = \begin{bmatrix} 0.6 & 0.4 \end{bmatrix}.$$

In order to find the percent of people who will buy American and the percent of people who will buy a Foreign vehicle on their third automobile we need to find the third state vector P_3. We know that $P_3 = P_0 T^3$. Therefore

$$P_3 = \begin{bmatrix} 0.6 & 0.4 \end{bmatrix} \left(\begin{bmatrix} 0.8 & 0.2 \\ 0.1 & 0.9 \end{bmatrix} \right)^3 = \begin{bmatrix} 0.4248 & 0.5752 \end{bmatrix}.$$

From the third state vector, we see that on their third automobile purchase, 42.48% will buy a foreign make and 57.52% will buy an American make.

17. The initial state vector for this problem is:

$$P_0 = \begin{bmatrix} 0.50 & 0.40 & 0.10 \end{bmatrix}.$$

To find the distribution of each major two years from now, we need to calculate the second state vector. This is:

$$P_2 = P_0 T^2$$

$$P_2 = \begin{bmatrix} 0.50 & 0.40 & 0.10 \end{bmatrix} \left(\begin{bmatrix} 0.87 & 0.05 & 0.08 \\ 0.05 & 0.85 & 0.10 \\ 0.15 & 0.03 & 0.82 \end{bmatrix} \right)^2$$

$$P_2 = \begin{bmatrix} 0.4516 & 0.34116 & 0.20724 \end{bmatrix}.$$

In two years, 45.16% of students will be business majors, 34.12% of students will be engineering majors, and 20.72% of students will be mathematics majors.

19. Let S represent state 1, the state of being immune to sulfasalazine, let C represent state 2, the state of being immune to Cipro®, and let N represent state 3, the state of not being immune to either antibiotic. The transition diagram for this problem is shown at the top of the next column.

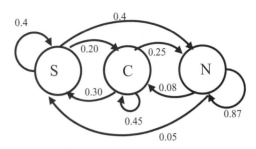

b. The transition matrix for the antibiotic resistance of this strain of bacteria is:

$$T = \begin{array}{c} \\ S \\ C \\ N \end{array} \begin{array}{c} \begin{array}{ccc} S & C & N \end{array} \\ \begin{bmatrix} 0.40 & 0.20 & 0.40 \\ 0.30 & 0.45 & 0.25 \\ 0.05 & 0.08 & 0.87 \end{bmatrix} \end{array}$$

c. The initial state vector for this strain of bacteria is:

$$P_0 = \begin{bmatrix} 0.20 & 0.30 & 0.50 \end{bmatrix}.$$

In order to fid the percent of this strain of bacteria that is immune to these antibiotics three generations from now, we need to calculate the third state vector.

$$P_3 = P_0 T^3$$

$$P_3 = \begin{bmatrix} 0.20 & 0.30 & 0.50 \end{bmatrix} \left(\begin{bmatrix} 0.40 & 0.20 & 0.40 \\ 0.30 & 0.45 & 0.25 \\ 0.05 & 0.08 & 0.87 \end{bmatrix} \right)^3$$

$$P_3 = \begin{bmatrix} 0.1559375 & 0.1683315 & 0.675731 \end{bmatrix}.$$

Three generations from now, 15.59% of this strain of bacteria will be immune to sulfasalazine, 16.83% of this strain of bacteria will be immune to Cipro®, and 67.57% of this strain of bacteria will not be immune to either of the antibiotics.

21. Using the matrix capabilities of a graphing calculator:

$$T^{-1} = \begin{bmatrix} 3.5 & -2.5 \\ -1.5 & 2.5 \end{bmatrix}$$

Therefore,

$$P_0 = P_1 \cdot T^{-1}$$

The solution is continued on the next page.

$$P_0 = \begin{bmatrix} 0.4 & 0.6 \end{bmatrix} \begin{bmatrix} 3.5 & -2.5 \\ -1.5 & 2.5 \end{bmatrix}$$

$$P_0 = \begin{bmatrix} 0.5 & 0.5 \end{bmatrix}$$

23. Using the matrix capabilities of a graphing calculator:

$$T^{-1} = \begin{bmatrix} \frac{-9}{67} & \frac{76}{67} \\ \frac{91}{67} & \frac{-24}{67} \end{bmatrix}$$

Therefore,

$$P_0 = P_1 \cdot T^{-1}$$

$$P_0 = \begin{bmatrix} 0.03 & 0.97 \end{bmatrix} \begin{bmatrix} \frac{-9}{67} & \frac{76}{67} \\ \frac{91}{67} & \frac{-24}{67} \end{bmatrix}$$

$$P_0 = \begin{bmatrix} \frac{88}{67} & \frac{-21}{67} \end{bmatrix}$$

The initial state vector is not valid. There is no way to use this transition matrix to get the first state vector.

25. Using the matrix capabilities of a graphing calculator:

$$T^{-1} = \begin{bmatrix} 0 & \frac{10}{9} & \frac{-1}{9} \\ 0 & 0 & 1 \\ \frac{5}{2} & \frac{-5}{9} & \frac{-17}{18} \end{bmatrix}$$

Therefore,

$$P_0 = P_1 \cdot T^{-1}$$

$$P_0 = \begin{bmatrix} 0.25 & 0.45 & 0.30 \end{bmatrix} \begin{bmatrix} 0 & \frac{10}{9} & \frac{-1}{9} \\ 0 & 0 & 1 \\ \frac{5}{2} & \frac{-5}{9} & \frac{-17}{18} \end{bmatrix}$$

$$P_0 = \begin{bmatrix} 0.75 & 0.1111 & 0.1389 \end{bmatrix}$$

27. We know that $P_n = P_0 T^n$, so the initial state vector $P_0 = \begin{bmatrix} 1 & 0 \end{bmatrix}$ we have

$$P_1 = \begin{bmatrix} 1 & 0 \end{bmatrix} \begin{bmatrix} 0.2 & 0.8 \\ 0.5 & 0.5 \end{bmatrix} = \begin{bmatrix} 0.2 & 0.8 \end{bmatrix}$$

$$P_2 = \begin{bmatrix} 1 & 0 \end{bmatrix} \left(\begin{bmatrix} 0.2 & 0.8 \\ 0.5 & 0.5 \end{bmatrix} \right)^2 = \begin{bmatrix} 0.44 & 0.56 \end{bmatrix}$$

$$P_3 = \begin{bmatrix} 1 & 0 \end{bmatrix} \left(\begin{bmatrix} 0.2 & 0.8 \\ 0.5 & 0.5 \end{bmatrix} \right)^3 = \begin{bmatrix} 0.368 & 0.632 \end{bmatrix}$$

$$P_4 = \begin{bmatrix} 1 & 0 \end{bmatrix} \left(\begin{bmatrix} 0.2 & 0.8 \\ 0.5 & 0.5 \end{bmatrix} \right)^4 = \begin{bmatrix} 0.3896 & 0.6104 \end{bmatrix}$$

$$P_5 = \begin{bmatrix} 1 & 0 \end{bmatrix} \left(\begin{bmatrix} 0.2 & 0.8 \\ 0.5 & 0.5 \end{bmatrix} \right)^5 \approx \begin{bmatrix} 0.3831 & 0.6169 \end{bmatrix}$$

$$P_6 = \begin{bmatrix} 1 & 0 \end{bmatrix} \left(\begin{bmatrix} 0.2 & 0.8 \\ 0.5 & 0.5 \end{bmatrix} \right)^6 \approx \begin{bmatrix} 0.3851 & 0.6149 \end{bmatrix}$$

$$P_7 = \begin{bmatrix} 1 & 0 \end{bmatrix} \left(\begin{bmatrix} 0.2 & 0.8 \\ 0.5 & 0.5 \end{bmatrix} \right)^7 \approx \begin{bmatrix} 0.3845 & 0.6155 \end{bmatrix}$$

$$P_8 = \begin{bmatrix} 1 & 0 \end{bmatrix} \left(\begin{bmatrix} 0.2 & 0.8 \\ 0.5 & 0.5 \end{bmatrix} \right)^8 \approx \begin{bmatrix} 0.3847 & 0.6153 \end{bmatrix}$$

$$P_9 = \begin{bmatrix} 1 & 0 \end{bmatrix} \left(\begin{bmatrix} 0.2 & 0.8 \\ 0.5 & 0.5 \end{bmatrix} \right)^9 \approx \begin{bmatrix} 0.3846 & 0.6154 \end{bmatrix}$$

$$P_{10} = \begin{bmatrix} 1 & 0 \end{bmatrix} \left(\begin{bmatrix} 0.2 & 0.8 \\ 0.5 & 0.5 \end{bmatrix} \right)^{10} \approx \begin{bmatrix} 0.3846 & 0.6154 \end{bmatrix}$$

To four decimal place accuracy, P_n is approximately equal to P_9 for all $n \geq 10$.

29. We know that $P_n = P_0 T^n$, so the initial state vector $P_0 = \begin{bmatrix} 0.4 & 0.6 \end{bmatrix}$ we have

$$P_1 = \begin{bmatrix} 0.4 & 0.6 \end{bmatrix} \begin{bmatrix} 0.2 & 0.8 \\ 0.5 & 0.5 \end{bmatrix} = \begin{bmatrix} 0.38 & 0.62 \end{bmatrix}$$

$$P_2 = \begin{bmatrix} 0.4 & 0.6 \end{bmatrix} \left(\begin{bmatrix} 0.2 & 0.8 \\ 0.5 & 0.5 \end{bmatrix} \right)^2 = \begin{bmatrix} 0.386 & 0.614 \end{bmatrix}$$

$$P_3 = \begin{bmatrix} 0.4 & 0.6 \end{bmatrix} \left(\begin{bmatrix} 0.2 & 0.8 \\ 0.5 & 0.5 \end{bmatrix} \right)^3 = \begin{bmatrix} 0.3842 & 0.6158 \end{bmatrix}$$

The solution is continued on the next page.

$$P_4 = \begin{bmatrix} 0.4 & 0.6 \end{bmatrix} \left(\begin{bmatrix} 0.2 & 0.8 \\ 0.5 & 0.5 \end{bmatrix} \right)^4 \approx \begin{bmatrix} 0.3847 & 0.6153 \end{bmatrix}$$

$$P_5 = \begin{bmatrix} 0.4 & 0.6 \end{bmatrix} \left(\begin{bmatrix} 0.2 & 0.8 \\ 0.5 & 0.5 \end{bmatrix} \right)^5 \approx \begin{bmatrix} 0.3846 & 0.6154 \end{bmatrix}$$

To four decimal place accuracy, P_n is approximately equal to P_4 for all $n \geq 5$.

31. We know that $P_n = P_0 T^n$, so the initial state vector $P_0 = \begin{bmatrix} 1 & 0 & 0 \end{bmatrix}$ we have

$$P_1 = \begin{bmatrix} 1 & 0 & 0 \end{bmatrix} \begin{bmatrix} 1 & 0 & 0 \\ 0.5 & 0 & 0.5 \\ 0 & 1 & 0 \end{bmatrix}$$

$$P_1 = \begin{bmatrix} 1 & 0 & 0 \end{bmatrix}$$

$$P_2 = \begin{bmatrix} 1 & 0 & 0 \end{bmatrix} \left(\begin{bmatrix} 1 & 0 & 0 \\ 0.5 & 0 & 0.5 \\ 0 & 1 & 0 \end{bmatrix} \right)^2$$

$$P_2 = \begin{bmatrix} 1 & 0 & 0 \end{bmatrix}$$

All state vectors are equal to P_0

33. We know that $P_n = P_0 T^n$, so the initial state vector $P_0 = \begin{bmatrix} 0.25 & 0.5 & 0.25 \end{bmatrix}$ we have

$$P_1 = \begin{bmatrix} 0.25 & 0.5 & 0.25 \end{bmatrix} \begin{bmatrix} 1 & 0 & 0 \\ 0.5 & 0 & 0.5 \\ 0 & 1 & 0 \end{bmatrix}$$

$$P_1 = \begin{bmatrix} 0.5 & 0.25 & 0.25 \end{bmatrix}$$

$$P_2 = \begin{bmatrix} 0.25 & 0.5 & 0.25 \end{bmatrix} \left(\begin{bmatrix} 1 & 0 & 0 \\ 0.5 & 0 & 0.5 \\ 0 & 1 & 0 \end{bmatrix} \right)^2$$

$$P_2 = \begin{bmatrix} 0.625 & 0.25 & 0.125 \end{bmatrix}$$

$$P_3 = \begin{bmatrix} 0.25 & 0.5 & 0.25 \end{bmatrix} \left(\begin{bmatrix} 1 & 0 & 0 \\ 0.5 & 0 & 0.5 \\ 0 & 1 & 0 \end{bmatrix} \right)^3$$

$$P_3 = \begin{bmatrix} 0.75 & 0.125 & 0.125 \end{bmatrix}$$

$$P_4 = \begin{bmatrix} 0.25 & 0.5 & 0.25 \end{bmatrix} \left(\begin{bmatrix} 1 & 0 & 0 \\ 0.5 & 0 & 0.5 \\ 0 & 1 & 0 \end{bmatrix} \right)^4$$

$$P_4 = \begin{bmatrix} 0.8125 & 0.125 & 0.625 \end{bmatrix}$$

$$P_5 = \begin{bmatrix} 0.25 & 0.5 & 0.25 \end{bmatrix} \left(\begin{bmatrix} 1 & 0 & 0 \\ 0.5 & 0 & 0.5 \\ 0 & 1 & 0 \end{bmatrix} \right)^5$$

$$P_5 = \begin{bmatrix} 0.875 & 0.0625 & 0.625 \end{bmatrix}$$

$$P_6 = \begin{bmatrix} 0.25 & 0.5 & 0.25 \end{bmatrix} \left(\begin{bmatrix} 1 & 0 & 0 \\ 0.5 & 0 & 0.5 \\ 0 & 1 & 0 \end{bmatrix} \right)^6$$

$$P_6 = \begin{bmatrix} 0.90625 & 0.625 & 0.03125 \end{bmatrix}$$

$$P_7 = \begin{bmatrix} 0.25 & 0.5 & 0.25 \end{bmatrix} \left(\begin{bmatrix} 1 & 0 & 0 \\ 0.5 & 0 & 0.5 \\ 0 & 1 & 0 \end{bmatrix} \right)^7$$

$$P_7 = \begin{bmatrix} 0.9375 & 0.03125 & 0.03125 \end{bmatrix}$$

$$P_8 = \begin{bmatrix} 0.25 & 0.5 & 0.25 \end{bmatrix} \left(\begin{bmatrix} 1 & 0 & 0 \\ 0.5 & 0 & 0.5 \\ 0 & 1 & 0 \end{bmatrix} \right)^8$$

$$P_8 = \begin{bmatrix} 0.953125 & 0.03125 & 0.015625 \end{bmatrix}$$

The solution is continued on the next page.

$$P_9 = \begin{bmatrix} 0.25 & 0.5 & 0.25 \end{bmatrix} \left(\begin{bmatrix} 1 & 0 & 0 \\ 0.5 & 0 & 0.5 \\ 0 & 1 & 0 \end{bmatrix} \right)^9$$

$$P_9 = \begin{bmatrix} 0.96875 & 0.015625 & 0.015625 \end{bmatrix}$$

$$P_{10} = \begin{bmatrix} 0.25 & 0.5 & 0.25 \end{bmatrix} \left(\begin{bmatrix} 1 & 0 & 0 \\ 0.5 & 0 & 0.5 \\ 0 & 1 & 0 \end{bmatrix} \right)^{10}$$

$$P_{10} \approx \begin{bmatrix} 0.9766 & 0.0156 & 0.0078 \end{bmatrix}$$

$$P_{11} = \begin{bmatrix} 0.25 & 0.5 & 0.25 \end{bmatrix} \left(\begin{bmatrix} 1 & 0 & 0 \\ 0.5 & 0 & 0.5 \\ 0 & 1 & 0 \end{bmatrix} \right)^{11}$$

$$P_{11} \approx \begin{bmatrix} 0.9844 & 0.0078 & 0.0078 \end{bmatrix}$$

$$P_{12} = \begin{bmatrix} 0.25 & 0.5 & 0.25 \end{bmatrix} \left(\begin{bmatrix} 1 & 0 & 0 \\ 0.5 & 0 & 0.5 \\ 0 & 1 & 0 \end{bmatrix} \right)^{12}$$

$$P_{12} \approx \begin{bmatrix} 0.9883 & 0.0078 & 0.0039 \end{bmatrix}$$

$$P_{13} = \begin{bmatrix} 0.25 & 0.5 & 0.25 \end{bmatrix} \left(\begin{bmatrix} 1 & 0 & 0 \\ 0.5 & 0 & 0.5 \\ 0 & 1 & 0 \end{bmatrix} \right)^{13}$$

$$P_{13} \approx \begin{bmatrix} 0.9922 & 0.0039 & 0.0039 \end{bmatrix}$$

$$P_{14} = \begin{bmatrix} 0.25 & 0.5 & 0.25 \end{bmatrix} \left(\begin{bmatrix} 1 & 0 & 0 \\ 0.5 & 0 & 0.5 \\ 0 & 1 & 0 \end{bmatrix} \right)^{14}$$

$$P_{14} \approx \begin{bmatrix} 0.9941 & 0.0039 & 0.0020 \end{bmatrix}$$

$$P_{15} = \begin{bmatrix} 0.25 & 0.5 & 0.25 \end{bmatrix} \left(\begin{bmatrix} 1 & 0 & 0 \\ 0.5 & 0 & 0.5 \\ 0 & 1 & 0 \end{bmatrix} \right)^{15}$$

$$P_{15} \approx \begin{bmatrix} 0.9961 & 0.0020 & 0.0020 \end{bmatrix}$$

35. We know that $P_n = P_0 T^n$, so the initial state vector $P_0 = \begin{bmatrix} 0 & 0 & 1 \end{bmatrix}$ we have

$$P_1 = \begin{bmatrix} 0 & 0 & 1 \end{bmatrix} \begin{bmatrix} 1 & 0 & 0 \\ 0.5 & 0 & 0.5 \\ 0 & 1 & 0 \end{bmatrix}$$

$$P_1 = \begin{bmatrix} 0 & 1 & 0 \end{bmatrix}$$

$$P_2 = \begin{bmatrix} 0 & 0 & 1 \end{bmatrix} \left(\begin{bmatrix} 1 & 0 & 0 \\ 0.5 & 0 & 0.5 \\ 0 & 1 & 0 \end{bmatrix} \right)^2$$

$$P_2 = \begin{bmatrix} 0.5 & 0 & 0.5 \end{bmatrix}$$

$$P_3 = \begin{bmatrix} 0 & 0 & 1 \end{bmatrix} \left(\begin{bmatrix} 1 & 0 & 0 \\ 0.5 & 0 & 0.5 \\ 0 & 1 & 0 \end{bmatrix} \right)^3$$

$$P_3 = \begin{bmatrix} 0.5 & 0.5 & 0 \end{bmatrix}$$

$$P_4 = \begin{bmatrix} 0 & 0 & 1 \end{bmatrix} \left(\begin{bmatrix} 1 & 0 & 0 \\ 0.5 & 0 & 0.5 \\ 0 & 1 & 0 \end{bmatrix} \right)^4$$

$$P_4 = \begin{bmatrix} 0.75 & 0 & 0.25 \end{bmatrix}$$

$$P_5 = \begin{bmatrix} 0 & 0 & 1 \end{bmatrix} \left(\begin{bmatrix} 1 & 0 & 0 \\ 0.5 & 0 & 0.5 \\ 0 & 1 & 0 \end{bmatrix} \right)^5$$

$$P_5 = \begin{bmatrix} 0.75 & 0.25 & 0 \end{bmatrix}$$

$$P_6 = \begin{bmatrix} 0 & 0 & 1 \end{bmatrix} \left(\begin{bmatrix} 1 & 0 & 0 \\ 0.5 & 0 & 0.5 \\ 0 & 1 & 0 \end{bmatrix} \right)^6$$

$$P_6 = \begin{bmatrix} 0.875 & 0 & 0.125 \end{bmatrix}$$

The solution is continued on the next page.

$$P_7 = \begin{bmatrix} 0 & 0 & 1 \end{bmatrix} \left(\begin{bmatrix} 1 & 0 & 0 \\ 0.5 & 0 & 0.5 \\ 0 & 1 & 0 \end{bmatrix} \right)^7$$

$$P_7 = \begin{bmatrix} 0.875 & 0.125 & 0 \end{bmatrix}$$

$$P_8 = \begin{bmatrix} 0 & 0 & 1 \end{bmatrix} \left(\begin{bmatrix} 1 & 0 & 0 \\ 0.5 & 0 & 0.5 \\ 0 & 1 & 0 \end{bmatrix} \right)^8$$

$$P_8 = \begin{bmatrix} 0.9375 & 0 & 0.0625 \end{bmatrix}$$

$$P_9 = \begin{bmatrix} 0 & 0 & 1 \end{bmatrix} \left(\begin{bmatrix} 1 & 0 & 0 \\ 0.5 & 0 & 0.5 \\ 0 & 1 & 0 \end{bmatrix} \right)^9$$

$$P_9 = \begin{bmatrix} 0.9375 & 0.0625 & 0 \end{bmatrix}$$

$$P_{10} = \begin{bmatrix} 0 & 0 & 1 \end{bmatrix} \left(\begin{bmatrix} 1 & 0 & 0 \\ 0.5 & 0 & 0.5 \\ 0 & 1 & 0 \end{bmatrix} \right)^{10}$$

$$P_{10} \approx \begin{bmatrix} 0.9688 & 0 & 0.0312 \end{bmatrix}$$

$$P_{11} = \begin{bmatrix} 0 & 0 & 1 \end{bmatrix} \left(\begin{bmatrix} 1 & 0 & 0 \\ 0.5 & 0 & 0.5 \\ 0 & 1 & 0 \end{bmatrix} \right)^{11}$$

$$P_{11} \approx \begin{bmatrix} 0.9688 & 0.0312 & 0 \end{bmatrix}$$

$$P_{12} = \begin{bmatrix} 0 & 0 & 1 \end{bmatrix} \left(\begin{bmatrix} 1 & 0 & 0 \\ 0.5 & 0 & 0.5 \\ 0 & 1 & 0 \end{bmatrix} \right)^{12}$$

$$P_{12} \approx \begin{bmatrix} 0.9844 & 0 & 0.0156 \end{bmatrix}$$

$$P_{13} = \begin{bmatrix} 0 & 0 & 1 \end{bmatrix} \left(\begin{bmatrix} 1 & 0 & 0 \\ 0.5 & 0 & 0.5 \\ 0 & 1 & 0 \end{bmatrix} \right)^{13}$$

$$P_{13} \approx \begin{bmatrix} 0.9844 & 0.0156 & 0 \end{bmatrix}$$

$$P_{14} = \begin{bmatrix} 0 & 0 & 1 \end{bmatrix} \left(\begin{bmatrix} 1 & 0 & 0 \\ 0.5 & 0 & 0.5 \\ 0 & 1 & 0 \end{bmatrix} \right)^{14}$$

$$P_{14} \approx \begin{bmatrix} 0.9922 & 0 & 0.0078 \end{bmatrix}$$

$$P_{15} = \begin{bmatrix} 0 & 0 & 1 \end{bmatrix} \left(\begin{bmatrix} 1 & 0 & 0 \\ 0.5 & 0 & 0.5 \\ 0 & 1 & 0 \end{bmatrix} \right)^{15}$$

$$P_{15} \approx \begin{bmatrix} 0.9922 & 0.0078 & 0 \end{bmatrix}$$

Exercises Section 11.3

1. Since all entries in the transition matrix T are positive, the transition matrix is regular.

3.

$$T = \begin{bmatrix} 0 & 0.5 & 0.5 \\ 0.2 & 0.4 & 0.4 \\ 0.8 & 0.1 & 0.1 \end{bmatrix}$$

$$T^2 = \begin{bmatrix} 0.5 & 0.25 & 0.25 \\ 0.4 & 0.3 & 0.3 \\ 0.1 & 0.45 & 0.45 \end{bmatrix}$$

All of the entries in T^2 are positive; therefore the transition matrix is regular.

5.

$$T = \begin{bmatrix} 1 & 0 & 0 \\ 0.5 & 0.5 & 0 \\ 0.2 & 0.3 & 0.5 \end{bmatrix}$$

$$T^2 = \begin{bmatrix} 1 & 0 & 0 \\ 0.75 & 0.25 & 0 \\ 0.45 & 0.30 & 0.25 \end{bmatrix}$$

The solution is continued on the next page.

We notice that all of the powers of T will have zero entries in the $t_{12}; t_{13}; t_{23}$ positions. Therefore, this transition matrix is not a regular transition matrix.

7. Multiplying the transition matrix T by itself repeatedly we see:

$$T^5 = \begin{bmatrix} 0.417 & 0.584 \\ 0.417 & 0.583 \end{bmatrix}$$

$$T^{10} \approx \begin{bmatrix} 0.416 & 0.583 \\ 0.416 & 0.583 \end{bmatrix}$$

Thus the limiting matrix to the nearest 0.01 is:

$$L = \begin{bmatrix} 0.42 & 0.58 \\ 0.42 & 0.58 \end{bmatrix}$$

9. Multiplying the transition matrix T by itself repeatedly we see:

$$T^5 = \begin{bmatrix} 0.723 & 0.277 \\ 0.555 & 0.445 \end{bmatrix}$$

$$T^{20} \approx \begin{bmatrix} 0.667 & 0.333 \\ 0.667 & 0.333 \end{bmatrix}$$

Thus the limiting matrix to the nearest 0.01 is:

$$L = \begin{bmatrix} 0.67 & 0.33 \\ 0.67 & 0.33 \end{bmatrix}$$

11. Multiplying the transition matrix T by itself repeatedly we see:

$$T^5 = \begin{bmatrix} 0.711 & 0.289 \\ 0.722 & 0.278 \end{bmatrix}$$

$$T^{10} \approx \begin{bmatrix} 0.714 & 0.286 \\ 0.714 & 0.286 \end{bmatrix}$$

The limiting matrix to the nearest 0.01 is:

$$L = \begin{bmatrix} 0.71 & 0.29 \\ 0.71 & 0.29 \end{bmatrix}$$

13. The stable state vector, $S = \begin{bmatrix} x_1 & x_2 \end{bmatrix}$, satisfies the following properties:

$$x_1 + x_2 = 1$$

$$\begin{bmatrix} x_1 & x_2 \end{bmatrix} \begin{bmatrix} 0.4 & 0.2 \\ 0.6 & 0.8 \end{bmatrix} = \begin{bmatrix} x_1 & x_2 \end{bmatrix}$$

This is the system:

$$x_1 + x_2 = 1$$
$$0.4x_1 + 0.2x_2 = x_1$$
$$0.6x_1 + 0.8x_2 = x_2$$

The solution to this system is $x_1 = 0.25$; $x_2 = 0.75$

Thus the stable state vector is:

$$S = \begin{bmatrix} 0.25 & 0.75 \end{bmatrix}.$$

This means that the limiting matrix is:

$$L = \begin{bmatrix} 0.25 & 0.75 \\ 0.25 & 0.75 \end{bmatrix}$$

15. The stable state vector, $S = \begin{bmatrix} x_1 & x_2 \end{bmatrix}$, satisfies the following properties:

$$x_1 + x_2 = 1$$

$$\begin{bmatrix} x_1 & x_2 \end{bmatrix} \begin{bmatrix} \frac{7}{8} & \frac{1}{8} \\ \frac{1}{6} & \frac{5}{6} \end{bmatrix} = \begin{bmatrix} x_1 & x_2 \end{bmatrix}$$

Thus, the following system must be satisfied:

$$x_1 + x_2 = 1$$
$$\tfrac{7}{8}x_1 + \tfrac{1}{8}x_2 = x_1$$
$$\tfrac{1}{6}x_1 + \tfrac{5}{6}x_2 = x_2$$

The solution to this system is:

$$x_1 = \tfrac{4}{7} \; ; \; x_2 = \tfrac{3}{7}$$

Thus the stable state vector is:

$$S = \begin{bmatrix} \frac{4}{7} & \frac{3}{7} \end{bmatrix}.$$

The solution is continued on the following page.

Using the stable state vector from the previous page, the limiting matrix is:

$$L = \begin{bmatrix} \frac{4}{7} & \frac{3}{7} \\ \frac{4}{7} & \frac{3}{7} \end{bmatrix}$$

17. The stable state vector, $S = \begin{bmatrix} x_1 & x_2 \end{bmatrix}$, satisfies the following properties:

$$x_1 + x_2 = 1$$

$$\begin{bmatrix} x_1 & x_2 \end{bmatrix} \begin{bmatrix} \frac{1}{7} & \frac{6}{7} \\ \frac{9}{10} & \frac{1}{10} \end{bmatrix} = \begin{bmatrix} x_1 & x_2 \end{bmatrix}$$

Thus, the following system must be satisfied:

$$x_1 + x_2 = 1$$
$$\tfrac{1}{7}x_1 + \tfrac{9}{10}x_2 = x_1$$
$$\tfrac{6}{7}x_1 + \tfrac{1}{10}x_2 = x_2$$

The solution to this system is:

$$x_1 = \tfrac{21}{41} \; ; \; x_2 = \tfrac{20}{41}$$

Thus the stable state vector is:

$$S = \begin{bmatrix} \frac{21}{41} & \frac{20}{41} \end{bmatrix}.$$

This means that the limiting matrix is:

$$L = \begin{bmatrix} \frac{21}{41} & \frac{20}{41} \\ \frac{21}{41} & \frac{20}{41} \end{bmatrix}$$

19. The stable state vector, $S = \begin{bmatrix} x_1 & x_2 & x_3 \end{bmatrix}$, satisfies the following properties:

$$x_1 + x_2 + x_3 = 1$$

$$\begin{bmatrix} x_1 & x_2 & x_3 \end{bmatrix} \begin{bmatrix} \frac{1}{5} & \frac{2}{5} & \frac{2}{5} \\ \frac{1}{2} & \frac{1}{4} & \frac{1}{4} \\ \frac{1}{3} & \frac{1}{6} & \frac{1}{2} \end{bmatrix} = \begin{bmatrix} x_1 & x_2 & x_3 \end{bmatrix}$$

The matrix equation becomes the system at the top of the next column.

$$x_1 + x_2 + x_3 = 1$$
$$\tfrac{1}{5}x_1 + \tfrac{1}{2}x_2 + \tfrac{1}{3}x_3 = x_1$$
$$\tfrac{2}{5}x_1 + \tfrac{1}{4}x_2 + \tfrac{1}{6}x_3 = x_2$$
$$\tfrac{2}{5}x_1 + \tfrac{1}{4}x_2 + \tfrac{1}{2}x_3 = x_3$$

The solution to this system is:

$$x_1 = \tfrac{1}{3} \; ; \; x_2 = \tfrac{4}{15} \; ; \; x_3 = \tfrac{2}{5}$$

Thus the stable state vector is:

$$S = \begin{bmatrix} \frac{1}{3} & \frac{4}{15} & \frac{2}{5} \end{bmatrix}.$$

This means that the limiting matrix is:

$$L = \begin{bmatrix} \frac{1}{3} & \frac{4}{15} & \frac{2}{5} \\ \frac{1}{3} & \frac{4}{15} & \frac{2}{5} \\ \frac{1}{3} & \frac{4}{15} & \frac{2}{5} \end{bmatrix}$$

21. The stable state vector, $S = \begin{bmatrix} x_1 & x_2 & x_3 \end{bmatrix}$, satisfies the following properties:

$$x_1 + x_2 + x_3 = 1$$

$$\begin{bmatrix} x_1 & x_2 & x_3 \end{bmatrix} \begin{bmatrix} 0.0 & 0.9 & 0.1 \\ 0.8 & 0.0 & 0.2 \\ 0.3 & 0.3 & 0.4 \end{bmatrix} = \begin{bmatrix} x_1 & x_2 & x_3 \end{bmatrix}$$

This is the system:

$$x_1 + x_2 + x_3 = 1$$
$$0x_1 + 0.8x_2 + 0.3x_3 = x_1$$
$$0.9x_1 + 0.0x_2 + 0.3x_3 = x_2$$
$$0.1x_1 + 0.2x_2 + 0.4x_3 = x_3$$

The solution to this system is:

$$x_1 = \tfrac{54}{139} \; ; \; x_2 = \tfrac{57}{139} \; ; \; x_3 = \tfrac{28}{139}$$

Thus the stable state vector is:

$$S = \begin{bmatrix} \frac{54}{139} & \frac{57}{139} & \frac{28}{139} \end{bmatrix}.$$

The limiting matrix is displayed at the top of the next page.

$$L = \begin{bmatrix} \frac{54}{139} & \frac{57}{139} & \frac{28}{139} \\ \frac{54}{139} & \frac{57}{193} & \frac{28}{139} \\ \frac{54}{139} & \frac{57}{139} & \frac{28}{139} \end{bmatrix}$$

23. Let A represent the state of owning an American-built car and F represent the state of owning a foreign-built car.

a. The transition diagram for the subscription rates is:

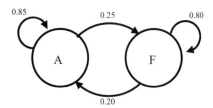

The transition matrix is:

$$T = \begin{bmatrix} 0.85 & 0.25 \\ 0.20 & 0.80 \end{bmatrix}$$

b. The stable state vector, $S = \begin{bmatrix} x_1 & x_2 \end{bmatrix}$, satisfies the following properties:

$$x_1 + x_2 = 1$$

$$\begin{bmatrix} x_1 & x_2 \end{bmatrix} \begin{bmatrix} 0.85 & 0.25 \\ 0.20 & 0.80 \end{bmatrix} = \begin{bmatrix} x_1 & x_2 \end{bmatrix}$$

This is the system:

$$x_1 + x_2 = 1$$
$$0.85x_1 + 0.20x_2 = x_1$$
$$0.25x_1 + 0.80x_2 = x_2$$

The solution to this system is:

$$x_1 = \tfrac{4}{7} \approx 0.571; \; x_2 = \tfrac{3}{7} \approx 0.429$$

Thus the stable state vector is:

$$S = \begin{bmatrix} \tfrac{4}{7} & \tfrac{3}{7} \end{bmatrix}.$$

Therefore, in the long run 57.1% of buyers will buy American-built cars and 42.9% of buyers will buy foreign-built cars.

25. Let state 1 represent the state of voting Democrat, state 2 represent the state of voting Republican, and state 3 represent the state of voting Independent. The stable state vector, $S = \begin{bmatrix} x_1 & x_2 & x_3 \end{bmatrix}$, satisfies the following properties:

$$x_1 + x_2 + x_3 = 1$$

$$\begin{bmatrix} x_1 & x_2 & x_3 \end{bmatrix} \begin{bmatrix} 0.80 & 0.15 & 0.05 \\ 0.08 & 0.90 & 0.02 \\ 0.20 & 0.15 & 0.65 \end{bmatrix} = \begin{bmatrix} x_1 & x_2 & x_3 \end{bmatrix}$$

This is the system:

$$x_1 + x_2 + x_3 = 1$$
$$0.80x_1 + 0.08x_2 + 0.20x_3 = x_1$$
$$0.15x_1 + 0.60x_2 + 0.15x_3 = x_2$$
$$0.05x_1 + 0.02x_2 + 0.65x_3 = x_3$$

The solution to this system is:

$$x_1 = 0.32 \; ; \; x_2 = 0.60 \; ; \; x_3 = 0.08$$

Thus the stable state vector is:

$$S = \begin{bmatrix} 0.32 & 0.60 & 0.08 \end{bmatrix}.$$

This means that the limiting matrix is:

$$L = \begin{bmatrix} 0.32 & 0.60 & 0.08 \\ 0.32 & 0.60 & 0.08 \\ 0.32 & 0.60 & 0.08 \end{bmatrix}$$

This means that if the transition probabilities continue, the Democrats will claim 32% of the vote, the Republicans will claim 60% of the vote, and the Independents will claim 8% of the vote.

27. Let state 1 represent the state of being right handed, state 2 represent the state of being left handed, and state 3 represent the state of being ambidextrous. The stable state vector, $S = \begin{bmatrix} x_1 & x_2 & x_3 \end{bmatrix}$, satisfies the following properties:

$$x_1 + x_2 + x_3 = 1$$

$$\begin{bmatrix} x_1 & x_2 & x_3 \end{bmatrix} \begin{bmatrix} 0.70 & 0.30 & 0 \\ 0.25 & 0.50 & 0.25 \\ 0.40 & 0.40 & 0.30 \end{bmatrix} = \begin{bmatrix} x_1 & x_2 & x_3 \end{bmatrix}$$

The solution is continued on the next page.

The system that is derived from the matrix equation on the previous page is shown below:

$$x_1 + x_2 + x_3 = 1$$
$$0.70x_1 + 0.25x_2 + 0.40x_3 = x_1$$
$$0.30x_1 + 0.50x_2 + 0.40x_3 = x_2$$
$$0.00x_1 + 0.25x_2 + 0.30x_3 = x_3$$

The solution to the system is:

$$x_1 = 0.488 \; ; \; x_2 = 0.390 \; ; \; x_3 = 0.122$$

Thus the stable state vector is:

$$S = \begin{bmatrix} 0.488 & 0.390 & 0.122 \end{bmatrix}.$$

b. Using the stable state vector from part (a), the limiting matrix is:

$$L = \begin{bmatrix} 0.488 & 0.390 & 0.122 \\ 0.488 & 0.390 & 0.122 \\ 0.488 & 0.390 & 0.122 \end{bmatrix}$$

The limiting matrix tells us in the long run that 48.8% of people will be right handed, 39% of people will be left handed, and 12.2% of people will be ambidextrous.

29. Let L represent the state of being low risk, M represent the state of being medium risk, and H represent the state of being high risk.

a. The transition diagram for the experiment is:

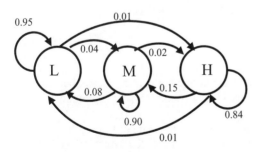

The transition matrix for this experiment is:

$$T = \begin{array}{c} \\ L \\ M \\ H \end{array} \begin{array}{ccc} L & M & H \\ \begin{bmatrix} 0.95 & 0.04 & 0.01 \\ 0.08 & 0.90 & 0.02 \\ 0.01 & 0.15 & 0.84 \end{bmatrix} \end{array}$$

b. The stable state vector, $S = \begin{bmatrix} x_1 & x_2 & x_3 \end{bmatrix}$, satisfies the following properties:

$$x_1 + x_2 + x_3 = 1$$

$$\begin{bmatrix} x_1 & x_2 & x_3 \end{bmatrix} \begin{bmatrix} 0.95 & 0.04 & 0.01 \\ 0.08 & 0.90 & 0.02 \\ 0.01 & 0.15 & 0.84 \end{bmatrix} = \begin{bmatrix} x_1 & x_2 & x_3 \end{bmatrix}$$

This is the system:

$$x_1 + x_2 + x_3 = 1$$
$$0.95x_1 + 0.08x_2 + 0.01x_3 = x_1$$
$$0.04x_1 + 0.90x_2 + 0.15x_3 = x_2$$
$$0.01x_1 + 0.02x_2 + 0.84x_3 = x_3$$

The solution to this system is:

$$x_1 = 0.573 \; ; \; x_2 = 0.348 \; ; \; x_3 = 0.079$$

Thus the stable state vector is:

$$S = \begin{bmatrix} 0.573 & 0.348 & 0.079 \end{bmatrix}.$$

We can determine from the stable state vector that in the long run, 57.3% of the drivers will be classified as low risk, 34.8% of the drivers will be classified as medium risk, and 7.9% of the drivers will be classified as high risk.

c. From the stable state vector in part (b), the limiting matrix is:

$$L = \begin{bmatrix} 0.573 & 0.348 & 0.079 \\ 0.573 & 0.348 & 0.079 \\ 0.573 & 0.348 & 0.079 \end{bmatrix}$$

The limiting matrix tells us that any driver, regardless of their initial classification, will be classified in the long run as a low-risk driver with a probability of 0.573, as a medium-risk driver with a probability of 0.348, and as a high-risk driver with a probability of 0.079.

Chapter 11 Summary Exercises

1. The transition diagram with all of the probabilities shown is:

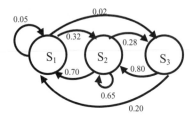

The transition matrix for this diagram is:

$$T = \begin{bmatrix} 0.05 & 0.32 & 0.63 \\ 0.70 & 0.02 & 0.28 \\ 0.20 & 0.80 & 0 \end{bmatrix}.$$

3. The state vector one transition later is:

$$P_1 = P_0 T$$

$$P_1 = \begin{bmatrix} 0.1 & 0.6 & 0.3 \end{bmatrix} \begin{bmatrix} 0.2 & 0.5 & 0.3 \\ 0.1 & 0.7 & 0.2 \\ 0.6 & 0.0 & 0.4 \end{bmatrix} =$$

$$P_1 = \begin{bmatrix} 0.26 & 0.47 & 0.27 \end{bmatrix}.$$

5. The stable state vector, $S = \begin{bmatrix} x_1 & x_2 \end{bmatrix}$, satisfies the following properties:

$$x_1 + x_2 = 1$$

$$\begin{bmatrix} x_1 & x_2 \end{bmatrix} \begin{bmatrix} 0.1 & 0.9 \\ 0.7 & 0.3 \end{bmatrix} = \begin{bmatrix} x_1 & x_2 \end{bmatrix}$$

This is the system:

$$x_1 + x_2 = 1$$
$$0.1x_1 + 0.7x_2 = x_1$$
$$0.9x_1 + 0.3x_2 = x_2$$

The solution to this system is:

$$x_1 = \tfrac{7}{16} \; ; \; x_2 = \tfrac{9}{16}$$

Thus the stable state vector is:

$$S = \begin{bmatrix} \tfrac{7}{16} & \tfrac{9}{16} \end{bmatrix}.$$

This means that the limiting matrix is:

$$L = \begin{bmatrix} \tfrac{7}{16} & \tfrac{9}{16} \\ \tfrac{7}{16} & \tfrac{9}{16} \end{bmatrix}.$$

7. The stable state vector, $S = \begin{bmatrix} x_1 & x_2 \end{bmatrix}$, satisfies the following properties:

$$x_1 + x_2 = 1$$

$$\begin{bmatrix} x_1 & x_2 \end{bmatrix} \begin{bmatrix} 0.5 & 0.5 \\ 0.6 & 0.4 \end{bmatrix} = \begin{bmatrix} x_1 & x_2 \end{bmatrix}$$

This is the system:

$$x_1 + x_2 = 1$$
$$0.5x_1 + 0.6x_2 = x_1$$
$$0.5x_1 + 0.4x_2 = x_2$$

The solution to this system is:

$$x_1 = \tfrac{6}{11} \; ; \; x_2 = \tfrac{5}{11}$$

Thus the stable state vector is:

$$S = \begin{bmatrix} \tfrac{6}{11} & \tfrac{5}{11} \end{bmatrix}.$$

This means that the limiting matrix is:

$$L = \begin{bmatrix} \tfrac{6}{11} & \tfrac{5}{11} \\ \tfrac{6}{11} & \tfrac{5}{11} \end{bmatrix}.$$

9. This matrix cannot be a transition matrix because it has a negative entry I the second row, second column.

11. Let state 1 represent the state of hitting the bulls-eye, and let state 2 represent the state of missing the bulls-eye. The transition matrix for this problem is:

$$T = \begin{matrix} & \begin{matrix} s_1 & \ s_2 \end{matrix} \\ \begin{matrix} s_1 \\ s_2 \end{matrix} & \begin{bmatrix} 0.82 & 0.18 \\ 0.73 & 0.27 \end{bmatrix} \end{matrix}$$

The solution is continued on the following page.

Given that the archer missed the bulls-eye on her first shot, this means that we are currently in the second state. Squaring the transition matrix on the previous page we see:

$$T^2 = \begin{matrix} & s_1 & s_2 \\ s_1 & \begin{bmatrix} 0.804 & 0.196 \\ s_2 & 0.796 & 0.204 \end{bmatrix} \end{matrix}$$

The above matrix says that two shots after the archers first miss, the probability that she will hit the bulls-eye and thus transition to state 1 is 0.796.

13. Multiplying the transition matrix T by itself repeatedly we see:

$$T^5 = \begin{bmatrix} 0.2641 & 0.4632 & 0.2727 \\ 0.2538 & 0.4734 & 0.2727 \\ 0.2579 & 0.4694 & 0.2727 \end{bmatrix}$$

$$T^{10} = \begin{bmatrix} 0.2576 & 0.4696 & 0.2727 \\ 0.2575 & 0.4697 & 0.2727 \\ 0.2576 & 0.4697 & 0.2727 \end{bmatrix}$$

$$T^{20} = \begin{bmatrix} 0.2576 & 0.4697 & 0.2727 \\ 0.2576 & 0.4697 & 0.2727 \\ 0.2576 & 0.4697 & 0.2727 \end{bmatrix}$$

Thus the limiting matrix to the nearest 0.0001 is:

$$L = \begin{bmatrix} 0.2576 & 0.4697 & 0.2727 \\ 0.2576 & 0.4697 & 0.2727 \\ 0.2576 & 0.4697 & 0.2727 \end{bmatrix}.$$

Cumulative Review

15. The augmented matrix is:

$$\begin{bmatrix} 1 & 2 & 1 & | & 4 \\ 2 & 1 & 3 & | & 11 \\ 3 & 5 & -2 & | & -2 \end{bmatrix}$$

Pivoting on the one in row 1, column 1 we obtain the matrix at the top of the next column.

$$\begin{bmatrix} 1 & 2 & 1 & | & 4 \\ 0 & -3 & 1 & | & 3 \\ 0 & -1 & -5 & | & -14 \end{bmatrix}$$

Next interchange rows 2 and 3

$$\begin{bmatrix} 1 & 2 & 1 & | & 4 \\ 0 & -1 & -5 & | & -14 \\ 0 & -3 & 1 & | & 3 \end{bmatrix}$$

Now we pivot on the negative one in row 2, column 3 to get:

$$\begin{bmatrix} 1 & 0 & -9 & | & -24 \\ 0 & 1 & 5 & | & 14 \\ 0 & 0 & 16 & | & 45 \end{bmatrix}$$

Finally, pivot on the 16 in row 3, column 3 to get:

$$\begin{bmatrix} 1 & 0 & 0 & | & \frac{21}{16} \\ 0 & 1 & 0 & | & \frac{-1}{16} \\ 0 & 0 & 1 & | & \frac{45}{16} \end{bmatrix}$$

The solution to the system is:

$$x = \frac{21}{16}; \, y = \frac{-1}{16}; \, z = \frac{45}{16}.$$

17. The Venn diagram for the set $A' \cap (B' - C')$ is:

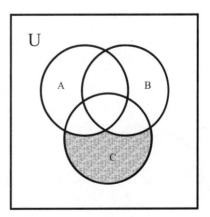

19. The initial simplex tableau is:

$$
\begin{array}{ccccc}
x & y & s_1 & s_2 & f \\
\end{array}
$$

$$
\left[
\begin{array}{ccccc|c}
1 & 1 & 1 & 0 & 0 & 20 \\
2 & 3 & 0 & 1 & 0 & 51 \\
\hline
-3 & -5 & 0 & 0 & 1 & 0
\end{array}
\right]
$$

Since negative five is the most negative entry in the last row, we select column 2 has the pivot column. The smallest quotient rule shows:

$R_1: \frac{20}{1} = 20$

$R_2: \frac{51}{3} = 17$

The pivot element is the three in row 2, column 2. Pivoting on this element results in the tableau:

$$
\begin{array}{ccccc}
x & y & s_1 & s_2 & f \\
\end{array}
$$

$$
\left[
\begin{array}{ccccc|c}
\frac{1}{3} & 0 & 1 & \frac{-1}{3} & 0 & 3 \\
\frac{2}{3} & 1 & 0 & \frac{1}{3} & 0 & 17 \\
\hline
\frac{1}{3} & 0 & 0 & \frac{5}{3} & 1 & 85
\end{array}
\right]
$$

Since there are no more negative entries in the last row, we have reached the optimal value of the objective function. The optimal solution is:

$x = 0; y = 17; s_1 = 3; s_2 = 0; f = 85$.

Chapter 11 Sample Test Answers

1. The transition diagram with all of the probabilities shown is:

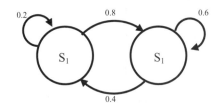

2. The transition matrix from problem 1 is:

$$
T = \begin{bmatrix} 0.2 & 0.8 \\ 0.4 & 0.6 \end{bmatrix}
$$

3. The completed transition diagram is:

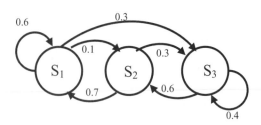

4. The transition matrix for the diagram in problem 3 is:

$$
T = \begin{bmatrix} 0.6 & 0.1 & 0.3 \\ 0.7 & 0.0 & 0.3 \\ 0.0 & 0.6 & 0.4 \end{bmatrix}
$$

5. The transition diagram for the transition matrix is:

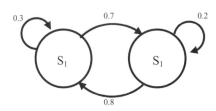

6. The transition diagram for the given matrix is:

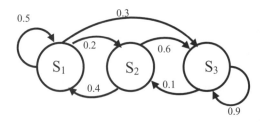

7. Let S represent the state of snow, let N represent the state of no snow.

The transition diagram for this problem is:

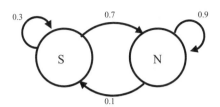

8. The transition matrix is:

$$T = \begin{bmatrix} 0.3 & 0.7 \\ 0.1 & 0.9 \end{bmatrix}$$

9. The state vector one transition later is:

$$P_1 = P_0 T$$

$$P_1 = \begin{bmatrix} 0.2 & 0.5 & 0.3 \end{bmatrix} \begin{bmatrix} 0.4 & 0.5 & 0.1 \\ 0.0 & 0.7 & 0.3 \\ 0.6 & 0.4 & 0.0 \end{bmatrix} =$$

$$P_1 = \begin{bmatrix} 0.26 & 0.57 & 0.17 \end{bmatrix}.$$

10. Since $P_2 = P_1 T$ and $P_1 = P_0 T$ then,

$$P_2 = P_0 TT = P_0 T^2.$$

$$P_1 = \begin{bmatrix} 0.2 & 0.5 & 0.3 \end{bmatrix} \left(\begin{bmatrix} 0.4 & 0.5 & 0.1 \\ 0.0 & 0.7 & 0.3 \\ 0.6 & 0.4 & 0.0 \end{bmatrix} \right)^2 =$$

$$P_2 = \begin{bmatrix} 0.206 & 0.597 & 0.197 \end{bmatrix}.$$

11. The transition matrix is not regular, because the entry in row 1, column 2 will always be zero regardless of how many times you multiply the matrix by itself.

$$T = \begin{bmatrix} 1 & 0 \\ 0.8 & 0.2 \end{bmatrix}$$

$$T^2 = \begin{bmatrix} 1 & 0 \\ 0.96 & 0.04 \end{bmatrix}$$

$$T^3 = \begin{bmatrix} 1 & 0 \\ 0.992 & 0.008 \end{bmatrix}$$

This means that we can never get to state 2 from state 1.

12. The transition matrix is a regular transition matrix. Notice that the square of the transition matrix is:

$$T^2 = \begin{bmatrix} 0.55 & 0.36 & 0.09 \\ 0.32 & 0.17 & 0.51 \\ 0.26 & 0.12 & 0.62 \end{bmatrix}$$

All of the entries in T^2 are positive, meaning that we can move from one state to any other state in at most two moves. Thus, the matrix is regular.

13. Raising the matrix to the 64th power gives us the limiting matrix to the nearest 0.01. The limiting matrix is:

$$L = T^{64} = \begin{bmatrix} 0.53 & 0.47 \\ 0.53 & 0.47 \end{bmatrix}$$

14. Raising the matrix to the 5th power gives us the limiting matrix to the nearest 0.01. The limiting matrix is:

$$L = T^5 = \begin{bmatrix} 0.57 & 0.43 \\ 0.57 & 0.43 \end{bmatrix}$$

15. The stable state vector, $S = \begin{bmatrix} x_1 & x_2 \end{bmatrix}$, satisfies the following properties:

$$x_1 + x_2 = 1$$

$$\begin{bmatrix} x_1 & x_2 \end{bmatrix} \begin{bmatrix} 0.47 & 0.53 \\ 0.26 & 0.74 \end{bmatrix} = \begin{bmatrix} x_1 & x_2 \end{bmatrix}$$

This leads to the system:

$$\begin{aligned} x_1 + x_2 &= 1 \\ 0.47x_1 + 0.26x_2 &= x_1 \\ 0.53x_1 + 0.74x_2 &= x_2 \end{aligned}$$

The solution to this system is:

$$x_1 = \tfrac{26}{79} \; ; \; x_2 = \tfrac{53}{79}$$

Thus the stable state vector is:

$$S = \begin{bmatrix} \tfrac{26}{79} & \tfrac{53}{79} \end{bmatrix}.$$

The limiting matrix is displayed at the top of the next page.

$$L = \begin{bmatrix} \frac{26}{79} & \frac{53}{79} \\ \frac{26}{79} & \frac{53}{79} \end{bmatrix}.$$

16. The stable state vector, $S = \begin{bmatrix} x_1 & x_2 \end{bmatrix}$, satisfies the following properties:

$$x_1 + x_2 = 1$$

$$\begin{bmatrix} x_1 & x_2 \end{bmatrix} \begin{bmatrix} 0.0 & 1.0 \\ 0.7 & 0.3 \end{bmatrix} = \begin{bmatrix} x_1 & x_2 \end{bmatrix}$$

This leads to the system:

$$\begin{aligned} x_1 + \quad x_2 &= 1 \\ 0.0x_1 + 0.7x_2 &= x_1 \\ 1.0x_1 + 0.3x_2 &= x_2 \end{aligned}$$

The solution to this system is:

$$x_1 = \tfrac{7}{17} \; ; \; x_2 = \tfrac{10}{17}$$

Thus the stable state vector is:

$$S = \begin{bmatrix} \frac{7}{17} & \frac{10}{17} \end{bmatrix}.$$

This means that the limiting matrix is:

$$L = \begin{bmatrix} \frac{7}{17} & \frac{10}{17} \\ \frac{7}{17} & \frac{10}{17} \end{bmatrix}.$$

Chapter 12
Game Theory:
Two–Player, Zero – Sum Games

Exercises Section12.1

1. The game is strictly determined because the two in row 2, column 2 is the least entry in row 2 and the greatest entry in column 2. Therefore, the value of the game is two. The game favors the row player. The best strategy for the row player is to play row 2. The best strategy for the column player is to play column 2.

3. The game is not strictly determined. The minimum entry in row 1 is three, which is not the maximum entry in column 2. The minimum entry in row 2 is two, which is not the maximum entry in column 1.

5. The game is strictly determined because the two in row 2, column 1 is the least entry in row 2 and the greatest entry in column 1. Therefore, the value of the game is two. The game favors the row player. The best strategy for the row player is to play row 2. The best strategy for the column player is to play column 1.

7. The game is strictly determined because the one in row 1, column 1 is the least entry in row 1 and the greatest entry in column 1. Therefore, the value of the game is one. This game favors the row player. The best strategy for the row player is to play row 1. The best strategy for the column player is to play column 1.

9. The game is not strictly determined, since there is no saddle point. The least entry in row 1 is negative two, but it negative two is not the greatest entry in column 3. The least entry in row 2 is negative five, but negative five is not the greatest entry in column three. The least entry in row 3 is negative 2, but negative two is not the greatest entry in column 4.

11. If x is any number, $0 < x < 3$, then x will be the least entry of row 1 and the greatest entry in column 1, and the game will be strictly determined. These are the only values of x for which the game will be strictly determined.

13. We construct a matrix in which the store is the row player and the economy is the column player. Let row 1 represent the strategy of remodeling (R) and row 2 the strategy on not remodeling (NR). Let column 1 be the condition that the economic situation is favorable, and let column 2 be the condition that a recession occurs. The payoff matrix is displayed at the top of the next column.

Economy

$$\text{Store} \begin{array}{c} \\ \text{R} \\ \text{NR} \end{array} \overset{\text{Favorable} \quad \text{Recesion}}{\begin{bmatrix} 100,000 & -40,000 \\ -50,000 & -45,000 \end{bmatrix}}$$

The game is strictly determined with a -40,000 as the saddle point. The best strategy for the store is row 1. Therefore the store should remodel.

15. Consider Nutone as the row player and Fabless as the column player. Let rows 1, 2 and 3 represent the strategies of offering sports-medicine, senior-citizen, and drug-therapy programs respectively. Let column 1 and 2 represent the strategy of offering sports-medicine or senior-citizen programs respectively.

a. The payoff matrix (where entries represent new customers for Nutone) is:

Fabless

$$\text{Nutone} \begin{array}{c} \\ \text{SM} \\ \text{SC} \\ \text{DT} \end{array} \overset{\text{SM} \quad \text{SC}}{\begin{bmatrix} 250 & 400 \\ 350 & 300 \\ 375 & 400 \end{bmatrix}}$$

b. The saddle point is the entry in row 3, column 1. 375 is the least entry in row 3, and the greatest entry in column 2.

c. The best strategy for Nutone is to play row 3, or open the drug-therapy program. The best strategy for Fabless is to open the sports-medicine program

17. Let Juan be the row player and let row 1 represent the red four and row 2 represent the black nine. Let Hector be the column player. Let column 1 represent the red five, column 2 represent the black seven, and column 3 represent the black 8.

a. The payoff matrix for this game is:

Hector

$$\text{Juan} \begin{array}{c} \\ \text{R4} \\ \text{B9} \end{array} \overset{\text{R5} \quad \text{B7} \quad \text{B8}}{\begin{bmatrix} 1 & -4 & -4 \\ -5 & 2 & 1 \end{bmatrix}}$$

b. The game is not strictly determined, because there is no saddle point.

19. Let Computer Software be the row player and Computer Hardware be the column player. Let rows 1 and 2 represent locating in Tallahassee and Austin respectively. Let columns 1 and 2 represent the same options for Computer Hardware.

a. Let the values of the payoff matrix represent profit of Computer Software in excess of Computer Hardware's profit. The payoff matrix is:

$$\begin{array}{cc} & \text{Computer Hardware} \\ & \begin{array}{cc} \text{T} & \quad\text{A} \end{array} \\ \text{Computer Software} \begin{array}{c} \text{T} \\ \text{A} \end{array} & \begin{bmatrix} 0 & 500,000 \\ -250,000 & 50,000 \end{bmatrix} \end{array}$$

b. This is a strictly determined game. The zero in row 1 is the saddle point, because it is the minimum entry in row 1 and the maximum entry in row 2.

c. The best strategy for Computer Software is row 1. That is, Computer Software should locate in Tallahassee. The best strategy for Computer Hardware is column 1. That is, Computer Hardware should also locate in Tallahassee.

Exercises Section 12.2

1. The expected value of the game is:

$$E(row) = \begin{bmatrix} 0.5 & 0.5 \end{bmatrix} \begin{bmatrix} 2 & -1 \\ 0 & 3 \end{bmatrix} \begin{bmatrix} 0.5 \\ 0.5 \end{bmatrix} = 1$$

3. The expected value of the game is:

$$E(row) = \begin{bmatrix} \frac{1}{3} & \frac{1}{3} & \frac{1}{3} \end{bmatrix} \begin{bmatrix} 2 & -1 & 4 \\ 0 & 3 & -2 \\ 1 & 2 & -5 \end{bmatrix} \begin{bmatrix} \frac{1}{3} \\ \frac{1}{3} \\ \frac{1}{3} \end{bmatrix} = \frac{4}{9}$$

5. The expected value of the game is:

$$E(row) = \begin{bmatrix} \frac{1}{2} & \frac{1}{2} \end{bmatrix} \begin{bmatrix} -1 & 5 & 4 \\ 2 & 3 & -6 \end{bmatrix} \begin{bmatrix} \frac{1}{3} \\ \frac{1}{3} \\ \frac{1}{3} \end{bmatrix} = \frac{7}{6}$$

7. The expected value of the game is:

$$E(row) = \begin{bmatrix} \frac{1}{3} & \frac{1}{3} & \frac{1}{3} \end{bmatrix} \begin{bmatrix} -3 & 2 & 5 \\ 1 & 4 & -3 \\ -5 & 1 & 3 \end{bmatrix} \begin{bmatrix} \frac{1}{3} \\ \frac{1}{3} \\ \frac{1}{3} \end{bmatrix} = \frac{5}{9}$$

9. The expected value of the game is:

$$E(row) = \begin{bmatrix} \frac{3}{8} & \frac{5}{8} \end{bmatrix} \begin{bmatrix} -3 & 5 & 7 \\ 0 & -2 & -4 \end{bmatrix} \begin{bmatrix} \frac{4}{11} \\ \frac{5}{11} \\ \frac{2}{11} \end{bmatrix} = \frac{-9}{88}$$

11. If the column player chooses strategy Q_1, the expected value of the game is:

$$E(row) = \begin{bmatrix} \frac{3}{5} & \frac{2}{5} \end{bmatrix} \begin{bmatrix} 0 & 2 \\ 1 & -3 \end{bmatrix} \begin{bmatrix} \frac{2}{3} \\ \frac{1}{3} \end{bmatrix} = \frac{4}{15} \approx 0.267 .$$

If the column player chooses strategy Q_2, the expected value of the game is:

$$E(row) = \begin{bmatrix} \frac{3}{5} & \frac{2}{5} \end{bmatrix} \begin{bmatrix} 0 & 2 \\ 1 & -3 \end{bmatrix} \begin{bmatrix} \frac{7}{11} \\ \frac{4}{11} \end{bmatrix} = \frac{14}{55} \approx 0.255 .$$

Since the expected value is the row players' value, the column player will prefer the lower of the two expected values. In other words, the column player's best strategy is Q_2.

13. If the column player chooses strategy Q_1, the expected value of the game is:

$$E(row) = \begin{bmatrix} \frac{3}{5} & \frac{2}{5} \end{bmatrix} \begin{bmatrix} \frac{3}{4} & \frac{1}{2} \\ \frac{1}{3} & \frac{-5}{4} \end{bmatrix} \begin{bmatrix} \frac{2}{3} \\ \frac{1}{3} \end{bmatrix} = \frac{29}{90} \approx 0.322 .$$

If the column player chooses strategy Q_2, the expected value of the game is:

$$E(row) = \begin{bmatrix} \frac{3}{5} & \frac{2}{5} \end{bmatrix} \begin{bmatrix} \frac{3}{4} & \frac{1}{2} \\ \frac{1}{3} & \frac{-5}{4} \end{bmatrix} \begin{bmatrix} \frac{7}{11} \\ \frac{4}{11} \end{bmatrix} = \frac{197}{660} \approx 0.298 .$$

Since the expected value is the row players' value, the column player will prefer the lower of the two expected values. In other words, the column player's best strategy is Q_2.

15. If the column player chooses strategy Q_1, the expected value of the game is:

$$E(row) = \begin{bmatrix} \frac{3}{5} & \frac{2}{5} \end{bmatrix} \begin{bmatrix} -3 & 4 \\ 2 & -1 \end{bmatrix} \begin{bmatrix} \frac{2}{3} \\ \frac{1}{3} \end{bmatrix} = 0.$$

If the column player chooses strategy Q_2, the expected value of the game is:

$$E(row) = \begin{bmatrix} \frac{3}{5} & \frac{2}{5} \end{bmatrix} \begin{bmatrix} -3 & 4 \\ 2 & -1 \end{bmatrix} \begin{bmatrix} \frac{7}{11} \\ \frac{4}{11} \end{bmatrix} = \frac{1}{11} \approx 0.091.$$

Since the expected value is the row players' value, the column player will prefer the lower of the two expected values. In other words, the column player's best strategy is Q_1.

17. The four in row 2, column 1 is the minimum value in row 2 and the maximum value in column 1; therefore, the saddle point is the four in row 2, column 1.

The row player should then always play row 2. The row matrix that represents the optimal strategy for the row player is $P_{opt} = \begin{bmatrix} 0 & 1 \end{bmatrix}$.

The column player should always play column 1. The column matrix that represents the optimal strategy for the column player is $Q_{opt} = \begin{bmatrix} 0 \\ 1 \end{bmatrix}$.

The value of the game is four. If the row player plays the optimal strategy, then the expected value of the game for any column strategy is:

$$E(row) = \begin{bmatrix} 0 & 1 \end{bmatrix} \begin{bmatrix} 3 & -5 \\ 4 & 6 \end{bmatrix} \begin{bmatrix} q_1 \\ 1-q_1 \end{bmatrix} = 6 - 2q_1$$

Since $0 \le q_1 \le 1$,

It must be that:

$$6 - 2q_1 \ge 4$$

If the column player plays the optimal strategy, then the expected value of the game for any row strategy is:

$$E(row) = \begin{bmatrix} p_1 & 1-p_1 \end{bmatrix} \begin{bmatrix} 3 & -5 \\ 4 & 6 \end{bmatrix} \begin{bmatrix} 1 \\ 0 \end{bmatrix} = 4 - p_1$$

Since $0 \le p_1 \le 1$,

It must be that:

$$4 - p_1 \le 4$$

Therefore we see that these strategies satisfy the fundamental theorem of game theory.

19. The negative seven in row 1, column 2 is the minimum value in row 1 and the maximum value in column 2; therefore, the saddle point is the negative seven in row 1, column 2.

The row player should then always play row 1. The row matrix that represents the optimal strategy for the row player is $P_{opt} = \begin{bmatrix} 1 & 0 \end{bmatrix}$.

The column player should always play column 2. The column matrix that represents the optimal strategy for the column player is $Q_{opt} = \begin{bmatrix} 0 \\ 1 \end{bmatrix}$.

The value of the game is negative seven. If the row player plays the optimal strategy, then the expected value of the game for any column strategy is:

$$E(row) = \begin{bmatrix} 1 & 0 \end{bmatrix} \begin{bmatrix} 4 & -7 \\ -3 & -8 \end{bmatrix} \begin{bmatrix} q_1 \\ 1-q_1 \end{bmatrix} = -7 - 3q_1$$

Since $0 \le q_1 \le 1$,

It must be that:

$$-7 - 3q_1 \ge -7$$

If the column player plays the optimal strategy, then the expected value of the game for any row strategy is:

$$E(row) = \begin{bmatrix} p_1 & 1-p_1 \end{bmatrix} \begin{bmatrix} 4 & -7 \\ -3 & -8 \end{bmatrix} \begin{bmatrix} 0 \\ 1 \end{bmatrix} = -8 + p_1$$

Since $0 \le p_1 \le 1$,

It must be that:

$$-8 + p_1 \le -7$$

Therefore we see that these strategies satisfy the fundamental theorem of game theory.

21. Let Tanya be the row player and Ilke be the column player. Let row 1 and row 2 represent the strategy of holding out one or two fingers respectively and similarly for columns 1 and 2. The payoff matrix (in cents) is:

$$\text{Tanya} \begin{array}{c} \\ 1 \\ 2 \end{array} \overset{\begin{array}{cc} 1 & 2 \end{array}}{\begin{bmatrix} 5 & -5 \\ -5 & 5 \end{bmatrix}}$$

Ilke

If each player holds out one finger twice as often as two fingers, then they hold out 1 finger $\frac{2}{3}$ of the time, and two fingers $\frac{1}{3}$ of the time. The expected value of the game is:

$$E(row) = \begin{bmatrix} \frac{2}{3} & \frac{1}{3} \end{bmatrix} \begin{bmatrix} 5 & -5 \\ -5 & 5 \end{bmatrix} \begin{bmatrix} \frac{2}{3} \\ \frac{1}{3} \end{bmatrix} = \frac{5}{9} \approx 0.56$$

The expected value is positive, which means that the row player Tanya is favored with this strategy.

23.
a. Consider Jack as the row player and Jill as the column player. Let row 1 represent Jack's strategy of showing a penny and row 2 represent Jack's strategy of showing a dime. Let column 1 represent Jill's strategy of showing a penny and column 2 represent Jill's strategy of showing a dime. The payoff matrix is:

$$\text{Jack} \begin{array}{c} \\ P \\ D \end{array} \overset{\begin{array}{cc} P & D \end{array}}{\begin{bmatrix} 1 & -10 \\ -1 & 10 \end{bmatrix}}$$

Jill

b. If Jack always plays the penny, and Jill always plays the dime, then the expected value of the game is:

$$E(row) = \begin{bmatrix} 1 & 0 \end{bmatrix} \begin{bmatrix} 1 & -10 \\ -1 & 10 \end{bmatrix} \begin{bmatrix} 0 \\ 1 \end{bmatrix} = -10.$$

c. If Jack always plays his penny and Jill plays her coins with an equal probability of $\frac{1}{2}$, then the expected value of the game is:

$$E(row) = \begin{bmatrix} 1 & 0 \end{bmatrix} \begin{bmatrix} 1 & -10 \\ -1 & 10 \end{bmatrix} \begin{bmatrix} \frac{1}{2} \\ \frac{1}{2} \end{bmatrix} = -4.5$$

d. The best strategy for Jill to play, since she is the column player, is the strategy that gives the most negative expected value. In this case, if Jack plays the penny with certainty, Jill should always play the dime or the strategy from part (b).

Exercises Section 12.3

1.
a. The one in row 2, column 2 is a saddle point. It is less than or equal to the one in row 2, column 1, so can be considered a minimum for the row, while at the same time it is a maximum for column 2. This is a strictly determined game.

The optimal row strategy is:

$$P_{opt} = \begin{bmatrix} 0 & 1 \end{bmatrix}.$$

The optimal column strategy is:

$$Q_{opt} = \begin{bmatrix} 0 \\ 1 \end{bmatrix}.$$

b. Since this is a strictly determined game, the value of the game is equal to the value of the saddle point, or in this case the value of the game is one.

3.
a. The matrix has no saddle point, so the game is not strictly determined. In order to use the techniques of this section, we need a payoff matrix with only positive entries, so we add six to each entry obtaining the new payoff matrix:

$$\begin{bmatrix} 4 & 9 \\ 10 & 1 \end{bmatrix}.$$

We need to remember that the value for this game will be six more than the value for the original game, so we will have to subtract six from the determined value to obtain the value of the original game. To determine the column strategies, we wish to solve the following linear programming problem:

Maximize $\qquad f = U + V$
subject to:

$$4U + 5V \le 1$$
$$10U + V \le 1$$
$$U, V \ge 0$$

The solution is continued on the next page.

On the other hand, we can determine the row strategies by solving the corresponding dual problem:

Minimize $g = X + Y$

subject to:

$$4X + 10Y \geq 1$$
$$9X + \quad Y \geq 1$$
$$X, Y \geq 0$$

We can solve both of these problems in one application of the simplex method of the column strategy linear programming problem. The initial simplex tableau is:

$$\left[\begin{array}{ccccc|c} 4 & 9 & 1 & 0 & 0 & 1 \\ 10 & 1 & 0 & 1 & 0 & 1 \\ \hline -1 & -1 & 0 & 0 & 1 & 0 \end{array}\right]$$

Pivoting on the nine in row 1, column 2 results in the tableau:

$$\left[\begin{array}{ccccc|c} \frac{4}{9} & 1 & \frac{1}{9} & 0 & 0 & \frac{1}{9} \\ \frac{86}{9} & 0 & \frac{-1}{9} & 0 & 1 & \frac{1}{9} \\ \hline \frac{-5}{9} & 0 & \frac{1}{9} & 0 & 1 & \frac{1}{9} \end{array}\right]$$

Pivoting on the $\frac{86}{9}$ in row 2, column 1 results in the tableau:

$$\left[\begin{array}{ccccc|c} 0 & 1 & \frac{5}{43} & \frac{-2}{43} & 0 & \frac{3}{43} \\ 1 & 0 & \frac{-1}{86} & \frac{9}{86} & 0 & \frac{4}{43} \\ \hline 0 & 0 & \frac{9}{86} & \frac{5}{86} & 1 & \frac{7}{43} \end{array}\right]$$

Thus, the minimum value of g and the maximum value of f is $g = f = \frac{7}{43}$.

This occurs when:

$$U = \frac{4}{43},\, V = \frac{3}{43},\, X = \frac{9}{86},\, Y = \frac{5}{86}.$$

The optimal row strategy is:

$$[vX \quad vY] = \left[\frac{9}{14} \quad \frac{5}{14}\right]$$

The optimal column strategy is:

$$\begin{bmatrix} vU \\ vV \end{bmatrix} = \begin{bmatrix} \frac{4}{7} \\ \frac{3}{7} \end{bmatrix}.$$

b. The value of this game is $v = \frac{1}{X+Y} = \frac{43}{7}$.

Remembering that this is the solution for the game whose entries in the payoff matrix had six added to each value, the value of the original game is $\frac{43}{7} - 6 = \frac{1}{7}$.

5.
a. The game is strictly determined. The four in row 1, column 2 is the minimum entry in row 1, as well as the maximum entry in column 2.

The optimal row strategy is:

$$P_{opt} = \begin{bmatrix} 1 & 0 \end{bmatrix}.$$

The optimal column strategy is:

$$Q_{opt} = \begin{bmatrix} 0 \\ 1 \end{bmatrix}.$$

b. Since this is a strictly determined game, the value of the game is equal to the value of the saddle point, or in this case the value of the game is four.

7.
a. This game is not a strictly determined game. We make the payoff matrix positive by adding three to each entry. The new matrix is:

$$\begin{bmatrix} 2 & 9 \\ 7 & 1 \end{bmatrix}.$$

We solve the associated linear programming problem as follows:

$$\left[\begin{array}{ccccc|c} 2 & 9 & 1 & 0 & 0 & 1 \\ 7 & 1 & 0 & 1 & 0 & 1 \\ \hline -1 & -1 & 0 & 0 & 1 & 0 \end{array}\right]$$

Pivoting on the seven in row 2, column 1 gives us:

$$\left[\begin{array}{ccccc|c} 0 & \frac{61}{7} & 1 & \frac{-2}{7} & 0 & \frac{5}{7} \\ 1 & \frac{1}{7} & 0 & \frac{1}{7} & 0 & \frac{1}{7} \\ \hline 0 & \frac{-6}{7} & 0 & \frac{1}{7} & 1 & \frac{1}{7} \end{array}\right]$$

The solution is continued on the following page.

Next pivoting on the $\frac{61}{7}$ in row 1, column 2 we get:

$$\left[\begin{array}{ccccc|c} 0 & 1 & \frac{7}{61} & \frac{-2}{61} & 0 & \frac{5}{61} \\ 1 & 0 & \frac{-1}{61} & \frac{9}{61} & 0 & \frac{8}{61} \\ \hline 0 & 0 & \frac{6}{61} & \frac{7}{61} & 1 & \frac{13}{61} \end{array}\right]$$

The optimal solution is:

$$f = g = \frac{13}{61}$$

when,

$$U = \frac{8}{61}, V = \frac{5}{61}, X = \frac{6}{61}, Y = \frac{7}{61}$$

The optimal row strategy is:

$$\left[\begin{array}{cc} vX & vY \end{array}\right] = \left[\begin{array}{cc} \frac{6}{13} & \frac{7}{13} \end{array}\right]$$

The optimal column strategy is:

$$\left[\begin{array}{c} vU \\ vV \end{array}\right] = \left[\begin{array}{c} \frac{8}{13} \\ \frac{5}{13} \end{array}\right].$$

b. Thus, $v = \frac{1}{U+V} = \frac{61}{13}$ and the value of the original game is $v - 3 = \frac{61}{13} - 3 = \frac{22}{13}$.

9.
a. The first row dominates the second row, so we can eliminate the second row from consideration. Since the resulting matrix has no saddle points, we proceed as in the previous problems. We delete the second row and then add three to each entry to make the payoff matrix have positive entries. The new payoff matrix is:

$$\left[\begin{array}{cc} 8 & 1 \\ 1 & 10 \end{array}\right].$$

We solve the associated linear programming problem with the initial tableau of:

$$\left[\begin{array}{ccccc|c} 8 & 1 & 1 & 0 & 0 & 1 \\ 1 & 10 & 0 & 1 & 0 & 1 \\ \hline -1 & -1 & 0 & 0 & 1 & 0 \end{array}\right]$$

Pivoting on the eight in row 1, column 1 gives us the matrix at the top of the next column.

$$\left[\begin{array}{ccccc|c} 1 & \frac{1}{8} & \frac{1}{8} & 0 & 0 & \frac{1}{8} \\ 0 & \frac{79}{8} & \frac{-1}{8} & 1 & 0 & \frac{7}{8} \\ \hline 0 & \frac{-7}{8} & \frac{1}{8} & 0 & 1 & \frac{1}{8} \end{array}\right]$$

Next pivoting on the $\frac{79}{8}$ in row 2, column 2 we get:

$$\left[\begin{array}{ccccc|c} 1 & 0 & \frac{10}{79} & \frac{-1}{79} & 0 & \frac{9}{79} \\ 0 & 1 & \frac{-1}{79} & \frac{8}{79} & 0 & \frac{7}{79} \\ \hline 0 & 0 & \frac{9}{79} & \frac{7}{79} & 1 & \frac{16}{79} \end{array}\right]$$

The optimal solution is:

$$f = g = \frac{16}{79}$$

when,

$$U = \frac{9}{79}, V = \frac{7}{79}, X = \frac{9}{79}, Y = \frac{7}{79}$$

The optimal row strategy for the new payoff matrix is:

$$\left[\begin{array}{cc} vX & vY \end{array}\right] = \left[\begin{array}{cc} \frac{9}{16} & \frac{7}{16} \end{array}\right]$$

However, we eliminated row 2 from the original game so the optimal strategy for the original game is:

$$\left[\begin{array}{ccc} \frac{9}{16} & 0 & \frac{7}{16} \end{array}\right]$$

The optimal column strategy is:

$$\left[\begin{array}{c} vU \\ vV \end{array}\right] = \left[\begin{array}{c} \frac{9}{16} \\ \frac{7}{16} \end{array}\right].$$

b. Thus, $v = \frac{1}{U+V} = \frac{79}{16}$ and the value of the original game is

$$v - 3 = \frac{79}{16} - 3 = \frac{31}{16}.$$

11. This game is not a strictly determined game. We make the payoff matrix positive by adding three to each entry. The new matrix is:

$$\left[\begin{array}{cc} 1 & 7 \\ 6 & 2 \end{array}\right].$$

The solution is continued at the top of the next page.

We solve the associated linear programming problem as follows:

$$\left[\begin{array}{cccc|c} 1 & 6 & 1 & 0 & 1 \\ 6 & 2 & 0 & 1 & 1 \\ \hline -1 & -1 & 0 & 0 & 0 \end{array}\right]$$

Pivoting on the six in row 2, column 1 gives us:

$$\left[\begin{array}{cccc|c} 0 & \frac{20}{3} & 1 & \frac{-1}{6} & 0 & \frac{5}{6} \\ 1 & \frac{1}{6} & 0 & \frac{1}{6} & 0 & \frac{1}{6} \\ \hline 0 & \frac{-2}{3} & 0 & \frac{1}{6} & 1 & \frac{1}{6} \end{array}\right]$$

Next pivoting on the $\frac{20}{3}$ in row 1, column 2 we get:

$$\left[\begin{array}{cccc|c} 0 & 1 & \frac{3}{20} & \frac{-1}{40} & 0 & \frac{1}{8} \\ 1 & 0 & \frac{-1}{20} & \frac{7}{40} & 0 & \frac{1}{8} \\ \hline 0 & 0 & \frac{1}{10} & \frac{3}{20} & 1 & \frac{1}{4} \end{array}\right]$$

The optimal solution is:

$$f = g = \tfrac{1}{4}$$

when,

$$U = \tfrac{1}{8}, V = \tfrac{1}{8}, X = \tfrac{1}{10}, Y = \tfrac{3}{20}$$

Thus, $v = \frac{1}{U+V} = 4$ and the value of the original game is $v - 3 = 4 - 3 = 1$.

The optimal row strategy is:

$$\left[\, vX \;\; vY \,\right] = \left[\, \tfrac{2}{5} \;\; \tfrac{3}{5} \,\right]$$

The optimal column strategy is:

$$\left[\begin{array}{c} vU \\ vV \end{array}\right] = \left[\begin{array}{c} \tfrac{1}{2} \\ \tfrac{1}{2} \end{array}\right].$$

13. The game is strictly determined. The one in row 1, column 2 is the minimum entry in row 1, as well as the maximum entry in column 2.

The optimal row strategy is:

$$P_{opt} = \left[\begin{array}{cc} 1 & 0 \end{array}\right].$$

The optimal column strategy is:

$$Q_{opt} = \left[\begin{array}{c} 0 \\ 1 \end{array}\right].$$

Since this is a strictly determined game, the value of the game is equal to the value of the saddle point, or in this case the value of the game is one.

15. The second row dominates the third row, so we can eliminate the third row. The reduced problem is:

$$\left[\begin{array}{cc} -2 & 6 \\ 8 & 0 \end{array}\right]$$

This game is not a strictly determined game. We make the payoff matrix positive by adding three to each entry. The new matrix is:

$$\left[\begin{array}{cc} 1 & 9 \\ 11 & 3 \end{array}\right].$$

We solve the associated linear programming problem as follows:

$$\left[\begin{array}{cccc|c} 1 & 9 & 1 & 0 & 1 \\ 11 & 3 & 0 & 1 & 1 \\ \hline -1 & -1 & 0 & 0 & 0 \end{array}\right]$$

Pivoting on the 11 in row 2, column 1 gives us:

$$\left[\begin{array}{cccc|c} 0 & \frac{96}{11} & 1 & \frac{-1}{11} & 0 & \frac{10}{11} \\ 1 & \frac{3}{11} & 0 & \frac{1}{11} & 0 & \frac{1}{11} \\ \hline 0 & \frac{-8}{11} & 0 & \frac{1}{11} & 1 & \frac{1}{11} \end{array}\right]$$

Next pivoting on the $\frac{96}{11}$ in row 1, column 2 we get:

$$\left[\begin{array}{cccc|c} 0 & 1 & \frac{11}{96} & \frac{-1}{96} & 0 & \frac{5}{48} \\ 1 & 0 & \frac{-1}{32} & \frac{3}{32} & 0 & \frac{1}{16} \\ \hline 0 & 0 & \frac{1}{12} & \frac{1}{12} & 1 & \frac{1}{6} \end{array}\right]$$

The solution is continued on the following page.

The optimal solution is:

$$f = g = \tfrac{1}{6}$$

when,

$$U = \tfrac{1}{16}, V = \tfrac{5}{48}, X = \tfrac{1}{12}, Y = \tfrac{1}{12}$$

Thus, $v = \tfrac{1}{U+V} = 6$ and the value of the original game is $v - 3 = 6 - 3 = 3$.

The optimal row strategy is:

$$\begin{bmatrix} vX & vY \end{bmatrix} = \begin{bmatrix} \tfrac{1}{2} & \tfrac{1}{2} \end{bmatrix}$$

However, since we eliminate row three from the original game, the optimal row strategy for the original game is:

$$\begin{bmatrix} \tfrac{1}{2} & \tfrac{1}{2} & 0 \end{bmatrix}.$$

The optimal column strategy is:

$$\begin{bmatrix} vU \\ vV \end{bmatrix} = \begin{bmatrix} \tfrac{3}{8} \\ \tfrac{5}{8} \end{bmatrix}.$$

17. The entries of the matrix are all positive. No row dominates another row and no column dominates another column, so we can not reduce the problem. Therefore, we solve the linear programming problem with the initial tableau:

$$\left[\begin{array}{ccccccc|c} 1 & 3 & 5 & 1 & 0 & 0 & 0 & 1 \\ 6 & 2 & 1 & 0 & 1 & 0 & 0 & 1 \\ 4 & 1 & 4 & 0 & 0 & 1 & 0 & 1 \\ \hline -1 & -1 & -1 & 0 & 0 & 0 & 1 & 0 \end{array} \right]$$

Pivoting on the six in row 2, column 1 results in the tableau:

$$\left[\begin{array}{ccccccc|c} 0 & \tfrac{8}{3} & \tfrac{29}{6} & 1 & \tfrac{-1}{6} & 0 & 0 & \tfrac{5}{6} \\ 1 & \tfrac{1}{3} & \tfrac{1}{6} & 0 & \tfrac{1}{6} & 0 & 0 & \tfrac{1}{6} \\ 0 & \tfrac{-1}{3} & \tfrac{10}{3} & 0 & \tfrac{-2}{3} & 1 & 0 & \tfrac{1}{3} \\ \hline 0 & \tfrac{-2}{3} & \tfrac{-5}{6} & 0 & \tfrac{1}{6} & 0 & 1 & \tfrac{1}{6} \end{array} \right]$$

Next we will pivot on the $\tfrac{10}{3}$ in row 3, column 3 to get the matrix at the top of the next column.

$$\left[\begin{array}{ccccccc|c} 0 & \tfrac{63}{20} & 0 & 1 & \tfrac{4}{5} & \tfrac{-29}{20} & 0 & \tfrac{7}{20} \\ 1 & \tfrac{7}{20} & 0 & 0 & \tfrac{1}{5} & \tfrac{-1}{20} & 0 & \tfrac{3}{20} \\ 0 & \tfrac{-1}{10} & 1 & 0 & \tfrac{-1}{5} & \tfrac{3}{10} & 0 & \tfrac{1}{10} \\ \hline 0 & \tfrac{-3}{4} & 0 & 0 & 0 & \tfrac{1}{4} & 1 & \tfrac{1}{4} \end{array} \right]$$

Next, pivot on the $\tfrac{63}{20}$ in row 1, column 2 to get:

$$\left[\begin{array}{ccccccc|c} 0 & 1 & 0 & \tfrac{20}{63} & \tfrac{16}{63} & \tfrac{-29}{63} & 0 & \tfrac{1}{9} \\ 1 & 0 & 0 & \tfrac{-1}{9} & \tfrac{1}{9} & \tfrac{-1}{9} & 0 & \tfrac{1}{9} \\ 0 & 0 & 1 & \tfrac{2}{63} & \tfrac{-11}{63} & \tfrac{16}{63} & 0 & \tfrac{1}{9} \\ \hline 0 & 0 & 0 & \tfrac{5}{21} & \tfrac{4}{21} & \tfrac{-2}{21} & 1 & \tfrac{1}{3} \end{array} \right]$$

Next pivot on the $\tfrac{16}{63}$ in row 3, column 6 to get:

$$\left[\begin{array}{ccccccc|c} 0 & 1 & \tfrac{29}{16} & \tfrac{3}{8} & \tfrac{-1}{16} & 0 & 0 & \tfrac{5}{16} \\ 1 & 0 & \tfrac{-7}{16} & \tfrac{-1}{8} & \tfrac{3}{16} & 0 & 0 & \tfrac{1}{16} \\ 0 & 0 & \tfrac{63}{16} & \tfrac{1}{8} & \tfrac{-11}{16} & 1 & 0 & \tfrac{7}{16} \\ \hline 0 & 0 & \tfrac{3}{8} & \tfrac{1}{4} & \tfrac{1}{8} & 0 & 1 & \tfrac{3}{8} \end{array} \right]$$

The optimal solution is:

$$f = g = \tfrac{3}{8}$$

when,

$$U = \tfrac{1}{16}, V = \tfrac{5}{16}, W = 0, X = \tfrac{1}{4}, Y = \tfrac{1}{8}, Z = 0$$

Thus, $v = \tfrac{1}{U+V+W} = \tfrac{8}{3}$. Since we did not add any values to the matrix, this is the value of the original problem.

The optimal row strategy is:

$$\begin{bmatrix} vX & vY & vZ \end{bmatrix} = \begin{bmatrix} \tfrac{2}{3} & \tfrac{1}{3} & 0 \end{bmatrix}$$

The optimal column strategy for the new problem is:

$$\begin{bmatrix} vU \\ vV \\ vW \end{bmatrix} = \begin{bmatrix} \tfrac{1}{6} \\ \tfrac{5}{6} \\ 0 \end{bmatrix}.$$

19. Column 3 dominates column 1 and therefore the column player should never play column 3. Thus we can delete column 3 from the problem. After deleting column 3, we see that row 1 dominates row 3, so the row player should never play row 3. So we can delete row 3 from the matrix. This gives us the new payoff matrix:

$$\begin{bmatrix} 1 & 0 \\ -1 & 1 \end{bmatrix}$$

We add two to each entry so that all entries are positive. The resulting payoff matrix is:

$$\begin{bmatrix} 3 & 2 \\ 1 & 3 \end{bmatrix}$$

We solve the associated linear programming problem as follows:

$$\left[\begin{array}{ccccc|c} 3 & 2 & 1 & 0 & 0 & 1 \\ 1 & 3 & 0 & 1 & 0 & 1 \\ \hline -1 & -1 & 0 & 0 & 1 & 0 \end{array}\right]$$

Pivoting on the three in row 1, column 1 gives us:

$$\left[\begin{array}{ccccc|c} 1 & \frac{2}{3} & \frac{1}{3} & 0 & 0 & \frac{1}{3} \\ 0 & \frac{7}{3} & \frac{-1}{3} & 1 & 0 & \frac{2}{3} \\ \hline 0 & \frac{-1}{3} & \frac{1}{3} & 0 & 1 & \frac{1}{3} \end{array}\right]$$

Next pivoting on the $\frac{7}{3}$ in row 2, column 2 we get:

$$\left[\begin{array}{ccccc|c} 1 & 0 & \frac{3}{7} & \frac{-2}{7} & 0 & \frac{1}{7} \\ 0 & 1 & \frac{-1}{7} & \frac{3}{7} & 0 & \frac{2}{7} \\ \hline 0 & 0 & \frac{2}{7} & \frac{1}{7} & 1 & \frac{3}{7} \end{array}\right]$$

The optimal solution is:

$$f = g = \tfrac{3}{7}$$

when,

$$U = \tfrac{1}{7}, V = \tfrac{2}{7}, X = \tfrac{2}{7}, Y = \tfrac{1}{7}$$

Thus, $v = \frac{1}{U+V} = \frac{7}{3}$. Since we added two to each entry of the reduced matrix, the value of the original game is
$$v - 2 = \tfrac{7}{3} - 2 = \tfrac{1}{3}.$$

The optimal row strategy for the reduced game is:

$$\begin{bmatrix} vX & vY \end{bmatrix} = \begin{bmatrix} \frac{2}{3} & \frac{1}{3} \end{bmatrix}$$

The optimal column strategy for the reduced game is:

$$\begin{bmatrix} vU \\ vV \end{bmatrix} = \begin{bmatrix} \frac{1}{3} \\ \frac{2}{3} \end{bmatrix}.$$

Recall that the third row and the third column were deleted from the original problem. The optimal strategies for the original game will be:

$$p_{opt} = \begin{bmatrix} \frac{2}{3} & \frac{1}{3} & 0 \end{bmatrix}$$

$$q_{opt} = \begin{bmatrix} \frac{1}{3} \\ \frac{2}{3} \\ 0 \end{bmatrix}.$$

21. Column 3 dominates column 2, so the column player should never prefer column 3 to column 2. Thus we can delete column 3, giving us the reduced payoff matrix:

$$\begin{bmatrix} 5 & 0 \\ -4 & 1 \end{bmatrix}$$

This game is not a strictly determined game. We make the payoff matrix positive by adding five to each entry. The new matrix is:

$$\begin{bmatrix} 10 & 5 \\ 1 & 6 \end{bmatrix}.$$

We solve the associated linear programming problem as follows:

$$\left[\begin{array}{ccccc|c} 10 & 5 & 1 & 0 & 0 & 1 \\ 1 & 6 & 0 & 1 & 0 & 1 \\ \hline -1 & -1 & 0 & 0 & 1 & 0 \end{array}\right]$$

The solution is continued on the following page.

Pivoting on the ten in row 1, column 1 gives us:

$$\left[\begin{array}{ccccc|c} 1 & \frac{1}{2} & \frac{1}{10} & 0 & 0 & \frac{1}{10} \\ 0 & \frac{11}{2} & \frac{-1}{10} & 1 & 0 & \frac{9}{10} \\ \hline 0 & \frac{-1}{2} & \frac{1}{10} & 0 & 1 & \frac{2}{10} \end{array}\right]$$

Next pivoting on the $\frac{11}{2}$ in row 2, column 2 we get:

$$\left[\begin{array}{ccccc|c} 1 & 0 & \frac{6}{55} & \frac{-1}{11} & 0 & \frac{1}{55} \\ 0 & 1 & \frac{-1}{55} & \frac{2}{11} & 0 & \frac{9}{55} \\ \hline 0 & 0 & \frac{1}{11} & \frac{1}{11} & 1 & \frac{2}{11} \end{array}\right]$$

The optimal solution is:

$$f = g = \frac{2}{11}$$

when,

$$U = \frac{1}{55},\ V = \frac{9}{55},\ X = \frac{1}{11},\ Y = \frac{1}{11}$$

Thus, $v = \frac{1}{U+V} = \frac{11}{2}$ and the value of the original

game is $v - 5 = \frac{11}{2} - 5 = \frac{1}{2}$.

Since we did not delete any rows from the original game, the optimal strategy for the original game and the reduced game is:

$$\left[\begin{array}{cc} vX & vY \end{array}\right] = \left[\begin{array}{cc} \frac{1}{2} & \frac{1}{2} \end{array}\right]$$

The optimal column strategy for the reduced game is:

$$\left[\begin{array}{c} vU \\ vV \end{array}\right] = \left[\begin{array}{c} \frac{1}{10} \\ \frac{9}{10} \end{array}\right].$$

Since we deleted the third column from the original game to create the reduced game, the optimal column strategy for the original game is:

$$q_{opt} = \left[\begin{array}{c} \frac{1}{10} \\ \frac{9}{10} \\ 0 \end{array}\right].$$

23. Let match be the row player and mismatch be the column player. Let row 1 represent placing the coin in the right hand and row 2 represent placing the coin in the left hand. Let column 1 represent guessing the right hand, and column 2 represent guessing the left hand. The payoff matrix for this game is:

$$\begin{array}{cc} & \text{Mismatch} \\ & \begin{array}{cc} \text{R} & \text{L} \end{array} \\ \text{Match} \begin{array}{c} \text{R} \\ \text{L} \end{array} & \left[\begin{array}{cc} -3 & 2 \\ 5 & -3 \end{array}\right] \end{array}$$

To make all the entries positive, we add four to each entry in the payoff matrix. The new payoff matrix is:

$$\left[\begin{array}{cc} 1 & 6 \\ 9 & 1 \end{array}\right]$$

We solve the associated linear programming problem as follows:

$$\left[\begin{array}{ccccc|c} 1 & 6 & 1 & 0 & 0 & 1 \\ 9 & 1 & 0 & 1 & 0 & 1 \\ \hline -1 & -1 & 0 & 0 & 1 & 0 \end{array}\right]$$

Pivoting on the nine in row 1, column 2 gives us:

$$\left[\begin{array}{ccccc|c} 0 & \frac{53}{9} & 1 & \frac{-1}{9} & 0 & \frac{8}{9} \\ 1 & \frac{1}{9} & 0 & \frac{1}{9} & 0 & \frac{1}{9} \\ \hline 0 & \frac{-8}{9} & 0 & \frac{1}{9} & 1 & \frac{1}{9} \end{array}\right]$$

Next pivoting on the $\frac{53}{9}$ in row 1, column 2 we get:

$$\left[\begin{array}{ccccc|c} 0 & 1 & \frac{9}{53} & \frac{-1}{53} & 0 & \frac{8}{53} \\ 1 & 0 & \frac{-1}{53} & \frac{6}{53} & 0 & \frac{5}{53} \\ \hline 0 & 0 & \frac{8}{53} & \frac{5}{53} & 1 & \frac{13}{53} \end{array}\right]$$

The optimal solution is:

$$f = g = \frac{13}{53}$$

when,

$$U = \frac{5}{53},\ V = \frac{8}{53},\ X = \frac{8}{53},\ Y = \frac{5}{53}$$

The solution is continued on the following page.

Thus, $v = \frac{1}{U+V} = \frac{53}{13}$ and the value of the original game is $v - 4 = \frac{53}{13} - 4 = \frac{1}{13}$.

The optimal row strategy for the original game is:

$$\begin{bmatrix} vX & vY \end{bmatrix} = \begin{bmatrix} \frac{8}{13} & \frac{5}{13} \end{bmatrix}$$

The optimal column strategy for the original game is:

$$\begin{bmatrix} vU \\ vV \end{bmatrix} = \begin{bmatrix} \frac{5}{13} \\ \frac{8}{13} \end{bmatrix}.$$

a. Match should adopt the strategy of placing the coin in her right hand $\frac{8}{13}$ of the time and in her left hand $\frac{5}{13}$ of the time.

b. Mismatch should choose the right hand $\frac{5}{13}$ of the time and the left hand $\frac{8}{13}$ of the time.

c. The game favors Match. She should win, on the average $\frac{1}{13}$ of a dollar per game. This is about 7.7 cents per game.

25. Let M-Mart be the row player and Q-Mart the column player. Let row 1 represent M-Mart featuring clothing and row 2 represent M-Mart featuring sporting goods. Let column 1 represent Q-Mart featuring electronics and let column 2 represent Q-mart featuring kitchenware. The payoff matrix for this game is:

Q-Mart

E   K

M-Mart $\begin{array}{c} C \\ S \end{array} \begin{bmatrix} 200 & -100 \\ -300 & 175 \end{bmatrix}$

To make the calculations easier, we rewrite the matrix to show the payoffs in hundreds of dollars:

Q-Mart

E   K

M-Mart $\begin{array}{c} C \\ S \end{array} \begin{bmatrix} 2 & -1 \\ -3 & \frac{7}{4} \end{bmatrix}$

Next we add four to each entry to obtain a positive payoff matrix displayed at the top of the next column.

Q-Mart

E K

M-Mart $\begin{array}{c} C \\ S \end{array} \begin{bmatrix} 6 & 3 \\ 1 & \frac{23}{4} \end{bmatrix}$

We solve the associated linear programming problem as follows:

$$\left[\begin{array}{cccc|c} 6 & 3 & 1 & 0 & 1 \\ 1 & \frac{23}{4} & 0 & 1 & 1 \\ \hline -1 & -1 & 0 & 0 & 0 \end{array}\right]$$

Pivoting on the 6 in row 1, column 1 gives us the matrix at the top of the next page.

$$\left[\begin{array}{cccc|c} 1 & \frac{1}{2} & \frac{1}{6} & 0 & 0 & \frac{1}{6} \\ 0 & \frac{21}{4} & \frac{-1}{6} & 1 & 0 & \frac{5}{6} \\ \hline 0 & \frac{-1}{2} & \frac{1}{6} & 0 & 1 & \frac{1}{6} \end{array}\right]$$

Next pivoting on the $\frac{21}{4}$ in row 2, column 2 we get:

$$\left[\begin{array}{cccc|c} 1 & 0 & \frac{23}{126} & \frac{-2}{21} & 0 & \frac{11}{126} \\ 0 & 1 & \frac{-2}{63} & \frac{4}{21} & 0 & \frac{10}{63} \\ \hline 0 & 0 & \frac{19}{126} & \frac{2}{21} & 1 & \frac{31}{126} \end{array}\right]$$

The optimal solution is:

$$f = g = \frac{31}{126}$$

when,

$$U = \frac{11}{126}, \, V = \frac{10}{63}, \, X = \frac{19}{126}, \, Y = \frac{2}{21}$$

Thus, $v = \frac{1}{U+V} = \frac{126}{31}$. The value of the original game is:

$$v - 4 = \frac{126}{31} - 4 = \frac{2}{31}$$

a. The optimal row strategy for the original game is:

$$\begin{bmatrix} vX & vY \end{bmatrix} = \begin{bmatrix} \frac{19}{31} & \frac{12}{31} \end{bmatrix}$$

The optimal column strategy for the original game is displayed at the top of the next page.

$$\begin{bmatrix} vU \\ vV \end{bmatrix} = \begin{bmatrix} \frac{11}{31} \\ \frac{20}{31} \end{bmatrix}.$$

The solution means that M-Mart should feature clothing $\frac{19}{31}$ percent of the time and it should feature sporting goods $\frac{12}{31}$ percent of the time. Q-Mart should feature electronics $\frac{11}{31}$ percent of the time and should feature kitchenware $\frac{20}{31}$ percent of the time.

b. The game is positive, so it favors the row player M-Mart. The value of the game is $\frac{2}{31}$ hundreds of customers. This means that M-Mart gains on average 6.45 customers each time they feature a special.

27. Let Bob be the row player and John be the column player. Let row 1 indicate that Bob played a nickel and row 2 indicate that Bob played a dime. Let column 1 indicate that John played a nickel and row 2 indicate that John played a dime. If the sum is even, Bob takes both coins, or Bob nets the value of John's coin. If the sum is odd, John takes both coins, or Bob loses the value of his coin. The payoff matrix for this game is:

John

$$\begin{array}{cc} & \text{N} \quad \text{D} \\ \text{Bob} \begin{array}{c} \text{N} \\ \text{D} \end{array} & \begin{bmatrix} 5 & -5 \\ -10 & 10 \end{bmatrix} \end{array}$$

We add 15 to both sides to obtain a positive payoff matrix. The new matrix is:

John

$$\begin{array}{cc} & \text{N} \quad \text{D} \\ \text{Bob} \begin{array}{c} \text{N} \\ \text{D} \end{array} & \begin{bmatrix} 20 & 10 \\ 5 & 25 \end{bmatrix} \end{array}$$

We solve the associated linear programming problem as follows:

$$\left[\begin{array}{ccccc|c} 20 & 10 & 1 & 0 & 0 & 1 \\ 5 & 25 & 0 & 1 & 0 & 1 \\ \hline -1 & -1 & 0 & 0 & 1 & 0 \end{array} \right]$$

Pivoting on the 20 in row 1, column 1 gives us the matrix at the top of the next column.

$$\left[\begin{array}{ccccc|c} 1 & \frac{1}{2} & \frac{1}{20} & 0 & 0 & \frac{1}{20} \\ 0 & \frac{45}{2} & \frac{-1}{4} & 1 & 0 & \frac{3}{4} \\ \hline 0 & \frac{-1}{2} & \frac{1}{20} & 0 & 1 & \frac{1}{20} \end{array} \right]$$

Next pivoting on the $\frac{45}{2}$ in row 2, column 2 we get:

$$\left[\begin{array}{ccccc|c} 1 & 0 & \frac{1}{18} & \frac{-1}{45} & 0 & \frac{1}{30} \\ 0 & 1 & \frac{-1}{90} & \frac{2}{45} & 0 & \frac{1}{30} \\ \hline 0 & 0 & \frac{2}{45} & \frac{1}{45} & 1 & \frac{1}{15} \end{array} \right]$$

The optimal solution is:

$$f = g = \frac{1}{15}$$

when,

$$U = \frac{1}{30},\, V = \frac{1}{30},\, X = \frac{2}{45},\, Y = \frac{1}{45}$$

Thus, $v = \frac{1}{U+V} = 15$. The value of the original game is:

$$v - 15 = 15 - 15 = 0$$

The optimal row strategy for the original game is:

$$\begin{bmatrix} vX & vY \end{bmatrix} = \begin{bmatrix} \frac{2}{3} & \frac{1}{3} \end{bmatrix}$$

The optimal column strategy for the original game is:

$$\begin{bmatrix} vU \\ vV \end{bmatrix} = \begin{bmatrix} \frac{1}{2} \\ \frac{1}{2} \end{bmatrix}.$$

Therefore, Bob should play his nickel two-thirds of the time and his dime one-third of the time. John should play each coin one-half of the time.

The vale of the game is zero. This indicates that if the players play their optimal strategies, the game is fair.

Chapter 12 Summary Exercises

1. The nine in row 1, column 2 is the smallest value in row 1 and the largest value in column 2. Therefore, nine is the saddle point. Thus, the value of the game is nine.

3. The two in row 2, column 1 is the smallest value in row 2 and the largest value in column 1. Therefore, two is the saddle point. Thus, the value of the game is two.

5. The expected value of the game if each player chooses strategies with an equal probability is:

$$E(row) = \begin{bmatrix} \frac{1}{2} & \frac{1}{2} \end{bmatrix} \begin{bmatrix} 0 & -3 \\ -2 & 1 \end{bmatrix} \begin{bmatrix} \frac{1}{2} \\ \frac{1}{2} \end{bmatrix} = -1.$$

7. The expected value of the game if each player chooses strategies with an equal probability is:

$$E(row) = \begin{bmatrix} \frac{1}{3} & \frac{1}{3} & \frac{1}{3} \end{bmatrix} \begin{bmatrix} 1 & 2 & 0 \\ 2 & 2 & 2 \\ -2 & 3 & 2 \end{bmatrix} \begin{bmatrix} \frac{1}{3} \\ \frac{1}{3} \\ \frac{1}{3} \end{bmatrix} = \frac{4}{3}.$$

9. The expected value of the game if each player chooses strategies with an equal probability is:

$$E(row) = \begin{bmatrix} \frac{1}{3} & \frac{1}{3} & \frac{1}{3} \end{bmatrix} \begin{bmatrix} -9 & 12 \\ 6 & -9 \\ 3 & -3 \end{bmatrix} \begin{bmatrix} \frac{1}{2} \\ \frac{1}{2} \end{bmatrix} = 0.$$

11. The four in row 2, column 1 is a saddle point. Therefore the value of the game is four.

The optimal strategy for the row player is:

$$P_{opt} = \begin{bmatrix} 0 & 1 \end{bmatrix}.$$

The optimal strategy for the column player is:

$$Q_{opt} = \begin{bmatrix} 1 \\ 0 \end{bmatrix}.$$

13. The two in row 1, column 2 is a saddle point. Therefore the value of the game is two.

The optimal strategy for the row player is:

$$P_{opt} = \begin{bmatrix} 1 & 0 \end{bmatrix}.$$

The optimal strategy for the column player is:

$$Q_{opt} = \begin{bmatrix} 0 \\ 1 \end{bmatrix}.$$

The optimal row strategy for the original game is:

15. Let Jamie be the row player and Twyla be the column player. Let row 1 represent Jamie's strategy of playing from the baseline and let row 2 represent Jamie's strategy of approaching the net. Let column 1 represent Twyla's strategy of playing from the baseline and let column 2 represent Twyla's strategy of approaching the net. Let the payoff matrix be the percentage of points won by Jamie. The payoff matrix is displayed below:

$$\begin{array}{cc} & \text{Twyla} \\ & \begin{array}{cc} \text{B} & \text{N} \end{array} \\ \text{Jamie } \begin{array}{c} \text{B} \\ \text{N} \end{array} & \begin{bmatrix} 60 & 30 \\ 45 & 65 \end{bmatrix} \end{array}$$

We solve the associated linear programming problem as follows:

$$\left[\begin{array}{ccccc|c} 60 & 30 & 1 & 0 & 0 & 1 \\ 45 & 65 & 0 & 1 & 0 & 1 \\ -1 & -1 & 0 & 0 & 1 & 0 \end{array} \right]$$

Pivoting on the 60 in row 1, column 1 gives us:

$$\left[\begin{array}{ccccc|c} 1 & \frac{1}{2} & \frac{1}{60} & 0 & 0 & \frac{1}{60} \\ 0 & \frac{85}{2} & \frac{-3}{4} & 1 & 0 & \frac{1}{4} \\ 0 & \frac{-1}{2} & \frac{1}{60} & 0 & 1 & \frac{1}{60} \end{array} \right]$$

Next pivoting on the $\frac{85}{2}$ in row 2, column 2 we get:

$$\left[\begin{array}{ccccc|c} 1 & 0 & \frac{13}{510} & \frac{-1}{85} & 0 & \frac{7}{510} \\ 0 & 1 & \frac{-3}{170} & \frac{2}{85} & 0 & \frac{1}{170} \\ 0 & 0 & \frac{2}{255} & \frac{1}{85} & 1 & \frac{1}{51} \end{array} \right]$$

The optimal solution is:

$$f = g = \frac{1}{55}$$

when,

$$U = \frac{7}{510}, V = \frac{1}{170}, X = \frac{2}{255}, Y = \frac{1}{85}$$

Thus, $v = \frac{1}{U+V} = 51$. Since we did not change the original payoff matrix, this is the value of the original game.

The solution is continued on the next page.

$$\begin{bmatrix} vX & vY \end{bmatrix} = \begin{bmatrix} \frac{2}{5} & \frac{3}{5} \end{bmatrix}$$

The optimal column strategy for the original game is:

$$\begin{bmatrix} vU \\ vV \end{bmatrix} = \begin{bmatrix} \frac{7}{10} \\ \frac{3}{10} \end{bmatrix}.$$

Jamie's optimal strategy is to stay at the baseline 40% of the time and to approach the net 60% of the time. Jamie will expect to win 51% of the points with this strategy.

17. If there is a saddle point for the game, then the game is strictly determined. This means that each player's best strategy is to make the same decision on each play of the game. The row player's strategy will be to choose the play corresponding to the row containing the saddle point and the column player's strategy will be to choose the play corresponding to the column that contains the saddle point.

19. The expected value of a two-person mixed strategy game is the row player's average value of the outcome of the game if both players randomly select their strategies according to some fixed set of probabilities.

Cumulative Review

21. The cost function is given by the line that passes through the points $(30, 400)$ and $(55, 600)$ where the x-value represent the number of items produced and the y-value represents the cost of producing x-items. Using the two points on the cost function, we determine the marginal cost or the slope of the line:

$$m = \frac{600 - 400}{55 - 30} = \frac{200}{25} = 8.$$

Using the point-slope formula of a line, the cost function is:

$$y - 400 = 8(x - 30)$$
$$y - 400 = 8x - 240$$
$$y = 8x + 160.$$

To use variables that better describe the equation, let $y = C(q)$ be the cost of producing q items, and let q represent the number of items produced. The descriptive cost function is:

$$C(q) = 8q + 160$$

23. The transportation matrix derived from the diagram is:

$$\begin{array}{c} \\ F \\ R \\ S \\ H \end{array} \begin{array}{c} \begin{array}{cccc} F & R & S & H \end{array} \\ \begin{bmatrix} 0 & 1 & 1 & 1 \\ 0 & 0 & 1 & 1 \\ 1 & 1 & 0 & 0 \\ 1 & 0 & 1 & 0 \end{bmatrix} \end{array}$$

25.

The means is $\frac{1+3+7+8+10+10+10}{7} = \frac{49}{7} = 7.0$.

The median is 8, the middle value.

The mode is 10, this is the value that occurs most often.

The range is 1 to 10.

Chapter 12 Sample Test Answers

1. The three in row 2, column 2 is a saddle point, so the game is strictly determined. The value of the game is three. Since the value of the game is positive, the game favors the row player.

2. The matrix has no saddle point. The five in row one is the minimum value of row 1, but it is not the maximum value of column 2. The negative two in row 2 is the minimum value of row 2, but it is not the maximum value of column 1. The game is not strictly determined.

3. The negative two in row 2, column 3 is a saddle point, so the game is strictly determined. The value of the game is negative two. Since the value of the game is negative, the game favors the column player.

4. If $x \leq 0$, then the zero in row 2, column 1 will be a saddle point and the game will be strictly determined.

5. If $-2 \leq x \leq 0$, then the x in row 2, column 2 will be the minimum entry in row 2, and the maximum entry in column 2. Therefore the game will be strictly determined

6. Consider the payoff to be the gain (or loss) from a 50% share. Let the row player be the First National Bank. Let row 1 represent the strategy of locating on campus and let row 2 represent the strategy of locating at the mall. Let the column player be Second National Bank. Let row 1 represent the strategy of locating on campus and let row 2 represent the strategy of locating at the mall.

The payoff matrix is displayed at the top of the next page.

2nd Nat. Bank

$$\begin{array}{cc} & \text{C} \quad \text{M} \end{array}$$

1st Nat. Bank $\begin{array}{c} \text{C} \\ \text{M} \end{array} \begin{bmatrix} 0 & 5 \\ -5 & 0 \end{bmatrix}$

7. The zero in row 1, column 1 is a saddle point. The best strategy for each bank is to locate on campus. The value of the game is zero.

8. The expected value of the game is:

$$E(row) = [0.5 \; 0.5]\begin{bmatrix} 3 & -5 \\ 0 & 8 \end{bmatrix}\begin{bmatrix} 0.5 \\ 0.5 \end{bmatrix} = \frac{3}{2} = 1.5$$

9. The expected value of the game is:

$$E(row) = \begin{bmatrix} \frac{1}{2} & \frac{1}{2} \end{bmatrix}\begin{bmatrix} -2 & 6 & 4 \\ 3 & 5 & -7 \end{bmatrix}\begin{bmatrix} \frac{1}{3} \\ \frac{1}{3} \\ \frac{1}{3} \end{bmatrix} = \frac{3}{2} = 1.5$$

10. The expected value of the game is:

$$E(row) = \begin{bmatrix} \frac{1}{3} & \frac{1}{3} & \frac{1}{3} \end{bmatrix}\begin{bmatrix} -1 & 3 & 7 \\ 2 & 3 & -2 \\ -4 & 1 & 5 \end{bmatrix}\begin{bmatrix} \frac{1}{3} \\ \frac{1}{3} \\ \frac{1}{3} \end{bmatrix} = \frac{14}{9} \approx 1.56$$

11. If the column player chooses strategy Q_1, the expected value of the game is:

$$E(row) = \begin{bmatrix} \frac{1}{3} & \frac{2}{3} \end{bmatrix}\begin{bmatrix} -1 & 2 \\ 3 & 0 \end{bmatrix}\begin{bmatrix} \frac{1}{4} \\ \frac{3}{4} \end{bmatrix} = \frac{11}{12} \approx 0.917$$

If the column player chooses strategy Q_2, the expected value of the game is:

$$E(row) = \begin{bmatrix} \frac{1}{3} & \frac{2}{3} \end{bmatrix}\begin{bmatrix} -1 & 2 \\ 3 & 0 \end{bmatrix}\begin{bmatrix} \frac{4}{13} \\ \frac{9}{13} \end{bmatrix} = \frac{38}{39} \approx 0.974$$

Since the expected value is the row players' value, the column player will prefer the lower of the two expected values. In other words, the column player's best strategy is Q_1.

12. If the column player chooses strategy Q_1, the expected value of the game is:

$$E(row) = \begin{bmatrix} \frac{1}{3} & \frac{2}{3} \end{bmatrix}\begin{bmatrix} 0.3 & -0.6 \\ -0.9 & 0.8 \end{bmatrix}\begin{bmatrix} \frac{1}{4} \\ \frac{3}{4} \end{bmatrix} = \frac{1}{4} \approx 0.25.$$

If the column player chooses strategy Q_2, the expected value of the game is:

$$E(row) = \begin{bmatrix} \frac{1}{3} & \frac{2}{3} \end{bmatrix}\begin{bmatrix} 0.3 & -0.6 \\ -0.9 & 0.8 \end{bmatrix}\begin{bmatrix} \frac{4}{13} \\ \frac{9}{13} \end{bmatrix} = \frac{1}{13} \approx 0.07.$$

Since the expected value is the row players' value, the column player will prefer the lower of the two expected values. In other words, the column player's best strategy is Q_2.

13. If the column player chooses strategy Q_1, the expected value of the game is:

$$E(row) = \begin{bmatrix} \frac{1}{3} & \frac{2}{3} \end{bmatrix}\begin{bmatrix} \frac{1}{3} & \frac{2}{5} \\ \frac{7}{8} & \frac{-1}{2} \end{bmatrix}\begin{bmatrix} \frac{1}{4} \\ \frac{3}{4} \end{bmatrix} = \frac{17}{720} \approx 0.0236.$$

If the column player chooses strategy Q_2, the expected value of the game is:

$$E(row) = \begin{bmatrix} \frac{1}{3} & \frac{2}{3} \end{bmatrix}\begin{bmatrix} \frac{1}{3} & \frac{2}{5} \\ \frac{7}{8} & \frac{-1}{2} \end{bmatrix}\begin{bmatrix} \frac{4}{13} \\ \frac{9}{13} \end{bmatrix} = \frac{44}{585} \approx 0.0752.$$

Since the expected value is the row players' value, the column player will prefer the lower of the two expected values. In other words, the column player's best strategy is Q_1.

14. This game is not a strictly determined game. We make the payoff matrix positive by adding two to each entry. The new matrix is:

$$\begin{bmatrix} 7 & 1 \\ 6 & 7 \end{bmatrix}.$$

We solve the associated linear programming problem on the following page.

$$\begin{bmatrix} 7 & 1 & 1 & 0 & 0 & | & 1 \\ 6 & 7 & 0 & 1 & 0 & | & 1 \\ \hline -1 & -1 & 0 & 0 & 1 & | & 0 \end{bmatrix}$$

Pivoting on the seven in row 1, column 1 gives us:

$$\begin{bmatrix} 1 & \frac{1}{7} & \frac{1}{7} & 0 & 0 & | & \frac{1}{7} \\ 0 & \frac{43}{7} & \frac{-1}{7} & 1 & 0 & | & \frac{1}{7} \\ \hline 0 & \frac{-6}{7} & \frac{1}{7} & 0 & 1 & | & \frac{1}{7} \end{bmatrix}$$

Next pivoting on the $\frac{43}{7}$ in row 2, column 2 we get:

$$\begin{bmatrix} 1 & 0 & \frac{7}{43} & \frac{-1}{43} & 0 & | & \frac{6}{43} \\ 0 & 1 & \frac{-6}{43} & \frac{7}{43} & 0 & | & \frac{1}{43} \\ \hline 0 & 0 & \frac{1}{43} & \frac{6}{43} & 1 & | & \frac{7}{43} \end{bmatrix}$$

The optimal solution is:

$$f = g = \tfrac{7}{43}$$

when,

$$U = \tfrac{6}{43}, V = \tfrac{1}{43}, X = \tfrac{1}{43}, Y = \tfrac{6}{43}$$

Thus, $v = \frac{1}{U+V} = \frac{43}{7}$ and the value of the original game is $v - 2 = \frac{43}{7} - 2 = \frac{29}{7}$.

The optimal row strategy is:

$$\begin{bmatrix} vX & vY \end{bmatrix} = \begin{bmatrix} \frac{1}{7} & \frac{6}{7} \end{bmatrix}$$

The optimal column strategy is:

$$\begin{bmatrix} vU \\ vV \end{bmatrix} = \begin{bmatrix} \frac{6}{7} \\ \frac{1}{7} \end{bmatrix}.$$

15. The three in row 2, column 2 is a saddle point. It is less than or equal to the three in row 2, column 1, so can be considered a minimum for the row, while at the same time it is a maximum for column 2. This is a strictly determined game. The value of the game is three.

The optimal row strategy is:

$$P_{opt} = \begin{bmatrix} 0 & 1 \end{bmatrix}.$$

The optimal column strategy is:

$$Q_{opt} = \begin{bmatrix} 0 \\ 1 \end{bmatrix}.$$

16. The six in row 2, column 1 is a saddle point. This is a strictly determined game. The value of the game is six.

The optimal row strategy is:

$$P_{opt} = \begin{bmatrix} 0 & 1 \end{bmatrix}.$$

The optimal column strategy is:

$$Q_{opt} = \begin{bmatrix} 1 \\ 0 \end{bmatrix}.$$

17. This game is not a strictly determined game. We make the payoff matrix positive by adding 11 to each entry. The new matrix is:

$$\begin{bmatrix} 1 & 27 \\ 23 & 7 \end{bmatrix}.$$

We solve the associated linear programming problem as follows:

$$\begin{bmatrix} 1 & 27 & 1 & 0 & 0 & | & 1 \\ 23 & 7 & 0 & 1 & 0 & | & 1 \\ \hline -1 & -1 & 0 & 0 & 1 & | & 0 \end{bmatrix}$$

Pivoting on the 23 in row 2, column 1 gives us:

$$\begin{bmatrix} 0 & \frac{614}{23} & 1 & \frac{-1}{23} & 0 & | & \frac{22}{23} \\ 1 & \frac{7}{23} & 0 & \frac{1}{23} & 0 & | & \frac{1}{23} \\ \hline 0 & \frac{-16}{23} & \frac{1}{23} & 0 & 1 & | & \frac{1}{23} \end{bmatrix}$$

Next pivoting on the $\frac{614}{23}$ in row 1, column 2 we get:

$$\begin{bmatrix} 0 & 1 & \frac{23}{614} & \frac{-1}{614} & 0 & | & \frac{11}{307} \\ 0 & 1 & \frac{-7}{614} & \frac{27}{614} & 0 & | & \frac{10}{307} \\ \hline 0 & 0 & \frac{8}{307} & \frac{13}{307} & 1 & | & \frac{21}{307} \end{bmatrix}$$

The solution is continued on the next page.

The optimal solution is:

$$f = g = \frac{21}{307}$$

when,

$$U = \frac{10}{307}, V = \frac{11}{307}, X = \frac{8}{307}, Y = \frac{13}{307}$$

Thus, $v = \frac{1}{U+V} = \frac{307}{21}$ and the value of the original game is $v - 11 = \frac{307}{21} - 11 = \frac{76}{21}$.

The optimal row strategy is:

$$\begin{bmatrix} vX & vY \end{bmatrix} = \begin{bmatrix} \frac{8}{21} & \frac{13}{21} \end{bmatrix}$$

The optimal column strategy is:

$$\begin{bmatrix} vU \\ vV \end{bmatrix} = \begin{bmatrix} \frac{10}{21} \\ \frac{11}{21} \end{bmatrix}.$$

18. The first row dominates the second row, so we can eliminate the second row from consideration. Since the resulting matrix has no saddle points, we proceed as in the previous problems. We delete the second row and then add five to each entry to make the payoff matrix have positive entries. The new payoff matrix is:

$$\begin{bmatrix} 15 & 1 \\ 1 & 19 \end{bmatrix}.$$

We solve the associated linear programming problem as follows:

The initial tableau is:

$$\begin{bmatrix} 15 & 1 & 1 & 0 & 0 & | & 1 \\ 1 & 19 & 0 & 1 & 0 & | & 1 \\ -1 & -1 & 0 & 0 & 1 & | & 0 \end{bmatrix}$$

Pivoting on the 15 in row 1, column 1 gives us:

$$\begin{bmatrix} 1 & \frac{1}{15} & \frac{1}{15} & 0 & 0 & | & \frac{1}{15} \\ 0 & \frac{284}{15} & \frac{-1}{15} & 1 & 0 & | & \frac{14}{15} \\ 0 & \frac{-14}{15} & \frac{1}{15} & 0 & 1 & | & \frac{1}{15} \end{bmatrix}$$

Next pivoting on the $\frac{284}{15}$ in row 2, column 2 we get the matrix at the top of the next column.

$$\begin{bmatrix} 1 & 0 & \frac{19}{284} & \frac{-1}{284} & 0 & | & \frac{9}{142} \\ 0 & 1 & \frac{-1}{284} & \frac{15}{284} & 0 & | & \frac{7}{142} \\ 0 & 0 & \frac{3}{142} & \frac{7}{142} & 1 & | & \frac{8}{71} \end{bmatrix}$$

The optimal solution is:

$$f = g = \frac{8}{71}$$

when,

$$U = \frac{9}{142}, V = \frac{7}{142}, X = \frac{9}{142}, Y = \frac{7}{142}$$

Thus, $v = \frac{1}{U+V} = \frac{71}{8}$ and the value of the original game is $v - 5 = \frac{71}{8} - 5 = \frac{31}{8}$.

The optimal row strategy for the new payoff matrix is:

$$\begin{bmatrix} vX & vY \end{bmatrix} = \begin{bmatrix} \frac{9}{16} & \frac{7}{16} \end{bmatrix}$$

However, we eliminated row 2 from the original game so the optimal strategy for the original game is:

$$\begin{bmatrix} \frac{9}{16} & 0 & \frac{7}{16} \end{bmatrix}$$

The optimal column strategy is:

$$\begin{bmatrix} vU \\ vV \end{bmatrix} = \begin{bmatrix} \frac{9}{16} \\ \frac{7}{16} \end{bmatrix}.$$

19. This game is not strictly determined, however, the third column dominates the second column, so the column player should never play column 3. Deleting column two, we reduce the payoff matrix to:

$$\begin{bmatrix} 9 & 0 \\ -2 & 1 \end{bmatrix}$$

Adding three to each entry in the payoff matrix we get the new positive payoff matrix:

$$\begin{bmatrix} 12 & 3 \\ 1 & 4 \end{bmatrix}$$

We solve the associated linear programming problem on the following page.

$$\begin{bmatrix} 12 & 3 & 1 & 0 & 0 & | & 1 \\ 1 & 4 & 0 & 1 & 0 & | & 1 \\ -1 & -1 & 0 & 0 & 1 & | & 0 \end{bmatrix}$$

Pivoting on the 12 in row 1, column 1 gives us:

$$\begin{bmatrix} 1 & \frac{1}{4} & \frac{1}{12} & 0 & 0 & | & \frac{1}{12} \\ 0 & \frac{15}{4} & \frac{-1}{12} & 1 & 0 & | & \frac{11}{12} \\ 0 & \frac{-3}{4} & \frac{1}{12} & 0 & 1 & | & \frac{1}{12} \end{bmatrix}$$

Next pivoting on the $\frac{15}{4}$ in row 2, column 2 we get:

$$\begin{bmatrix} 1 & 0 & \frac{4}{45} & \frac{-1}{15} & 0 & | & \frac{1}{45} \\ 0 & 1 & \frac{-1}{45} & \frac{4}{15} & 0 & | & \frac{11}{45} \\ 0 & 0 & \frac{1}{15} & \frac{1}{5} & 1 & | & \frac{4}{15} \end{bmatrix}$$

The optimal solution is:

$$f = g = \tfrac{4}{15}$$

when,

$$U = \tfrac{1}{45},\ V = \tfrac{11}{45},\ X = \tfrac{1}{15},\ Y = \tfrac{1}{5}$$

Thus, $v = \frac{1}{U+V} = \frac{15}{4}$ and the value of the original game is $v - 3 = \frac{15}{4} - 3 = \frac{3}{4}$.

The optimal row strategy for the new payoff matrix is:

$$\begin{bmatrix} vX & vY \end{bmatrix} = \begin{bmatrix} \frac{1}{4} & \frac{3}{4} \end{bmatrix}$$

The optimal column strategy for the new game is:

$$\begin{bmatrix} vU \\ vV \end{bmatrix} = \begin{bmatrix} \frac{1}{12} \\ \frac{11}{12} \end{bmatrix}.$$

We did not change the rows to the original game, so the optimal row strategy for the original game is:

$$p_{opt} = \begin{bmatrix} \frac{1}{4} & \frac{3}{4} \end{bmatrix}.$$

However, since we deleted the third column from the original game, the optimal column strategy for the original game is:

$$q_{opt} = \begin{bmatrix} \frac{1}{12} \\ \frac{11}{12} \\ 0 \end{bmatrix}.$$

20. Let Mia be the row player and Elena be the column player. Let row 1 indicate that Mia played a dime and row 2 indicate that Mia played a quarter. Let column 1 indicate that Elena played a dime and row 2 indicate that Elena played a quarter. If the coins match, Mia gets the product of the two coins. If the coins do not match, Elena gets ten times the sum of the two coins. The payoff matrix for this game is:

$$\begin{array}{cc} & \text{Elena} \\ & \begin{array}{cc} D & Q \end{array} \\ \text{Mia} \begin{array}{c} D \\ Q \end{array} & \begin{bmatrix} 1.00 & -3.50 \\ -3.50 & 6.25 \end{bmatrix} \end{array}$$

We add 3.75 to both sides to obtain a positive payoff matrix. The new matrix is:

$$\begin{array}{cc} & \text{Elena} \\ & \begin{array}{cc} D & Q \end{array} \\ \text{Mia} \begin{array}{c} D \\ Q \end{array} & \begin{bmatrix} 4.75 & 0.25 \\ 0.25 & 10.00 \end{bmatrix} \end{array}$$

We solve the associated linear programming problem as follows:

$$\begin{bmatrix} 4.75 & 0.25 & 1 & 0 & 0 & | & 1 \\ 0.25 & 10.00 & 0 & 1 & 0 & | & 1 \\ -1 & -1 & 0 & 0 & 1 & | & 0 \end{bmatrix}$$

Pivoting on the 4.75 in row 1, column 1 gives us:

$$\begin{bmatrix} 1 & \frac{1}{19} & \frac{4}{19} & 0 & 0 & | & \frac{1}{19} \\ 0 & \frac{759}{76} & \frac{-1}{19} & 1 & 0 & | & \frac{18}{19} \\ 0 & \frac{-18}{19} & \frac{4}{19} & 0 & 1 & | & \frac{1}{19} \end{bmatrix}$$

The solution is continued on the next page.

Next pivoting on the $\frac{759}{76}$ in row 2, column 2 we get:

$$\begin{bmatrix} 1 & 0 & \frac{160}{759} & \frac{-4}{759} & 0 & \Big| & \frac{52}{253} \\ 0 & 1 & \frac{-4}{759} & \frac{76}{759} & 0 & \Big| & \frac{24}{253} \\ \hline 0 & 0 & \frac{52}{253} & \frac{24}{253} & 1 & \Big| & \frac{76}{253} \end{bmatrix}$$

The optimal solution is:

$$f = g = \frac{76}{253}$$

when,

$$U = \frac{52}{253}, \; V = \frac{24}{253}, \; X = \frac{52}{253}, \; Y = \frac{24}{253}$$

Thus, $v = \frac{1}{U+V} = \frac{253}{76}$. The value of the original game is:
$$v - 3.75 = \frac{253}{76} - 3.75 = \frac{-8}{19}$$

The optimal row strategy for the original game is:

$$\begin{bmatrix} vX & vY \end{bmatrix} = \begin{bmatrix} \frac{13}{19} & \frac{6}{19} \end{bmatrix}$$

The optimal column strategy for the original game is:

$$\begin{bmatrix} vU \\ vV \end{bmatrix} = \begin{bmatrix} \frac{13}{19} \\ \frac{6}{19} \end{bmatrix}.$$

Mia should play a dime $\frac{13}{19}$ of the time and a quarter $\frac{6}{19}$ of the time. Elena should also play a dime $\frac{13}{19}$ of the time and a quarter $\frac{6}{19}$ of the time.

21. The value of the game is $\frac{-8}{19}$, which is negative, hence the game favors the column player Elena. Elena will win on average $\frac{8}{19}$ of a dollar per play, or slightly over 42 cents per play.

Appendix I Exercises

1. a. No. **b.** Yes. **c.** Yes. **d.** No.

2. a. No. **b.** Yes. **c.** Yes. **d.** No.

3. a. No. **b.** No. **c.** Yes. **d.** No.

4. a. No. **b.** No. **c.** No. **d.** Yes.

5. a. Yes. **b.** Yes. **c.** Yes. **d.** No.

6.

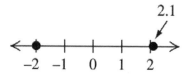

7.

8.

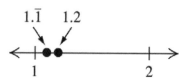

9.

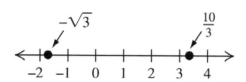

10.

11.
$$2(3-7)=$$
$$(2)(3)-(2)(7)=$$
$$6-14=$$
$$-8$$

12.
$$2 \div \left[1+1(3\bullet2-6)\right]=$$
$$2 \div \left[1+(6-6)\right]=$$
$$2 \div \left[1-0\right]=$$
$$2$$

13.
$$-3(2-x)=$$
$$(-3)(2)-(-3)(x)=$$
$$-6+3x$$

14.
$$-4(x-8)-30=$$
$$(-4)(x)-(-4)(8)-30=$$
$$-4x+32-30=$$
$$-4x+2$$

15.
$$-2(2-3^2)=$$
$$(-2)(2)-(-2)(9)=$$
$$-4+18=$$
$$14$$

16.
$$-2(2-3)^2=$$
$$-2(-1)^2=$$
$$-2(1)=$$
$$-2$$

17.
$$3x+2=11$$
$$3x=11-2$$
$$3x=9$$
$$x=3$$

The equation is conditional. The only solution is $x=3$.

18.
$$2-2x=4$$
$$-2x=4-2$$
$$-2x=2$$
$$x=-1$$

The equation is conditional. The only solution is $x=-1$.

19.

$$3x + 7 = x - 5$$
$$3x + 7 - x = -5$$
$$2x = -5 - 7$$
$$2x = -12$$
$$x = -6$$

The equation is conditional. The only solution is $x = -6$.

20.

$$0.25x + 200 = 0.3x + 320$$
$$-0.05x = 120$$
$$x = -2400$$

The equation is conditional. The only solution is $x = -2400$.

21.

$$0.25x + 100 = 0.25(x + 400)$$
$$0.25x + 100 = 0.25x + 100$$
$$0 = 0$$

The equation is consistent. Any value of x will satisfy the equation.

22.

$$0.25x + 100 = 0.25(x + 100)$$
$$0.25x + 100 = 0.25x + 25$$
$$0 \neq -75$$

The equation is inconsistent. There are no values of x that satisfies this equation.

23.

$$0.25x + 100 = 0.25(400 - x)$$
$$0.25x + 100 = 100 - 0.25x$$
$$0.5x = 0$$
$$x = 0$$

The equation is conditional. The only solution is $x = 0$.

24.

$$2(x - 3) + 4 = 2x - 1$$
$$2x - 6 + 4 = 2x - 1$$
$$2x - 2 = 2x - 1$$
$$0 \neq 1$$

The equation is inconsistent. There are no values of x that satisfies this equation.

25.

$$2(x - 3) + 4 = 2x - 2$$
$$2x - 6 + 4 = 2x - 2$$
$$2x - 2 = 2x - 2$$
$$0 = 0$$

The equation is consistent. Any value of x will satisfy the equation.

26.

$$\left(3^2\right)\left(3^{-1}\right) = 3^{2+(-1)} = 3^1 = 3.$$

27.

$$\frac{2^4}{2^5} = 2^{4-5} = 2^{-1} = \frac{1}{2}.$$

28.

$$\frac{4 \cdot 3 \cdot 2 \cdot 1}{(2 \cdot 1)(2 \cdot 1)}\left(\frac{1}{3}\right)^2\left(\frac{2}{3}\right)^2 =$$
$$6\left(\frac{1}{9}\right)\left(\frac{4}{9}\right) =$$
$$\frac{8}{27}.$$

29.

$$\frac{3 \cdot 2 \cdot 1}{(2 \cdot 1)(1)}\left(\frac{1}{4}\right)^2\left(\frac{3}{4}\right)^1 =$$
$$3\left(\frac{1}{16}\right)\left(\frac{3}{4}\right) =$$
$$\frac{9}{64}.$$

30.

$$4\left(\frac{1}{3}\right)^0\left(\frac{2}{3}\right)^4 =$$

$$4(1)\left(\frac{16}{81}\right) =$$

$$\frac{64}{81}.$$

31.

$$2^x = 3$$

$$\ln(2^x) = \ln(3)$$

$$x\ln(2) = \ln(3)$$

$$x = \frac{\ln 3}{\ln 2}$$

32.

$$3^x = 2$$

$$\ln(3^x) = \ln(2)$$

$$x\ln(3) = \ln(2)$$

$$x = \frac{\ln 2}{\ln 3}$$

33.

$$12\cdot 4^x = 36$$

$$4^x = 3$$

$$\ln(4^x) = \ln(3)$$

$$x\ln(4) = \ln(3)$$

$$x = \frac{\ln 3}{\ln 4}$$

34.

$$100(1.01)^x = 500$$

$$(1.01)^x = 5$$

$$\ln\left((1.01)^x\right) = \ln(5)$$

$$x\ln(1.01) = \ln(5)$$

$$x = \frac{\ln 5}{\ln 1.01} \approx 161.7472$$

35.

$$\frac{100\left[1-(0.01)^x\right]}{1-0.01} = 101$$

$$\frac{100\left(1-(0.01)^x\right)}{0.99} = 101$$

$$100\left(1-(0.01)^x\right) = 99.99$$

$$1-(0.01)^x = 0.9999$$

$$(0.01)^x = 0.0001$$

$$\ln(0.01)^x = \ln(0.0001)$$

$$x\ln(0.01) = \ln(0.0001)$$

$$x = \frac{\ln(0.0001)}{\ln(0.01)} \approx 2$$

36.

$$S_5 = \frac{2\left[1-3^5\right]}{1-3}$$

$$S_5 = \frac{2\left[-242\right]}{-2}$$

$$S_5 = 242.$$

37.

$$S_{10} = \frac{3\left[1-\left(\frac{1}{2}\right)^{10}\right]}{1-\frac{1}{2}}$$

$$S_{10} = \frac{3\left[\frac{1023}{1024}\right]}{\frac{1}{2}}$$

$$S_{10} = 6\left[\frac{1023}{1024}\right]$$

$$S_{10} = \frac{3069}{512}.$$

38.

$$S_{20} = \frac{1\left[1-(0.1)^{20}\right]}{1-0.1}$$

$$S_{20} \approx \frac{1[1]}{0.9}$$

$$S_{20} \approx 1.1111.$$

39.

$$S_{25} = \frac{200\left[1 - (0.01)^{25}\right]}{1 - 0.01}$$

$$S_{25} \approx \frac{200[1]}{0.99}$$

$$S_{25} \approx 202.0202 \, .$$

40.

$$S_5 = \frac{50\left[1 - (1.01)^{5}\right]}{1 - 1.01}$$

$$S_5 = \frac{50\left[-0.5101005\right]}{-0.01}$$

$$S_5 = 255.05025$$